MUNK'S ROLL

VOL. VIII

LIVES OF THE FELLOWS OF THE ROYAL COLLEGE OF PHYSICIANS OF LONDON

continued to 1988

Edited by

GORDON WOLSTENHOLME

Kt, OBE, LLD, DTech, FRCP, FIBiol

Harveian Librarian

assisted by

Valérie Luniewska BA

IRL PRESS

at

OXFORD UNIVERSITY PRESS

Oxford New York Tokyo

1989

Royal College of Physicians of London
11 St Andrews Place
London NW1 4LE
1989

British Library Cataloguing in Publication Data
Munk's roll.
Vol. 8: Lives of the fellows of the Royal College of Physicians of London.
1. England. Physicians. Organisations.
Royal College of Physicians of London.
Fellows – Biographies – Collections
I. Wolstenholme, G.E.W.
(Gordon Ethelbert Ward),
1913–
610.92'2

ISBN 0-19-963132-8

The engraving of the College arms on the title-page was made by Reynolds Stone

Printed by Information Press Ltd, Oxford, England

FOREWORD

The growth of the College can be gauged by the rate at which *Munk's Roll* appears. The first volume, published in 1878, contained 'biographical sketches of all the eminent physicians whose names are recorded in the annals from the Foundation of the College in 1518 to its removal in 1825 from Warwick Lane to Pall Mall East.' Whereas a single volume was all that was needed to span the first 307 years of the College's existence, four more were necessary to take us from 1825 to 1965. Since then the pace of our expansion has been even greater so that Sir Gordon Wolstenholme, who has occupied the post of Harveian Librarian since 1979, has found it necessary to produce three volumes to cover the years since 1965 - a prodigious effort but one of great importance to the College. We are most grateful to him and Valerie Luniewska for this further publication.

R HOFFENBERG

President

PREFACE

This eighth volume of *Munk's Roll* contains biographies of 321 Fellows whose deaths occurred between 1 January 1984 and 31 December 1988, and of a few other Fellows where notification of death was received too late for their inclusion in previous volumes.

For a third time the present editors have included photographs of Fellows, where available. This applies to 272 (84.7%) of the entries in the volume.

We have thought it useful to include on this occasion a combined index to Volumes IV to VIII, since with the more frequent appearance of the later volumes it is not easy to remember within just what period a Fellow died. One of us (VL) has devoted much of her own time to its careful compilation.

We are deeply grateful to the many authors for the time, care and affection they have put into the writing of the biographies. We have encouraged them to write with greater freedom than in the past about their friends' personalities and experiences, and to give more information of the direction of research where this became part of the fabric of medical science. As a result, there is here much of the history of clinical science in the past half-century. The biographies are not uncritical, and praise and admiration have all the more merit.

This is the third and last volume of *Munk's Roll* for which we have been responsible. We hope it proves a source of happiness and pride to relatives, colleagues and friends. In particular, we wish it to be a tribute to those many Fellows who have upheld the best traditions of the College in a period of ever increasing involvement in education and standards of conduct. It is to their example and instruction that succeeding generations of Fellows and Members are for ever indebted.

Readers of many of these biographies will be reminded not only of their debt to the past but also their own responsibilities for the future.

Sir Gordon Wolstenholme
Valérie Luniewska

June 1989

The editors dedicate Volumes VI, VII and VIII to the Presidents of the Royal College of Physicians during the period of their preparation:

Sir Douglas Black
Sir Raymond Hoffenberg

With respect and admiration.

AAMC	Australian Army Medical Corps
Aberd	Aberdeen
ADMS	Assistant Director of Medical Services
AFC	Air Force Cross
AIF	Australian Imperial Force
AKC	Associate of King's College
AMA	American Medical Association
ASTMS	Association of Scientific Technical and Managerial Staff
BA	Bachelor of Arts
BAO	Baccalaureus Artis Obstetricae (Bachelor of Obstetric Art)
BAOR	British Army of the Rhine
BCh	Baccalaureus Chirurgiae (Bachelor of Surgery)
BChir	Baccalaureus Chirurgiae (Bachelor of Surgery)
BEF	British Expeditionary Force
Belf	Belfast
Birm	Birmingham
BM	Bachelor of Medicine
BMA	British Medical Association
Br Col	British Colombia
BRCS	British Red Cross Society
BS	Bachelor of Surgery
BSc	Bachelor of Science
BSc Med	Bachelor of Science in Medicine
Bt	Baronet
CAMC	Canadian Army Medical Corp
CB	Companion of the Order of the Bath
CBE	Commander of the Order of the British Empire
CC	Companion of the Order of Canada
CD	Canadian Forces Decoration
CH	Companion of Honour
CIE	Companion of the Order of the Indian Empire
CM	Chirurgiae Magister (Master of Surgery)
CMG	Companion of the Order of St Michael and St George
CSI	Companion of the Order of the Star of India
CVCP	Committee of Vice-Chancellors and Principals
CVO	Companion of the Royal Victorian Order

CVRI	Cardiovascular Research Institute (University of California)
Cantab	Cambridge
Capt	Captain
ChB	Chirurgiae Baccalaureus (Bachelor of Surgery)
ChM	Chirurgiae Magister (Master of Surgery)
Co	County
Col	Colonel
DADMS	Deputy Assistant Director of Medical Services
DC	District of Columbia
DCH	Diploma of Child Health
DCL	Doctor of Civil Law
DChO	Diploma in Ophthalmic Surgery
DCP	Diploma in Clinical Pathology
DD	Doctor of Divinity
DDMS	Deputy Director of Medical Services
DDS	Doctor of Dental Surgery
DFC	Distinguished Flying Cross
DGMS	Director General of Medical Services
DHMSA	Diploma of the History of Medicine
DHSS	Department of Health and Social Security
DIH	Diploma in Industrial Health
Dip Bact	Diploma in Bacteriology
DJMK	Datoship Paducka Mahkuta Kelantan
DKSM	Diploma of the Kitchener School of Medicine
DL	Deputy Lieutenant
DLitt	Doctor of Letters
DM	Doctor of Medicine
DMR	Diploma in Medical Radiology
DMRD	Diploma in Medical Radio-diagnosis
DMRE	Diploma in Medical Radiology and Electrology
DMS	Director of Medical Services
DO Dubl	Diploma of Ophthalmology Dublin
DPH	Diploma of Public Health
DPM	Diploma in Psychological Medicine
DSM	Distinguished Service Medal
DSO	Distinguished Service Order
DSc	Doctor of Science
DTM	Diploma in Tropical Medicine
DTM&H	Diploma in Tropical Medicine and Hygiene
Dubl	Dublin, Trinity College

Durh	Durham
DVA	Department of Veterans Affairs
Edin	Edinburgh
EMS	Emergency Medical Service
Eng	(following Diploma) Royal Colleges of Physicians of London, and Surgeons of England
ENT	Ear, Nose and Throat
ERD	Emergency Reserve Decoration (Army)
FAA	Fellow of the Australian Academy of Science
FAAP	Fellow of the American Academy of Pediatrics
FACE	Fellow of the Australian College of Education
FACMA	Fellow of the Australian College of Medical Administration
FACP	Fellow of the American College of Physicians
FACS	Fellow of the American College of Surgeons
FAMSI	Fellow of the Academy of Medical Sciences of India
FANZPsych	Fellow of the Australian and New Zealand College of Psychiatrists
FCGP	Fellow of the College (later Royal College) of General Practitioners
FCOG	Fellow of the British (later Royal) College of Obstetricians and Gynaecologists
FCP Philad	Fellow of the College of Physicians of Philadelphia
FCPS Pak	Fellow of the College of Physicians and Surgeons of Pakistan
FCPath	Fellow of the College (later Royal) of Pathologists
FCRA	Fellow of the College of Radiologists of Australia
FCST	Fellow of the College of Speech Therapists
FEIS	Fellow of the Educational Institute of Scotland
FFCM	Fellow of the Faculty of Community Medicine
FFOM	Fellow of the Faculty of Occupational Medicine
FFR	Fellow of the Faculty of Radiologists

FICS	Fellow of the International College of Surgeons
FKC	Fellow of King's College London
FLEX Lic.USA	Federal Licensing Examination USA
FOM	Faculty of Occupational Medicine
FPAS	Fellow of the Pakistan Academy of Sciences
FPS	Fellow of the Pharmaceutical Society
FRACP	Fellow of the Royal Australasian College of Physicians
FRACS	Fellow of the Royal Australasian College of Surgeons
FRAHS	Fellow of the Royal Australian Historical Society
FRAS	Fellow of the Royal Astronomical Society
FRCGP	Fellow of the Royal College of General Practitioners
FRCOG	Fellow of the Royal College of Obstetricians and Gynaecologists
FRCP	Fellow of the Royal College of Physicians of London
FRCPA	Fellow of the Royal College of Pathologists of Australia
FRCPath	Fellow of the Royal College of Pathologists
FRCPC	Fellow of the Royal College of Physicians of Canada
FRCPE	Fellow of the Royal College of Physicians of Edinburgh
FRCPG	Fellow of the Royal College of Physicians of Glasgow
FRCPI	Fellow of the Royal College of Physicians Ireland
FRCPsych	Fellow of the Royal College of Psychiatrists
FRCPS	Fellow of the Royal College of Physicians and Surgeons of Glasgow
FRCR	Fellow of the Royal College of Radiologists
FRCS	Fellow of the Royal College of Surgeons of England
FRCSE	Fellow of the Royal College of Surgeons of Edinburgh
FRCSI	Fellow of the Royal College of Surgeons of Ireland
FRFPS	Fellow of the Royal Faculty(later College) of Physicians and Surgeons of Glasgow
FRIBA	Fellow of the Royal Institute of British Architects
FRIC	Fellow of the Royal Institute of Chemistry
FRS	Fellow of the Royal Society

FRS(C)	Fellow of the Royal Society of Canada
FRSC	Fellow of the Royal Society of Chemistry
FRSE	Fellow of the Royal Society of Edinburgh
FRSH	Fellow of the Royal Society of Health
FRSS	Fellow of the Royal Statistical Society
FRSSAfr	Fellow of the Royal Society of South Africa
FSA	Fellow of the Society of Antiquaries
FSc(Med)	Fellow in Medical Sciences
FZS	Fellow of the Zoological Society
GCVO	Knight Grand Cross of the Royal Victorian Order
GLC	Greater London Council
GM	George Medal
GOC	General Officer Commanding
Glasg	Glasgow
HAC(RHA)	Honourable Artillery Company (Royal Horse Artillery)
HMS	His/Her Majesty's Ship
Heidlb	Heidelberg
Hon	Honorary;Honourable
HP	House Physician
HS	House Surgeon
IATA	International Air Transport Association
IBP	International Biological Programme
ICI	Imperial Chemical Industries Limited
IMS	Indian Medical Service
Irel	(following Diploma)Royal College of Physicians and Surgeons in Ireland
JP	Justice of the Peace
KBE	Knight Commander of the Order of the British Empire
KC	King's Counsel
KCB	Knight Commander of the Bath
KCMG	Knight Commander of St Michael and St George
KCSI	Knight Commander of the Star of India
KCSG	Papal Knight of the Order of St Gregory
KCVO	Knight Commander of the Royal Victorian Order
KG	Knight of the Order of the Garter
KLJ	Knight St Lazarus of Jerusalem
KSG	Papal Knight of the Order of St Gregory

Kt	Knight
LCC	London County Council
LFPS	Licentiate of the Faculty (later Royal Faculty) of Physicians and Surgeons of Glasgow
LM	Licentiate in Midwifery
LMCC	Licentiate of the Medical Council of Canada
LMS	Licentiate in Medicine and Surgery
LMSSA	Licentiate in Medicine and Surgery, Society of Apothecaries
LRCP	Licentiate of the Royal College of Physicians of London
LRCPE	Licentiate of the Royal College of Physicians of Edinburgh
LRCPI	Licentiate of the Royal College of Physicians of Ireland
LRCSE	Licentiate of the Royal College of Surgeons of Edinburgh
LRCSI	Licentiate of the Royal College of Surgeons in Ireland
LRFPS	Licentiate of the Royal Faculty (later College)of Physicians and Surgeons of Glasgow
LSA	Licentiate of the Society of Apothecaries
Liverp	Liverpool
LLD	Doctor of Laws
Lond	London
Lt-Cdr	Lieutenant-Commander
Lt-Col	Lieutenant-Colonel
Lt-Gen	Lieutenant-General
Luck	Lucknow
MA	Master of Arts
Manch	Manchester
MB	Bachelor of Medicine
MBE	Member of the Order of the British Empire
MC	Military Cross
MCh	Magister Chirurgiae(Master of Surgery)
MCCM(NZ)	Member of the New Zealand College of Community Medicine
MCPSA	Member of the College of Physicians,South Africa
MD	Doctor of Medicine
MD(Psychol Med)	Doctor of Medicine in Psychological Medicine
MD(State Med)	Doctor of Medicine in State Medicine

MEF	Middle East Forces
Melb	Melbourne
MICE	Member of the Institute of Civil Engineers
Minn	Minnesota
MMR	Mass Miniature Radiography
MMSA	Master of Midwifery of the Society of Apothecaries
MOIC	Medical Officer-in-charge
MP	Member of Parliament
MPS	Member of the Pharmaceutical Society
MRCP	Member of the Royal College of Physicians of London
MRCPE	Member of the Royal College of Physicians of Edinburgh
MRCPI	Member of the Royal College of Physicians of Ireland
MRCS	Member of the Royal College of Surgeons of England
MRIA	Member of the Royal Irish Academy
MS	Master of Surgery
MSc	Master of Science
NAPT	National Association for the Prevention of Tuberculosis
NAS	National Academy of Sciences (USA)
New Bruns	New Brunswick
NIMR	National Institute of Medical Research
NRC	National Research Council(USA)
NUI	National University of Ireland
NZ	New Zealand
OBE	Officer of the Order of the British Empire
OC	Officer of the Order of Canada
OM	Order of Merit
OMA	Ontario Medical Association
OTC	Officers' Training Corps
Oxon	Oxford
PC	Privy Councillor
PJG	Pingat Jasa Gemilang
POW	Prisoner of War
PRCP	President of the Royal College of Physicians of London
PhD	Doctor of Philosophy
Punj	Punjab

Queen's	Queen's University, Kingston, Ontario, Canada
QAIMNS/R	Queen Alexander's Indian Medical Nursing Service, Reserve
QARNS	Queen Alexander's Royal Nursing Service
RAAF	Royal Australian Air Force
RAAMC	Royal Australian Army Medical Corps
RACP	Royal Australasian College of Physicians
RAF	Royal Air Force
RAFVR	Royal Air Force Volunteer Reserve
RAMC	Royal Army Medical Corps
RAMC(SR)	Royal Army Medical Corps(Special Reserve)
RAMC(T)	Royal Army Medical Corps(Territorial)
RCAMC	Royal Canadian Army Medical Corps
RCVNR	Royal Canadian Volunteer Naval Reserve
RFA	Royal Fleet Auxiliary
RNH	Royal Naval Hospital
RRC	Royal Red Cross
RSCI	Royal College of Surgeons of Ireland
RFA	Royal Field Artillery
RMO	Regimental Medical Officer;Resident Medical Officer
RN	Royal Navy
RNVR	Royal Naval Volunteer Reserve
RSA	Royal Society of Arts
RSM	Royal Society of Medicine
RUI	Royal University of Ireland
Revd	Reverend
Rev	Reverend
Roch	Rochester
S.Afr	South Africa
St And	St Andrews
Sask	Saskatchewan
ScD	Scientiae Doctor(Doctor of Science)
Sgt	Sergeant
Sheff	Sheffield
Sqdn Ldr	Squadron Leader
TA	Territorial Army
TC	Trinity Cross of Republic of Trinidad and Tobago
TD	Territorial Decoration

UC Dubl	University College,Dublin
UCLA	University of California,Los Angeles
UNRRA	United Nations relief and Rehabilitation Administration
VAD	Voluntary Aid Detachment
VD	Colonial Auxiliary Forces Officers' Decoration (Volunteer Decoration,later VRD)
WAust	Western Australia
WHO	World Health Organization
West Ont	Western Ontario
Wits	Witwatersrand

Abbreviations of journal titles follow the system used in *World Medical Periodicals* and *World List of Scientific Periodicals.*

An asterisk(*) against FRCP means that the Fellow was elected under the special bye-law which provides for the election to the Fellowship of 'Persons holding a medical qualification, but not Members of the College, who have distinguished themselves in the practice of medicine or in the pursuit of Medical or General Science or Literature...'. Where the list of honorary degrees is too lengthy to include in entirety in the heading a list is available in the library.

ADDIS, Harold St Clair Campbell *b*.2 June 1913 *d*.9 Mar 1986
MB BChir Cantab(1939) MRCP(1947) FRCP(1972)

Campbell, as he was known to his friends, was born in Edinburgh, the son of William Robert Addis FRCOG who, after the 1914-18 war, became consultant obstetrician and gynaecologist at St Mary's Hospital, Manchester. His mother was Rhoma Ethel, née Hutchinson, daughter of Thomas Megisson Hutchinson, an engineer.

Campbell's father was trained as an artist at the Slade School,and later in Paris, before he took up medicine; he won the MC in the first world war. His brother, Campbell's uncle, was Thomas Addis MD FACP, professor of medicine at Stanford University, USA. Thomas Addis was well known for the 'Addis Count' and was author of the textbook *Glomerular nephritis: diagnosis and treatment*, New York, MacMillan & Co.,1948. Apart from his father and uncle, Campbell had many connexions with medicine. His sister Caroline qualified in Manchester and later did cancer research in London; she was married to Roger Warwick, professor of anatomy at Guy's Hospital and co-editor of the 35th British edition of Gray's *Anatomy*, London, Longman, 1973. One of Campbell's sons, two daughters-in-law, a nephew and a niece, were all medical.

Campbell began his education at Moor Allerton Preparatory School in Manchester and then, at age 14, went to Shrewsbury School. He was a very good cross country runner, and in his final year at school organized the sports day with great success. At Cambridge he was on the cross country running first team and did some rowing for his College. He also played some rugby. After qualification and two house posts at Manchester Royal Infirmary, he joined the Navy. He was on the cruiser *Birmingham* when it was torpedoed in the Mediterranean. He was mentioned in despatches. After the war he returned to Manchester, to Ancoats and later Salford Royal hospital.

After obtaining his MRCP, Campbell went to work for Kellgren in the department of rheumatic research at Manchester Royal. In 1953 he was appointed consultant chest physician in Sheffield. He was never very happy in this post and, in 1956, at the suggestion of former colleagues he applied for and was appointed to the post of consultant in geriatric medicine in the Wigan and Leigh Hospitals in central Lancashire. This he found very much to his liking; he had a great concern for the old and disabled and was a tower of strength in the rehabilitation of the chronic sick.

As a colleague Campbell was absolutely dependable and ever considerate in all he did. he worked closely with consultants in all specialties - several of whom were friends from his student days. Serving on numerous hospital committees, he was one of two consultants on the first new area boards which were formed in 1973-74.

He retired in 1977 and greatly enjoyed his retirement. His hobbies were gardening, reading and contract bridge, and he adored the post retirement cottage he bought in Wales. His wife, Sally, was theatre staff nurse on Geoffrey Jefferson's neurosurgical unit at Manchester Royal Infirmary. Campbell did his first junior post on this unit and met her there. They were married in 1940. He was very much a home-loving man and his five grandchildren were a great joy. He was survived by his wife and three sons, one of whom is a general practitioner in Auckland.

TBSD

[*Brit.med.J.*,1986,**292**,1403]

ALERGANT, Cecil David *b.*15 May 1915 *d.*22 July 1988
LRCP MRCS(1939) MBChB Liverp(1939) MD(1948)
MRCP(1949) FRCP(1973)

Cecil Alergant was born in Liverpool and educated at the Liverpool Institute and in Paris, where he developed fluency in French. He studied medicine at Liverpool University and distinguished himself by winning the Holt medal for physiology. He qualified in 1939 and, following clinical attachments, went into general practice.

He served in the RAMC from 1943-47, during which time he developed an interest in venereology. After the war he returned to England to take up duties as senior registrar in venereology at Newsham General Hospital, rapidly obtaining his MD in 1948, and his membership of the College the following year. After a brief spell as senior registrar at the Liverpool Royal Infirmary he was appointed consultant venereologist at that hospital in 1950, thus commencing a distinguished career which spanned three decades.

Cecil Alergant's kindness, warmth, diplomacy and clinical acumen were appreciated greatly by staff and patients alike, and his marvellous intellect contributed significantly to research in the department. His medical opinions continued to be sought long after his retirement due to the depth of his knowledge and its contemporary nature.

'CDA', as he was affectionately known, was a great teacher at both undergraduate and postgraduate level and was instrumental in initiating lectures in venereology for nurses and midwives in many hospitals in Liverpool and west Lancashire. He was appointed clinical lecturer in venereology at Liverpool University in 1950, and became lecturer in charge of the department in 1965. His many and varied qualities led to his appointment as departmental head in 1966, a post which he held until his retirement.

In 1967, 'CDA' initiated the Liverpool course for the diploma in venereology and to this day it is the only course in which the diploma is awarded by a university. It has worldwide recognition.

He was an active and enthusiastic champion of the Medical Society for the Study of Venereal Diseases, being treasurer from 1969-75 and president from 1975-77. Regular attendances, with his wife Joan, both in England and abroad, established strong professional and lifelong friendships with many members. His active participation continued long after retirement; his last presentation being at the Society's UK meeting in Cambridge, shortly before he died. His paper was received with warm enthusiasm and this, and the social events he attended with his wife, will long be remembered by those fortunate enough to have been there.

As a father, grandfather and father-in-law, Cecil Alergant had few peers. His generosity, humour and capacity for love, marked him as a great family man and his home in Liverpool was never short of visitors; a fact that was not lost on his devoted wife who spent many hours in the kitchen as a result. He was fond of crosswords and mathematical problems and derived much pleasure from teasing his family with both. His intellect and general knowledge complemented his unofficial position as the family arbiter and counsellor. Many have been the richer for his advice.

Retirement to Llandegla, North Wales, did not diminish family contact - long weekends and holidays allowing him time with his several grandchildren, whom he adored. Although deteriorating health limited his physical activity, his mental agility remained intact until his sudden death at home in July, 1988. Medicine had lost a wonderful physician and teacher; his family lost a patriach and gentleman. PBC

[*Brit.med.J.*,1988,**297**,1328; Photo]

ALTOUNYAN, Roger Ernest Collingwood

*b.*24 Oct 1922 *d.*10 Dec 1987
AFC(1945) BChir Cantab(1956) MB(1957) MRCP(1974)
FRCP(1982) DM Soton(1983)

Roger Altounyan was born in Aleppo, Syria. He was the third generation of a famous medical family originally based in the Middle East. His grandfather, an American orphan, was born in Turkey, educated by a group of American missionaries, and eventually sent to medical school at Columbia University. After qualifying he returned to Turkey via Europe, where he met his wife, an Irish nurse.

In Turkey, he and his wife travelled from village to village by horseback, conducting surgeries and operations, and eventually settled in Aleppo, Syria, where they established the Altounyan Hospital.

The first Dr Altounyan sent his wife to England to have her children and consequently the Altounyans became British citizens.

Roger Altounyan and his four sisters were educated at boarding school in England and spent many holidays on their own in the family home at Lake Coniston in the Lake District. In the summer of 1927 they were joined by J Arthur Ransome and provided him with the inspiration for his novel *Swallows and Amazons*.

After completing his school education in England, Roger returned to Syria. His education was interrupted by the outbreak of war, which forced him to flee from Syria to Palestine where he joined the Middle East Command in Jerusalem. He initially trained as a pilot in Bulawayo, Rhodesia, before returning to England. At first he flew bombers and then joined the pilots advanced flying unit, followed by the flying instructors school and the battle school. He was awarded the AFC in 1945.

At the end of the war Roger entered Emmanuel College, Cambridge, to study medicine and went on to Middlesex Hospital, London, to complete his clinical training. It was while at medical school that he developed bronchial asthma for the first time.

After qualifying at the Middlesex he returned to the family hospital in Aleppo, where he worked as assistant physician and surgeon to his father until 1955. As a result of the Suez crisis Roger returned to England in late 1955 where he took up a post in the research department of Benger Laboratories. However, he soon began to miss contact with patients, a situation which was rectified by the late Bill Robinson, a chest physician at Monsall Hospital, who allowed him to conduct an unofficial outpatient clinic in the disused dental department of Monsall Hospital. There - with a spirometer and two chairs - he held three outpatient sessions a week, which rapidly became very busy. Ever mindful of the patients' requirements, his outpatient sessions were run late in the afternoon and into the evening, so that patients did not have to take time off work. The clinics were carried out without nursing assistance, and the patients themselves were frequently driven to their homes by Roger at the end of the clinic.

While working at Bengers he became interested in a research project looking for smooth muscle relaxants among some synthetic derivatives of Khellin. He suggested that he could use himself as a guinea pig, and he used an extract of guinea pig hair, to which he was sensitive, as a standard challenge. Over the next two years he tested two compounds a week on himself and it became clear that some Khellin derivatives were active in preventing muscle constriction. Despite this he was moved from the project into the laboratory, and the research team was ordered to cease work on the project. But the team were so sure that their work was leading somewhere that they continued with the project in secret, with twice weekly challenge tests being carried out during the outpatient sessions at Monsall Hospital. This work led to the development of di-sodium cromoglycate. Coincident with the identification of this bis-chromone Bengers laboratory was taken over by Fisons, and the project was reinstated. Di-sodium cromoglycate proved to be active only by inhalation, and then in doses of several milligrams. The solution to this problem proved to be the spinhaler, a prototype of which Roger Altounyan had already conceived and manufactured at home. He had noticed after starting work at Monsall Hospital that some asthmatic patients were unable to use pressurised inhalers. He then designed, using metal Tyrozet tubes, a crude inhaler from which patients administered an isoprenaline/lactose combination from gelatin capsules. It was from the principles incorporated in the design of this original inhaler that the spinhaler was developed.

Over the next 20 years Roger continued in active research looking for new compounds which would be more active in the treatment of bronchial asthma.

In 1951 he married Hella Christel, daughter of Reinhold Schumacher, a teacher. They had two sons and three daughters. Throughout his life Roger retained a deep affection for the Lake District, which he visited regularly and could usually be found either sailing on Lake Coniston or char fishing.

His questioning mind was active in all aspects of his life, from his specially ventilated sitting-room fireplace to the oxygen concentrator which he was using during his final illness, to which he added a bag in circuit to increase flow rate during exertion. His friends and colleagues who visited him at home remember with some trepidation his home-made wines produced from a variety of sources ranging from elderberries to tea. This last product was always highly potent but of very variable palatability. He was a quick thinker, and always an excellent conversationalist. Throughout his life he maintained an intense interest in people and would spend hours helping with difficult patients. He was indeed an inspiration to all those with whom he had contact.

CACP

[*The Times*,12 Dec 1987; *Lancet*,1988,1,193; Doctor,10 Nov 1988; Photo]

ANDERSON, Ian McAllister *b.*20 Oct 1911 *d.*10 Dec 1986
MB ChB Glasg(1934) MRCP(1939) MD(1945) FRCP(1957)

Ian McAllister Anderson was a proud Scot who was born in Kelvinside, Glasgow. His father, William Donald Anderson, was a qualified doctor but practised dentistry. He died when Ian was eight years old, but the family was relatively prosperous and did not suffer financially. Ian was educated at the Glasgow and Kelvinside Academies, entered the medical school of Glasgow University in 1929 and after graduation was successively house surgeon and house physician at the Royal Infirmary, and obstetric house officer at the Maternity Hospital.

Early in his career he came in contact with Leonard Findlay [*Munk's Roll*,Vol.V,p.130], an incandescent red-headed man who was the second holder of the first established chair of child health in Britain. Findlay had a histrionic manner, and was much interested in rickets which was then prevalent among indigent Scots in Glasgow. He denied the value of cod liver oil, but Ian Anderson always defended

Findlay despite this. When Findlay moved down to the Princess Elizabeth of York Hospital, Shadwell, Ian followed him and became his resident there. The move to London did not alter Findlay's authoritarian manner or his florid style, and Ian was subjected to his notes being corrected in red ink in full view of the students and staff.

At that time, on ward rounds, Ian found that gastroenteritis was prevalent in the slums of London and he went to work as assistant resident at the St Louis Children's Hospital in the USA, where Hartman was a pioneer in the management of acidosis from salt and water depletion. He returned to Britain to become resident medical officer at Shadwell, and in 1939, when Findlay moved to Oxford, Ian was appointed as first assistant (i.e.registrar) to the children's department at the London Hospital where he worked with Bellingham-Smith [*Munk's Roll*,Vol.VI, p.41] and Tallerman [*Munk's Roll*,Vol.VII,p.567]. It was there that he met his wife, Eileen Nolan-Neylan, who was a staff nurse. He was attracted to her pretty appearance, and she was attracted to his handsome presence and his clinical diagnostic skill.

From 1940-42, during the second world war, he was a physician in the Emergency Medical Service and then joined the RAMC as a medical specialist with the rank of major, seeing service in the Middle East, India, Scotland, and subsequently in Germany, where he attended Field Marshall Montgomery. They got on well. When Ian wished to go to Copenhagen to visit a friend, Montgomery made his personal plane available for him; when Ian got off at Copenhagen the expectant throng waiting at the airport were disappointed to see only a lieutenant colonel walking down the steps. Ian enjoyed this moment nevertheless.

After discharge from the Army, Ian became children's physician at the Prince of Wales Hospital and ran a cardiac clinic there, mainly for those with rheumatic heart disease, and was later appointed consultant physician to the Queen Elizabeth Hospital for Children, both at Shadwell and Hackney Road. He worked at all these hospitals until he retired. Because of his experience with heart disease and his contact with Clarence Gavey [*Munk's Roll*,Vol.VII,p.207], who was cardiologist at Westminster Hospital, he was also appointed to the staff there and, from 1947, to the Westminster Children's Hospital. At some time he held appointments at Ilford Public Health Clinic, St George's Hospital, Ilford, and he was honorary paediatrician to St Stephen's Hospital, Fulham, until 1964.

Ian had his own mind; he opposed the random use of antibiotics in the management of gastroenteritis; when the MRC was funding the unit at the Queen Elizabeth Hospital for Children, Hackney, he disagreed with their use and was outspoken in committee. He was therefore well ahead of his time, but was excluded from the gastroenteritis ward by the medical committee whilst he was on holiday. He took holidays infrequently after that.

He put all his energies into developing a unit at the Westminster with, at first, the late Sir Clement Price-Thomas [*Munk's Roll*,Vol.VI, p.382] and later with Charles Drew, for the management of children with congenital heart disease. It was very demanding work, Drew having perfected the technique of profound hypothermia; it meant detailed investigations and meticulous post-operative care. He was assisted by Joseph Humble [*Munk's Roll*,Vol.VII,p.284], the haematologist at the Westminster, and by juniors who subsquently made their way in paediatric cardiology. They included C G Newman, his successor, and Michael Tynan, professor of paediatric cardiology at Guy's Hospital. Charles Ayoub was a paediatric cardiologist who learnt a great deal from him and subsequently practised at the American Hospital in Beirut.

Patients might be referred by friends or colleagues, but were often 'picked up' at the various hospitals that Ian attended, so that at different times of the week he would be in Tottenham, Ilford, Hackney, Shadwell, or even Banstead in Surrey, as well as at St Stephen's. He travelled around London on public transport, being picked up at Guildford every evening by his wife, who waited there for him. Weekends were utilized for the Saturday ward round and playing golf, which was his great love. His other love was going to and entertaining at the Caledonian Club, where his reminiscences were often succinct, powerful and pointed.

Apart from the original work, which he directed, in congenital heart disease - in which he was assisted by Handley Coles (q.v.) who did all the catheterizations for him - Ian collaborated with Joe Humble and John Hobbs in bone marrow transplantation. Humble and Pegg had pioneered this technique at the Westminster Hospital and the first patient in the world to receive a successful bone marrow transplant was one of Ian's patients. The work was subsequently taken up by Kenneth Hugh-Jones and the department now has a national, if not international, reputation.

At different times, Ian was secretary of the section of paediatrics at the Royal Society of Medicine, and examiner for the diploma of child health. He obtained his membership of the College in 1939 and was later elected a Fellow. He was stern as an examiner, but generous as a marker. He could be frightening and demanded a great deal of his juniors. Those who survived did well and there are many who are grateful to him for putting them in line.

Retirement was irksome to Ian Anderson and he undertook a few locums. Eventually he moved up to Meriden, near Coventry, with his wife, so that they could be close to other members of the family. He suffered a very unhappy terminal illness, in which he was devotedly supported by his family.

LS

[*Daily Telegraph*,18 Dec 1986]

ANDREWES, ***Sir*** **Christopher Howard** *b.*7 June 1896 *d.*31 Dec 1988 Kt(1961) MRCS LRCP(1921) MB B Lond(1921) MD(1922) MRCP(1923) FRCP(1935) FRS(1939) Hon LLD Aberd(1963) Hon MD Lund(1968)

Christopher Andrewes was an internationally known medical virologist who spent most of his working life in full-time research as a member of the staff of the Medical Research Council.

He was born in London, the son of Sir Frederick William Andrewes, a pathologist at St Bartholomew's Hospital, and a Fellow of the College [*Munk's Roll*,Vol.IV.p.376]. His mother, Phyllis Mary Hamer, came from a publishing family.

Christoper Andrewes was educated at Highgate School and studied medicine at Bart's, where he received gold medals on graduation, and also for his MD two years later. He was on the medical professorial unit until 1927, with a two year break as an assistant physician at the Rockefeller Institute Hospital, New York.

He served in the RNVR during the first world war, and in the second he worked in the EMS laboratory service, the forerunner of our public health laboratory service. However, most of the time he worked at the National Institute for Medical Research, first at Hampstead and then at Mill Hill, where he became head of the division of virology and bacteriology. He was deputy director of the Institute from 1953 until he retired in 1961. In 1946 he founded the common cold unit at Salisbury, which he directed as part of his division. The unit became internationally known scientifically and was widely supported by the British public, so that by the time it closed its doors in 1989 over 18,000 volunteer inoculations had been made in the course of over 1000 separate 10-day trials. During his time as director important observations were made on the spread of colds, and the first steps in culturing and identifying the causal viruses.

Andrewes' opinion on a range of virological topics was much valued. He was a member of various scientific committees and of the governing body of the Animal Virus Research Institute, Pirbright. He gave an Oliver-Sharpey lecture on viruses in relation to tumours, 1934, and a Leeuwenhoek lecture for the Royal Society on the place of viruses in nature (1952). He was also invited to lecture abroad.

In 1939 he was elected a fellow of the Royal Society and was on the council from 1945-47. He was knighted in 1961.

Christopher Andrewes was characteristically seen in a tweed jacket, white hair rather dishevelled, his ruddy face expressing interest, puzzlement, concentration, or fun - never boredom. In spite of his medical training his research was in many ways an expression of his lifelong interest in natural history. He cited this as his main hobby, but in fact he reached a professional standard as an entomologist. He, with Smith [*Munk's Roll*,Vol.V.p.385] and Laidlaw [*Munk's Roll*,Vol.V, p.236], discovered the human influenza A virus in 1933, and in the following years he studied the spread of the virus around the world and, through the work of the World Influenza Centre in his department, began to work out the properties of each of the new viruses as they appeared and swept across the world. He also made important contributions to virus taxonomy and was involved in international discussions to get agreement on the proper approach to the subject, and then on the grouping and naming of viruses.

Although Andrewes was never an official teacher he had great influence as a communicator. He gave excellent scientific papers and lectures, and generated lively and penetrating discussions both in international fora and at the bench. Members of his staff, and many visitors to the laboratory, were benefited. For many years he enlivened the meetings of the Medical Research Club in London and was a founder of the informal virus group, which subsequently became one of the most dynamic and influential sections of the Society of General Microbiology. His department was probably the best of the small group of virus research laboratories in action after the war and it made an important contribution to the later flowering of British virology. He also wrote books on viruses for the general public, who received them well.

Christopher Andrewes was a warm family man and took great care of and interest in his three sons, two of whom became general practitioners. As well as taking his children on country walks, he would invite his staff out too. Indeed it was expected that the recently recruited science graduates would accompany him each morning on his fortnightly visits to the common cold unit along the nearby drove road. They learned to recognize the birds and the flowers and, if they were wise, absorbed something of his schoolboy enthusiasm for the life history of respiratory viruses and the biology of their reactions with cells. Until he was 90 he continued to be interested in his laboratory 'family', who came down one by one to visit him and his wife at his chalet bungalow in a village near Salisbury. His wife, Kathleen Lamb, whom he had married in 1927, predeceased him. He was survived by his three sons.

DAJT

[*The Times*,2 Jan 1989; *The Independent*,4 & 11 Jan 1989; *The Guardian*,9 Jan 1989; *Brit.med.J.*,1989,**298**,180; Photo]

ANGUS, Joan *b.*30 Mar 1924 *d.*17 Aug 1985
MB ChB Glasg(1947) MRCP(1953) FRCP(1974)

Joan Angus was born and bred in Glasgow, her father Lawrence Angus a business man, her mother Agnes a nurse. She was educated at Glasgow High School for Girls and then studied medicine at the Western Infirmary. Following tradition she moved southwards after qualifying and, since she had early made up her mind that she wanted to be a paediatrican, after a variety of appointments in Birmingham and

London she obtained her MRCP and was appointed house physician to Sir Wilfrid Sheldon [*Munk's Roll*,Vol.VII,p.531] at the Hospital for Sick Children in Great Ormond Street. Further paediatric appointments followed at the Central Middlesex Hospital and the Sorrento Maternity Hospital and then, quite on impulse, she decided to take a sabbatical and set out with three friends in a small boat, planning to sail through the Mediterranean and down the east coast of Africa. Alas, the Suez crisis prevented their passage through the Canal so they turned back and spent several months in Malta where Joan worked in the professorial unit at St Luke's Hospital, living much of the time on her boat and supporting her friends. After a year they returned to England but, following a couple of consultant locum appointments, Joan was off again; this time to become lecturer in paediatrics in the University of the West Indies where she spent several very happy years. She was appointed consultant paediatrician to the Teeside Hospitals in 1966 and elected to the Fellowship of the RCP in 1974.

These are the bare facts, but they do not tell the full story of this courageous woman's life. Whilst still in her 'teens she developed the first symptoms of chronic rheumatoid disease which was never to leave her.She was seldom free from pain, and as the years went by became physically more and more disabled. During her years in Middlesborough she underwent several major surgical operations, but always returned quickly to her work with its large clinical load in a busy and scattered department - a department she led quietly, smoothly and effectively. At the end of 1983, and now resigned to a more or less wheelchair existence, she retired. Happily this retirement, which she had dreaded, proved to be throughly enjoyable. A grateful District Health Authority had allowed her to keep a couple of outpatient sessions, but she soon found her days so filled that they began to intrude and she was glad to give these up. She joined the Open University and was very proud to have completed the first two years of the course. Ever a gregarious person she had time to see her many devoted friends, to enjoy her home and garden and care for her aged father.

Joan was an excellent clinician, well read and with wide experience. She published very little but she took great pleasure in teaching her junior staff, and in their success. She had a truly remarkable flair for getting instantly on to good terms with parents and children, and cared deeply about her child patients. Similarly, with her happy, vivacious and outgoing personality, she had an infinite capacity for forming lasting friendships with a wide variety of people. Intelligent and cultured, despite her illness she loved living and lived her life fully. Countless friends found her brilliant personality and selflessness an inspiration.

She collapsed suddenly one evening at her home in Stokesley and died the next day in the hospital she had served for 17 years. RGW

[*Brit.med.J.*,1985,**291**,1212-3; Photo]

ARMSTRONG, Thomas George *b.*30 Sept 1910 *d.*9 June 1985
BA Cantab(1931) MRCS LRCP(1934) MB BChir(1935)
MRCP(1936) MD(1939) FRCP(1970)

Tom Armstrong was born in Krugersdorp, South Africa, where his father was a resident magistrate. He came to Britain for his university education and graduated with a BA from Trinity College, Cambridge, in 1931. In 1934 he qualified in medicine with the conjoint, followed by the Cambridge degree in 1935 and his membership of the College in 1936. He was greatly influenced by John A Ryle [*Munk's Roll*,Vol.IV,595] who became his close friend.

During the second world war Tom Armstrong was among the first of the RAMC doctors to be sent to the Middle East. He was based at a Cairo Military Hospital and later a casualty clearing station, and was subsequently promoted to lieutenant colonel. On demobilization in 1946 he returned to South Africa and set up a private practice in Durban. He was also appointed visiting physician to St Aidan's Hospital and the King Edward VIII Hospital. While at the latter hospital he did considerable research on amoebic dysentery, a common complaint in South Africa, and developed an interest in cardiology which led him to study under Velva Schrire of Cape Town [*Munk's Roll*,Vol.VI,p.40] and later, in London, under Paul Wood [*Munk's Roll*,Vol.V,p.456]. On his return to South Africa again, he was instrumental in setting up the sub-department of cardiac surgery and cardiology in Durban.

In 1970 he gave up private practice and joined the staff of Wentworth Hospital as a full time consultant physician, and at the same time was appointed a lecturer in cardiology in the sub-department of cardiology at the University of Natal. He became interested in the haemodynamics of constrictive pericarditis, and made a valuable contribution in the field of valvular heart disease; describing the left atrial lift as being distinctive from the right ventricular heave. He also wrote papers in connexion with the genesis of the various components of the first heart sound.

Tom spent a considerable amount of time with his patients and was much loved by them; he was a particularly kind man and not only concerned himself with their medical problems but also helped many with their social or economic difficulties. An excellent teacher, he contributed a great deal to the training of medical postgraduates at the University of Natal and assisted in the training of senior registrars, more particularly in the area of non-invasive cardiology. He retired from full time practice in 1981, due to ill health, but continued to work part-time at the Wentworth Hospital, finally retiring at the end of 1984.

In his younger days Tom was interested in gliding and obtained a certificate of qualification, accompanying many gliding teams to various parts of Europe. He enjoyed outdoor life, especially trout fishing, and was interested in horticulture, being well acquainted with the indigenous flora of South Africa. He had a small farm on the outskirts of Durban where he constantly experimented with seed cultivation, being particularly interested in proteas. After a visit to Italy, he found a new interest in theatre and opera.

His wife, Yvonne, survived him.

ASM/YA/GW

[Photo]

ASTON, Howard Campbell *b.*29 Aug 1910 *d.*9 Oct 1986
MB ChB Birm(1933) MRCP(1939) FRCP(1973)

Howard Aston was born in Birmingham, where his father Bernard Aston was a sales manager. He was educated at the King Edward VI School, Birmingham, going on to Birmingham University where he graduated in medicine in 1933 with distinction in forensic medicine and toxicology. After house appointments at the Midland Nerve Hospital and Queen's Hospital, he was appointed medical officer in the London County Council Hospital Service in 1936. He was senior resident physician at Lewisham Hospital in 1942, but joined the RAMC as a medical specialist with the rank of major. After demobilization he returned to Lewisham, and was later appointed consultant physician to the Bethnal Green Hospital in 1948, where he spent the rest of his professional life.

Howard was a well read, up to date and very sound general physician, with an open mind always ready to encourage new developments and techniques. At Bethnal Green he was in charge of staff health, of a children's ward, and of adult wards - being one of the last of a generation of general physicians to have the care of neonates and also of elderly patients. He was in his 50s before the hospital had a paediatricain or geriatrician on the staff.

Being a quiet man, of few words, it was not easy to get to know him. He was not a man who threw his weight about, but his occasional interventions in committee were respected since they were usually put forward to prevent something unfair or unwise.

He married Kathleen Marguerite Mutch, daughter of a headmaster, in 1942 and they had three daughters, all of whom became distinguished either academically or musically. Howard was an accomplished musician on stringed instruments and there was a tradition of family oriented chamber music, as well as orchestral playing. Holidays were often spent at the Dartington music school.

Howard was an observant traveller and also a keen gardener, but the last few years of his working life were marred by a succession of serious organic illnesses: diabetes mellitus with ketosis during gangrenous appendicitis, bronchial asthma with status asthmaticus during a flight when above Heathrow Airport, coronary artery disease and gastrointestinal haemorrhage. He coped manfully, and came back to work after each episode, reluctantly retiring at the age of 63 because of ill health - and then only because of the insistence of his wife and his physician. Fortunately, after retirement his health improved at first, possibly due to less exposure to atmospheric pollution or to less stress.

RHB

BAIN, Cyril William Curtis *b.*5 June 1895 *d.*14 Aug 1987
MC(1918) MA BM BCh Oxon(1922) MRCP(1924) DM(1933)
FRCP(1938)

Curtis Bain was senior physician to Harrogate General Hospital, a distinguished cardiologist, son of a remarkable father, and president of the BMA in 1949 - the first year of the National Health Service.

He was born at Thornfield, near Manchester. His father, William Bain, spent twelve years in general practice at Heaton Chapel, where he married Ellen Curtis. He gained the membership of the College, was elected FRCS Edinburgh, and settled in Harrogate as a spa physician, being elected a Fellow of the College in 1912 [*Munk's Roll*,Vol.IV, p.522]. Curtis Bain grew up in an atmosphere of medical research and practice in Harrogate, and was educated at Wellington College, Christ Church Oxford and St Thomas's Hospital, London.

His studies were interrupted by the first world war, when he was gazetted to the Duke of Wellington's Regiment in August 1914 and saw active service in France and Flanders with the West Riding Regiment attached to the Machine Gun Corps. He ended the war with the rank of major, was mentioned in despatches and awarded the Military Cross for conspicuous gallantry and devotion to duty.

In 1917, with a shortage of doctors, he was asked to resume his studies and qualified in 1922. After early appointments at St Thomas's he became RMO at the National Heart Hospital before being appointed physician at Harrogate General Hospital, and to the cardiographic department at the Royal Bath Hospital. He obtained the membership of the College in 1924, the DM in 1933, and was elected a Fellow in 1938.

Like his father before him, he combined a busy practice in a Yorkshire spa town with active research, and wrote several original papers on the electrocardiograph and conduction defects. A walking holiday in the Black Forest in the spring of 1928 started a fruitful collaboration with Terence East of King's College, with whom he wrote

Recent Advances in Cardiology, London, J&A Churchill, 1929, which went to five editions before passing to Wallace Brigden. He also served on the council of the British Cardiac Society.

A loyal BMA member, he was honorary secretary of the Harrogate division from 1925-29, chairman from 1939-46, and as president welcomed the British Medical Association to its annual meeting in Harrogate in 1949. Meetings of the scientific sections were extended to four days, and the handbook included an advertisement for the fourth edition of *Recent Advances...* . His presidential address was on the theme 'Man and the machine', in which he warned his colleagues against making machines their masters. Reliance upon tests, he said, engendered a habit of mind in which more confidence was felt in a figure than in an opinion based upon clinical judgement.

He was a tall, precise, commanding figure. On his consulting room wall he had a map of Harrogate indicating the distribution of rheumatic fever cases, and his electrocardiogram became a feature of domiciliary practice in the town.

He retired to St Ives, Cornwall, in 1958 and, already a keen walker and trout fisher, taught himself to sail. For 15 years - based on Falmouth - he explored the roads, and the Helford River in his cruiser *Zest*, with his customary determination. *Who's Who* gave gardening as his later recreation. He renewed his friendship with Terence East as they investigated the Cornish saints together.

In 1930 he married Diana Pease, who survived him together with their daughter, a general practitioner in Portsmouth, and their two sons; one of whom is an ENT surgeon in Ontario. Fortunately, his enquiring mind and forthright views remained intact until his death, which followed an operation for a fractured hip.

AGS

[*Brit.med.J.*,1987,**295**,790; *The Times* 21 Aug 1987; Photo]

BAKER-BATES, Eric Tom *b.*18 Sept 1905 *d.*20 May 1986
MB ChB Liverp(1928) MRCP(1931) MD(1933) FRCP(1950)

Eric Baker-Bates (B-B) was born at St Helens, Lancashire, the son of Henry Baker-Bates, a physician and surgeon, and his wife Edith Pennington, the daughter of Henry Pennington who was also a physician and surgeon. Throughout four centuries his family tree included physicians - and some surgeons - and his father was four times mayor of St Helens.

Eric was educated at Heaton Moor College, Merchant Taylors, and the University of Liverpool, graduating with honours and distinction in medicine and pathology. After junior house appointments he obtained his membership of the College and became consultant physician to several hospitals including Southport General Infirmary, St Helens, Warrington, and Clatterbridge. In all, he was part-time consultant to some 11 hospitals in what is now known as Merseyside. He gave a lifetime of dedicated service to the people and hospitals of Liverpool. An outstanding student, he went on to develop a passion for teaching. When he was transferred from the Southern General Hospital to the professorial unit at the Royal Infirmary he became Henry Cohen's, later Lord Cohen of Birkenhead [*Munk's Roll*,Vol.VII,p.106], junior consultant colleague and they formed a high powered teaching unit; their teaching methods may well have caused a few raised eyebrows but they were enormously effective. No one who participated in them will ever forget the Friday 'circus' of those days, with B-B introducing and presenting splendid clinical material and Henry dissertating on B-B's themes. Were a student's spirits to sink and medicine seem pointless and tedious, a few weeks under their wing would inevitably raise morale and restore enthusiasm. They may not have been able to follow B-B's example and give their whole life to medicine, but they certainly received a strong incentive to attempt to do so.

As a teacher B-B was down to earth, simple to understand, and quick to get to the nub of a problem; he always talked about 'the patient', never 'the case'. He frequently identified scurvy in many of the underprivileged patients he saw, and he would put sick elderly patients into his car and take them to hospital himself if time was important.

Early every morning, Sundays included, he would set off in his Rolls Royce accompanied by as many students as the vehicle would hold, and would work through the host of patients encountered from Southport to St Helens, and from Clatterbridge to Broadgreen, with unflagging energy and inexhaustible enthusiasm. He taught practical clinical medicine, spiced with an endless flow of anecdotes and aphorisms. Many will recall the cautionary gambit (allegedly biblical), 'What does it profit a man if his wife is a widow?', and 'The Almighty is the best physician even if he hasn't got his plate in Rodney Street.'

As an extrovert he could be infuriating and many students had phone calls at 7 a.m., always asking them to do something for a patient. They learnt to walk round the hospital at night searching for an empty bed for someone he might wish to admit.

He married Norah Stewart Kirkham in 1928 and they had five children, four sons and a daughter. The marriage was dissolved in 1954.

Although a lifelong Protestant, B-B was essentially ecumenical in outlook and a requiem mass was held in his honour at St Catherine of Siena Church, Lowton Leigh. There were troughs as well as peaks in his temperament but he was entirely without any pretensions, was proud of his St Helens background, always supported his juniors and gave unstinted service to Liverpool. His hobbies were motoring, especially sports cars, and collecting silver and furniture. GW/VL

[*Brit.med.J.*,1986,**292**,1679;**293**,145; Liverpool med.Inst.,*Trans.and Rep.*,1975; *Evening Standard* 26 Feb 954; Photo]

BALLANTYNE, Diamond Allan *b.*1 Sept 1911 *d.*5 June 1984
MB ChB NZ(1937) MRCP(1946) MRACP(1952) FRACP(1958) FRCP(1969)

Diamond Allan Ballantyne was born in New Guinea of New Zealand parentage. His father, Andrew Ballantyne, was a clergyman and his mother, May, née Jenness, was the daughter of a watchmaker and jeweller. His early education was at Eastern Hutt, Lower Hutt, and he went on to Wellington College and the University of Otago. His clinical studies were pursued at Dunedin Hospital.

Immediately after qualification he held house appointments at the Napier Hospital, New Zealand, and with the advent of the second world war he enlisted in the 2nd New Zealand Expeditionary Force in 1940, where he served until 1945. In 1941 he was captured in Crete, on two separate occasions, being classified as British POW 23918. He was twice mentioned in despatches.

After the war he came to the UK and spent a year in London as a house physician at the Hammersmith Hospital and at the Brompton. He returned to New Zealand in 1948 and was appointed visiting physician at the Hastings and District Fallen Soldiers Memorial Hospital, where he served until his retirement.

Ballantyne had a firm social commitment; in 1959 he became brigade surgeon to the Hastings divison of St John's Ambulance Brigade, he was officer in charge of the medical unit of the Hastings Civil Defence Organization in 1961, and a member of the council of Hastings Marriage Guidance in 1963. He was also a member of council of the New Zealand branch of the British Medical Association.

He married Sybil Joyce, née Clinick, the daughter of a farmer, in 1939. There were no children of the marriage.

His leisure pursuits included trout fishing and farming; he was a commercial asparagus grower. He also had a love of music and enjoyed playing the piano. GW/VL

BALME, Harold Wykeham *b.*5 Nov 1918 *d.*24 June 1988
MB BChir Cantab(1943) MA(1947) MRCP(1947) MD(1948) FRCP(1959)

By the death of Wykeham Balme medicine lost one of its most colourful characters who can never be replaced. He was born in 1918 in northern China, the second son of a medical missionary, Harold Balme MD FRCS, who wrote a book on the relief of pain and subsequently settled as a general practitioner in Henley on Thames. Wykeham was educated at Windlesham House and Marlborough College, and then went on to King's College, Cambridge, where he studied medicine. Whilst up at university he played the French horn in the Cambridge Orchestra.

After graduation he became house physician to the medical unit at Bart's, at the Sector Hospital at Hill End, St Albans. A year later he joined the RAMC, serving in France and Germany. After demobilization and return to civilian life, he obtained his membership of the College in 1947 and was appointed medical registrar to Leslie Cole [*Munk's Roll*,Vol.VII,p.109] at Addenbrooke's Hospital, Cambridge, in the same year. He returned to Bart's a year later where he remained for the next six years, first under Eric Scowen, later Sir Eric, and then under Ronald Christie (q.v.) in the medical professorial unit.

He was appointed consultant physician to Whipps Cross Hospital in 1954 and to St Bartholomew's Hospital two years later. In 1959 he was elected to the Fellowship of the College. He wrote several papers on general medicine during his registrar days, being particularly interested in the genetics of diabetes. Later he wrote on more general topics such as 'Advice to a house physician' and 'Rheumatoid arthritis from the patient's point of view.'

He married his wife Joan Lund in 1940, while still an undergraduate. She was a teacher and the daughter of a missionary. They had five children; three boys and two girls. The eldest son, Nicholas, died tragically at the age of 18, and it seemed to many of his friends that Wykeham never fully recovered from his death. He was a great family man and very hospitable at his lovely home in Loughton. Although he did not holiday abroad, he visited American centres about which he gave a somewhat acidulous account. He later went to Lamborene, Albert Schweitzer's hospital. After his appointment at Bart's as a general physician, he realized that he should have a sub-specialty and chose rheumatology. He became the hospital's first consultant in the subject in the 1960s. He visited several rheumatology centres and, although he did not produce any major original contribution on the subject, he founded a school of rheumatology at Bart's from which emanated many consultants in rheumatology, including Paul Dieppe of Bristol Royal Infirmary, Ted Huskinson and David Doyle at Bart's, Tom Corverman at Bath, and P A Bacon at Birmingham; thus providing an excellent opportunity for young men to develop the specialty. He also took over rheumatology at Whipps Cross Hospital, in the absence of the consultant in physical medicine, being followed by David Doyle.

Wykeham Balme was one of the founders of the medical education centre at Whipps Cross and became its second tutor. He also founded the North East Metropolitan Physicians Club, which still meets twice a year at hospitals in the region. He was sub-dean at Bart's and president of the rugby club; his name has been perpetuated there in a room in the new Robin Brooke Centre and this must have been a source of satisfaction to him.

He retired early, at the age of 62, after having staphylococcal septicaemia. Although he was itching to retire, immediately he did so he started doing hospital locums as consultant physician; first at Hackney Hospital, commuting from Muker in North Yorkshire for four days each week, and then at other hospitals in England - including the Isle of Wight and Darlington.

Wykeham was full of repartee and wise sayings: 'Taken on a new lease of wife', 'Its sex, God bless it', 'If I miss a committee meeting at Bart's I find I have lost my one bed, but if I miss one at Whipps Cross I find they have given me an extra ward to look after.' He was outspoken and perhaps offended the ultrasensitive, but to his friends he was a true delight, and he looked after his colleagues and their relatives with devotion. He was a very able physician and quick to assess the genuine case. At times he seemed to have little sympathy with neurotics and he had no time for malingerers.

He had several interesting hobbies, but being a somewhat secret person these often only became apparent to his friends accidently. He was an excellent photographer and concentrated particularly on flowering plants and birds; he had a wonderful collections of slides of English wild flowers, which he hoped to publish. He also had a lifelong interest in music; he played the piano well but preferred to do so in private. After retirement he joined the Muker silver band, playing the trombone, and indeed they were rehearsing to perform at the Durham Festival at the moment when he died. He was a keen gardener, and brewed his own wine. He was one of the few physicians to own a pub, the *Farmers Arms* at Muker, which was managed by his brother-in-law. This pub is on the Pennine Way and serves lunches to hikers and other wayfarers. He had a lovely house on a hilltop overlooking the village, known as Kisdon House. And during his days as a registrar he was the proud possessor of two ancient Rolls Royce cars.

Wykeham Balme will long be remembered for his bushy eyebrows, quick responses, wit and somewhat earthy humour, and for his great skill as a diagnostician and a deflator of pomposity. He was conservative in his views, especially regarding medical treatment, being sceptical of the value and potential hazards of some modern drugs. It is somewhat ironic that he had suffered from angina pectoris intermittently for two years prior to his death and had never sought the advice of a cardiologist.

KWGH

[*Lancet*, 1989, 1, 56; Photo]

BANNERMAN, Robin Mowat *b.*2 Feb 1928 *d.*8 Mar 1985
BA Oxon(1949) BM BCh(1952) MA(1960) DM(1960)
MRCP(1955) FRCP(1973) FACP(1966)

Robin Bannerman was born in Alton, Hampshire, the son of a Scottish pathologist, Robert George Bannerman,and his wife Charlotte Sinclair Mowat.He was educated at Westminster School and went up to Christchurch, Oxford, as a Westminster Scholar in 1946, obtaining an honours degree in animal physiology. His clinical years were spent at St Thomas's Hospital where, after qualifying in 1952, he served successively as casualty officer, house physician, senior house physician, demonstrator in clinical pathology and medical registrar.

Bannerman early developed an interest in disorders of the blood and it was very appropriate that in 1957, having been awarded the Radcliffe travelling fellowship at University College, Oxford, he should spend a year in Carl V Moore's department in St Louis, USA, which had become a major centre of haematological research. His work with Moises Grinstein on the defect of haem synthesis in thalassaemia provided the basis for his prize winning monograph *Thalassaemia*, New York, Grune and Stratton, 1961. A second year in the United States was spent with Victor McKusick at Johns Hopkins. His friendship with McKusick was an important factor in his decision to return subsequently to the United States.

Back in England in 1960 he joined the Nuffield Department of Medicine in Oxford, where he collaborated with the writer and others in studies of iron deficiency and iron absorption. We developed a friendship which survived a move by both of us to the USA in 1963. Robin Bannerman was greatly attracted to the excellent department of medicine which Evan Calkins was developing in Buffalo and decided to become one of the mainly young band of physician investigators at the University of Buffalo, where he remained until his death. He established a medical genetics unit at Buffalo General Hospital. In 1975 he became director of the divisions of medical genetics and human genetics in the respective departments of medicine and genetics. He created a first class clinical and teaching programme in clinical genetics. He continued to investigate haem and porphyrin metabolism in the haemoglobinopathies and thalassaemia and contributed significantly to knowledge of globin and iron metabolism. His unit attracted many young investigators who have subsequently distinguished themselves. Though he had known that he had a lethal malady long before his death, Robin continued to work right up to the end.

Robin married Franca Angela Eleonora Vescia in 1953 and they had three daughters, one of whom is now a physician in Montreal. Their homes were happy and welcoming ones from the time when they lived in a cottage in Oxfordshire, which lacked even the rudiments of civilization in its physical resources, to the house in Buffalo memorable to the writer for the night of the great blackout of 1966; when the electric power failed throughout the whole of the State of New York.

Robin Bannerman will be remembered for his gentleness and shyness, and for the determination and firmness of purpose which those qualities concealed. He was a man totally committed to maintaining the highest standards of precision in thinking and expression. On first contact he seemed an unusually serious person but increasing friendship revealed an attractive, pawky sense of humour which, together with his unfailing kindness, endeared him to all who knew him. He sometimes gave the impression that he would have been happier as a farmer, in the steps of his maternal grandfather, but had he so chosen, medicine would have lost a scholar.

KBT/GW

[*Lancet*,1985,**1**,1114; *Buffalo News*,9 Mar 1985; Photo]

BASSADONE, Enrique Teofilo *b.*15 Mar 1919 *d.*23 Dec 1986
LMC Madrid(1933) MRCS LRCP(1942) MRCP(1951)
FRCP(1970)

Enrique Bassadone was born and went to school in Huevla, Southern Spain, though as his parents were from Gibralter he was actually a British subject. He chose to study medicine at the University of Madrid, where he qualified Licenciado de medicina y cirugía in 1933, the equivalent of our first medical degrees. After qualification he first held a house physician post and then served as a junior pathologist before being appointed junior assistant to Jiménez Díaz, the professor of medicine in Madrid. He kept this post for three years.

On the outbreak of the Spanish Civil War in 1936 he joined the Government faction and served with great distinction, rising to the rank of brigadier general. He saw much active service during the early stages of the war and when Madrid fell to General Franco's forces he was fortunate to escape to France, despite the problems of his subsequent internment under difficult conditions.

His good fortune returned when he was allowed to take up residence in Britain. Despite the fact that his Spanish qualification was deemed acceptable for practice in this country, he worked to obtain the Conjoint with typical industry and application, qualifying in 1942 in the middle of the second world war. He was working at this time at St Olave's as senior resident physician and was establishing a reputation both for clinical work and for nutritional research on the development of the wartime national loaf, carried out under J A Ryle [*Munk's Roll*,Vol.IV,p.595] at Guy's.

For Enrique Bassadone, 1947 was a very important year. He was appointed assistant physician at St Alfege's and married Mary Kirkland, a nurse working at St Olave's. Both events were to prove pivotal in his life, exchanging the chaos and confusion of his early years for the stable framework of hospital and home life which so characterized his later years. He remained at St Alfege's, which became the new Greenwich District Hospital, for 27 years until his retirement in 1974. His achievements were many. He set up one of the very first intensive care units in the country and became physician in charge of the unit after his appointment to the Miller Hospital, and the new Greenwich District Hospital. He was very influential during the planning and commissioning of the new hospital. Always interested in politics, he was chairman of the medical staff committee for several years, but first and foremost he was a physician with great diagnostic acumen and even greater compassion.

Enrique was an extremely likeable man. Dapper and lithe is appearance, he always had a ready smile, tempered by a thoughtful - and often thought provoking - comment. He had a wide knowledge of life and its vicissitudes, as well as of medicine, art, literature and politics A well informed man, he never made you a present of his opinions; he was not reserved, radiating a quiet, restrained buoyancy and optimism. Needless to say, his patients adored him and his colleagues relied heavily on his advice and wisdom.

After retirement he went back to live in Southern Spain, with his wife Mary, where they spent several happy years now that he was free from the risk of political chastisement. Eventually they returned to England. In these latter years, before he suffered the stroke that so gravely affected the final period of his life, he maintained his interest in medicine and people, often popping in to postgraduate meetings to 'keep his brain alive', as he put it. He was survived by his wife Mary and their three sons.

PM/GW

[*Brit.med.J.*,1987,**294**,654; Photo]

BEDRI, Ali *b.26* Nov 1903 *d.13* Jan 1987
MBE DKSM Sudan(1928) *FRCP(1952)

Ali Bedri was the son of Sheikh Babikir Bedri, an outstanding Sudanese educationalist. He was educated in Rufaa Blue Nile province where his father founded his schools, going on to the Gordon Memorial College in Khartoum where he graduated as a teacher in 1923. He subsequently decided to study medicine and was one of the first group of students to enter the Kitchener School of Medicine, graduating with distinction in 1928. Following house appointments at Khartoum Hospital he served in Singa, Dongola, the Nuba Mountains and Sennar, demonstrating outstanding ability. Later he became the first Sudanese to hold the post of senior medical officer at Omdurman Hospital and Khartoum Hospital, a post previously held only by Syrian or Lebanese doctors. In 1937 he came to Britain for postgraduate training at Hammersmith.

On his return to the Sudan he became the first Sudanese doctor to hold the post of assistant deputy director of medical services under Sir Eric Pridie [*Munk's Roll*,Vol.VII.p.479] an international authority on tropical medicine. Bedri's previous experience as a medical inspector in various parts of the Sudan had brought him into contact with the many health problems of the country and the need to build an efficient medical service. He faced the difficult task of deciding priorities in view of the scanty budget available and the exacting conditions of the second world war, and carried out his responsibilities with wisdom, moderation and civility. Under the guidance of his British teachers and senior colleagues, he developed the foresight which made him the first architect of an independent Sudanese medical profession. He was appointed a member of the Advisory Council for Northern Sudan and was a member of the Sudanization Committee. Although his status as a civil servant prevented him from overt indulgence in politics, he was a staunch believer in Sudanese independence. In 1949 he became a member of the first legislative assembly and was appointed Minister of Health.

While in office as Minister of Health, during the period before independence, he initiated a 10-year plan for building an efficient health service. He foresaw the need to cater for postgraduate studies and training so that Sudanese doctors could shoulder their responsibilities in both preventative and curative medicine, and for the training of paramedicals, nursing staff, maternity and child welfare, and health workers. An able administrator, he was wise, patient and a stickler for perfection. In return he gave his staff the freedom, trust and support they needed.

Ali Bedri maintained close contacts with the Royal Colleges in Britain and other British medical institutions. He fostered relations with the RCP both at home and abroad. He never failed to attend Comitia when he was in London during his summer visits. In Khartoum, he made a point of meeting any visiting members of the British Colleges to discuss with them medical standards and postgraduate studies; a practice he continued when president of the Sudan General Medical Council.

Bedri voted for self-government of the Sudan and with its advent in 1954 he relinquished his post as Minister of Health, returning to private medical practice. He continued to take an active role as president of the Sudan Medical Council, as well as working with the Red Crescent.

Ali Bedri's ability made him the doyen of medicine in the Sudan. He had a brilliant career, building the structure of medical training, exemplary bedside manners and sound traditions at a time when few diseases had specific treatment, and there were no antibiotics even for endemic diseases. He always welcomed his colleagues, particularly his juniors, who came to him frequently for advice. He never disappointed them and was always courteous and helpful. His compassion, his smile and his insight were greatly appreciated by his patients.

On his 81st birthday he invited all his colleagues who had graduated from the Kitchener School of Medicine and greatly enjoyed talking to those of a younger generation. His 'family' of colleagues retained a real warmth and affection for him, as he did for them.

He also kept a very special affection for and attachment to all the members of the large family of Bedri, reliving his life and youth in his own children. He was survived by his son and four daughters, and twelve grandchildren.

AMH/GW

[*The Times*,29 Jan 1987; Photo]

BEHRMAN, Simon *b.*14 Dec 1902 *d.*9 Dec 1988
BSc Lond(1925) MRCS LRCP(1928) MRCP(1932) FRCP(1964)

Simon Behrman was consultant emeritus in neurology to the Guy's group of hospitals when he died.

He was born in Imperial Russia, the son of Leopold Behrman, an influential timber merchant, and his early education was at the gymnasium in St Petersburg. The family was forced to flee from St Petersburg at the time of the revolution, and managed to escape only hours before the Bolsheviks seized their home. Simon Behrman was a refugee in Europe for over a year, eventually reaching Glasgow in 1919 to stay with medical relatives. Although unable to speak a word of English, within 18 months of his arrival he had secured a scholarship to University College London, where he obtained a BSc with first class honours. He proceeded to do his clinical training at St Bartholomew's Hospital medical school and qualified in 1928. His first house appointment was at Princess Beatrice Hospital, following which he worked at Queen's (now Queen Elizabeth) Hospital for Children in Hackney Road, from where he wrote the first of many scientific papers. In 1932 he obtained his membership of the College, and then went to the National Hospitals for Nervous Diseases, Queen Square, where his enthusiasm for neurology was fired by Sir Charles Symonds [*Munk's Roll*,Vol.VII,p.563] whom he always regarded as his mentor. He was house physician and registrar at Maida Vale Hospital for Nervous Diseases from 1930-33, and later registrar at Queen Square and Guy's Hospital.

Behrman's special interest in neuro-ophthalmology led to his appointment as consultant physician to the Central London Ophthalmic Hospital. in 1939, and he remained on the staff of Moorfields Eye Hospital where he was held in the highest regard. He retired in 1967.

During the second world war, Simon Behrman worked in the EMS, based at Guy's Hospital and covering the south east metropolitan region. At the inception of the NHS, he was appointed consultant neurologist to Bromley Hospital and in 1950, with Geoffrey Knight, established the south east metropolitan regional neurosurgical and neurological unit at the Brook Hospital, Woolwich, the first of its kind, which served as a model for similar units being developed elsewhere in the country. He was also consultant in neurology to St Olav's, Dulwich; St Francis, Farnborough, Kent; St Leonard's, Shoreditch; St Giles and Queen Mary's (Stratford) Hospitals. He was elected to the fellowship of the College in 1964.

Behrman was widely known for his exceptional clinical skills and outstandingly detailed history taking. He always paid considerable attention to letters written by general practitioners and in particular to any case reports from psychiatric social workers - no doubt influenced by his marriage in 1940 to Doris Engelbert who had been senior psychiatric social worker at Guy's Hospital.

His close association with ophthalmologists at Moorfields brought recognition of the differences in approach to medical problems between ophthalmologists and physicians. It prompted him to found the Eye Physic Club in the 1960s, to bring together the two disciplines, with which he remained actively involved up to the time of his death.

Simon Behrman was a prolific writer, the author of over 300 scientific publications. His fascination with all aspects of medicine led to papers on such diverse topics as trauma and orthopaedics, psychiatry, medical history, and medicine in literature. All his writings were characterized by the most beautiful English prose, and he continued to write articles until a few months before he died. He had an encylopaedic knowledge of medicine and neurology; one of his special interests was papilloedema.

Outside medicine, as a young man he enjoyed walking and climbing. His appreciation of antique furniture stemmed from his early association with his father's timber trade. He had a wide knowledge of architecture and took pleasure in studying many of the old buildings in South London, particularly around Blackheath. He enjoyed Russian literature, which he could read in the original.

Simon lived in the same house in Harley Street for over 50 years and often said that his reason for staying there was its proximity to the Royal Society of Medicine, in whose library he could often be found late into the evening.

He was survived by his wife and four children, one of whom is a physician.

JC

[*The Times*,27 Dec 1988; *Brit.med.J.*,1989,298,520; Photo]

BENSON, Robert Charles Stirling *b.*22 Aug 1915 *d.*16 May 1986
MRCS LRCP(1939) MB BS Lond(1939) MRCP(1946)
FRCP(1969)

While on the golf course at Blakeney, Norfolk, Robert Benson felt faint, lay down, and quickly passed away as a result of a heart attack. If he could have chosen the manner of his going that is roughly what he would have opted for; golf was one of his favourite pastimes. Blakeney, a beautiful flint and brick village on the fine but austere north coast of Norfolk, was the place he chose for his retirement, a place where he had made some close friends and passed several exceptionally happy years. A heart attack was the old enemy he had grappled with in the lives of others for most of his professional career; an enemy that had had a silent dig at him some ten years previously.

Robert Benson was born in South Australia, where his father was a doctor. He was educated at St Peter's College, Adelaide, but his entire medical education took place in England and he never returned to Australia for any spell of medical practice. The London Hospital was the scene of his medical training, both as student and as junior doctor, and of that great institution he remained an enthusiastic and lifelong admirer. He graduated in 1939, obtained his MRCP London in 1946, and was elected to the Fellowship in 1969. 1947 was a seminal year: he became cardiac registrar to Sir John Parkinson [*Munk's Roll*, Vol.VII,p.443] and William Evans (q.v.) and was awarded the Paterson Research Scholarship. One of the most endearing of Bob's qualities was the warmth of his feelings, and his loyalty to those whom he liked and admired; his regard for William Evans amounted almost to veneration. William Evans was internationally renowned and admired as one of the greatest teachers of his generation. There grew up between him and Bob a relationship almost like that of father and son, a relationship which endured for the rest of his life. One of Bob's delights was to talk about visits he paid to the great man in his retirement in South Wales. Jeeva Raj, once Evan's registrar at Rush Green Hospital, writes: 'In his book *Journey to Harley Street*, [London, Rendal, 1968] William Evans describes the warm friendship and spontaneous loyalty extended to him by his younger colleagues, one of whom was Robert Benson. William Evans writes that they met to reminisce of earlier days, regale him with their diverse activities in the practice of cardiology, and announce their plans and aspirations for the future. These meetings gladdened his heart.' For Benson's retirement party Evans wrote a letter which included these words: 'Among the many registrars who helped me none gave greater service to me and my patients than Robert Benson, none showed greater loyalty to the fundamental clinical data which we collected from our patients, whom we examined together.'

In 1949 Benson became chief clinical assistant to William Evans at the National Heart Hospital, and in the same year took over from him as consultant cardiologist at Oldchurch Hospital. A year later he was appointed consultant physician to Rush Green and Victoria Hospitals.

Bob was responsible for introducing at a local level the early mobilization of patients after myocardial infarction, and later the concept of specialized coronary care. In the 1960s, Ward 5 at Rush Green Hospital became a coronary care unit under his directorship. It increased to seven beds, making it one of the largest in the NE Thames Region, and it included a laboratory in which right heart catheterization was carried out. The unit attracted excellent technical, nursing and junior medical staff, some of whom went on to careers in cardiology, and it remains an enduring monument to Bob's professional achievements. He was open to new ideas and was quick to introduce worthwhile ones, such as continuous ECG monitoring, treadmill ECG testing, and 72 hour ECG tape recording and analysis. But as Jeeva Raj says: '...he always reminded us not to be carried away by technological advances in cardiology lest scientific curiosity should reach a point of sin'. He was elected a member of the British Cardiac Society in 1952, and was a founder member of the Junior Cardiac Club in London. He was a regular attender at cardiac meetings. In 1956, in the *British Heart Journal*, **18**, p.529-543, he published jointly with J F Smith a paper on cardiac amyloidosis.

Benson had admirable teaching and training abilities, and when he knew that a young doctor or other member of his staff was working reliably and loyally, under his direction, he became an enthusiastic adviser and supporter. He felt the same about his patients. But he reacted in a somewhat different fashion to anyone displaying what were to him less desirable characteristics. He demanded obedience as if he were commander of a ship; which may have been part of his nature. This was strongly reinforced by his wartime experience in the Royal Naval Volunteer Reserve. He was at sea in the Western Approaches Atlantic during the worst period of the war, from 1940-41, and later he was principal medical officer on an armed merchant cruiser in the Pacific. At the time of demobilization he held the rank of surgeon lieutenant commander, RNVR.

He undoubtedly had qualities of leadership; he was respected and admired by a loyal band of medical, nursing and paramedical colleagues. But there was a darker side to those same qualities: he could not bear to be thwarted, or even disagreed with. In this respect his inflexibility resulted in some hurtful, even broken, human relationships which he could not or would not seek to mend.

In 1948 he married Inger Kristine, a Norwegian by birth, and there were two sons and a daughter of the marriage. The marriage was dissolved in 1974.

JRB/GW

[*Brit.med.J.*,1986,**292**,1679; Photo]

BENSON, Trevor Baylor *b.*22 Sept 1925 *d.*5 July 1984
MB ChB Manch(1949) MRCGP(1964) MRCP(1968) DPhys Med(1970) FRCP(1982)

Trevor Baylor Benson was born in Rodney Street, Liverpool, - an auspicious start to a medical career. His father was a bank manager, and the young Benson had a somewhat peripatetic life. Most of his childhood was spent in the Lancashire cotton towns and he attended Bury Grammar School before proceeding to the Manchester Medical School and Manchester Royal Infirmary. He qualified in 1949 and was appointed house surgeon to Sir Harry Platt - a clinician who influenced him greatly. He saw national service with the Royal Air Force at Nocton in Lincolnshire. He was designated junior orthopaedic specialist and obtained a wide range of clinical experience. Two important events occurred while he was stationed in Lincolnshire: he met Ruth Oliver, a Bart's trained nurse, and then quite inexplicably he sustained a massive pulmonary embolism. He was regarded as fortunate to survive this illness and thereafter bowed to medical advice that he would never be strong enough to pursue a career in orthopaedic surgery.

He married Ruth in 1953 and sought an opening in general medical practice. He became a principal in Helsby the following year, 1954. This was, and still is, a sleepy Cheshire village acting as a dormitory area for the heavy manufacturing localities of Ellesmere Port, Runcorn and Warrington. He was happy there. He and Ruth brought up three children and grew the most splendid tomatoes.

He was a widely respected local family doctor and founder member of the Royal College of General Practitioners. He continued however to retain his interest in musculo-skeletal problems and returned to hospital based medicine with a clinical attachment at the Manchester Royal Infirmary. Remarkably and characteristically, he decided that he would obtain the membership of the Royal College of Physicians. That he should even contemplate such study more than two decades after qualification raised a number of eyebrows, but he became MRCP(UK) at the first attempt; following it, in 1970, with the diploma in physical medicine.

Services in rheumatology and rehabilitation in the Mersey region had been notoriously poor. In 1975 the Regional Health Authority appointed him its first consultant in this specialty. It was a wrench for him to leave general practice, but having done so he devoted himself wholeheartedly to the development of services for rheumatology and the younger physically disabled. His paper on 'The younger disabled unit at Fazakerley Hospital' in the *British Medical Journal*, 11 August 1979, in collaboration with E Williams, has become a classic reference on the subject. He was greatly surprised when invited in 1981 to join the standing medical advisory committee at the Department of Health. Shortly afterwards he was elected FRCP.

A kindly, courteous and quiet man, he wore a benign smile and often appeared bumbling and absent-minded. He had a rigorous intellect but found great difficulty in organizing himself or his workload for maximum efficiency. It was this mixture of friendliness, intellectual purity and gentle eccentricity, which endeared him to family, patients and colleagues. He was survived by his wife and three children. EW/GW

[*Brit.med.J.*,1984,**289**,707; Photo]

BLACK, Kenneth Oscar *b.*10 Nov 1910 *d.*5 Aug 1987
MA Cantab(1932) MRCS LRCP(1935) BChir(1936) MB(1937) MRCP(1937) MD(1942) FRCP(1946)

Kenneth Black was born at Scarborough, the son of George Barnard Black, a solicitor and company director, and his wife Annie Elizabeth Hirst, daughter of John Hirst, minister of religion. He received his early education at Bootham School, York, and then studied natural science at Leeds University for a year before going up to King's College, Cambridge, as an exhibitioner. He gained a first class in the natural sciences Part I, and also excelled in athletics. In 1931 he won the Cambridge University Athletic Club's senior sports one mile in four minutes 38 seconds, and was awarded his blue for cross country running.

In 1932 Kenneth Black entered the medical college of St Bartholomew's Hospital, with which he was to remain associated for the rest of his life. He qualified with the conjoint in 1935, obtaining his Cambridge MB in 1937 and his membership of the College in the same year. He was appointed house physician to George Graham [*Munk's Roll*,Vol.VI,p.205] and later became his chief assistant. At an early stage of his career he developed an interest in biochemical disorders, especially in diabetes. In 1940 he became a consultant with the Emergency Medical Service, before being commissioned in the RAMC in 1942. He served in West Africa and India as a medical specialist, and later as officer in charge of a medical division with the rank of lieutenant colonel.

Shortly after demobilization in 1946 he was appointed consultant physician at Bart's and was elected to the Fellowship of the College that same year. He took charge of the diabetic clinic at Bart's, and under his leadership it became a model for developments eleswhere, with an accent on team work and cooperation between dieticians, chiropodists, social workers, and his ophthalmological and biochemical colleagues. He was also involved with the early activities of the British Diabetic Association.

He served on the staff of St Luke's Hospital for the Clergy, and was an examiner in medicine for London University, the Society of Apothecaries and the College.

By nature he was a shy man, avoiding all forms of ostentation. He enjoyed many friendships, and his qualities of gentleness and kindness were combined with tremendous strength of mind when defending principles in which he believed. Throughout the years he retained his youthful appearance and continued to enjoy athletic pursuits.

His family, on both sides, were Methodists and he was brought up under strict Methodist traditions. While living in London he became greatly involved with Hinde Street Methodist Church, where he held many offices, and developed a lasting interest in theology. He studied the New Testament in depth and read widely on the subject. His Christian faith was the greatest force in his life, and his medical work and contact with patients were entirely fulfilling for him.

In 1959 he married Virginia, daughter of Herbert Lees, stockbroker, and they had two daughters, Anne and Clare. He was a devoted family man and in retirement they enjoyed family life to the full; at home with long walks on the South Downs when his old love of botany was shared by all, and opportunities for travel abroad where he could pursue his interest in archaeology. His wife and daughters survived him.

WEG

[*Brit.med.J.*,1987,**295**,1493; *Lancet* 1987,**2**,753-4]

BLAIR, Leslie George *b*.13 Oct 1901 *d*.1 Nov 1984
MRCS LRCP(1923) DMRE Cantab(1933) MRCP(1948)
FFR(1954) FRCP(1954) FRCR(1975)

George Blair was born at Birkenhead where his father, Charles James Blair, was a schoolmaster. His mother, Louisa, was the daughter of John Kynsey, a civil servant. 'Billy' as he was known to his friends, received his early education in Birkenhead and went on to the University of Edinburgh and the London Hospital, where he qualified with the conjoint in 1923. His early clinical experience as a house physician at the Brompton Hospital led to his interest in chest diseases, from which he himself suffered, and by 1933 he had decided to specialize in radiology, and initially obtained an unpaid post as clinical assistant in the X-ray department at the London Hospital. He was appointed radiologist to Harefield County Hospital in 1936, and began his successful career in radiology. He continued to work at the Harefield after being appointed as director of the X-ray departments at the Brompton and Great Ormond Street in 1946. He also gave considerable service to local institutions in Buckinghamshire. He became one of the leading chest radiologists of his day, finally retiring from the NHS in 1967, although he continued to practice in Harley Street until well after his 80th year.

In the post war years, Blair was one of the three leaders in thoracic radiology. It was a new and controversial discipline and there were times when he needed to call upon all his experience and knowledge to contest the challenges of Sir Peter Kerley [*Munk's Roll*,Vol.VII,p.317] and George Simon [*Munk's Roll*,Vol.VII,p.536], and radiological meetings tended to be very lively occasions.

Seeking a sounder basis for interpretation, Blair was among the first to use the Gough-Wentworth total lung section method for pathological-radiological correlation, with the assistance of his colleagues in pathology. He rarely disappointed radiologists who sought his opinion on difficult cases, and his papers were meticulously and elegantly delivered. His powerful visual memory enabled him to remember many patients by their lung patterns. As a teacher he was outstanding in reporting sessions and covered a large volume of work, particularly in mass surveys, yet he never lifted a finger - so that his self-winding watch kept stopping.

Billy was an authority on thoracic conditions in children, especially tuberculosis, at a time when such conditions were more common, and his first article in *The British Journal of Radiology* was on the interpretation of children's chest X-rays. He also wrote lucidly and with authority on thoracic trauma, lung abscess and other infections.

Based on the work of Foster-Carter and Hoyle, he established the importance of pulmonary anatomy in relation to disease, best exemplified by the chapters on segmental lesions in the lung in *Modern Trends in Radiology*, 1948. Although less well documented, he also wrote papers on bone disease, e.g. in the reticuloses.

Blair possessed great organizational and administrative ability and was genuinely director of his two main departments; his skill in committee management being largely due to his meticulous advance preparation. He was elected to the Faculty of Radiologists in 1954, becoming vice-president in 1956, and was honorary president of the European Society of Paediatric Radiology during the London congress in 1961. In 1946, he was awarded La Medaille Vermeil de la Reconnaissance Française for his services to the French community in Britain during the war.

Billy married Olive Stewart Brown in 1927. There were no children of the marriage and his wife predeceased him by a few years. He had relied heavily on her unfailing support and after her death, being himself in failing health, he withdrew from medical and social activities, apart from his private practice. He ran a very successful practice in Harley Street where he was greatly dependent on his radiographer, Maureen Wild. And in his hospital departments he always ensured that his juniors did all the physical work while he himself was truly a consultant; this inactivity possibly resulting from his early ill health.

Blair had a quiet, reserved personality, controlled and dignified, but when challenged he could produce reserves of sardonic wit. He was at his best during small gatherings of colleagues, such as club meetings, when he was greatly encouraging to younger members.

His favourite leisure pursuits were fishing and shooting, and he had his own pheasant shoot near Amersham; he was a very good host and enjoyed sharing it with friends.

TL

[*Clinical Radiology*,Nov 1985,p.552; Photo]

BLANEY, Patrick Joseph Aloysius *b.*12 Aug 1917 *d.*5 Aug 1984
MB BCh BAO Belf(1939) MRCP(1947) MD(1952) FRCP(1974)

Patrick Blaney was born in Belfast, where his father was a wine merchant. He studied medicine at Queen's University Belfast and after graduation he worked in general practice, and in various hospital posts in Belfast and Nottingham. He was senior registrar at the General Hospital, Nottingham, between 1946-52. He married Mary Catherine McNeill in 1951 and they had nine children; five sons and four daughters.

Blaney moved to Dublin in 1953 as one of the first whole time consultants appointed by the then Dublin Board of Assistance. He joined four other consultants (one physician, two surgeons and one obstetrician) at Kevin's Hospital. This small group of dedicated men faced the mammoth task of caring for over 2,000 patients. They brought modern medicine to the sick poor of the city, and they changed the image of the hospital from that of the South Dublin Union to that of an active municipal hospital.

Over the years many young doctors followed Patrick Blaney on his wardrounds. The length of these rounds and the number of patients seen on them are already legendary. Those who worked for him describe his great skills as a bedside clinician. Equally impressive was his humanity. He showed an awareness of the importance of social factors in the genesis of disease and in this he was a man very much ahead of his time. His acumen in diagnosis and treatment of acute illness was matched by his compassion and care for those with chronic disability. He was in every respect 'a complete physician', a doctor of the old school of bedside medicine who was well acquainted with the developments and possibilities of modern technology.

Students and young doctors benefited from his teaching. Most of his registrars went on to obtain higher degrees and they are now scattered around the world in consultant and professorial posts. Many of the county physicians worked on Patrick Blaney's wards. He was an acute observer, and one of the earliest clinical reports on hypothermia came from his unit.

The achievement of Blaney and his colleagues in developing St Kevin's Hospital contributed in a major way to the evolution of the present day St James's Hospital. The decision to develop St Kevin's as a major teaching and regional hospital and to amalgamate it with three other voluntary hospitals in the Federated Group (Mercer's, Sir Patrick Dun's and Baggot Street) was a momentous one. It called for long hours of planning and negotiation. Blaney became an active and committed member of the project team for the new hospital. He showed great generosity during the years of transition from St Kevin's to St James's, sharing his facilities with colleagues from other hospitals and encouraging the expansion of the consultant staff by the appointment of new young consultants. Thirty years ago, when Patrick Blaney arrived, there were five consultants; today there are 65 consultants attached to St James's Hospital.

Blaney inspired all those who worked for him - junior doctors, nurses and paramedical staff - and he also unconsciously set standards for his colleagues in the hospital.

He had a very happy family life. He was a keen yachtsman and a member of the Royal St George Yacht Club, Dunlaoghaire; an interest which he shared with his children. He bore his final illness with bravery and continued to see patients up to a short time before he died. A Blaney gold medal for the best final year medical student has been established to commemorate this noble and wise physician in the hospital he served selflessly for so many years.

DC/GW

[Photo]

BLOOM, Harris Julian Gaster *b.*30 June 1923 *d.*21 Dec 1988
MRCS LRCP(1947) MB BS Lond(1947) MD(1949) MRCP(1950) DMRT(1954) FFR(1956) FRCP(1971) FRCR(1975) FRCS(1985) Hon FACR Hon FAAP

Julian Bloom was a consultant radiotherapist at the Royal Marsden Hospital for the 30 years between 1958 and 1987, and subsequently became an honorary visiting professor at the Institute of Cancer Research. Also, he was founder and co-chairman of the Radiotherapy and Oncology Centre at the Cromwell Hospital, London, where he continued to care for patients until shortly before his death.

He was born at Sheffield where his father, Arthur Bloom, was an engineer, and educated at Fleet School and the Middlesex Hospital, University of London. After house posts at the Middlesex, he was assistant pathologist for two years and was then called up for National Service and appointed a senior specialist in pathology in the RAMC from 1950-52. From 1952-56 he trained in radiotherapy in the Meyerstein Institute of the Middlesex Hospital; his first major contribution to oncology was at this early stage of his career with a seminal study, together with W W Richardson, on 'Histological grading and prognosis in breast cancer: a study of 1409 cases of which 359 have been followed for 15 years.' This grading of breast cancer remains in widespread use today.

Julian Bloom joined the department of radiotherapy at the Royal Marsden Hospital in 1958, at a time when high energy linear accelerators were becoming available for clinical practice, and he was one of the leaders of a generation of radiotherapists who developed current techniques of external beam radiation. His practice of radiotherapy was characterized by an exacting attention to detail; and proposed radiotherapy fields were frequently adjusted by him with millimetre precision. His trainees regarded approval of their initial radiotherapy plans as a high accolade. His dedication to patient care was uncompromising and his hours of work within the Royal Marsden Hospital soon became legendary. A 12-hour working day was routine, his afternoon clinic rarely finishing before 9pm and occasionally extending past midnight. The same attention to detail was brought to bear on administrative matters and he was twice elected chairman of the department of radiotherapy and oncology. His academic abilities, and his dedication to all aspects of his medical career, were more than matched by the warmth and charm of his personality; those many physicians who devoted long hours accompanying him in clinical or academic pursuits were willing victims of his enthusiasm.

Bloom specialized in the treatment of urological tumours and tumours of the brain, and he made major improvements in the management of both groups. In a series of articles in the 1960s he reported the hormone responsiveness of advanced renal carcinoma and, again, this pioneering work has stood the test of time and remains one of the first approaches to treatment for patients who cannot be cured by surgery. He was an authority on the radiotherapy of bladder cancer and linked the work of the Royal Marsden hospital and the Institute of Urology in the analysis of combined modality therapies for locally advanced tumours.

The major enterprise of Julian Bloom's career was the establishment of a specialized brain cancer unit at the Royal Marsden Hospital, with the support of surgical colleagues at Atkinson Morley's Hospital, the Hospital for Sick Children, Great Ormond Street, and the National Hospitals for Nervous Diseases, Queen Square. He saw the need for broadbased investigation of these rare and devastating diseases and his unit incorporated both child and adult patients, and medical specialists in endocrinology, psychology and rehabilitation. Additionally, the unit had strong links with the Institute of Cancer Research, and the cancer research fund which bears his name supported the investigation of brain tumour chemotherapy in a laboratory setting. Clinical investigations were based on the radiotherapy of brain tumours but characteristically encompassed also hypothermia, radiosensitizers, immune modulation and high dose chemotherapy in a series of collaborative research protocols. His personal series of patients with brain tumours validated the use of radiotherapy for pituitary tumours, craniopharyngioma and meningioma. He saw gratifying improvements in outcome for children with brain tumours, and as chairman of the brain tumour study group of the International Society of Paediatric Oncology (SIOP) he developed the role of adjuvant chemotherapy in medulloblastoma. His reported results in the management of childhood brain tumours have become benchmarks against which future progress must be measured, and his 30 years in the field were honoured by an International Neuro-oncology Conference held at the Royal Marsden Hospital in September 1988.

Amongst many professional honours were fellowships of the

American Academy of Paediatrics,1981; the Belgian Radiotherapy Society, 1983; the Royal College of Surgeons of England,1985, and the Venezuelan Cancer Society, 1978. The principal theatre in the new Royal Marsden Conference Centre was named 'The Julian Bloom Lecture Theatre'.

AH²/GW

[*The Times*,22 Dec 1988; *The Independent*,2 January 1989; *Lancet*,1989,1,171]

BOLTON, John Henderson *b.*9 Mar 1907 *d.*14 Nov 1988
MB BS Melb(1932) MD(1936) MRCP(1939) FRCP(1957) FRACP(1964)

Born in Bendigo, the son of a bank manager, John Bolton was educated at Wesley College and the University of Melbourne, graduating in medicine in 1932. He became a resident medical officer at the Royal Melbourne Hospital, and was appointed its medical superintendent in 1936. Concerned by the morbidity amongst the hospital's nurses, he recorded and weighed the food entering and leaving their dining room; he subsequently reported a significant decline in morbidity, not so much by improving the diet as enhancing its palatability. His concern with nutrition, so important in his war years, lasted throughout his life. In 1938 he undertook postgraduate training in London, interrupted by the outbreak of war. Rather than return to Australia to enlist, he joined the RAMC as a lieutenant and went to France in January 1940 with the 17th General Hospital. A fortunate billeting during the cold war greatly advanced his appreciation of French red wines, but in May 1940 he was one of six of the hospital's medical officers to stay behind to care for the wounded after the evacuation of France. He was repatriated to the United Kingdom in 1943 with nephritis, which he characteristically studied in himself to the extent that the camp facilities allowed. With improvement in his health, John Bolton voluntarily returned to active service in Belgium and Germany, a brave move as further capture of a repatriated prisoner carried dire penalties. He was mentioned in despatches in 1946, when he was discharged with the rank of lieutenant colonel. In February 1944 he contributed a paper on his experience of diet and nutrition as a prisoner of war to the Inter-Allied Conferences on War Medicine, and later reported on an epidemic of diphtheria.

Returning to Melbourne with the WRNS officer, Elizabeth Rodwell, née Nankivell, whom he had married in 1944, he established a consultant medical practice which extended over 40 years, indeed virtually to his death. He was appointed to the medical staff of his old hospital, serving as an outpatient and later inpatient physician until his retirement in 1966, with periods as dean of the clinical school and as chairman of the honorary medical staff. He was also a physician to the

Repatriation General Hospital, and for some years a research scholar in the university department of obstetrics and gynaecology. For many years he was a member of the Walter and Eliza Hall Institute of Medical Research and a board member, representing the medical staff of the Royal Melbourne Hospital, from 1958-66. He was able to ensure that the clinical research units of the Institute and the Hospital as a whole maintained a successful relationship.

As a postgraduate teacher John Bolton was outstanding. At a time when the teaching of clinical medicine tended to be dogmatic and didactic, he was ever questioning. The appearance or disappearance of a physical sign was, for example, a challenge to explain the mechanism or to explore the pathophysiological basis for the change; expertise in clinical medicine was to be gained not merely by observing but by thinking at the bedside. Treatment was a mini-experiment to be followed by quantitative data or objective evaluation wherever possible, tested often by statistical appraisal. His use of statistics in 'simple' situations as well as in research projects was well in advance of the times; his statistical method was not always appreciated by senior physicians, as when he recorded the number of times individual members of the honorary medical staff spoke at their business meetings.

Perhaps Bolton's most important contribution to nutritional evaluation was the progressive grading of tongue changes in nutritional deficiency, but equally important was his regular indication of minor signs to students, signs such as diminution of vibration sense, follicular keratosis and angular cheilosis in patients with other major problems; statistical data from a continuing survey of all his patients provided the basis for the interpretation and nutritional significance of these signs. Later work related nutritional factors to the outcome of pregnancy as well as to coronary disease and the health status of the elderly. He introduced a quantitative response to the assessment of capillary fragility, thus allowing its application in a variety of clinical situations not immediately obvious. Another major interest was haematology, notably reflected in research on red cell fragility, but also apparent in his papers on nutritional subjects. The stimulus he provided to juniors with different interests is reflected in collaborative work published locally on asthma and diaphragmatic behaviour. Towards the end of his life he was preparing or assisting in papers on fat distribution and on cancer chemotherapy - the latter another collaborative interest with a younger colleague. Few physicians in full time private practice can have achieved his research output, but for the most part it was based on his own patients and directed towards their collective welfare. Not surprisingly, he devoted no attention to medical politics or economics.

John was the kindest of men and the most humanistic of physicians; academe, which he might well have adorned, he eschewed as too restrictive of his interest in people. Reticent rather than shy in his personal relationships - he declined entry in *Who's Who in Australia*, and never added to the basic entry in medical directories - he was yet a stimulating and provocative conversationalist and a great club man; with Elizabeth, he was a superb host, and withal he was a golfer and classical violinist of great ability; he established an orchestra in Germany, and played jazz for pocket money as a student. He also exploited an esoteric theory of chance in numbers with such sureness as to be requested not to play roulette at a London club, and to extract a regular return in later life from the local government lottery.

Truly a most remarkable, original and lovable man, revered by his patients and students, John Bolton left an indelible mark on the history of medicine in Melbourne, He was survived by his wife Elizabeth; their only child died in childhood.

BG/MGW

[RACP History of Medicine Library:papers & bibliography; Photo]

BORRIE, Peter Forbes *b.*26 Mar 1918 *d.*6 Apr 1984
MB BChir Cantab(1942) MRCP(1948) MA(1950) MD(1951)
FRCP(1960)

Peter was born in Basra, Iraq, where his father, who was a New Zealander, was a surgeon and founder of the British Hospital. It was, therefore, at an early stage in Peter's life that he learnt to appreciate that the world is composed of different races and different types of people. His education at Rugby School and Clare College, Cambridge, added to his ability to rub shoulders with people of different intellectual, racial and educational backgrounds. He completed his undergraduate training at St Bartholomew's Hospital, London, in 1942, and for the rest of his life he took an interest in the hospital and medical school, which were very close to his heart in affection. Following a house physician appointment to George Graham [*Munk's Roll*,Vol.VI,p.205] at Bart's, he served in the Royal Army Medical Corps as regimental medical officer to the Royal Engineers, in both England and north western Europe, and in 1947 he was appointed as a dermatologist and worked in Egypt and Greece. He obtained his membership of the College in 1948 and 12 years later was elected to the Fellowship.

After demobilization he took up an appointment as senior registrar in dermatology at St Mary's Hospital, Paddington, and in 1949 returned to work at his favourite London Hospital, St Bartholomew's, first as senior registrar, then as chief assistant, and later as assistant physician and consultant physician in charge of the dermatology department, and he remained there until his retirement in 1978. Like many physicians of his era, he did a round of other smaller hospitals within the home counties and was, at times, dermatologist to St John's Hospital for Diseases of the Skin, St Andrew's Hospital Dollis Hill, Barnet General Hospital, and Moorfields Eye Hospital.

At a time when lupus erythematosus was not a very well known disease, Peter worked on the clinical aspects and patterns of this disease and his Cambridge MD was awarded for a thesis on it in 1951. For the rest of his days as a clinician he continued to take a lively interest in the cutaneous vasculitis aspect of lupus, and of other related diseases, and also on the different clinical manifestations of xanthomatosis.

Although he was not a prolific writer of original papers a number of his reports are repeatedly quoted today, and they reflect the accuracy and importance of his original observations and contributions to medical literature. As editor of *Roxburgh's Textbook of Common Diseases of the Skin,* from 1959, his name was well known to dermatologists throughout the world, and it is interesting that this textbook remains in popular demand despite the plethora of texts from almost every university hospital in the United Kingdom.

Peter's main dermatological interest was, however, directed to the clinical care of patients. He enjoyed making people better, and seeing that the patient left his presence with satisfaction and a hopeful outlook. He was always willing to learn about new drugs and techniques and his good relationship with his registrars, and other junior staff, enabled them to pass on to him their concepts of modern trends in therapy and understanding of disease, and he looked upon this communication as a major source of his continuing education. He read his journals with enthusiasm and felt time spent attending prolonged medical meetings was time wasted, particularly if the meeting proceedings were subsequently published.

Medical students seemed to have a particular rapport with Peter. He enjoyed teaching them, and made a great effort to learn their names at a time when there were so many students coming through the dermatology department for tuition that the weaker brethren considered both the tuition and getting to know individual students to be an impossibilty. He stimulated a significant interest in the dermatological part of the curriculum and a considerable number of Bart's Hospital graduates who worked in the skin department as junior medical staff later became consultant dermatologists in teaching hospitals in London, and in other parts of the country. Their interest was, to a great extent, stimulated by Peter's enthusiasm for his subject.

As he became more senior his colleagues looked upon him as one of their elder statesmen. He was elected to the board of governors of Bart's, was chairman of the medical council of the Hospital and president of the section of dermatology of the Royal Society of Medicine. He was a most successful chairman with a directness of approach to subjects, and to his colleagues, which crystallized his feelings about situations in a refreshing way. His accuracy and objectivity were respected by his friends and associates.

Peter was most relaxed and happy in his own home with Helen, a soldier's daughter and his devoted wife, and his four children of whom he was very proud. He loved to play golf, to garden, and to entertain friends, and the hospitality of his home was shared by many whatever their social status in life.

He had that special ability of being able to disagree violently and forcefully with his colleagues and still remain close friends with them. He was outspoken about those aspects of life which concerned him and, some years ago, he felt compelled to voice his reluctance to appoint a South African anaesthetist to a consultant post when there were many United Kingdom trained individuals clamouring to obtain the prestigious appointments on a London teaching hospital. This type of racial bias, so unpopular in today's world, was stimulated not by a sense of superiority of one group over another but by the belief that such appointments should not be given to foreign graduates when the home market was over-subscribed with well trained candidates.

Ill health forced Peter to retire early but he was able to enjoy six years in Cornwall following his hobbies, including golf and gardening. Having walked out of St Bartholomew's Hospital after his retirement party he never again came to London, nor to the hospital which he loved so dearly; it was as if he felt that his own era at that institution had passed. A service of thanksgiving for the life and work of this whole man was held in the parish church of St Bartholomew the Less, within the walls of St Bartholomew's Hospital, on 6 June 1984. It was attended by a large number of friends and colleagues from every aspect of his former life who came to pay a tribute to this independent, loyal and hardworking man, to whom accuracy and an objective approach to problems was of supreme importance. He was a complete and loveable person, and his family looked back on a life well fulfilled.

DDM

[*Lancet*,1984,2,415; *The Times* 16 Apr,26 May 1984]

BOUTOURLINE-YOUNG, Harben Jay *b.*16 Mar 1916 *d.*4 Sept 1982
MRCS LRCP(1942) MB BS Lond(1943) DCH(1943) MD(1948)
MRCP(1948) MD Florence(1954) MD Harvard(1949)
FRCP(1970)

Pat Young, as he was known for the first half of his life, was born in Paeroa, New Zealand, the son of Harben Robert Young, a consulting engineer, and his wife Lena Isabella Jay. It is of interest that Pat was a cousin of Austen and Neville Chamberlain. He went to Westport Primary School and St Andrew's College, Christchurch, winning a gold medal and 'excellence' on entry to Canterbury College. Here he studied for a degree in commerce, but switched to University College London to study medicine.

After a house appointment in Hemel Hempstead, Pat was RMO at the Queen Elizabeth Hospital for Children under George Newns (q.v.),and had served four months as an outpatient registrar at Great Ormond Street when he was called up into the RAMC, in which he worked from 1944-46, reaching the acting rank of major. The end of the war found Pat in Trieste, where he was responsible for service families in the hospital commanded by the writer (GW).

Pat had been married and had one daughter, but in Italy he fell in love with a very young and beautiful girl of aristocratic Russian family, Elizabeth Mary Boutourline (Lily). They married in 1948 and Pat assumed the name of Boutourline-Young. They had five children. Lily then qualified brilliantly in medicine in Florence but as a woman was completely frustrated in her ambition to be a surgeon.

He had returned to England after the war and was a senior registrar in paediatrics at St Mary's, obtaining his MD and membership. In 1949 and again in 1952 Pat worked in the United States as research assistant in neonatal paediatrics at the Boston Lying-in hospital, at Harvard University as assistant professor of physiology and at Yale as associate professor of paediatrics. His life work was in connexion with the Harvard-Florence Research Project, 1952-69, continued in the Harvard-Yale Research Project from 1970, of both of which he was director. The projects were mainly concerned with the impact of poverty and malnutrition on child development and the influence of the environment on adolescent behaviour; both were studies of children of Italian stock in Florence, Sicily and Boston.

Pat Boutourline-Young had all the ease and geniality of a New Zealander, combined with a simple modesty which obscured his acute and logical intelligence. He enjoyed the contrast between his New Zealand background and American democracy on the one hand, and his seigneurial state in Florence. It was unfortunate that serious problems arose between him and his wife, although in fact there was never any actual separation and he continued his work and practice from his office. Towards the end his escape into overindulgence handicapped him in recovery from a car accident and other illnesses, but he could still be a superb host. His many scientific papers provided an important and solid foundation for studies of child development in many countries.

JB[1]/GW

[Photo]

BRAY, Percival Thomas *b.*17 July 1912 *d.*13 June 1988
BSc Wales(1933) MB BCh(1936) MRCP(1945) DCH(1946) FRCP(1968)

Percy Bray was born in Newport, when it was in Monmouthshire, and died at Cardiff where he had spent most of his professional life. He was the son of Arthur Ward Bray, a commercial traveller, and his wife Florence Isobel Edwards.

Percy Bray went to Cardiff to read medicine and soon showed his true academic colours. A man of great intellect, it was no surprise to those who knew him that he graduated with first class honours and obtained the gold medals in surgery, medicine and pathology. He spent two years at the Cardiff Royal Infirmary before joining the RAMC for the duration of the war.

After demobilization he obtained his membership of the College, followed a few months later by the DCH, and later becoming a Fellow. On obtaining a Nuffield fellowship in child health, he joined the staff of the Hospital for Sick Children in Great Ormond Street to prepare himself for a career in paediatrics. Many of the staff there were to become his lifelong friends.

On his return to Cardiff, as assistant paediatrician, he was founder member of the newly formed department of child health. He was a staunch supporter of the academic department, and his appointment took him to many hospitals in the former county of Glamorgan. He took a great interest in cystic fibrosis, lecturing about it both at home and abroad and travelling to Israel, the Philippines, and East Germany; becoming a well known and loved leader in a multidisciplinary research and service group in Cardiff.

Percy was a man of great charm; not only did he have an acute puckish wit but also a dry sense of humour which came through his many anecdotes; he would be in a room only a few minutes before generating outbursts of laughter. One was always left wondering how much was fact and how much was poetic licence; but the story stuck. He was also a man of scholarship and would read the latest journals while others were taking their lunch. It was little wonder that he had the edge on them in any departmental grand round in the later afternoon. He was by nature a man of logic - and blossomed after taking a university course on the subject; he confounded all, and left his sparring partners speechless.

Before specializing in paediatrics, he had had an interesting life as a general practitioner in Port Talbot. On one occasion, when he was delivering a lady of a child, in her farmhouse, he found it was necessary to use forceps and became aware of an interested observer, who was sympathetic to the mother's expressed distress and was grunting in unison. He could not distract his attention from the mother, and assumed that the observer was the grandmother. He was somewhat startled to find that it was the domestic goat which had joined them to ensure fair play.

Bray was one of the last paediatricians to deal with an outbreak of smallpox in Great Britain. He also took charge of any paediatric work which was required after the Aberfan disaster.

His interests, outside medicine and his family, were ornithology and cricket. He married Betty Wyn Hughes, the daughter of a clerk in holy orders, in 1937, and was inseparable from her. They had two children, a son and a daughter. He was survived by his wife, children and three grandchildren.

OPG

[*Brit.med.J.*,1988,**297**,612; *Lancet*,1988,**2**,407-8; Photo]

BRIGGS, Edmund James Seeds Nixon *b*.30 June 1921 *d*.25 May 1987
MB BS Lond(1945) DCH(1947) MD(1948) MRCP(1948)
FRCPC(1972) FRCP(1975)

James Nixon Briggs died suddenly and unexpectedly at his home in Winnipeg, shortly before his 66th birthday.

Jim Briggs, as he was known to his friends and colleagues, was born in Dublin of Irish parents. His father was a general practitioner in Wallasey, Cheshire, and it was there and in Belfast that he spent his childhood. He was educated at Cheltenham College and King's College, London. He did his medical training at Charing Cross Hospital, graduating with honours in 1945. After house appointments at Charing Cross Hospital and at Hillingdon County Hospital, he moved to the Sheffield Children's Hospital in 1946, where he was to come under the influence of R S Illingworth, the first full time professor of child health at the University of Sheffield. As clinical assistant to the professorial unit, Jim had a heavy clinical and teaching load but despite this he obtained his DCH, MD, and membership of the College during the next two years. His career in paediatrics was temporarily interrupted between 1948-50 when he served as a medical officer in the Royal Air Force, becoming medical specialist at the RAF hospital in Aden with the rank of squadron leader.

Following demobilization he returned to Sheffield as senior registrar to Illingworth, who was in the process of building a strong academic paediatric unit. Jim was caught up in the many clinical studies that were being conducted into the children's diseases of the day and developed particular interests in rheumatic fever, nephritis and epilepsy; several important research papers on these topics appeared under his name.

In 1952 Jim accepted a position in the paediatric department of the Winnipeg Clinic and emigrated to central Canada with his family. He rapidly established a large practice in Winnipeg, where his unique clinical skills became recognized by the medical community throughout Manitoba. He was a brilliant diagnostician and spared no effort to get to the bottom of the most complex clinical problem. As a staff member of the Children's Hospital of Winnipeg and the Winnipeg General Hospital, and on the part-time faculty of the University of Manitoba, his skills as a teacher and investigator were quickly appreciated by Harry Medovy, and during the late 1950s and '60s he became much involved in improving the level of perinatal care, to which he made a major contribution. Working closely with obstetricians and pathologists, he became interested in the then poorly understood condition of the respiratory distress syndrome that resulted in the deaths of many prematurely born infants. He developed the hypothesis that this condition might be successfully treated with an aerosolized surface tension lowering agent and in 1953 he conducted a controlled clinical trial to test his hypothesis. The treatment was unsuccessful, but what is remarkable is that the published report in 1955 shows that his thinking about the pathogenesis of respiratory distress was far ahead of his time and antedated the discovery of pulmonary surfactant by some four years. Jim loved to teach and he sustained the respect of his students throughout his career. In recognition of his contribution to medical education and research the University of Manitoba promoted him to full professor in 1974. He was elected a Fellow of the RCP in 1975.

While continuing his busy and exacting paediatric practice in Winnipeg, Jim became concerned about the special problems of Canada's native children in Northern Manitoba. He accepted a consultant appointment to St Anthony's Hospital at The Pas in 1954, and from then until his death he made regular visits to this northern town, as well as to several isolated Indian communities, to see patients referred to him and to teach local health professionals about the many problems, medical, cultural and social, that were responsible for the greater morbidity and mortality rates of Indian children as compared to those of Canada as a whole. The much better health these children now enjoy is undoubtedly due in part to Jim's pioneering work and to the example he set to others who later became involved in the solution of the health care problems of native children in Canada.

Jim had a flair for administration and his talents were put to good use at the hospitals he served. For many years he was on the council of the Winnipeg Clinic and was involved in its administrationa and in recruitment of new staff. For the last few years of his life he was also heavily involved with the affairs of the College of Physicians and Surgeons of Manitoba. Appointed a member of council in 1974, he served at one time or another on all its committees and became its president in 1985.

Jim had a very stimulating influence on his professional colleagues; things could never be dull when he was around. He usually had some story or some quip that made one laugh and he talked easily to all types of people, especially the parents of his patients. He was forthright and never hesitated to criticize when something was not to his liking, especially something he felt was injurious to the highest standards of patient care. Criticism, however, was constructive, never malicious, and he never bore a permanent grudge.

His life was lived to the full. His professional life was extremely time consuming and demanding but he seemed to thrive on hard work. Yet with his family, of which he was inordinately proud, and with his friends he could relax completely. Many enjoyed the hospitality of Jim and his wife Judy in their attractive house and beautiful garden on the banks of the Assiniboine River. Apart from medicine, his interests were catholic. He was particularly interested and knowledgeable about the history of the part of the country in which he lived, and he enjoyed good literature and classical music. In his youth he was a strong swimmer and he regularly played tennis until shortly before his death.

Jim Briggs was survived by his wife Judy, his son Rodney - in the Canadian Diplomatic Service -, his daughter Jane, and three grandchildren.

JCH

[*Lancet*,1987,2,641; Photo]

BRINTON, Wilfred Danvers *b.*19 July 1911 *d.*10 Apr 1985
BA Oxon(1933) LMSSA(1938) BM BCh(1938) MRCP(1946)
DM(1950) FRCP(1967)

Wilfred Brinton was of a distinguished medical family. His paternal grandfather was William Brinton MD FRCS FRCP who died in 1867 [*Munk's Roll* Vol.IV p.76]. His maternal great-grandfather was Sir William Bowman Bart.,FRS FRCS (1816-1892) of capsular fame. His brother was Denis H Brinton DM FRCP (q.v.) former dean of St Mary's, with whom Wilf kept up a weekly correspondence.

Wilfred's mother suffered from rheumatoid arthritis and he held that he was conceived because her physician advised that another pregnancy might at least alleviate her affliction. He was born and educated at Eton College, where his father was an esteemed housemaster (retired 1924, died 1941) whose pupils continued to meet annually until 1979. His education continued at New College, Oxford, and Guy's Hospital. He had many additional interests: archaeology, early Scottish history, Italian art, an intense and rather pedantic interest in the use of words, and natural history - one holiday, before the war, he spent largely dissecting a stranded whale. These interests may have delayed his progress as a medical student but they enriched the remainder of his life.

Wilf married Barbara Lyall, a fellow Oxford graduate, in 1940. They both had strong personal attributes, but the fundamental bond was firm and lasting. Their two adored, and indulged, daughters have matured to responsible professional careers.

Wilf joined the RAMC in 1939 and was posted to Cairo late in 1940 where he was first MO to CSDIC, and then to hospitals in Tripoli and Southern Italy. Barbara was also in the Middle East from 1942 and worked for three years there in civil supplies. In 1945 Wilf returned to Guy's as senior registrar. He was particularly interested in congenital heart disease and heart lesions. On Tuesday afternoons he conducted a teaching round at one of the satellite hospitals, which was very well attended by junior staff who appreciated his interest in his patients and his subject, and his fresh and lively presentations. His publications were principally on congenital heart disease. In 1950 he was appointed consultant physician to the Winchester Group of Hospitals, with the special interest of cardiology. He served town, gown and country with equal dedication. His rounds were extremely well attended. His duties included visits each week to Basingstoke, Alton, and Andover hospitals. For many years he served the College as examiner in pathology for the Membership examination.

Though both husband and wife were 'good mixers' neither had time, nor inclination, for the social life. Real friends became members of an extended family. To this, before my time, belonged Angus the fisherman at Morar who taught Wilfred everything he knew about the sea. Those in Winchester certainly included Mr Benham the gardener, and Mrs Harris who skilfully ran her own home and also managed the Brinton's. 'Old Nan', Wilf's nannie, was of the family in England and in Scotland. There was also his old friend and neighbour at Traigh: VSS in Arisaig.

Devoted as he was to his birthplace, Wilf was more attached to Morar and Arisaig on the west coast of Scotland. Here, from 1885-1924 the Bowman family had rented, but never bought, a house. Their many guests were never expected to stay for less than a fortnight in view of the arduous journey to get there. Barbara has transcribed and edited the 'Journals' of those summers. It was here that Wilf spent his boyhood, his last nine years, and many times in between.

In his youth Wilf was active in climbing mountains. He also sailed the unpredictable seas of the Hebrides to Eigg, Canna and Loch Brittle. In his fishing he followed the advice of his old mentor, Angus: 'Any time I being dead, Master Wilfred, this is where you'll get the fish.' The trick was to transform red lugworms into red raspberries in 12 hours. We rose early to dig the lugworms at low tide and, after a quick kipper breakfast, hurried to the boat house to thread the bloody worms onto the murderous hooks of the long line. The line was placed so that it could be played out from the boat without a tangle. Then a long row to the mouth of the Morar river where the line would be played out on the bank indicated many years before by Angus. Back to the boat house for marmalade sandwiches, usually in pouring rain, then the long row back to the line.Hauling in the line requires the skill of the hauler and yet more skill of the oarsman to keep the boat in suitable trim. The fish came in: plaice, starfish, skate, and hopefully no congers. Back to the cottage to gut the fish to the cries of raucous hungry gulls. Then a platter (aschet) of fresh fish to VSS, which she rewarded with a plate of raspberries. Hey presto! Red lugworms to red raspberries in 12 hours.

On retirement Wilf made no delay in returning to a skilfully rebuilt 'Old Mary's Cottage'. Too little time was given for these last happy and contented years. In his final brief illness Wilf made light of his constraints. He even said that he enjoyed hospital food. His skilful and compassionate physician, however, allowed him to return home to Arisaig. He watched the ever changing colours and lights over the sea and the islands, he fingered his books, he heard again his favourite recordings. So his last days had a tranquillity comparable to that of the Strauss 'Last Four Songs'.

HEH

[*Lancet*,1985,1,1172; *Guy's Hosp.Gaz.*,May 1985; Photo]

BRINTON, Denis Hubert *b*.9 Dec 1902 *d*.13 Mar 1986
MRCS LRCP(1927) BM BCh Oxon(1928) MRCP(1929)
DM(1937) FRCP(1938)

Denis Hubert Brinton was born at Eton, where his father was classics master and later a housemaster. His mother, Dorothea Bowman, was the daughter of a solicitor. He died at his home in Burley on the edge of the New Forest which he loved, cared for in his last illness by his wife Rosemary and a devoted family doctor; exactly as he wished.

As befitted his paternal genes, his birth place and his *alma mater*, he was given a classical education in two senses, style and content. His preparatory school was Summerfields at Oxford, followed by a return to Eton, thence back to Oxford, New College, where he read classics before turning to medicine to follow his grandfather, William Brinton, also a Fellow of the College in his time [*Munk's Roll*,Vol.IV, p.74]. Sadly, this broad high road to doctoring has become a rarity, by-passed by pragmatic narrower lanes. He went to St Mary's Hospital Medical School with a university scholarship in 1925, to win student prizes in hygiene, psychological medicine and forensic medicine. He took his Oxford final degrees in 1928, became a member of the College a year later, gained his doctorate in 1937 and was elected Fellow in 1938.

His younger brother Wilfred (q.v.), who died before him, became a consultant physician in Winchester after training at Guy's. So Eton, not a traditional nursery for medical science, reared two distinguished doctors from the same Brinton generation.

Brinton married twice; by his first wife Joan Violet Hood, daughter of a South Wales mine owner, he had two sons. The elder became Conservative member of Parliament for Gravesham. The younger died as a schoolboy at Eton during the last war. A second bitter blow struck when his first wife came to an untimely death. He was fortunate in finding great comfort and happiness in a second marriage, to someone he knew well and had worked with as dean, Rosemary Cockerill. He was a man who needed love and companionship to supplement an innate privacy. Both wives gave it to the full.

His junior hospital appointments included house physician to the medical unit at St Mary's under its first professor, Langmead, and at the National Hospital for Nervous Diseases, Queen Square. Later he became registrar to both hospitals in their turn. He was appointed to the consultant staff of both hospitals in the same year - 1935. This greatly prized double appointment gave the widest experience in neurology; the intrinsic and rare at the Neurological Institute and the more commonplace, associated with disease outside the system, at the general teaching hospital.

By the outbreak of war in 1939 it might have seemed that here was a fully finished young neurologist but, as with so many of his peers, war service added unusual professional experience and a deeper understanding of his fellows that was to be invaluable. He served in the Royal Air Force Volunteer Reserve from 1939-45, as a so-called neuropsychiatrist, a title with as many interpretations as it has syllables, which always brought a wry smile to his lips. Most of his time in the Air Force was spent in the Middle East where he was mentioned in despatches. He gained the rank of air commodore. It was typical of the quality of his mind and his great interest in natural science, particularly ornithology, that he grasped two incidents that came his way by chance during those years. The first, an epidemic of meningococcal meningitis which later was to be the basis of a book published by E&S Livingstone, *Cerebrospinal Fever,* 1941. The other, an outburst of a mysterious acute neurological disease the cause of which he traced to the adulteration of wheat by poisonous tares. He rediscovered a phenomenon that had been known in those lands since Biblical times.

Within months of return to civilian life he was invited by his colleagues at St Mary's to undertake a task that needed delicacy at first, mediation later and resource throughout; an appointment for which he had a remarkable amalgam of gifts, unusual not so much in their quality and number as that they were collected in one man. He was asked to become dean. There was need for change. The transition from brilliantly successful servant to despot takes time. Brinton's predecessor had had that, perhaps in part due to an understandable academic inertia in time of war. Praiseworthy steadfastness had become stubborn agnosia of the senses that tell a man when he has outlived his office. The depth of the tyranny to be overcome was measured by the fact that a plot was necessary for its removal.

Denis Brinton's particular contribution to the School and hospital was forged during his five years as dean. He was pressed to stay longer but was more mindful of the circumstances of his election than those who put him there. He stood by his own dictum, five years at least, and no more than seven. His introduction to the job was made more formidable by the presence of a complacent student body, used to undisciplined ease by an administration believing in self determination. There was much to do. Replacement of administrative muddle by method, the introduction of democracy, decision by consensus, open government, stiffer discipline, a tutorial system, a refurbished curriculum and other devices toward balanced excellence were hurtful to those who had to shift. Luckily the entry of mature students from war service saw the virtue of the changes and made them prosper. Liking for order made him a good administrator, not a love of bureaucracy of which he had none. Distaste for unnecessary and repetitive propositioning of the University led him to be a founder member of the Association of Metropolitan Deans, where medical schools' linen might be washed in decent privacy and ironed for understanding and approval. His experience in choice of medical students, and his delight in a worldwide follow up, led to the 1951 Bradshaw Lecture at the College, 'The selection of medical students'.

The School's affairs were not the only target for concern and renovation. The National Health Service was born a year after the new dean took office. It brought as many recalcitrant curses as good wishes. Doctors are renowned for seeing the unique relationships in their work as an inviolate recommendation for the *status quo*. The preservation of the teaching hospital autonomy by the survival of the old Board of Governors did little to ease their new irksome harness and inflamed resentment from local hospitals without such cover; hospitals that had been commanded by the Department of Health to come under St Mary's wing.

Brinton's qualities then became apparent. His clear mind, eye and ear for excellence, logical exposition, firm but diplomatic resolve and, most of all, his example, allowed him to choose the best even of a bad lot, to propose, debate, find agreement and implement. No one could doubt his altruism, lack of prejudice except for quality, and overt honesty. He had humour without malice. People were persuaded not against their better judgement but because of it. St Mary's were lucky at that time in having two men working in harmony and mutual regard in positions that then, and even sometimes now, were thought to be divided by a fence of prejudice; Brinton and George Pickering [*Munk's Roll*,Vol.VII p.464] the second holder of the University chair of medicine. By virtue of their individual excellence, each could see in the other the need for both aspects of medical practice and teaching by the full-time academic and the part-time servant of the NHS. They saw that they were complementary, and the absurdity of separating them. This view percolated to the staff generally and brought a benignity that enhanced the School's well-being and reputation. After Pickering's departure to Oxford his place in this productive team was amply filled by Stanley Peart.

Brinton's personal taste was not for what is usually implied by medical science or even clinical science. Both these are given a too narrow interpretation. If science is thought to represent careful and precise obervation, then he applied it to bedside neurology as is shown by his publications, modest in number but not in content. His exactitude did not prevent a warm sympathy and understanding for his patients.

Brinton might persuade with elegance and charm but he was a reluctant communicant about himself. One had to learn about him from observation, by piecing together occasional scraps of revelation and gradually working them into a portrait; of a man with high standards of living and behaviour from which he never departed, and of practice which he taught by modest precept and few words. The picture of an old fashioned gentleman, but not outmoded. One met people who disagreed with him in the particular, those whose ethics might look shabby by his, some who envied him for the apparent ease of a patrician career, others made uneasy by his several exemplary achievements - but no one who disliked him. He sought intimate friendship rarely outside his family but gained strength from it, of the kind he found in unostentatious religious belief and practice.

His services outside the common run were invited widely. He accepted few because he never saw himself in the rare image he truly was. However, Archbishop Fisher pressed him into membership of his Commission on Divine Healing, 1953-58. What more appropriate person? An open mind capable of logical argument in the handling of evidence, paying tribute to science basic to medical orthodoxy and yet needing no reminder that man does not live by bread alone, too wise to denounce the cure but contemptuous of pseudo-scientific claims, preferring a belief in the power of mind.

Brinton had the gift to procure the best out of those he worked with and for, as well as calling for a high level of probity from himself. CHE

[*Brit.med.J.*,1986,**292**,1743;**293**,214; *Lancet*,1986,**1**,1047; *The Times*,31 Mar 1986; *St Mary's Hosp.Gaz.*,1972,**78**,7;1964,**70**; Photo]

BRISCOE, William Alexander *b.*26 May 1918 *d.*21 Dec 1985
BM BCh Oxon(1942) DM(1951) MRCP(1974) FRCP(1975) FACP(1976)

With the death of William (Bill) Briscoe,at the age of 67, pulmonary physiology and medicine lost a distinguished research worker and physician and a man whose charming idiocyncrasies were nearly as well known as his important scientific contributions.

Bill Briscoe was born in London, the son of Henry V A Briscoe, professor of inorganic chemistry at Imperial College in the University of London, and Rebecca Stevenson Briscoe, a graduate in mathematics from London University, much interested in education. Bill was educated at the Royal Grammar School, Newcastle-upon-Tyne, and later in London at St Paul's School. He was a good student, excelling particularly in zoology and mathematics. Classics, however, was not exactly his forte and Bill required some gentle tutoring in Latin to ensure success at his Oxford examination.

During his late teens, both in Newcastle and in London, he became a prolific painter in watercolours. For those who knew him, it came as no surprise that his favourite subjects were industrial scenes with factories belching black smoke, although he also executed quiet, charming paintings where cottage and rolling landscape dominated the scene. He won all the art prizes at St Paul's but after leaving for Oxford with an open scholarship to New College, and for reasons that will always be obscure, he almost never painted again. An occasional poster, during his student days at Oxford, wittily advertising the coming of a seminar speaker was the sum total of his artistic output.

He did well at Oxford but was always more interested in the laboratory side than on gaining honours. He did not get a first and it was strange that, even in late life, this able physician and original clinical scientist would refer to this irrelevant disappointment. After a junior appointment at the Brompton, he went off to the war, serving three years in the RAMC in North Africa, Italy and India. During the war, while serving in the Italian theatre, he was asked by his commanding officer to obtain a dozen or so pints of blood. To show the troops how safe and easy it was to donate blood, he became the first volunteer. This worked so well that Bill repeated the demonstration over and over again until his Hb was about 7.0 gm % and his distraught commanding officer - now Sir Gordon Wolstenholme, and currently Harveian Librarian and Editor of *Munk's Roll* - ordered him to cease and desist.

After demobilization, he became RMO at Hammersmith Hospital where, under the inspiring leadership of Sir John McMichael, he blossomed as a scientist and respiratory physiologist. Never one to embrace modern technology if string and sealing wax would suffice, his Heath Robinson type laboratory probed basic questions in intrapulmonary gas mixing.

Although he was an eccentric, he was recognized by his colleagues as a deep thinker and a truly original research worker. He continued his training at Bellevue, where he came under the lasting influence of André Cournand and Dickinson Richards, and also at Philadelphia where he was much influenced by Julius Comroe [q.v.]. He returned to the UK in 1953 and for two years continued his studies at the Pneumoconiosis Research Unit at Cardiff. But when an invitation from Comroe arrived, inviting him to return to Philadelphia, he accepted and returned to the United States where he was to continue his academic career in physiology and medicine.

In 1956 the pull of the Cournand lab, and his affection for New York City, led him to accept an invitation to work at Bellevue and to join the faculty of medicine at the College of Physicians and Surgeons of Columbia University. For 12 years he was to remain in that legendary laboratory which was a dynamic focus for all that was best in pulmonary medicine and physiology. He contributed steadily to his field with thoughtful articles illuminating many basic questions in distribution of lung gases in health and disease. He took great care over the preparation of his papers and would not publish unless he thought he had something important to say.

In 1968 he joined the faculty of the New York Hospital-Cornell Medical Center as a full professor of medicine and head of the pulmonary division. He led the division with great humour and intellectual vigour, and was much beloved by his colleagues. Genuinely modest, he hated pretension and was a most witty and skilful pricker of the bubbles of pomposity. For as long as I can remember, his battle-scared wallet was a pocket hardware store - razor blades, paper clips, and miniscule screwdrivers; his pockets a battery of pens and pencils of all shapes and sizes and many hues. Those who knew his office on the 4th floor at New York Hospital will have an ineradicable memory of a Merlin's cave overstuffed with the unexpected.

He was not much of a general clinician but enjoyed contact with all his patients, who implicitly trusted him. What he may have lacked in expert knowledge was made up by his warmth and his ready willingness to admit ignorance. Even when he had a pretty good idea of what was going on he would take pains to defer to experts in the field. He was an excellent and clear minded lecturer, and even a lifelong mild stammer did not prevent his lectures from being original and informative.

A heavy and lifelong smoker, it was in 1971 that he noticed increasing hoarseness. Cancer of the larynx was diagnosed in 1972 and a laryngectomy was performed. With indomitable spirit and good humour, he quickly mastered laryngeal speech, taught the technique to others in his spare time, and was soon back giving lectures to his students who frequently gave him a standing ovation. From now on, however, increasing illness dogged his life. He required two hip operations for crippling arthritis, nearly died from a pulmonary embolism, and in 1980 a new tumour was found at the back of the tongue. His speech became more difficult and secondary growth of his original carcinoma became apparent. From then on he communicated more and more by writing. His notes, cryptic, witty and to the point, achieved with Bill a new art form. A year after his bilateral hip replacement in 1983, a third tumour arose - this time in the pharynx. Despite heroic surgery, swallowing rapidly became difficult and Bill's life became far from easy.

It was remarkable that without the power of oesophageal speech, in constant pain and discomfort and the development of dysphagia which required a gastroscopy, he was still able during the closing days of his life to remain cheerful. Broken in body but never in spirit, this eccentric Englishman, who at one time roamed the streets of New York looking into trash cans for interesting objects which he could use in the laboratory, died at his home in New York City.

Bill's many contributions to respiratory medicine were recognized by his election to the Association of American Physicians and membership on several advisory committees to the National Institutes of Health and National Aeronautical Space Administration. Although his work was difficult to follow and few could discuss his work on Bohr's integral isopleths with intelligence, he was able to write with his colleagues Arthur DuBois, Robert Forster and Julius Comroe, the classic textbook on *The Lung*, Chicago, Year Book Publishers, 1955.

He was correcting the proofs of the third edition a week before he died.

In 1955 he married Dr Anne E Milman of New York City, a graduate of Vassar College and Yale University, who is on the faculty of Colombia University College of physicians where she served as associate professor. Bill Briscoe was buried in a pre-Revolution cemetery at Haddam Neck, deep in the Connecticut countryside, where he and his wife, who survived him, had spent so many happy summers. There were no children.

AGB

[*Lancet*,1986,1,110; *New York Times* 22 Dec 1985; Photo]

BROCKBANK, William *b.*28 Jan 1900 *d.*12 Mar 1984
TD BA Cantab(1921) MRCS LRCP(1924) MB BChir(1925)
MRCP(1926) MD(1929) FRCP(1940) Hon MSc Manch(1972)

William Brockbank was the archetypal Manchester medical man of his day and generation. Although he served the city of his birth in many ways his outstanding contribution was to the Manchester clinical school. Significantly, his father E M Brockbank FRCP [*Munk's Roll*, Vol.V, p.51] was a physician to Manchester Royal Infirmary, and also for many years chairman of the medical library committee of the University. Both of the interests implied in these appointments held by the father were to be cultivated and extended by the son.

Brockbank was educated at Bootham's School, York, and Caius College, Cambridge, and spent the clinical years of his medical course in Manchester. He held junior clinical appointments at Manchester Royal Infirmary and also at the Brompton Hospital, London. He returned to Manchester to take up what was then the key post of resident medical officer at Manchester Royal Infirmary. As RMO he was responsible for all admissions to the medical wards of the hospital; a responsibility which ensured that every physician on the staff would have ample opportunity to assess the calibre of the current holder of the post. Thus the appointment of William Brockbank to the staff of the Infirmary at the age of 32 speaks for itself. He maintained a wide general competence as a physician, together with a particular interest in chest diseases - a field in which he contributed to the establishment of a practical regime for the treatment of asthma with steroids. He was a most enthusiastic and effective teacher, with a flair for inculcating a sound clinical approach during the student's first clinical appointment. He favoured those areas of medicine in which a didactic approach was appropriate: for example, his popular lecture on worms prepared the student to renew acquaintance with them in the 'pots' section of the final examination.

Brockbank's keen interest in both the instruction and welfare of students led to his appointment as dean of clinical studies in 1939. He remained in this post until his retirement from the hospital in 1965. It was interrupted by a period of military service in the Middle East, during which he rose to the rank of lieutenant colonel in charge of a medical division.

William Brockbank was a central figure for successive generations of clinical students and was constantly available to help with their problems. Although his clinical and decanal work took up most of his time he continued to maintain a lively interest in medical history, in the medical library, and in the general medical life of the city. As a medical historian he gave the FitzPatrick Lectures at the College in 1950 and 1951, the Vicary lecture at the Royal College of Surgeons in 1956, and the Gideon de Laune Lecture of the Society of Apothecaries in 1963. The medical library was under the care of a joint committee formed by the University and the Manchester Medical Society and Brockbank was a member for many years, including a period as chairman. He took a keen interest in the library and combined this with his interest in medical history by setting out, week by week, an exhibition of documentary historical material in the hope of stimulating the interest of students and staff. In the immediate post war years there were still a number of separate societies in Manchester for medicine, surgery, pathology etc., and between 1948 and 1950 William Brockbank set himself a task which is always potentially difficult: that of amalgamating them all into one strong Manchester Medical Society. He was a trustee and committee member of the new joint society and later became its president. To commemorate the bicentenary of his beloved Infirmary he wrote a history of the hospital, characteristically well illustrated, entitled *Portrait of a hospital,1752-1948,* London 1952. He also wrote a collection of biographies of the MRI staff, starting in 1830 but prudently stopping short at 1948:*The honorary medical staff of the MRI 1830-1948*, Manchester, 1965. He was the obvious choice to give the paper on the history of the local medical school, and he did so with wit and erudition.

Outside medicine his interests lay in cricket, water colours, the Manchester Luncheon Club, and the Scout movement - in which he became a district commissioner and was awarded the Silver Acorn. Both physically and mentally he was more of an oak than an acorn, and his contribution to the medical school was stalwart and unequalled. He always greatly enjoyed a dinner party, which he would enliven with a flow of apposite anecdotes. His occasional directness of manner was totally ineffective in concealing a very kind heart and a genuine desire to be helpful. Sadly, he remained unmarried.

DAKB

[*Brit.med.J.*,1984,**288**,1169; *Lancet* 1984,1,693; *The Times*,19 Jan 1984; Photo]

BROD, Jan *b.*19 May 1912 *d.*10 Feb 1985
MD DSc Prague(1937) *FRCP(1969)

Jan Brod was born in Novy Jicin, Czechoslovakia. His father was a business manager and his mother, Marta Pachnerova, also came from a family in business. He went to school in the town of his birth and then to Charles University, Prague. After graduation he worked in hospitals in Prague and Vienna. He left Czechoslovakia at the time of the German occupation and held a French government research fellowship in Paris for almost a year. On the invasion of France he volunteered for the Free Czech Army and served in a first aid station caring for wounded civilians in Compiégne near the French front. On the collapse of France he became attached to a Czech field ambulance in Britain, and for a year was seconded as a house physician to the Warneford hospital in Leamington Spa. In 1942 he volunteered for the RAMC and served in Britain, North Africa and Italy with field ambulances, and later as a graded physician in a military hospital. He was mentioned in despatches.

On demobilization Jan became an assistant at the 1st Medical Clinic at Charles University, and spent a year as a Rockefeller fellow under Homer Smith at New York University. In 1951 he joined the Institute of Cardiovascular Research in Prague, becoming its director in 1963. He was the chief initiator of the 'Manifesto of 2000 words' in support of Dubcek's liberalization policy in Czechoslovakia and had to leave Prague when the Russian tanks occupied the country later, in 1968. He was invited as visiting professor to Mainz, and in 1969 accepted the offer of the second chair of medicine at the Hanover Medical School, a position he held until his retirement in 1981.

Brod was a true European; he lectured extensively in many European countries and was conversant with some eight languages, in most of which he was fluent. In 1971 he delivered the Lilly Lecture at the College. He had an international reputation in the fields of nephrology and hypertension. He was a founding member of the International Societies of Nephrology and of Hypertension, and was president of the 2nd International Congress of Nephrology in Prague. He was a member of many medical societies and academies throughout Europe, and in Argentina. In Britain, he was a member of the Medical Research Society and the Renal Association, and shortly before his death was made a fellow of Green College, Oxford, and an honorary member of the British Hypertension Society. In 1961 he received the Purkinje medal of the Czechoslovakian Academy of Science. He wrote several monographs including a momumental work *The Kidney*, London, Butterworth, c.1973, which was translated into three languages from the original Czech, and over two hundred scientific papers.

Scientifically, his early work was in the field of renal diseases, particularly chronic pyelonephritis, and he will be remembered for introducing the endogenous creatinine clearance test into clinical practice. He progressed to the study of the disturbed homeostasis of water and electrolytes in heart failure and then, in 1954, he turned his attention to the pathogenesis of hypertension, a subject which was to remain his dominant interest right up to the time of his death. At first he was concerned with the autonomic control of blood pressure and the manner in which this is disturbed by external influences in normotensive and hypertensive subjects. With his research team in Prague he demonstrated that the response to the cold pressor test was enhanced in normotensive subjects to a greater extent in those with, than in those without, a family history of hypertension. His researches were extended to the study of the response to mental stress and to a more precise definition of the manner in which flow to various territories was selectively affected. He provided clear evidence that the pattern of haemodynamic response to emotional stress in normal subjects, involving renal and skin vasoconstriction and muscle vasodilatation, was similar to that in unstimulated hypertensive subjects. The peripheral response to such stress in hypertensives was shown to be greater and more protracted than in normotensives. These studies have largely been the stimulus to subsequent research workers in the field and have provided strong support for the existence of a genetic-environmental interaction in the pathogenesis of essential hypertension. With his research group in Hanover he continued his studies of haemodynamic changes in different forms of hypertension. In renal disease he demonstrated an early rise in cardiac output preceding the development of hypertension and possibly attributable to reduced venous distensibility. The rise in output was followed by a return to normal associated with a rise in peripheral resistance in more advanced hypertension. These findings, together with the observation that whereas vascular resistance in muscles is lowered in essential but unchanged in renal hypertension, provided good evidence for a difference in pathogenesis in the early stages of these two forms of hypertension. Brod continued his researches after his retirement and, within a few weeks of his death, he communicated recent original work to a meeting of the Medical Research Society, This was concerned with studies of the response of subjects with and without early renal disease to body fluid volume expansion and led him to the conclusion that renal hypertension was a consequence of altered sodium homeostasis, resulting in an increase of resting blood volume and increased susceptibility to fluid overload. Furthermore, he observed that subjects with early renal disease were unable to adjust to fluid overload and the associated increase in cardiac output by increasing arterial and venous distensibility and in consequence suffered a rise in blood pressure.

Brod was a bedside clinician in the English tradition and his ward rounds in Hanover were a unique experience for his undergraduate students. Many of these rounds were conducted in English and were attended by medical staff from the British Military Hospital in Hanover.

He was a deeply caring doctor and taught respect for patients and avoidance of their discomfort. He wrote critically in later years about the substitution of clinical judgement by technology.

Brod regarded England as his second motherland, his beloved Czechoslovakia being barred to him. From the war years he retained a keen interest in the RAMC and it gave him great satisfaction to be made an honorary member of the officers' mess both in Hanover and Millbank. He had a particular attachment to Leamington Spa from the time he was a house physician there during the war and, shortly before his death acquired a house there in which he planned eventually to retire. He was a staunch supporter of the College, and his record of attendances at meetings would put to shame many a Fellow resident in England.

His very many friends both in Britain and abroad speak of the warmth of his personality and of his unwavering loyalty. He had a quite phenomenal memory and could recall details of encounters with his friends long after they had forgotten. He often seemed more at home in each country he visited than did the natives themselves, and usually knew as much or more than they about their country's history, topography, architecture, music and art. He was an accomplished pianist and a great lover of music, particularly Czech opera and Wagner. He was an invariable optimist, for had he not escaped in turn from Nazi persecution in Czechoslovakia, from the German Army in France, from the Coventry bombing in 1940, from the sinking of the *Windsor Castle* in the Mediterranean in 1942, and finally from the Russian communists in 1968. Although he could show anger, he never spoke with malice of anyone. He tended to converse too fast and not infrequently left his audience gasping in pursuit of his wide ranging ideas. He inspired good fellowship among physicians and research workers. Many throughout the world remember with pleasure attending the annual International Nephrological Symposia which he organized in Hanover and at which he created the perfect atmosphere for both the formal and informal exchange of knowledge. A visit to the opera was mandatory and even recalcitrant opera goers on these occasions were converted overnight by Jan's enthusiastic interpretations of the complexities of the librettos. His private hospitality was unfailingly generous and spontaneous.

A visit by Jan to one's own department would be an event to be remembered. He would pull out his notebook and record obsessionally the details of the exchanges, making apt comments and suggestions throughout. Thoughts expressed at such visits would be recalled unerringly years later.

Jan Brod married, at the age of 48, Elisabeth Staffenova, known to all as Ula, who had been a medical student of his at Charles University. It was an ideal marriage and gave him support through good times and bad, lifting his spirits and making it possible for him to achieve so much. A touching tribute to her appears in the dedication of his great book on the kidney. They had one son, whom Jan sent to be educated in England and who was an undergraduate at Magdalen College, Oxford, at the time of Jan's death. Tragedy befell many of his family, for he lost 40 relatives in Nazi concentration camps, including his mother.

Tributes to him were paid at memorial services in both Hanover and Leamington Spa. His ashes were brought from Germany, at his request, to be deposited at Leamington Spa.

JML

[*Brit.med.J.*,1985,**290**,719-720; *Lancet* 1985,**2**,470-471; *The Times*,26 Feb 1985; *J.Hypertens.*,Oct 1985,3(5),527-8; Photo]

BRODY, Hyman Philip *b.*18 Oct 1905 *d.*13 Mar 1985
MB BS Sheff (1929) MRCP (1932) FRCP (1947)

Hyman Brody was born and died in Sheffield, where he had spent all his practising life. He was educated at the Central School in Sheffield and gained the Lancasterian scholarship in 1920, matriculating a year later. He then registered as a dental student but after two years changed to medicine, enrolling at Sheffield University Medical School. After qualifying he obtained house appointments at the Royal Infirmary, Sheffield, first as assistant casualty officer, then medical registrar and subsequently as first assistant to A E Barnes [*Munk's Roll*, Vol.V,p.28]. He was appointed consultant physician to the Sheffield Infirmary in 1945.

As a general physician with an interest in gastroenterology his opinion was widely sought throughout South Yorkshire, and for many years he maintained additional outpatient sessions at, at least, one district general hospital in the area. His diagnostic acumen became a byword and his practice grew accordingly. A quiet, modest man, he always had time for his patients despite the pressures of a busy professional life. He was essentially a clinician and many generations of students and junior hospital staff attached to his firm had the invaluable experience of witnessing clinical excellence combined with compassion.

Brody was president of the Sheffield Medico-Chirurgical Society in 1970. He retired from active practice in the same year and was appointed postgraduate dean, a post he held for the next two years. For many years he was an active member of the Jewish community in Sheffield and was a wise counsellor to many who sought his help.

He was survived by his wife Gertrude who had always given him invaluable support throughout his professional life and in particular during the many months of his illness which he bore with notable courage. They had two sons: one an author and sociologist, the other a consultant cardiologist practising in the United States.

JDD

[*Brit.med.J.*,1985,**290**,1437]

BROTHERSTON, *Sir* John Howie Flint *b.*9 Mar 1915 *d.*12 May 1985
Kt(1972) MA Edin(1935) MB ChB(1940) DPH(1947) MD(1950) Dr PH Johns Hopkins(1952) FRCPE(1957) FRCPSG(1964) FFCM(1972) FRCP(1978) Hon LL Aberd(1971) Hon MD Bristol(1981)

John Brotherston was a leading figure in health promotion and health services in the UK and abroad for more than 30 years. He was born into an Edinburgh legal family and had a pioneering medical mother, Margaret Brotherston MBE. Throughout his life he managed to combine a strong Scottishness, even a minor Scottish Nationalist orientation, with an international outlook reflected in an abundance of overseas friends and travel. His first degree at Edinburgh University was in the arts faculty, where he became president of the Students' Representative Council. His undergraduate medical career, also at Edinburgh, was not particularly distinguished academically, no doubt because of his polymathic interests in undergraduate affairs.

John was remarkably reticent about his war service in the RAMC, 1941-46; part of his abhorrence of hierarchical organizations. He was wounded in Sicily and stubbornly refused amputation of his leg. Within a few years after demobilization John had established lifelong relationships with the London School of Hygiene and Johns Hopkins University, USA. He regarded his own prime mentor as James Mackintosh, from whom he acquired an interest in the effects of housing on health.

The chair of public health at Edinburgh University was a predictable goal, and within a decade he had transformed the department by his own energies and by those of others whom he recruited, including a few from London. John tended to lean on individuals he knew and transmute them, rather than reform them. During this period he was elected dean of the faculty of medicine and embarked on a major revision of the undergraduate curriculum. He began to organize weekend retreats where participants were expected to work from dawn to dusk and beyond, and then carry on working in a social setting. These tactics reached their apotheosis after 1964 when he became chief medical officer, Scottish Home and Health Department, and threw his energies into no less a task than the reorganization and integration of Scotland's health services. Within a few years there was no doctor in Scotland worth his salt who could say that he had not had the opportunity to express his views to the chief medical officer.

John worked closely with the World Health Organization, and at a time when most schools of public health were still plodding along their well worn paths - and the Hunter report and wider vision of the Faculty of Community Medicine had yet to make their impact - he established a diploma in medical services administration at Edinburgh University; the first international course of its kind, to which WHO sent doctors with scholarships from many countries. He had a fine perception of what was going to be useful for future development; as assessor and, later, member of the Medical Research Council he made valuable contributions to its discussions. He gave 21 years of continuous service to the General Medical Council, and was an outstanding chairman of the education committee from 1970-75.

After so energetic a career some men might have welcomed retirement, but in 1977 John Brotherston returned to the university chair that he left in 1964, as emeritus professor. From 1978-81 he was president of the UK Faculty of Community Medicine, a body he had worked so hard to establish in earlier years. He was also chairman of ASH (Scotland).

Politically John was eclectic, but as he aged he became progressively more interested in Scotland, culminating in his presidency of the unsuccessful campaign for a Scottish Assembly. In later years John developed his interest in sailing, where he favoured the west coast of Scotland. Of longer duration was his indifferent golf, and his skilled photography, where he was early adept with a miniature camera. Although not a drinking man his interest in wine making, and even the drinking of it, developed in his later years.

John published sparingly, but to great effect. At first concerned with housing and health, the scope of his publications widened to include health and health services. Perhaps his finest mark was 'Inequality: is it inevitable?', Galton Lecture 1975. When he recognized he had a terminal illness John redoubled the pace of his editorial activity on the history of the Scottish health service from 1900 to the present time, to complement his MD thesis of 1950 on *Observations on the early public health movement in Scotland,* London School of Hygiene and Tropical Medicine, memoir 8, 1952.

John was survived by his wife, the former Irene Low (Bunty), also a doctor, whom he married in 1939. They had two sons and two daughters, one of whom followed in his footsteps in his chosen medical specialty. He was a devoted grandfather to nine grandchildren. GDF/GW

[*Brit.med.J.*,1985,**290**,1670; *Lancet*,1985,**1**,1229;
The Times 16 May 1985; *Daily Telegraph*,15 May 1985;
Glasgow Herald,14 May 1985; *Community Medicine* (1985)**7**,155-6;
Photo]

BROWN, Ian Keith *b.*5 April 1929 *d.*28 March 1988
MB ChB Liverp(1954) MD(1963) MRCP(1965) FRCP(1975)

Keith Brown was born in Wallasey, Lancashire, where his father was a medically qualified dental surgeon. He was educated at Wallasey Grammar School and the University of Liverpool. His long association and love for Walton Hospital started when he was a student and discovered the wealth of clinical material it had to offer, and he encouraged friends to accompany him to staff teaching sessions. He was appointed house physician in 1956. He subsequently served at the Whiston, the David Lewis Northern Hospitals, and on the professorial unit at the Liverpool Royal Infirmary. In 1965 he became senior registrar at Broadgreen Hospital, and was appointed consultant physician to the Walton in 1968.

At that time the Walton Hospital was an outlying hospital whose existence was hardly recognized at the centre. Over the next 20 years the hospital gained in stature and this was largely due to Keith's own efforts. He had a reputation as an able and inspiring teacher, and he was delighted when teaching firms were established at the hospital in 1974. He became the second clinical sub-dean shortly afterwards. Over four years he established undergraduate teaching in the hospital and then, characteristically, resigned because he thought new blood was needed. The structure he developed has remained largely unchanged.

Keith was a superb clinician. His practice was founded on sound general medicine and common sense. His ability to calm and comfort was limitless, and generations of junior doctors sought his guidance and advice. He never sought personal advancement, yet his ability and ease with people ensured that he became a major influence on medicine in Liverpool. He was regional adviser for the College, and chairman of the board of faculty of the university. He was also appointed admissions sub-dean for the faculty of medicine, but was unable to take up this post owing to illness. The appointment which gave him most pleasure was that of president of the Liverpool Medical Institution. He died during his term of tenure, but managed to give an erudite inaugural talk, and presided over the annual dinner with joy and humour.

Keith said his main interest, outside medicine, was golf - but in fact it was people. It was his love of people that illuminated both his professional and private life. Although diffident by nature he was also gregarious and struck an immediate rapport with people regardless of age or status. He married Gwyneth Carrington, daughter of a colliery manager, in 1954, and they had a son and a daughter, Roly and Laura. His happy and supportive home life provided a solid base for his many achievements.

RW/GW

[*Brit.med.J.*,1988,**296**,1339; *Lancet*,1988,**1**,1061]

BROWNE, Stanley George *b.8.*Dec 1907 *d.20.*Jan 1986
CMG(1976) OBE(1965) MB BS Lond(1933) MRCP(1934)
FRCS(1935) DTM(1936) MD(1954) FRCP(1961) FKC(1976)
KLJ(1984)

Stanley Browne, a world authority on leprosy and a lifelong medical missionary, was born in New Cross, London, the second son of Arthur

Browne, a Post Office clerk who was also secretary to the local Baptist church. Stanley's early education was outstanding, at Brockley Central School he obtained many of the school prizes. He left school at the age of 15 and obtained the post of junior clerk in the Town Clerk's department at Deptford Town Hall. His desire, however, was to become a medical missionary and go to university. He matriculated at 18 having gained the first non-vocational scholarship of the London County Council for two years at King's College. He succeeded in intermediate BSc and second MB, and gained a three-year scholarship at King's College Hospital, qualifying in 1933. In his student days at King's he was awarded the Barry, Leathes and Warneford prizes and the Warneford gold medal, and at King's medical school he was awarded the Jeff medal, and prizes in obstetrics, urology, orthopaedic surgery, surgical pathology and hygiene. He obtained distinctions in surgery, forensic medicine and hygiene in the graduation examinations. He combined house appointments at King's with postgraduate study; obtaining his MRCP in 1934 and his FRCS in 1935, and the Murchison scholarship of the college as the outstanding medical graduate. After being accepted by the Baptist Missionary Society for work in the Belgian Congo, he went to Antwerp to study French and tropical medicine, obtaining the DTM diploma in 1936. His MD thesis was on onchocerciasis.

In April 1936 he arrived in Yakusu, on the upper reaches of the River Congo, and threw himself into the work of a medical missionary. This involved playing an active part in the control of trypanosomiasis and onchocerciasis over a wide area. His rural surveys showed a high incidence of leprosy, particularly in the Kombe district, and he endeavoured to find the cause and cure for this disease and built a leprosarium across the River Congo at Yalisombo. On his first furlough he met Marion (Mali), eldest daughter of the Rev Dr H R Williamson, who had been a missionary in China. They became engaged prior to Browne's return to Yakusu in April 1940 and, six months later, Mali joined him and they were married in Yakusu on 15 November. They had three sons: Derek, Alastair and Christopher, all born in Yakusu. Two of them became doctors.

By this time the Yalisombo leprosarium had been established and Browne's interest in leprosy was increasing. He was one of the first to use Dapsone in the treatment of leprosy and continued his research in the use of Chaulmoorgra oil and Dapsone from 1947-58. Other interests while he was in Yakusu was the filming of the leprosy scene of the *Nun's Story* with Audrey Hepburn, Dame Edith Evans and Peter Finch, and discovering the breeding place of the larvae of the fly *Simulium Damsonum* as the host for onchocerciasis. As principal of the School for Medical Auxiliaries at Yakusu he helped develop a rural health service which later become a model. Whilst on leave in England in 1958 he resigned from the Baptist Missionary Society. He had no clear idea of what his next step would be. The deciding factor was an invitation from the government of the Eastern Region of Nigeria to succeed Frank Davey as senior leprologist at Uzuakoli, and one month later he arrived in Nigeria. He was to make a contribution second to none in his fight against leprosy, and was known in West Africa as 'Mr Leprosy', and sometimes as 'Bonganga'.

From this time onwards he devoted all his energies to combating leprosy. At Uzuakoli, he pioneered the use of a new anti-leprosy drug B663 now known as clofazamine, and used in multidose treatment of leprosy. He published numerous research papers on leprosy, and continued his interest in rural health work and the use of rural dispensaries for field surveys.

Stanley Browne returned to England at the end of 1965 to assume the directorship of the Leprosy Study Centre in London where he succeeded Robert Cochrane (q.v.). From his base in London he was able to advise many governments on their leprosy control programmes. He was invited by Sir George Godber to succeed Robert Cochrane as consultant adviser to the Department of Health and Social Security in leprosy, an appointment he relinquished in 1979. He was the medical consultant to the Leprosy Mission from 1965-78, and for several years was medical secretary to LEPRA and editor of the *Leprosy Review*.

His organizational abilities were widely sought and he was appointed secretary and treasurer of the International Leprosy Association from 1966-84. He was president of the Royal Society of Tropical Medicine and Hygiene from 1977-79, and WHO consultant on the training of medical auxiliaries, and adviser on its onchocerciasis programme in West Africa. Being fluent in French, he was visiting professor at the Antwerp Tropical Medicine Institute and the universities of Strasbourg and Rome. He was a member of the Leprosy Expert Panel of the World Health Organization and chairman of the medical commission of the International Federation of Anti-Leprosy Associations.

As a highly respected Christian doctor, Stanley Browne was president of the Christian Medical Fellowship of Great Britain from 1969-71, chairman of its overseas committee 1966-77, president of the International Congress of Christian Physicians 1982-86, president of the Ludhiana British Fellowship, and became the first doctor president of the Baptist Union of Great Britain and Ireland in 1980.

He was a gifted teacher and writer, adding over 500 learned articles to scientific journals, mainly on leprosy, tropical dermatology and tropical medicine. He beame president of the International Association of Physicians to the Overseas Services from 1982-84, and his wide interest in ecumenical matters was shown in his appointment to the executive committee of the International Schweitzer Fellowship, and adviser to Mother Teresa's leprosy hospital in Calcutta.

Stanley Browne was an outstanding advocate of cooperation between missions and governments on health matters and was responsible for initiating programmes in which leprosy was integrated into general health services. His important contributions to tropical medicine have been recognized by the award of the Stewart prize for epidemiology from the British Medical Association, the medal of the Royal Africa Society, the Ambuj Nath Bose prize for tropical medicine from the Royal College of Physicians of Edinburgh, and the JN Chaudry gold medal from the Calcutta School of Tropical Medicine. He was also awarded the Damien Duton award for outstanding contributions to leprosy. As a Fellow of King's College London, he became one of the first three Fellows of King's College Hospital medical school in 1978. He was elected a Fellow of the College in 1961. Among the many other honours he received, apart from the CMG and OBE, there were three from the King of the Belgians: the Chevalier de l'Ordre Royal du Lion, Officer de l'Ordre de Leopold II, and the insignia of Commander of the Order of Leopold - the highest honour accorded in Belgium to a civilian.

His wife Mali, and sons, survived him. DSB/WHJ

[*Brit.med.J.*,1986,**292**,491,705; *Lancet*1986,1,455; *Baptist Times* 17 Apr 1980; *Kent & Sussex Courier* 26 June 1970; *Brit.med.J.*,1966,1,305; *Medical News* 14 Jan 1960; *Battle against leprosy...*,Nancy Martin,1985; *Mister Leprosy...*,Phyllis Thompson,1980; *Bonganga*,Sylvia & Peter Duncan,1956; *Heralds of health:the saga of Christian missions*, ed.S.G.Browne et al,CMF Publications,1985; Research papers (Zaire) to 1985 at Wellcome Trop.Inst.,London; Photo]

BULL, James William Douglas *b.*23 Mar 1911 *d.*5 July 1987
CBE(1976) BA Cantab(1932) LMSSA Lond(1934) MA(1936) MB BChir(1936) MRCP(1937) DMR Lond(1938) MD(1947) FRCP(1951) FFR(1954) FRCS(1977) FRCR(1978) Hon FACR

James Bull, an only son, was born and brought up in Stony Stratford, Buckinghamshire, where his father, Douglas William Anderson Bull, MD Cantab JP, was a general practitioner. His mother was Eileen Comfort, née Waddell. Recording his own biographical details, James, in the manner of the time to which he whole-heartedly belonged, wrote as the profession of his mother's father, James Waddell, the single word 'Gentleman'.

Educated at Dumpton House preparatory school, Repton and Gonville and Caius, Cambridge, where he obtained a second class honours degree in the natural sciences tripos, Part 1, he won an entrance scholarship to St George's Hospital in 1932. Jobs as casualty officer, house surgeon and house physician were followed by that of assistant curator of the pathology museum, and then medical registrar. His use of pathological specimens, and his special interest in the anatomy that they revealed, lasted all his life. He began to specialize in radiology and, fired by Wylie McKissock, he won a Rockefeller travelling fellowship to Sweden in 1938, to study the new techniques of neuroradiology under Erik Lysholm at the Seraphimer Hospital in Stockholm. Lysholm had made great progress in the technical precision and interpretation of skull radiography and ventriculaography. He and the Lysholms became great friends.

The war intervened before he considered that his studies were complete, though he was appointed radiologist at Maida Vale Hospital in 1939. In 1940 he joined the RAMC and was sent to the head injury hospital at Oxford. There he met and married Edith Burch, the undergraduate chemistry student elder daughter of Charles Schiff Burch, chairman of Halesowen Steel. Almost immediately afterwards he was posted to Singapore. With the fall of Singapore there began three and a half years of very difficult life for James, and of loneliness for his young bride. James was never afterwards able to accept even casual social contact with the Japanese, but was unwilling to talk about what he had seen and suffered himself. He did, however, leave a short memoir that he allowed to be passed on to the keeper of the department of documents at the Imperial War Museum in which he briefly describes the savagery of the invasion of his hospital, and something of the starving years in Changi POW camp, where he had managed to import a fluoroscope. When the X-ray film stock ran out, he used fluoroscopy to good effect in spite of causing erythema on some patients' backs and his own chest.

Following his release and repatriation, he was on the first ship back to Stockholm. After Erik Lysholm's death, Erik Lindgren had taken over and work was proceeding at the Karolinska Hospital on percutaneous angiography and on pneumoencephalography. In 1946 James Bull was appointed to the staff at St George's, where he worked at the Atkinson Morley as neuroradiologist for Wylie McKissock. In 1947 he joined Hugh Davis at the National Hospitals for Nervous Diseases, Queen Square. At both centres his energy, determination and ability in negotiation and persuasion were instrumental in the introduction of the new methods that he had learned in Sweden.

James had played tennis for Buckinghamshire in his youth. Later he played golf. He was a little over six feet tall, handsome and with a strong and commanding voice. It came naturally for him to be first, and for the whole of his active career, up to his retirement from the health service in 1975 following a coronary artery thrombosis, from which he made a good recovery, he led the enlargement, teaching and changes in neuroradiology in Britain. From his position as second consultant to Hugh Davis, and then as director of the Lysholm department of radiology at the National, named after his teacher and friend, he was able to train many of the next generation of neuroradiologists for the world at large; those who had spent time at Queen Square played a particularly prominent role in North America.

From his 1946 visit to Sweden he brought back with him the rediscovered technique of percutaneous angiography. The first percutaneous carotid angiogram in England was performed by him in a small department near outpatients in the National Hospital, Queen Square. He was a competitive man and a stimulating rivalry was instigated and stoked by Sir Geoffrey Jefferson, [*Munk's Roll*,Vol.V., p.213], between him and Reginald Reid who worked at the Manchester Royal Infirmary.

By 1950, if not earlier, he was experimenting with vertebral angiography both at Queen Square and St George's. At about the same time he initiated Myodil ventriculography, which proved to be a very benign and much used method of investigating posterior fossa tumours for many years. On the other hand, he refused to employ water-soluble contrast media for intrathecal use until Metrizamide came along, a conservatism that was justified by events.

The next innovation for Britain, seen by James Bull in America and introduced to Queen Square in 1963 - with his customary passionate, practical drive, creating not only the funds but also the building alterations and necessary assistants - was radionuclide brain scanning, which maintained its place in neuroradiology until the EMI scanner displaced it. His part in the CT revolution was also characteristic. First he made sure that Queen Square obtained a preproduction prototype, then - worried at the slow progress of EMI Ltd towards quantity production and finding himself in the company of Zena Powell, the managing director's wife, at a reception, he convinced her of the importance of the new imaging method and *she* convinced her husband, John.

James Bull, a member of the councils of both the Royal College of Physicians and of the Royal College of Surgeons, was president of almost everything including, at the age of 44, of the 4th Symposium Neuroradiologicum - the ultimate honour for a neuroradiologist. He was once described, slightly enviously, by a neurologist who did not know him very well, as 'Such a very *grand* man'. His generous support for his colleagues, especially younger ones, had to be contrasted with his public presence; splendid, dominating, English. Naturally he was foundation president of the British Society of Neuroradiologists. He was successfully wooed by the very innovative young French also to become president of the European Society. He had a great respect for them. As he said, aside, at an early meeting in Strasbourg, in that magnificant carrying voice of his: 'No flies on these frogs, George.'

His scholarly enthusiasm for another subject, the history of the building of the London-Birmingham railway, impelled him on to the footplate of the locomotive drawing the first morning train for that journey, shepherded, naturally, by a senior manager of British Rail who had been got out of bed for the purpose. And his enthusiasm for opportunities of scholarship for students drove him to great efforts for the creation of radiological chairs.

His wife Edith had been his closest friend and ally and her sudden death in 1979 was a very hard blow. He tried to recover from his loneliness by marrying again; a disaster which was of very short duration. He taught himself then to live alone but his children and the friends of his more active years, loved as they were and loving, never replaced the companionship that he had had from Edith. She, in her own right a leader, knew how to care for the demands of leadership in him. On Friday, 3 July, 1987 while driving at twilight in Henley's Fair Mile, he struck the back of a lorry that had pulled out into the road. He suffered, as was shown by the CT scan, an irreversible brain injury.

GH du B

[*Brit.med.J.*,1987,**295**,337; *The Times*,9 Oct 1987; Photo]

BULL, *Sir* Graham MacGregor *b.*30 Jan 1918 *d.*14 Nov 1987
Kt(1976) MB ChB Cape Town(1939) MD(1947) MRCP(1947) FRCP(1954)

Graham MacGregor Bull was born in Burma, where his father Arthur Barclay Bull, was a medical officer to an oil company. He was brought up in South Africa, where he attended the Diocesan College in Cape Town and subsequently graduated in medicine, with distinction, at the University of Capetown. After a short spell in general practice, he was appointed tutor and assistant in the department of medicine at the Groote Schuur Hospital, where his students - some of them now medical men of distinction in the UK - still recall his talents as a memorable teacher even in those early days.

In 1947 he was awarded a travelling fellowship by the South African Council for Scientific and Industrial Research, which he held at the Postgraduate Medical School at Hammersmith. His MD thesis on postural proteinuria had drawn him to the attention of John McMichael, later Sir John, who appointed him to a lectureship in his department of medicine. W J Kolff had already left the school but his pioneering work on the artifical kidney was being continued by E G L Bywaters and A M Joekes. Bull joined this team and, when Bywaters returned to his original interest in rheumatology, he took over the renal unit. With Joekes and K G Lowe, he evaluated renal function in acute tubular necrosis and proceeded to develop a conservative form of treatment, the 'Bull regimen', for use in the acute phase of renal failure. In simple terms this involved measuring and replacing fluid and electrolytes lost by the patients to keep them in balance until natural recovery could take place. This resulted in an immediate reduction in what had previously been a high mortality rate. As postgraduate students from all over the world flocked to Hammersmith in the postwar years, Bull's teaching had a profound influence on the future practice of renal medicine.

In 1952, at the early age of 34, Bull was appointed to the first full-time chair of medicine at Queen's University, Belfast. After surviving, with good humour, a gruelling initiation ceremony by the students, 'The Bull' - as he came to be known - set about introducing a modern approach to medical teaching and practice, which swept like a breath of fresh air through the whole of Ulster. His outstanding expertise at the bedside and his encyclopaedic knowledge of the medical field, which included such neglected subjects as psychiatry and geriatrics, earned him the respect of senior colleagues, assistants and students alike. His early experience in general practice helped to ensure the cooperation of general practitioners in establishing record linking systems, and his support gave an impetus to Belfast's flying ambulance service for the early treatment of cases of myocardial infarction. He brought to the medical school an appreciation of the importance of mathematical accuracy in both clinical and research work, and he was influential in establishing a department of medical statistics. As chairman of the Northern Ireland Hospital Authority's education and research committees he played a dominant role in developing postgraduate education and promoting research over the whole field of medicine in the province.

In 1966, following the untimely death of the director-designate, John Squire [*Munk's Roll*,Vol.VI,412], Bull was appointed director of the Medical Research Council's new clinical research centre at Northwick Park, with Richard Doll, later Sir Richard, as his deputy director. The centre had been sited deliberately at a district general hospital rather than in a university environment so that emphasis could be placed on studies of human disease as it presented in a defined community. Bull took over when the plans for the centre and the hospital were complete and building had begun. He promptly turned his attention to recruitment and operational policies, with a steady resolve and dedication, cloaked by a relaxed, informal and good-humoured manner. By the opening day, in 1970, he had in place a happy and well organized force of research workers ready to occupy their new laboratories and carry on their work without loss of momentum.

Bull was a dominant figure both in the centre and in the hospital; he himself had an imaginative and highly original approach to any problem and he was always available for advice. He stood behind his divisional heads, never seeking to direct them but encouraging them to develop their research programmes in their own way. He played a major role in ensuring that the Northwick Park complex, including the research centre with its own wards, was first accepted and then highly regarded by the health authorities, the local community, and the patients treated there. He himself was widely respected and rightly trusted, but he had no easy task. The Northwick Park scheme was from the beginning the subject of frequent and often carping criticism in university circles. The staff at the centre were employed predominantly for their research prowess, without recourse to the other outlets open to university employees, and this selective approach made recruitment more difficult. The initial concept of joint NHS/MRC appointments to integrate more fully hospital and research staff met with insuperable problems. Decisions taken by the MRC itself closed various options for the newly established institute and its director. The strong emphasis on technology in the early days may have reflected in part the pressures emanating from the Rothschild proposals. Nevertheless, in spite of these and other difficulties, Bull succeeded in commissioning from scratch a burgeoning centre for clinical research which conformed with the MRC's intention that the programme should be orientated to those conditions common in the community which are so often neglected as subjects for research in traditional university departments - such as infectious diseases, nutrition and metabolic disorders. Bull's achievements received national recognition when he was knighted in 1976. He retired in 1978.

Bull served as a member of the Medical Research Council from 1962-66. His continuing interest in tropical medicine made him a distinguished chairman of the Council's tropical medicine research committee, the medical advisory committee of the Overseas Development Ministry, and the clinical investigation committee of the SEATO Cholera Research Laboratory. He became a member of the executive council of the Ciba Foundation in 1970, and was a trustee from 1979 until his death. He was second vice-president of the Royal College of Physicians from 1978-79.

Graham Bull was essentially a kind man. He was humane and compassionate, as well as highly ethical, in all his dealings with patients and their kin. He was loyal and trusting in relation to his staff, who had not only a great respect but also a deep affection for him. He was a congenial, buoyant and rational colleague. As a committee member he was full both of original ideas and benign amusement, and only occasionally expressed exasperation. And he was unostentatiously faithful as a friend. He had many interests outside medicine and was a wide ranging traveller; his knowledge of the Stock Exchange would have put many professionals to shame, and he was also an expert cook, keen wine-maker, and a genial host.

He married in 1947 Megan, née Jones, who had been a fellow student in Cape Town. Her career in the prison medical service culminated in her appointment in 1973 as governor of Holloway Prison, a post she held for nine years, being awarded the OBE. She shared her husband's wide interests and his pleasure in entertaining colleagues and friends. There were four children of the marriage; a daughter who is a consulting paediatric cardiologist, and three sons - an accountant, a zoologist and a musician - whose interests reflect the broadly based interests of their father.

In 1988 the Sir Graham Bull Memorial Prize was founded at the College to commemorate the achievements of a fine man. The prize is awarded annually for meritorious research carried out by a scientist under the age of 45 in the field of clinical research in its broadest sense, to include such areas as surgery, psychiatry, pharmacology, molecular biology etc.

SH

[*Lancet*,1987,**2**,1346; *Brit.med.J.*,1988,**296**,; *The Times*,18 Nov 1987; *The Independent*,4 Dec 1987; *The Daily Telegraph*,1 Dec 1987; *MRC News*,Mar 1988,28-29; *Medical News*,5 Mar 1966; *Belfast News Letter*,5 May 1954; Photo]

BURNET, *Sir* Frank Macfarlane *b*.3 Sept 1899 *d*.31 Aug 1985 OM(1958) AK(1978) KBE(1969) Kt(1951) MB BS Melb(1922) MD(1924) PhD Lond(1928) FRS(1942) FRACP(1948) FRCPE(1953) *FRCP(1953) Hon DSc Cantab(1946) Hon DSc Oxon(1968) Hon FRCS(1969) Nobel Prize(1960)

Frank Macfarlane Burnet was born at Traralgon, Victoria, a country town where his father was manager of the bank. He attended the local school, where he was happy, often wandering in the woods collecting beetles and identifying the birds. This awakened his love for biology and he was delighted to read Darwin's *Origin of Species.* Darwin remained his hero. Later he boarded at Geelong College and excelled at science. He then entered the medical school at the University of Melbourne, gaining second place in his final year in 1922. During his year as house physician at the Royal Melbourne Hospital he decided to become a neurologist but was posted to the pathology department, and so began a lifelong committal to the bench.

He was slim, of medium height, fine featured, shy and an ardent worker. His spark of genius impressed Charles Kellaway, director of the adjoining Walter and Eliza Hall Institute of Medical Research. Kellaway appointed him to his team, which included Neil Hamilton Fairley [*Munk's Roll*,Vol.VI,p.171] and Gordon Cameron [*Munk's Roll*,Vol.VI,p.80]. In 1926, Kellaway sent Burnet to London to work with C J Martin at the Lister Institute. After gaining his PhD he returned to Melbourne.

Under Kellaway's guidance he received public acclaim: he solved the mystery of the 'Bundaberg Tragedy', when twelve children died following diphtheria immunization - he found virulent staphylococci in the injected fluid.

He returned to London in 1931 and worked for three years under Henry Dale at the National Institute of Medical Research. He developed Goodpasture's method of culturing a virus on the membrane of a developing chick embryo. He was deeply impressed by that method. By good fortune, in a neighbouring laboratory P P Laidlaw, C Andrewes (q.v.) and Wilson Smith made history: they transmitted the influenza virus to ferrets.

Inspired by his London experiences, Burnet hastened back to Melbourne where there was an outbreak of influenza among the nursing staff of the hospital. He injected sputum into his chick embryos and colonies of influenza virus appeared on the membranes. He was over-joyed - a world first! Now a vaccine could be made. Indeed, it was the first step in his famous researches into the unfathomed depths of virology.

By a series of brilliant experiments, he and his colleagues unravelled the obscurities of G K Hirst's haemagglutination test for the presence of influenza virus. They discovered that virus possessed a mucolytic enzyme and the substrate was neuraminic acid which coated the red blood cells. They called the enzyme 'neuraminidase'. They also found that neuraminidase was present in cholera toxin and its mucolytic action contributed to the destruction of bowel epithelium.

During those days Burnet also led his team in studies outside his laboratory.In 1930, 312 cases of poliomyelitis occurred in Victoria. He transmitted the causal virus to monkeys, and was the first to discover there were two specific strains with no cross-immunity. This was of vital importance for John Enders in Boston when he produced the first successful vaccine and was awarded the Nobel Prize. At that time a colleague wrote of Burnet: 'His enthusiasm and meticulous method of study were most impressive. Slim, handsome and of medium height, his pensive expression, penetrating gaze and apparent shyness, would blossom into a lucid description of his current research'.

In 1936 Burnet discovered that the cause of Q fever, which was attacking abattoir workers, was a rickettsial organism. It was called *Rickettsia burneti* in his honour.

In 1942 he was elected a Fellow of the Royal Society and treasured the many friendships he made there. He succeeded Kellaway, as director of the Hall Institute and research professor of experimental medicine in the University of Melbourne, in 1944.

In 1949, Burnet and P Lind discovered 'recombination'. They found that if two different strains of influenza virus were grown together under suitable conditions they formed a new strain with characteristics of both parents. Burnet wrote: 'It would not be a fantasy to say that in the influenza virus we have a tool - a uniquely suitable tool - with which to break some of the codes that lie at the basis of cellular life'. In 1950 Burnet guided Alick Isaacs in his discovery of *interference* between 'inactive' and 'active' strains of influenza virus when they were grown on the chick embryo membrane. Later, Isaacs continued that research in London and, collaborating with J Lindenmann, in 1957 they discovered the famous biological product *interferon*.

The virus causing an epidemic of severe encephalitis in the River Murray area was discovered in Burnet's laboratory by E French in 1951 and the epidemiology was unravelled by G Anderson: the virus was harboured by humans in New Guinea, carried south by migrating waterbirds and transmitted by mosquitos to animals and man.

For some time Burnet had experienced an increasing urge to investigate the immunity system. Why did man tolerate his own tissues, yet reject foreign tissues? That urge reached its climax in 1957. He proclaimed to his colleagues that virus research would cease and research into the body's immunity system would begin. He vigorously discussed antibody production and action, and formulated his clonal selection theory: antibodies were directed against invaders such as foreign tissue grafts whereas antibodies directed against the host's own tissues were eliminated.

He was intrigued by the autoimmune diseases, where that tolerance of the host's own tissues failed and tissue destroying antibodies appeared, causing a disease such as systemic lupus erythematosus. He visited the hospital to see cases of autoimmune disease and had heated debates with the physicians. His casting into the future was a stimulus and a joy to all. In 1963 he joined with Ian Mackay to write *Autoimmune Diseases: Pathogenesis, Chemistry and Therapy*, Springfield, Illinois, Thomas, c.1963. It was a pioneer publication.

In his laboratory he studied the 'New Zealand black' mice which developed autoimmune disease.He found changes in the thymus gland and postulated that lymphocytes from the gland might be producing damaging antibodies. Under Burnet's guidance, Gustav Nossal researched with J Lederberg, a Nobel Prize winner visiting from America. In 1958 they discovered a single lymphocyte released only one strain of antibody even when exposed to two strains. Then, led by Burnet, N Warner and A Szenberg first discovered there were two populations of lymphocytes, each with a different function, either 'graft rejection' or 'antibody production to an injected antigen'; later they were called 'T' and 'B' cells. This sparked off worldwide research.

During those successful years of immunity research Burnet's personality changed. He emerged from the cloistered existence which had served him so well during his virus research to become more extrovert. He was convincing in his pronouncements on the progress of science and the future of the human race. He took an increasing interest in his fellow men. He maintained his love of the Australian bush by joining the 'Wallaby Club' for long walks, to enjoy good company and the birds he knew so well.

In 1949 Burnet and Frank Fenner had predicted the phenomenon of 'immunological tolerance'. Now, in 1960, Burnet was awarded the Nobel Prize which he shared with Peter Medawar (q.v.) who had skilfully confirmed their prediction. It was awarded 'For the discovery of immunological tolerance'.

HM Queen Elizabeth II awarded him the Order of Merit in 1958, and in the following year he received the highest award of the Royal Society - the Copley Medal - previous recipients being Charles Darwin and Captain James Cook of HMS *Endeavour*. He later delivered the Croonian lecture to the Royal College of Physicians, entitled 'The Autoimmune Diseases'; Lord Adrian being in the chair.

Burnet accepted the honours with pride and modesty, retaining a refreshing trace of his early shyness which was matched by his still youthful appearance.

In 1965 he retired as director of the Walter and Eliza Hall Institute and was succeeded by Gustav (later Sir Gustav) Nossal. It was an occasion on which his colleagues showed their admiration: they were deeply grateful for his gifts to mankind, and the inspiration and guidance he had given to so many young scientists. During his retirement Burnet was constantly engaged in writing, lecturing and travelling. He enjoyed staying with Sir Gordon Wolstenholme at Ciba Foundation House in London, with John Enders in America, and with Konrad Lorenz in Austria to see his 'new light on animal ways' as told in *King Solomon's Ring*.

Burnet's 70th birthday was celebrated in the Great Hall of the Royal Australasian College of Surgeons. His virtues and achievements were extolled by Frank Fenner and by Gustav Nossal who said: 'No one who loves Australian science will ever forget his example'.

Amidst those happy events came sorrow for Mac Burnet. In 1973 his beloved wife Linda died from leukaemia. Charming, intelligent and amusing, she had been a pillar of strength both at home and during their travels. Later, in 1976, he married Hazel Foletta who, as a librarian, had helped him with his writings.

He wrote several books, including *Immunity, ageing and cancer : medical aspects of mutation and selection,* San Francisco. Freeman, 1976, in which he stated: 'Cancer occurs when a key enzyme of the group handling DNA repairs mutates to an exceptionally error-prone form. Advancing age advances the degree of proneness of these enzymes.' In 1985 he developed cancer with wide spread metastases. He knew his days were numbered but, true to form, he followed his downward course with astonishing interest - almost with glee. He had elected to spend his last days at the home of his son, Ian, at Port Fairy on the south coast of Victoria, the land of his predecessors. After a state funeral in Melbourne, he was buried near Port Fairy: 'close to where he had studied nature as a child'. His daughter Deborah wrote: 'It was a perfect Spring day with brilliant blue sea in the distance and skylarks singing overhead. I think Pa would have heartily approved.' Yes, it was a fine ending to a brilliant career.

IW/GW

[*Brit.med.J.*,1985,**291**,747,980; *Lancet* 1985,**2**,620; *The Times* 2 Sept & 17 Sept 1985; *Brit.med.J.*,1960,**2**,1296,1306; *Triangle* Vol.6 Apr 1963; *Lancet*,1965,**1**,897-8; *Sante Publique*,Feb 1961; *Australian News*,13 Sept 1873; *Univ.of London Gaz.*,July 1960; *Aust.Ann.Med.*,1969,**18**,293; Walter & Eliza Hall Inst.of Med.Research,*Annual Review* 1978/9 'Tribute...'; *Changing Patterns: an atypical autobiography*,Melbourne and London, Heinemann,1968; Photo]

CAMERON, James Kenneth *b.*20 June 1905 *d.*30 Apr 1985
MRCS LRCP Lond(1927) MB ChB Liverp(1927) MRCP(1929) FRCP(1969)

Kenneth Cameron - known to his friends as Jake - was a local celebrity on Merseyside for 20 years; he made an impact from the day of his appointment as medical superintendent of Leasowe Open Air Hospital on the Wirral in 1947 until his retirement to North Wales in 1969. His fame spread far beyond the confines of the medical profession into the music world of the region. His renown as a raconteur and conversationalist made him very welcome at concerts, meetings and dinners where he displayed his brilliant wit.

Kenneth was born at Rockferry, Cheshire, the son of James Cameron, an analytical chemist, and his wife Alice Jones whose father, Henry, was a boot manufacturer. He was educated at Birkenhead School and the University of Liverpool. After the usual house appointments at the David Lewis Northern Hospital, Liverpool, (now the Royal Northern), he became medical registrar and tutor. While working as a registrar he gained his membership of the College. It was said that he was the youngest man to have passed this examination and he seemed set for a teaching career at the Northern. But Jake Cameron fell in love and in those days, without private means, general practice provided the only way in the medical profession to support a wife and family. He married Sheelah Wilkinson in 1931, and they had one daughter, Jane. It was at this time that he moved to Milnethorpe, Westmorland. He had been in general practice some four years when, at the age of 30, he became crippled by severe ankylosing spondylitis. He was bed-ridden for the next four years, nursed by Sheelah in a cottage in Anglesey.

By the outbreak of the second world war in 1939 Jake had started to get about on crutches and had a petrol driven three-wheel invalid carriage. The Emergency Medical Service appointed him superintendent of the Ormskirk EMS Hospital, Lancashire, and when hostilities ended Jake found that his own long illness and his years at Ormskirk enabled him to return to specialist work as a resident doctor at the North Wales Sanatorium, for the incidence of tuberculosis had greatly increased during the war years. The hospital at Llangwyfan stood high and remote in the Clwyd range, near Denbigh, and in the winter it was often snowbound. There were fabulous tales of the manpower and spade-work needed at this time of the year in order to get Jake and his carriage there, so that he could work in the wards. In 1947 he was appointed chest physician to the Leasowe (Wirral) Open Air Hospital, where he spent 22 happy years, concentrating on the treatment of lung tuberculosis in children. It led to his paper on the role of surgery in the treatment of primary pulmonary tuberculosis in children, based on a long term study of 409 children, in collaboration with John Hay [*Munk's Roll*,Vol.V,p.178] and Leslie Temple. In 1969 he was elected a Fellow of the College.

It was in Ormskirk that he took up the violin while reclining in his invalid carriage. He became accomplished on this instrument and as a musicologist, and was the anchor-man in an amateur quartet. He was also an ardent supporter of the Rodewald Music Society in Liverpool. His arms and hands had escaped serious involvement with arthritis, so that he apprenticed himself to Mellor Riding, a well known cabinet maker in the northwest of England, and became adept at cabinet making. He was no mean opponent at croquet, in which his mallet took the place of crutches, and in later years he became an expert embroiderer and contributed many pieces of work to Exeter Cathedral. Fortunately he was able to continue with his 'manual hobbies' until the end of his life, and could keep up a lengthy correspondence with friends all over the world. He was extremely well read and his knowledge of Chaucer, Shakespeare and Donne was encyclopaedic.

In his 70s, after retirement to a bungalow on the moors above Ruthin, he designed and built himself a weaving loom and made cloth from which he had a suit tailored. He also made 'lengths' for Sheelah and Jane. Jane (Hobson) became a marine biologist of considerable fame.

Under the guidance of the Forestry Commission, Jake's Welsh hillside was planted with specimen trees and he pursued gardening day by day, with tools he designed and made himself which enabled him to do what might have appeared to be impossible jobs. Inevitably, there came a day when he overbalanced and fell, fracturing his hip, but he survived the accident and returned to many of his previous activities. As the years passed, his arthritis and ankylosis worsened but he stayed on his crutches until the end. His determination in the face of adversity was characteristic of him, and he retained his sparkling intellect and conversation until the day, when living with his daughter in the Isle of Man after the death of his wife in 1983, a haemorrhage from a vocal cord tumour brought about his death.

HKL

[*Brit.med.J.*,1985,**290**,1255; Photo]

CAPEL, Leslie Henry *b.*27 Dec 1917 *d.*18 Dec 1988
MB BS Lond(1950) MRCP(1953) MD(1961) FRCP(1970)

Leslie Capel was born in Cardiff, South Wales, the son of a local business man, Percy Capel, and his wife Fanny Ballinson. He was educated at Cardiff High School. On the outbreak of war he volunteered for Army service, but was rejected because he was severely short-sighted. He became a cowman on a dairy farm and liked to recount how he began to read widely on bovine diseases and came to be regarded as a local expert. One day he was asked to examine a bull which he judged to be particularly friendly; he was promptly tossed over its head with resulting serious leg injuries, and that was the end of his farming career. Undaunted, he decided to take up veterinary studies at Reading University, where he proved to be an outstanding student. Because of his special aptitude for physiology it was suggested that he should make a further change of career and study medicine. In 1945 he joined St Mary's Hospital, where he won several prizes in physiology, pharmacology and histology, and from where he graduated in 1950.

During his junior house jobs at St Stephens, the Whittington and the London Chest Hospital, he developed an interest in thoracic medicine, particularly in the then growing study of the physiology of lung function. Some of his early ideas, influenced by his training at the Trudeau Sanatorium in America, where he was a Fulbright Scholar in 1953-54, helped to form a solid basis for the understanding of the mechanisms of airflow limitation.

He became registrar at the London Chest Hospital in 1954, and senior registrar in 1956, and he was then given the task of setting up the respiratory function department and of training staff to run it. He did it with great efficiency. He became highly skilled in analysing the problems causing respiratory incapacity, and his colleagues came to rely on him a great deal for his assessment of their patients' fitness for cardiothoracic surgery in borderline cases. He also became part-time senior registrar at Highwood Hospital for Children, which was then situated at Brentford, Essex. His sympathetic manner made him a great favourite with the children who named him 'the Professor' because of his likeness to a character in a then popular television series called 'Thunderbirds'.

At Highwood he developed his interest in respiratory allergy, and this led to his appointment as consultant in that department at the Royal National Throat, Nose and Ear Hospital, in February 1963.

In 1962 he was appointed consultant physician to the London Chest Hospital, and it was there that he devoted so much of his life.

Leslie Capel was a man of vision. His early ideas on lung sounds, on lung function, on specialist nurses, on rehabilitation for those suffering from cardiothoracic disease, and on the need for recognition and help for those who were handicapped by the invisible disability of severe respiratory disease, were all years ahead of his time. His pioneering efforts helped to bring many of his ideas to fruition, but with his customary modesty he would never take any credit for them.

He was a gentle, compassionate physician who, by his example, demonstrated the importance of treating the patient rather than the disease. His warmth and humanity endeared him to his patients and to his colleagues. His mind was an original and enquiring one, and he inculcated in his students a healthy scepticism for time honoured practices and for fashionable innovations which were not supported by good evidence of their value.

He was also a gifted teacher and clinician; his weekly discussion meeting in his respiratory function department was one of the most popular features at the London Chest Hospital, and junior and senior doctors came from far and wide to listen. Even when he was close to retiring age he still possessed that youthful energy and infectious enthusiasm which, for so many years, inspired his staff and students.

Leslie Capel's manner was courteous and deliberate, and his speech and writing were models of precision. He loved words, and he wrote with great ease and skill; this was apparent when he was joint editor of the *British Journal of Diseases of the Chest,* and associate editor of the *Drugs and Therapeutics Bulletin,* and also in his many medical contributions to both journals and textbooks, on lung function, smoking and chronic bronchitis. His favourite crusade was the correct use of the English language. He preferred the short sentence, and the shorter word, and he instilled clarity of thought and word in colleagues and students.

Capel became an effective broadcaster on both television and radio and it is said that, following one of his anti-smoking programmes, shops in some parts of the country ran out of chewing gum. His interests were wide; he collected old medical instruments, and he built his own telescope and became much absorbed in the study of astronomy.

Just a few weeks before his sudden death in December, he attended a reception at the London Chest Hospital for the formal acceptance of the 'Capel Collection' of historical medical instruments, in the newly opened museum, and it was touching to see the real affection and respect shown to him by former colleagues and staff. They regarded him as a lifelong friend and counsellor, as indeed he was.

Leslie Capel came late to that reception - but it was expected. He was nearly always late for meetings, so much so that it could be confidently relied on. His other vice was to mislay important papers, generally his own but sometimes those belonging to colleagues whom he had just visited and whose papers he would gather up with his own. Whenever important papers were missing at the hospital, the cry would go up 'Has Leslie been in here recently?'.

Leslie was a sensitive, cultured and tolerant man, and always ready to help, but he could be firm and incisive when he saw injustice and wrongdoing. Over the past 20 years his contribution to thoracic medicine and to the London Chest Hospital had been considerable; his death produced in many a real sense of personal loss.

In 1953 he married Eileen Grace Hots, daughter of a veterinary surgeon, and they had one son, Jonathan. They both survived him. MC

[*Brit.med.J.*,1989,**298**,249-50 1520; Photo]

CARD, Wilfrid Ingram *b.*13 Apr 1908 *d.*12 Jan 1985
MB BS Lond(1931) MD(1933) MRCP(1934) FRCP(1944)
MRCPE(1951) FRCPE(1953) FRCPG(1967)

The son of Henry Charles Card, Wilfrid was educated at Tonbridge School, leaving with an exhibition in mathematics. His medical training was at St Thomas's Hospital, where he had a distinguished undergraduate career, obtaining the Seymour Graves Toller prize and the Mead medal. He was a house physician at the London Chest Hospital and also to Sir Francis Fraser [*Munk's Roll*,Vol.V,p.141] at the Postgraduate Medical School; then he returned to St Thomas's Hospital as medical registrar to Sir Maurice Cassidy [*Munk's Roll*,Vol.IV,p.528]. He obtained the Perkins scholarship in 1936 and the Louis Jenner research fellowship in 1938. In 1939 he was appointed physician to outpatients at St Thomas's and was elected to a Beit research fellowship. He had already studied gastroscopy in Leipzig, and gastrointestinal radiology in Hamburg, and while in Germany also gained an understanding of and bitter loathing for the Nazi movement. This last gave him a sense of justice in a war which he felt, with Churchill, was unnecessary and the result of British unpreparedness. Remembering what he called his 'wasted years' he was anxious that Britain should never make this mistake again.

From the outbreak of war until 1942 he was in charge of the medical division at Botley's Park War Hospital. From 1942-46 he served in the Royal Army Medical Corps as a lieutenant colonel, and saw service in West Africa, India, Burma, Ceylon, Malaya and Siam. In 1945 he was appointed to a research group and organized a field trial of a new scrub typhus vaccine.

After the war he felt that the opportunity for his research at St Thomas's Hospital was not sufficient and he accepted an invitation to move to Edinburgh where Sir Stanley Davidson [*Munk's Roll*,Vol.VII, p.136] was introducing new blood into Edinburgh medicine. As physician in charge of the gastrointestinal unit at the Western General Hospital Wilfrid was able to give full rein to his skills as an innovator despite, at that time, primitive facilities. With Sir John Bruce, later the professor of clinical surgery, he initiated a joint medical and surgical gastrointestinal unit which acquired national and international renown. The medical research building, available to whole and part-time staff alike, was his brain child. He was the moving force behind the establishment of a university department of medicine at the Western General Hospital, and as a true democrat he managed to persuade first his colleagues and then the faculty of medicine that the department should have an elected chairman in charge rather than an appointed professor - an arrangement common in other countries but almost unknown in Britain. He was a natural leader, though he looked on himself rather as *primus inter pares*, and as his new colleagues became his friends he was able to guide them towards objectives which he recognized, often much earlier than they, to be of merit. He was aware of the importance of social meetings of members of the staff, and he founded a staff dining club where his knowledge of wines and enjoyment of companionship proved catalytic in the achievement of the rapport which he fostered. On leaving Edinburgh he gave his colleagues a magnificent silver loving-cup of modern design which bore William Harvey's words *'Concordia res parvae crescunt'* and this now serves as a fine memorial to him on occasions when the staff dine together. His colleagues in their turn founded a memorial lecture in his name and, by then in retirement, he was to give the first lecture. He generously described this as the greatest honour of his medical career.

He established a close partnership with his nursing colleagues and was influential in the appointment, as matron, of Estelle Adamson, also from St Thomas's, where she was later also to be matron. They both warmly subscribed to Florence Nightingale's aphorism that 'our patients are our honoured guests'. (He also liked to quote another of Florence Nightingale's sayings: 'A nurse should wash her hands every day; if her face too so much the better.')

While in Edinburgh he and his colleagues published a classic paper on the relationship between the parietal cell mass and the acid output of the stomach, among a constant stream of publications relating mainly to gastrointestinal physiology in man. At an early stage in the development of the GI unit he enlisted the closest support of clinical psychologists and social workers to develop his own version of holistic medicine.

Those who knew Wilfrid Card best were usually impressed most by his vision. This was manifested particularly in his second major interest in what was to prove a classic contribution, in 1967, entitled 'Towards a Calculus of medicine', *Medical Annual* 1967,**85**,9-21. He often returned to the thesis *Ars Longa Vita Brevis;* 'since all doctors, without exception, are mortal, their hardly earned skill and wisdom die with them.' [ibid.p.9]. His principal aim became to work towards '...a clear concept of disease, a rigorous definition of all clinical and laboratory data, an assessment of their discriminatory value, and an analytic process whereby a set of data from the patient can be assigned to a diagnostic, prognostic or treatment class. We need nothing less than a calculus of medicine.' He quoted George Boole's observation of 1847: 'We might justly assign it as the definitive character of a true Calculus, that is a method resting upon the employment of Symbols, whose laws of combination are known and general, and whose results admit of consistent interpretation.' [ibid.p.9]. As Henrik Wulff said at a Memorial Festschrift held at the RCP London on 17 June 1985: 'Card proposed a new paradigm of clinical thinking...' ['Measuring Gut feelings: The scientific basis for clinical medicine.'].

As these mathematical interests became dominant he was to move in 1966 to a newly created chair of medicine with relation to mathematics and computing at the University of Glasgow. Relieved now of the administrative and committee work for which he had become so much in demand in Edinburgh, this phase of his career became the most productive in research publications. His renewed energy may have derived in part from the atrophy of his social life with the loss of his beloved wife Tilly. After his formal retirement he maintained his active research life right up to the end; on his sudden and unheralded death there was a new manuscript in his brief case.

'Even those of us who never had the opportunity to work with Professor Card recognize the remarkable diversity of his talents. We recognize him as a scientist who did very important physiological research, we respect him as the statistician and mathematician who applied the ideas of decision theory to clinical problems, we see him as our teacher who had the gift of explaining complex issues in simple terms, and we appreciate that all his activities were carried forward by his ability to inspire his colleagues and by his talents for cooperation, organization and administration. However, Professor Card's achievements do not only reflect his diversity of talents. They also reflect the fact that he used his talents to serve a single purpose. Wilfrid Card will be remembered as a clinician and a humanist because everything he wrote, including the most theoretical papers, aimed at improving the care of the individual patient.' (H R Wulff, Memorial Festschrift).

Add to this that he was also physician to the Queen in Scotland; secretary to the Association of Physicians; a skilful navigator and sailor; a talented carpenter - he built the superstructure of his cabin cruiser; that he was blessed with an infectious laugh; that he was a much beloved husband, parent and grandparent, and that as a person of integrity and a privately devout man he was an embodiment of Christian virtues. One can understand the comment of Gerry Crean, a colleague who was to work longer than anyone with him in Edinburgh and in Glasgow: 'He never knew how much we loved him.' JS

[*Brit.med.J.*,1985,**290**,474; *Lancet*.1985,**1**,293; *The Times* 18 Jan 1985; *The Scotsman* 21 Mar 1966; *Medical News* 1 Apr 1966,2 Apr 1966; Photo]

CARTER, Cedric Oswald *b.*26 Sept 1917 *d.*12 Mar 1984
BM BCh Oxon(1942) MRCP(1948) FRCP(1964) MA(1965) DM(1966)

Cedric Carter was a highly respected medical geneticist. He revered intelligence; perhaps not shrewdness nor physical insight, but the sort that would help to score 160 points on the IQ scale and help to solve intricate puzzles. He believed, I think, in the concept of a 'central intelligence' that could be measured and he thought and wrote a good deal about the inheritance of that quality. If he had a quiet passion then it was for positive eugenics and nothing gave him more pleasure than to note the tendency for Social Class I families to have more children. Indeed. he thought it was the duty of those with higher intelligence to pass these genes on to the next generation and this, combined with his love for children, contributed to his own family of seven.

Cedric understood and made contributions to many of the difficult concepts in genetics. Polygenic or multifactorial inheritance particularly intrigued him. His work on the clinical consequences of polygenic inheritance has been universally accepted and the 'Carter effect' - the risk of recurrence in sibs is higher after the birth of the rarely affected sex - has been included in textbooks on genetics. This gave him enormous pleasure.

He founded the UK Clinical Genetic Society and joined Fraser Roberts (q.v.) to run the first genetic counselling clinic in this country centered at the Hospital for Sick Children. He pioneered, with Kath Evans, the use of family studies to ascertain recurrence risks in the common congenital malformations, and he directed the MRC genetics unit, situated in the Institute of Child Health, between the years 1964 and 1982. He wrote three books: *Human heredity,* Baltimore,Penguin Books,1962; *The ABC of medical genetics*,Boston, Little Brown,1969; and in conjunction with T J Fairbank *The Genetics of locomotor disorders*, Oxford, Oxford Univ.Press,1974, as well as hundreds of articles. It was in the nature of C O Carter to do all this unostentatiously. He never gave hasty judgements and his collected works will stand as an accurate reference for many decades. His unemotional attitude exasperated many who visited the unit, and it could be disconcerting to have a conversation with him on a difficult subject - many of his sentences would be interspersed with long silences, leaving the recipient unsure whether to wait patiently or to interject. What emerged in the end was usually accurate and to the point. Although he contributed enormously to the field of congenital malformations he often stated that he would have preferred to have devoted his life to understanding normal variations rather than abnormalities.

Carter was the son of a captain in the Royal Navy and he had an 'old-fashioned' idea about duty, especially in time of war. He expected others to have it too.

Almost as a preparation for retirement, he later developed an interest in palaeontology. Some evenings he would rush off to give, or attend, a lecture at the Natural History Museum. These spurts of physical energy were in evidence throughout his life. Although physically small he loved sport and was immensely proud of his Oxford blue for boxing. Towards the end of his life he cycled from the station to his home in Keston and frightened his family and colleagues by falling off twice, once landing in hospital with concussion. No amount of persuasion would convince him that he should stop, and on retirement he turned with similar gusto to long distance running.

A scientist in everything he did, Carter charted his course, his pulse, and anything else he could find to measure. He died soon after completing his first marathon, suddenly and silently. He would have wished for no better end to his life.

MB

[*The Times*,17 Mar 1984; *Lancet*,1984,**1**,692; Photo]

CHALMERS, Theodore Moir *b*.10 July 1919 *d*.3 Aug 1984
MB ChB Edin(1941) MRCP(1948) FRCP(1965) MD Cantab(1966) MRCPE(1967) MD Edin (1970) FRCPE (1970)

The foundation of the Clinical Medical School at the University of Cambridge, after years of discussion and against seemingly insurmountable opposition, owed much to the vision and quiet determination of Theodore Chalmers. This was one of many achievements in a distinguished medical career.

Chalmers was born in Inverness, the youngest of three sons of a Highland family with a strong medical tradition. He was educated at Inverness Royal Academy and received his medical training in Edinburgh, where he graduated in 1941. After an appointment as house physician in Inverness, he served in the Royal Air Force for five years, performing unit and hospital duties as a squadron leader in the combat zones of North Africa and Italy. He spoke little of his war experiences in later life but they no doubt contributed to his passionate opposition to nuclear weapons and his disinterest in chauvinism. He seriously considered settling in Italy, but on return to England he decided to resume his career in civilian life, first at the Brompton Hospital and then as registrar at the Hammersmith Hospital. His research interests began at the Middlesex Hospital where he studied the effects of autonomic drugs on sweat secretion, which formed the basis of his MD thesis. This took him to the fields of electrolyte and water metabolism and neurohypophyseal function, which led him to describe the use of nicotine as a diagnostic test in diabetes insipidus [*Clin. Sci*.1951,**10**,137-144]. At that time considerable advances were being made into the effects of adrenal corticosteroids on electrolyte metabolism, and he pursued this line of research whilst a Nuffield Foundation fellow at the University of Pennsylvania Hospital in Philadelphia. On his return to this country he became senior lecturer in medicine at the Welsh National School of Medicine in Cardiff and during his short stay there he described one of the earliest cases of Conn's syndrome [*Lancet*,1956,**1**,127-132]. He then returned to London as first assistant to the professorial medical unit at the Middlesex Hospital in 1954. There he made studies of adipose tissue metabolism and obesity, and in particular of the fat mobilising activity of fasting urine, a subject in which he later collaborated with the University department of biochemistry at Cambridge [*Lancet*, 1958,**1**,866-869]. His research into adipose tissue metabolism formed the basis of the Malthe lecture to the Norwegian Medical Association which he delivered in 1964.

Chalmers moved to Cambridge in 1962 on his appointment as consultant physician to the then United Cambridge Hospitals and East Anglian Regional Hospital Board. His main duties were at Addenbrooke's Hospital but he also had clinical responsibilities at Fulbourn Hospital and the County Hospital at Huntingdon. From the start he set up a first class clinical service and organized postgraduate clinical teaching sessions which he maintained throughout his career. He retained his broad interest in endocrinology but now turned his attention to bone and mineral metabolism. These new interests were relatively untapped areas of research in Cambridge and over the next 20 years he and his co-workers published a series of papers on hyperparathyroidism, renal stones, metabolic bone disease and vitamin D metabolism. All this coincided with major advances in the understanding of vitamin D metabolism by a group of scientists directed by Egon Kodicek at the MRC Dunn Nutritional Laboratory in Cambridge, which culminated in their discovery of the unique steroid hormone 1,25-dihydroxycholecalciferol. Chalmers fostered links with this group and their friendly collaboration produced nearly 20 years of fruitful research.

In 1965 he was elected Fellow and director of medical studies at St John's College, Cambridge, where he taught physiology and guided the careers of countless medical students. Although not schooled in Cambridge he was proud of his association with St John's and had a high regard for its tradition, style and excellence. He was appointed dean of the School of Clinical Research and Postgraduate Teaching at Cambridge in the same year. It was natural, with his long academic background and deep interest in medical education, that he should take on this role and he quickly recognized the potential for extending medical education and clinical research in Cambridge, and that these goals could best be realized with the establishment there of an undergraduate clinical medical school. He espoused this cause despite the widely known and long held opposition to such a proposal from many within the University. This did not deter him and, with the favourable attitudes expressed in the Todd report of 1968 and the redevelopment of Addenbrooke's on the 'new site', the time was ripe for such a venture. He was appointed to the Clinical School planning committee in 1969 and in that role travelled widely in the USA studying medical curricula. After prolonged and sometimes difficult negotiations, which he conducted with customary patience and determination, the clinical school was established in 1975 and Chalmers became the foundation dean. He held this post for four years before returning to full-time clinical practice in 1979. During that time a novel curriculum was introduced, incorporating in-depth topic teaching and extending the Cambridge preclinical supervision system into the clinical years.

Chalmers was a member of the Association of Physicians and served at various times on the executive committees of the Medical Research Society and Renal Association; as honorary secretary of the Middlesex Hospital Research Society, and of the section of experimental medicine and therapeutics of the Royal Society of Medicine. He was a member of the Council of the Royal College of Physicians, and won election to the General Medical Council in 1979 which enabled him to continue his interest in general medical education. He was held in high regard as an examiner in the final MB and periodically served in this capacity for the universities of Cambridge, Belfast, Wales, London, St Andrews and Manchester. He was also external examiner for the first final MB examination to be held at the medical school at Enugu, Nigeria. He established a lasting association with this institution, so that many students from Nigeria were enabled to take their electives in Cambridge.

As a physician, Chalmers was a man of the highest intelligence and integrity. His junior medical and nursing colleagues developed a strong sense of loyalty to him and the things he stood for, and strove hard to match his high standards. He brought the best out of people and was particularly good with medical students. An excellent and decisive clinician, with a wide knowledge of medicine, he consistently took superb clinical notes and was an avid reader of journals. His modest, calm and phlegmatic manner was an example to us all. An honest and straightforward man with a light, dry sense of humour, he was never harsh or unduly critical but he despised dogma, arrogance and incompetence. He practised transcendental meditation to an advanced level and was particularly interested in physiological responses to this technique, and its application to the treatment of anxiety and mild hypertension. As a founder member of the British Association for Medical Applications of Transcendental Meditation he did much to promote interest in the subject both within and outside the medical profession.

At a personal level, Theo was a quiet and unassuming family man. His interests were highly intellectual. He was especially fond of music and, with his wife, sang in the Cambridge Philharmonic Choir, often giving snippets of rehearsals to his juniors in the lifts between ward rounds. He also enjoyed gardening and camping holidays in France with his family.

He married Doreen Boorman in 1951, by whom he had three daughters and two sons, one of whom also qualified in medicine.

PHA/MRC

[*Brit.med.J.*,1984,**289**,506,773; *Lancet*,1984,**2**,474,533,994; *The Times* 14 Aug 1984; Photo]

CHANDRA, Harish *b.*30 Nov 1919 *d.*[? Nov 1984]
MB Bombay(1942) DCH(1945) DCH Eng(1948) MRCP(1950) FRCP(1971) FIAP India(1978)

Harish Chandra was born in Hyderabad, India, the son of the late Rai Harbans Chandra, who was in government service, and his wife the late Brij Rani. He married Pratab Kishore, alias Meena Rani, in 1944, and they had one son, Rohit. Harish Chandra died in the USA, following an operation for a stenosed bile duct, at the age of 65.

His early education was at Madrasai Aliya and St George's Grammar School and his medical studies were carried out at the Grant Medical College, Bombay, from 1937-32. After qualification he held resident appointments at Jerbai Wadia Children's Hospital, passing his DCH examination in Bombay in 1945. He came to London in 1947 and studied at the Institute of Child Health and other institutions, obtaining his membership of the College in 1950. He was elected a Fellow of the Indian Academy of Medical Sciences in 1969, and of the Indian Academy of Paediatrics in 1978.

Chandra was appointed a lecturer in paediatrics at the Osmania Medical College, Hyderabad, and consultant in paediatrics. He was also appointed superintendent and chief paediatrician at the Nuloufer Hospital for Woman and Children, Hyderabad, in 1953. In 1957 he became professor of paediatrics at Osmania Medical College, officer in charge of the Institute for Child Health, Hyderabad in 1958, and later professor and head of the upgraded department of paediatrics in 1967. For a brief time he was dean of Osmania Medical College.

He was a member of the Association of Paediatricians of India from 1951, president of the Indian Academy of Paediatrics in 1969 and from 1957-58 was president of the Hyderabad branch of the Indian Medical Association. He was also a member of the board of studies, academic council and faculty of medicine at Osmania University.

Harish Chandra was also a member of the advisory committee of maternal and child health, Government of India, temporary adviser to the World Health Organization, south-east Asia region, and a short-term WHO consultant. In 1972 he was appointed an adviser to WHO, south-east Asia region, New Delhi, a post which he held until 1979. He played an active role in forming the Indian Academy of Paediatrics in 1963. For some time he was a member of the Medical Council of India.

Chandra was an undergraduate and postgraduate teacher in paediatrics and examiner for the doctorate of medicine in paediatrics at many universities. He published many papers in national medical journals.

Outside medicine, he was fond of racing and swimming and was steward of the Hyderabad Race Club and chairman of the Sailing Club,

Hyderabad. He was a connoisseur of food and music, and all the good things of life.

SCS

[*Indian Paediatrics*,Nov.1984,Vol.**21**,no.11,p.831; Photo]

CHINNER, Melville Ernest *b.*17 June 1902 *d.*30 Nov 1985
MB BS Adelaide(1925) MD(1938) MRACP(1938) MRCP(1939) FRACP(1947) FRCP(1961) Hon FACP (1969)

Melville Chinner was the son of Alfred Chinner, a horticulturist, and Edith,née Batten, whose father was also a horticulturist. A combination of genetic and environmental influences undoubtedly played an important part in Melville's many interests outside medicine. He was an expert in the growing of camellias and old-fashioned roses and president of the Orchid Club of South Australia. His interests extended beyond the floral kingdom. His garden in Adelaide abounded in beautiful blooms, but also contained aviaries where he bred many varieties of Australian finches; he was a recognized authority on these birds. His love of beautiful things embraced fine porcelain, paintings, wine and glassware. He was a true connoisseur who could converse on many subjects and shared his enjoyment and interest in diverse ideas with all who knew him. He would have made a fine, though conservative, director of an art gallery. In 1941 he married Louise Furner Blake, the daughter of a dentist, who bore him two daughters.

Despite his outside interests, his main purpose in life was dedicated to the care of the sick, to the practice of internal medicine as an excellent consultant physician, and to the teaching of undergraduate and postgraduate medicine. Yet he did not achieve the status of a member of the Board of Censors of the Royal Australasian College of Physicians, possibly by reason of his early professional background.

To those who did not know him, the impression he gave on first acquaintance was misleading. A quiet and rather self-effacing manner hid a strong underlying personality with fairly fixed ideas and beliefs. Further acquaintance revealed a character of dedication and compassion. His was not the rumbustious, extrovert humour which is often associated with the Australian prototype in many walks of life, but that of an incisive mind and a dry wit. He taught in a structured manner based on logic and a deep knowledge of all branches of internal medicine. Although there was a lack of flamboyance with which to excite and stimulate the fancies of his students, those who wished to learn respected his approach and manner of teaching. Throughout his clinical life he served as a superb example to his colleagues and his juniors by his dedicated, kindly care for the sick as a complete and caring doctor. It was his habit to see things as either black or white with only occasional shades of grey, and thus he was in demand as a chairman, committee member, and adjudicator in many areas of medical politics and even outside his professional sphere. He was a true conservative whose emotions only became really aroused when discussing left wing politicians, bureaucrats or big government. Nevertheless, he was usually able to supress these feelings and communicate amicably with those who held opposing views. Only rarely did his beliefs interfere with his deep sense of justice. He believed strongly in the future of his country, of which he was extremely proud. A considerable competence in many areas of active sport in his earlier years was continued with an above average excellence in golf as the years advanced.

Chinner's secondary schooling was at St Peter's College, Adelaide, followed by undergraduate medical training at the University of Adelaide and the Royal Adelaide Hospital. After graduation he entered general practice but his ambition was to become a physician. As with many physicians of his generation, apprenticeship in general practice gave him a deep understanding of the problems of his patients and his general practitioner colleagues. Consultant status was achieved in 1938, followed by a year of postgraduate training in London. The war of 1939-1945 interrupted his consultant career and he served as a physician medical officer with the Australian Imperial Forces in New Guinea and the Pacific zones. With the cessation of hostilities he entered private consultant practice in Adelaide and served firstly as an honorary physician to the Adelaide Children's Hospital and then for 20 years as honorary physician to the Royal Adelaide Hospital, until his retirement at the age of 65 years. His breadth of knowledge in all areas of internal medicine ensured his success in private practice, and recognition as a teacher. He was devoted to the interests and careers of many of his younger colleagues and did much to support their aspirations.

Despite the heavy demands on his time, he was able to serve his profession and the community in a wide variety of other responsibilities. He was president of the South Australian branch of the British Medical Association (as it was in those years) in 1956 and 1957, and chairman of the Medical Staff Society of the Royal Adelaide Hospital, as well as being a member of many committees associated with both these bodies. However, his abiding love was for the Royal Australasian College of Physicians; service as a member of Council,from 1957 to 1968, culminated in his election as president in 1969 and 1970.

Men of lesser commitment might then have called it a day, but Chinner continued a lifetime of dedicated service. A member of the Medical Board of South Australia, he was its president from 1971 until 1975. He was holding office when a severe left-sided stroke ended his active career. His intellect remained unimpaired, but he was obliged to depend on others and this depressed him. He forced himself to develop new interests and bore the limitations in his life with fortitude. The many friends who visited him found him a continuing source of stimulation. His depth of experience and wisdom was invaluable to those who sought his advice in the remaining 10 years of his life. The sudden death of his wife was a loss from which he never really recovered. Those close to him knew that he would have welcomed the end when it occurred several years later.

Melville Chinner was a man of integrity whose life was largely devoted to the service and interests of others. A respect for his fellow beings and a love of fine things ensured a full and complete life, as well as many happy memories for his friends and colleagues both in Australia and overseas.

AKG/GW

[RACP *College Newsletter*,Feb 1969,2(3),3; Photo]

CHRISTIE, Ronald Victor *b.*4 June 1902 *d.*27 Sept 1986
MB BCh Edin(1925) MSc McGill(1933) MD Edin(1937) DSc Lond(1938) MRCP(1938) FRCP(1942) FRCPC(1955) Hon DSc Dublin(1962) Hon FRCPE(1969) Hon DSc Edin(1970) Hon DSc McGill(1978) Hon LLD Otago(1975)

Ronald Christie was born in Edinburgh, son of Dugald Christie CMG who, as a medical missionary, had founded the medical school at Mukden in Manchuria. Ronald spent the first eleven years of his life at Mukden, returning to George Watson's College in Edinburgh in 1913 to continue his education. He never forgot the Mandarin Chinese he had learned as a boy.

After medical graduation and a year of internships, he sailed for New York to become a research assistant at the Rockefeller Institute in 1926. Here he worked with Carl Binger and D D Van Slyke. Although he was offered a staff position in 1928, he joined J C Meakins, physician in chief at the Royal Victoria Hospital, Montreal. In 1929 he won a scholarship and spent that year in Aschoff's pathology laboratory in Freiburg, Germany, becoming fluent in German. From 1930-35 he was the first full-time research associate in medicine at the Royal Victoria. Here he did his pioneering work on lung mechanics in emphysema. He returned to Britain in 1935, accepting the position of reader in medicine with A W M Ellis [*Munk's Roll*,Vol.VI,p.162],later Sir Arthur, at the London Hospital. In 1938 he was appointed professor of medicine at St Bartholomew's Hospital.

During the second world war Ronald became involved, through the Medical Research Council, with the early use of penicillin; and in the years immediately after the war he visited a number of countries on behalf of WHO, speaking on its manufacture and use. He also published important clinical trials on penicillin in the treatment of subacute bacterial endocarditis. From 1946-55, he built up a respiratory research group at Bart's in which many young investigators were trained. In 1955 he accepted an invitation to return to the Royal Victoria and McGill as physician in chief. Here he was able to implement many of his ideas on the structure of an academic department of medicine, and its integration with part-time physicians. Residency training was considered an important priority. The department, under his leadership, became a leading research and advanced training centre in North America.

In the London and Canadian Royal Colleges of Physicians he played an important role in restructuring the specialty examinations in medicine.

He was full-time dean of the McGill Faculty from 1964-68, when he retired. However, there was no period of inactivity since he was supported by the Commonwealth Foundation to investigate the relationship of different patterns of health care financing to medical school activities. This work led to his 1969 Harveian Oration on 'Medical education and the State', which contained many original perceptions. At the time of his death Ronald was working on a paper on Erasistratus whose physiological experiments, in the third century BC, Ronald felt had been unjustly criticized by Galen.

Apart from his honorary degrees, Ronald received many honours - including the Trudeau medal of the American Thoracic Society in 1975, the gold medal of the American College of Chest Physicians in 1969, and the Duncan Graham award of the Royal College of Physicians and Surgeons of Canada in 1983.

To these factual details much might be added. Ronald was an excellent and sympathetic bedside teacher, both to students and patients; his absolute integrity was immediately evident. He had an acute perception of the untested assumptions of contemporary medicine but no interest in its 'haute politique'. He had no particular liking for public speaking. His unrivalled experience of medical practice on both sides of the Atlantic made him a shrewd critic of the vices and virtues characteristic of Britain and North America. His dedication to the ideals of academic medicine was apparent from his career.

He was the most 'unpompous' of men, readily approachable and with a detachment from and insight into the pyschology of medical practice. He was an acute judge of other people and built his successful career by encouraging others to make the best use of their potential. He never lost his interest in China and startled the performers of the Peking Opera, who came to the Royal Victoria Hospital to commemorate Norman Bethune, by thanking them in Chinese.

His love of dry-fly fishing was proverbial. He had had a croft in Scourie in Sutherlandshire since the mid-1930s, and three months before he died he went there for three weeks. His friends wondered whether the countries he chose to study for the Commonwealth Foundation, and the timing of his visits, did not bear some relationship to the trout fishing.

Whatever patterns of health care financing, or of organized medical research, may develop in the future, Ronald's qualities will always remain central to excellence.

Ronald married Joyce Ervine in 1933; she predeceased him in 1968. Their children were Janet (Dr Janet Seely of Ottawa) and Dugald, a lawyer, living in Vancouver BC. In 1976 he married Manette Loomis, widow of his long time friend Dr Alfred Loomis of New York.

DVB/GW

[*Lancet*,1986,**2**,988; Photo]

CLARK-KENNEDY, Archibald Edmund *b.*23 Apr 1893 *d.*2 Sept 1985
MA Cantab(1914) MRCS LRCP(1918) MRCP(1921) MD(1923) FRCP(1930)

Archibald Clark-Kennedy, formerly consultant physician to the London Hospital,1928-58; dean of the London Hospital medical college 1937-53, and fellow of Corpus Christi College, Cambridge, 1919-1985, died in Addenbrooke's Hospital after a few days' unconsciousness. As Chris Bartlett wrote in the obituary in the *British Medical Journal*, 'His great span of life, lived out in fullness to the very end, matched his physical, intellectual and moral stature.'

The son of the Revd A E Clark-Kennedy, he was educated at Wellington College and Corpus Christi College, Cambridge, where he graduated with first class honours in the natural sciences tripos. Elected a fellow of Corpus Christi in 1919, he was director of medical studies for many years.

On the outbreak of war in 1914 he volunteered for the Army and was commissioned into the Queen's Royal West Sussex Regiment (5th Territorial Battalion) and served in India and Mesopotamia. On his return to England in 1917 he completed his medical studies at the London. He qualified with the conjoint and gained a commission in the RAMC, serving in France as medical officer to 158 Army RFA Brigade. On demobilization his higher medical qualifications were acquired rapidly. After working on the medical unit under Sir Arthur Ellis [*Munk's Roll*,Vol.VI,p.162] he was appointed a physician to the London Hospital in 1928, and in 1937 dean of the London Hospital medical college.

His 17 years as dean of the London Hospital medical college included the 1939-45 war, throughout which he ran his medical firm in Whitechapel, while maintaining his school, for the first few years, scattered over a consortium of hospitals in East Anglia, mainly linked by his own personal efforts. He placed the preclinical departments in Cambridge and was able - as he put it - to keep an eye on them and on Corpus at the same time. From 1939-45 Archie carried much of the burden of running the London's large section of the Emergency Medical Service. Almost single-handed, as far as the management of the London college was concerned, he kept it alive and functioning efficiently, disseminated as it was over four counties. Throughout the war he was one of the few, at one period there were only two, physicians with beds in Whitechapel. After the war he had to cope with changes when the London Hospital became part of the National Health Service, and its medical school became an autonomous college of the University of London, financially dependent on the University Grants Committee. Problems of the long term future jostled with immediate difficulties of every kind, all of them calling for urgent action from the dean.

Archie later became acknowledged as an outstanding educationalist. In 1947 the first volume of his book *Medicine: The patient and his disease*, Edinburgh, E&S Livingstone, was published. The second, *Medicine: Diagnosis, prevention and treatment*, appeared in 1953, and a third, *Medicine: medicine in a human setting*, in 1954.

As a clinician his great skill was in identifying the true nature of a problem. He relied little on the emotional stimulus of new discoveries. For this reason some considered his practice of medicine unscientific and his teaching too theoretical, but although he constantly asserted the importance of showing students how to learn rather than conveying the fact, he never despised the latter. Neither was he reluctant to apply new discoveries provided he was convinced logically of their usefulness.

Besides articles in medical journals he was the author of *Patients as people*, London, Faber & Faber, 1957; *Human disease*, England, Harmondsworth Penguin Books, 1957; *How to learn medicine*, London, Faber & Faber, 1959 and *Clinical Medicine, the modern approach*, with C W Bartley, London, Pitman, 1960. His interest in history produced *Stephen Hales, DD FRS: an 18th century biography*, Cambridge University Press, 1929. In retirement he undertook a two-volume *History of the London Hospital*, London, Pitman Medical Publications, c.1963. This was followed by *Edith Cavell: pioneer and patriot*, London, Faber & Faber, 1965, and *Man, medicine and morality*, London, Faber & Faber, c.1969. Undeterred by advancing years, he wrote *Attack the colour*, London, Research Publishing Company, 1975; *A Victorian soldier*, 1980, and finally *Cambridge to Botany Bay*, 1983, the story of an infamous first cousin, J R Mortlock, on which was based an entertaining television programme.

Archie's physical achievements became almost legendary. His great height, long legs and inexhaustible stamina made him a fine long distance runner. He ran for Cambridge Hare and Hounds against Oxford before the first world war, and was still delighting students in the 1950's running cross country. Towards the end of the second world war he lost the sight of one eye, and a cataract obscured his vision in the other. For some 26 years he could only see with difficulty. For many years after the war he rode to hounds, relying on the visual acuity of his horse. On holiday he would hunt with a foot pack in Cumbria, and also went salmon fishing in the Solway Firth. He loved to beagle and sail, but his greatest joy was walking the fells. He completed the Pennine Way in 14 days at 78 years old. He celebrated his 80th birthday at the top of Skiddaw. In his 80's bicycle tours covered most parts of Britain, beginning usually on a Sunday when it was easier to cycle across London from Liverpool Street to some other terminus, and then take the train to a point from which he would move from one youth hostel to the next.

His wife Phyllis, who died in 1978, had been trained as a nurse at the London. His daughter and one grand-daughter nursed at the same hospital. His son was a fellow and steward of estates at Corpus. Archie continued to live an independent life until shortly before his death. This continued activity was a triumph of thought and courage over differing kinds of disability. Finding his moped, and later his bicycle, both of which had served him faithfully for decades, were now too insecure, he maintained his mobility on a tricycle, a racing model with 10 gears. This enable him to dine regularly at Corpus right up to the end, and he remained one of the best conversationalists at High Table, fully abreast of all current affairs. Towards the end of his life he became a Unitarian and among his later publications was *Basic belief, or,the faith of an agnostic*, Cambridge, Cambridge University Press, c.1980, an expression of what he called 'basic relief'. Above all he valued his independence, which he retained until the end, curtailed by only a few days in hospital before his death.

The British Medical Journal obituary pertinently remarked that his greatness arose partly from a '...touch of eccentricity about him. It imparted an orginality of thought that demolished traditional attitudes but only to expose truth.' Archie had a spark of the divine fire which makes the elect of any profession.

BLH

[*Brit.med.J.*,1985,**291**,828; *Lancet*,1985,**2**,621; *The Times*,3 Sept 1985; *Letters of the Corpus Assoc.***65**,Michaelmas 1986 (Corpus Christi College); *Lond.Hosp.Gaz.*,Dec 1974,**77**(4),3-7; July 1972,**75**(3),8-12; July 1958,**61**(3),76-78; Oct 1953,**56**(5)143-44; Photo]

CLAYTON, *Sir* Stanley George *b.*13 Sept 1911 *d.*12 Sept 1986
Kt 1972 MRCS LRCP(1934) MB BS Lond(1934) FRCS(1936) MRCOG(1939) MD(1941) MS(1942) FRCOG(1951) *FRCP(1977) Hon FCOG S.Afr(1975) Hon FRACOG (1984)

Stanley George Clayton was the son of the Rev George and Florence Clayton. He was born at Hankow in China where his father was serving as a Methodist missionary. With his typical sense of humour he often recounted the story of how, at the age of two, when on a journey on the Trans-Siberian railway with his parents, his inquisitiveness caused him to fall out of the train - fortunately, into thick snow. He came to regard this as evidence of his ability to survive.

At the age of eight he was sent to Kingswood School, near Bath, to begin his formal education. From there he went to King's College London with a Sambrooke scholarship in 1929. Before going on to King's College Medical School, as a clinical medical student, such was his reputation as a tough forward on the rugby field that he was prematurely recruited into the Hospital's First 15 at a time when there were several international players in the team. As a medical student he soon made his mark, gaining the Hallet prize of the Royal College of Surgeons in 1931 and the Jelf medal and Todd prize in 1934, the year in which he qualified. After two years of house appointments he became registrar at King's in 1936, and at Queen Charlotte's Hospital in 1938. During this time he also served as a lecturer in anatomy at King's College.

The war years saw him serving as a general surgeon as well as a gynaecologist in the EMS and, in spite of a heavy clinical load, he sat and passed successfully both the MD and the MS. Later in the war he was recruited into the RAMC, where he served in Germany with the rank of major. Returning to civilian life and his chosen specialty he was soon appointed to the consultant staff at King's, Queen Charlotte's and the Chelsea Hospital for Women.

Until 1963 he filled these appointments with distinction, gaining a reputation as a superb teacher, a sound clinician, and a first class surgeon. But what he really wanted was a full-time academic career and when the opportunity arose to become professor at the Postgraduate Institute at Chelsea and Queen Charlotte's, he welcomed it in spite of his regret at leaving King's. Yet his first loyalty was always to King's and when, after prolonged negotiations, it became possible to establish a professorial chair at King's, Clayton was an automatic choice. He was one of the most outstanding professors in the post-war era. Though not himself an original scientific research worker, he possessed those intellectual attributes which enabled him to see quite clearly where research was needed to enhance knowledge and elucidate the causes of disease and illness and thus improve the prospects for better treatment of patients. He gathered round him a team of young research workers, all of whom under his guidance made significant contributions, especially in relation to techniques for monitoring the well-being of the foetus during pregnancy and labour. He laid the foundations of an academic department which has become one of national and international repute.

All those who were taught by Clayton testify to his pre-eminence as a teacher. He could present any subject, however complex, with a clarity that few teachers are able to achieve. He was impatient of irrelevancies and always went straight to the core of a problem.

Clayton contributed many original articles to journals, wrote two short textbooks, and edited *Medical disorders during pregnancy*, London, Churchill, 1951, with Samuel Oram. It was as an editor that he excelled: from 1963-72 he edited the *British Journal of Obstetrics and Gynaecology*, having been assistant editor to James Young for many years previously. Under his guidance the *Journal* more than doubled its circulation worldwide, by improving its quality. Every paper that was submitted was most carefully vetted and edited. His Joseph Price oration to the American Association of Obstetricians and Gynaecologists in 1975, should be compulsory reading for any would-be medical author. His title was 'Printed communications'; being filled with much characteristic humour and command of the English language, he also revealed great understanding. He recalled how, whenever it was necessary to reject an article, he would write to the author in longhand which he felt would make the unsuccessful writer feel that the article had really been given careful consideration. These letters often contained helpful advice. Editing involved many hours of tedious reading but, being a voracious reader, Clayton was quite happy to do this. His reading was by no means confined to medical literature and he was reputed to have read the *Encyclopaedia Britannica* from cover to cover. His ability to bring out quotations from so many literary sources was a tribute both to his reading and to his memory.

As time passed and his career blossomed it was obvious that he was destined for many high offices. In turn he was vice-president and president of the RCOG, chairman of the conference of Colleges and president of the Association of Family Planning Doctors. He served on the medical advisory committee on medical training in the EEC, and for seven years filled that most onerous of posts: chairman of the distinction awards committee. His administrative skills and obvious integrity enabled him to gain the respect and admiration of all those who worked with him. One of his outstanding achievements was to guide Australian gynaecologists through a minefield of legal and professional problems to a successful outcome when they decided to separate from the parent College and found one of their own. So grateful were they that, at their own expense, they sent over an Australian artist to paint his portrait - which now hangs in the College at Regents Park - and later made him an honorary Fellow of their new Australasian College.

'No man is wiser for his learning; wit and wisdom are born with a man' (*Table-talk*, John Selden, 1584-1654); Clayton was born with a good measure of both. He was not an easy man to know, being a rather shy and private person. At times he appeared abrupt and direct in speaking his mind but he always spoke with conviction. He was somewhat impatient; in a hurry to get on with the next job as soon as the previous one was finished. He was totally loyal to his hospital and medical school, to the Royal College, and to his colleagues and friends. Although not naturally fond of social gatherings, whenever his presence was required by virtue of his office, as on many occasions, he was always charming, whether host or guest, and sincere in his speeches.

Clayton was fortunate to have a happy home life with Kathleen, née Willshire, whom he married in 1936, and their two children. He left his friends in no doubt about the enormous companionship he received from her throughout his professional life, and especially in the years before she died when she was not always in the best of health.

When the Claytons settled in their home at Leatherhead he had little time, and no great enthusiasm, for hobbies; but he had a garden and that garden had to be kept in order. As time passed, he not only did this but also became somewhat of an expert. Relaxing with his friends gave him great pleasure and it was then that his quiet sense of humour was apparent. It was a regular sight to see him, with three other colleagues, take half-an-hour off in the lunch interval to enjoy a game of cribbage in the staff common room at King's. There was often much hilarity among both onlookers and players. Living in Leatherhead by choice, he appeared to enjoy driving a series of fast Jaguar cars to and from home and hospital every day, and he was greatly saddened when he had to confess that the pleasure of night driving had gone.

Clayton suffered a mild coronary but made light of it. He must have had a good deal of angina but, in spite of this, he was hesitant about worrying his doctors. Following an operation, from which he was recovering quite satisfactorily, he suffered another coronary, this time severe, and died as he would have wished - in the hospital to which he had given devoted service for more than 50 years.

JHP

[*Brit.med.J.*,1986,**293**,**829**; *Lancet*,1986,**2**,759; *The Times*,16 Sept 1986; Photo]

CLEMENT-JONES,(née YIP) **Vicky Veronica**

*b.*23 Dec 1948 *d.*30 July 1987

MA Cantab(1971) MB BChir(1974) MRCP(1976) FRCP(1986)

Vicky was born in Hong Kong, the daughter of Teddy and Susie Yip. She moved with her family to East Grinstead at the age of seven, and was educated at the Notre Dame Convent, Lingfield, and East Grinstead County Grammar School. She gained an exhibition to Girton College and graduated in 1971 with a double first in the medical sciences tripos, archaeology and anthropology. She then studied medicine at St Thomas's Hospital medical school and obtained her MB in 1974. She continued working at St Thomas's, progressing to senior house physician in neurology and thoracic medicine.

In 1976 Vicky moved to St Bartholomew's Hospital to take up the post of registrar in the medical professorial unit, and was later awarded the Aylwen bursar appointment as a research fellow in the department of endocrinology. There she carried out innovative research into enkephalins and endorphins, and contributed to more than 30 published papers, including one showing the scientific basis for acupuncture.

In September 1982 Vicky learned that she had advanced ovarian cancer and, in her own words, she 'crossed the divide from doctor to patient'. For the first time she fully understood the emotions and problems faced by patients and their families, their feelings of uncertainty and isolation and their need for information which combined accurate medical knowledge with the practical aspects of coping, which most patients learned by experience. She became aware of the importance of better communication between patients with cancer and their families and friends, and the health professionals working with them. After briefly resuming her medical career, she decided to found a charity which would provide information, practical advice and emotional support to all cancer patients and their families. Vicky described her own experiences as a cancer patient and her reasons for founding BACUP in an article entitled 'Cancer and beyond: the formation of BACUP' which was published in the *British Medical Journal*, 1985,**291**,1021-23.

Vicky set about this task with characteristic energy, single-mindedness and professionalism, gaining the support of senior medical colleagues, other cancer organizations, and individual patients, their relatives and friends. She also visited America to see the services provided by the National Cancer Institute and the American Cancer Society. BACUP (the British Association of Cancer United Patients) was registered as a charity in October 1984, with Vicky as chairman and honorary director. Just twelve months later, the National Cancer Information Service was launched, with specially trained senior cancer nurses providing information and emotional support by telephone and letter. By the time of Vicky's death, a year and a half later, there were six full-time nurses responding to more than 1,500 enquiries per month; more than 20 booklets and leaflets about cancer had been published, and BACUP News, cancer patients' own newspaper, was in its 4th issue.

Throughout this time Vicky was receiving treatment but made the most of the time that she had, when necessary directing BACUP's development from her hospital bed with a mobile phone and always enthusiastic to present BACUP's latest achievements in interviews with the press and TV. She reassured everyone that despite her increasing debility and constantly changing expectations she enjoyed a good life, taking the greatest pleasure not only in her successful establishment of BACUP as a service to cancer patients and their families, but also through sharing her personal delight in music, food and gardening, with her own family and friends. The memory of her remains a source of inspiration and encouragement to all who knew her.

Her husband, Tim, a company solicitor, is a Trustee of BACUP. There were no children of the marriage.

MLS

[*Brit.med.J.*,1987,**295**,677; *Lancet*,1987,**2**,407; *The Times*,1 Aug 1987; Photo]

COCHRANE, Archibald Leman *b.*12 Jan 1909 *d.*18 June 1988 CBE(1969) MBE(1946) MA MB Chir Cantab(1938) MRCP(1961) FRCP(1964) FFCM(1972) Hon Dr York Univ(1975) Hon DSc Rochester USA(1977)

All fellows are unique, but none (in his generation) was as unique as Archie Cochrane. Born at Galashiels into a family that had become moderately wealthy by manufacturing textiles, his early life was traditional enough, taking him via Uppingham to King's College, Cambridge, where he obtained first class honours in both parts of the natural sciences tripos and nurtured the ambition of becoming a don. Fortunately for medicine, the problem he began to tackle proved to be a sterile aspect of tissue culture and he soon became dissatisfied with its triviality and his own technical incompetence. When, therefore, he developed a functional symptom, which no English physician could cure, he abandoned his project and sought advice at the Kaiser Wilhelm Institute in Berlin, then the fount of academic medicine. Again obtaining no relief and becoming intrigued, like so many other young people at the time, with the ideas of Freud, he turned to psychoanalysis. The next two years in Berlin, Vienna and Amsterdam, following his Jewish analyst as he fled from Hitler, gave him a hatred of fascism, fluent German, and a sceptical attitude to all theories not validated by experiment.

In 1935 Archie returned to Britain and started clinical medicine at University College Hospital, where he was greatly influenced by the enthusiasm and scientific rigour of Sir Thomas Lewis [*Munk's Roll*, Vol.VI,p.531] and George Pickering, later Sir George [*Munk's Roll*, Vol.VII,p.464] and by the social concern of Philip D'Arcy Hart. It was four years, however, before he qualified, as the outbreak of the Spanish Civil War led him to join an ambulance unit in support of the Republic, in which he must have been the only member with neither political nor religious affiliation. He returned to Britain after a year, in which he had gained experience of triage on the Madrid front, and qualified just in time to complete a house job at the West London Hospital and to obtain a research appointment at UCH - before resigning to join the Army. Eighteen months later he found himself a prisoner of war, having been captured with an ill-starred commando unit in Crete. For most of the rest of the war he was the British officer in charge of prison hospitals, in which he had first hand experience of Nazi brutality, carried out his first controlled experiment to find a cure for famine oedema, became an expert on tuberculosis, and exposed himself to such a large dose of x-rays that he developed, 20 years later, a squamous carcinoma on the back of his hand.

On demobilization, Archie obtained a Rockefeller fellowship in preventive medicine, which involved a course at the London School of Hygiene and Tropical Medicine, where he fell under the influence of Austin Bradford Hill, later Sir Austin, and then a year at the Henry

Phipps clinic in Philadelphia. At the latter, he came across the problem of inter- and intra-personal variation in the reading of x-rays and developed his interest in the scientific study of diagnostic and prognostic error.

Back in Britain, Archie joined the MRC's pneumoconiosis research unit under Charles Fletcher and worked steadily for the next 40 years, first as a member of the unit, then from 1960 as David Davies professor of tuberculosis and diseases of the chest and director of the MRC's new epidemiology unit in Cardiff, and finally, after official retirement, as the first president of the Faculty of Community Medicine, 1972-75, and an honorary member of the unit he had previously directed. Employed, at first, to study the factors that determined the development and progression of pneumoconiosis in coal mines, Archie imaginatively sought to test the hypothesis that progressive massive fibrosis was determined by tuberculous infection, by making intensive efforts to eliminate tuberculosis from one valley while leaving another to standard practice. His plan was thwarted however by the introduction of streptomycin, which almost eliminated tuberculosis from both. Turning to investigate the role of dust, he demonstrated for the first time a quantitative correlation between the development of the disease and exposure to dust (rather than silica) by studying 'random colliers' whose exposure was measured by members of his team who 'shadowed' them at work.

When the National Coal Board took over the work on pneumoconiosis, Archie began to use the technique of population surveys that he had developed in the study of coal miners, to study the natural history and causes of a wide range of common diseases from anaemia to glaucoma. In the course of this work, he helped to set new standards for the burgeoning science of epidemiology by refusing to accept lapse rates of more than five per cent in surveys and follow-up studies and by checking the reproducibility of any measurement or item of the subject's history.

Just as important were his efforts to establish medical and social practice on a sound basis by persuading clinicians, medical administrators, and members of other professions (including magistrates and schoolteachers) to base their practice on the results of randomized controlled trials rather than on personal impression, precept, theory or tradition. These were epitomized in his short book on *Effectiveness and efficiency: random reflections on health services*, London, Nuffield Provincial Hospitals Trust, 1972, which was based on his own experience in testing the value of domiciliary, outpatient and inpatient treatment in the National Health Service, and won him international acclaim.

Despite frequently working a 12-hour day, Archie somehow found time to create a garden at his home at Rhoose Farm that came to be included in the national garden scheme, to collect with discrimination modern paintings and sculpture, and to offer hospitality to friends and students on a scale more reminiscent of Edwardian than Elizabethan times. Unusually for someone with so much concern for communal welfare, he was equally concerned for the welfare of his family, his staff and his friends. When his sister was admitted to a psychiatric hospital, he refused to accept the diagnosis of dementia and sought a fresh opinion in Edinburgh, which led to the discovery that she, and he, were both porphyriacs. Concerned that relatives might, unknown, endanger their lives by exposing themselves to barbiturates or sulphonamides, he sought urinary and faecal specimens from 153 of them scattered as far afield as New Zealand and obtained specimens from 152.

To talk with Archie might occasionally be tiresome, if one was not immersed in the topic, as he found small talk difficult and preferred to discourse at length on his current enthusiasm. To work with Archie was to obtain a lifelong friend, on whom one could rely for unfailing support in times of need.

RD

[*Brit.med.J.*,1988,**297**,63,355,419,614,1704,1989,**298**,454; *Lancet*,1988,**2**,57-58; *The Times*,22 June 1988; Video; Photo]

COCHRANE, Robert Greenhill *b.*11 Aug 1899 *d.*3 Aug 1985
CMG(1969) MB ChB Glasg(1924) MRCS LRCP(1924)
DTM&H(1924) MRCP(1927) MD(1928) FRCP(1945)

Robert Greenhill Cochrane ('Bob' to his many friends) was born in Pei-Tei-Ho, China, of missionary parents. What was to be an eventful career almost ended before it began when, a few months after his birth, the family was captured by Boxers while trying to escape. They were released after a faithful servant had successfully pleaded with their captors. His father Thomas Cochrane, also a physician, was a forceful character who, on returning to China in 1901 after leave in England, founded the Peking Union Medical College and Hospital with money from the Chinese reparations to the great powers and a personal donation from the Empress Dowager herself.

During his early years Robert was taught in China by his mother but his formal education began when, at the age of nine, he was sent to the School for the Sons of Missionaries at Blackheath, England, and then to Eltham when the School moved there. He graduated in medicine in Glasgow in 1924, with commendation. After a few months at St Bartholomew's Hospital, London, and a short course at the London School of Hygiene and Tropical Medicine, he gained the conjoint diploma and the London diploma in tropical medicine. That same year he sailed for India under the auspices of the Mission to Lepers. He could not have had a better introduction to leprosy, for he sat at the feet of the renowned Ernest Muir in Calcutta for three months. His first term was spent in Purulia and Bankura in West Bengal. He returned to

Britain in 1927 and was admitted MRCP. He had also collected enough data on leprosy, and had enough diligence, to prepare a thesis for his MD.

After making a somewhat critical report in his survey of the Mission to Lepers' homes and hospitals in India, he rather abruptly severed his connexion with the Mission. In 1929 he was appointed medical and general secretary of BELRA (the British Empire Leprosy Relief Association) which, having been formed some four years previously, needed a man of Cochrane's standing and drive to guide its future course. In 1931 he was invited to a consultation of leading leprologists in Manila, called by the Leonard Wood Memorial Foundation, and the International Leprosy Association was formed. Bob was appointed its first secretary-treasurer. This period set the pattern for his future rather restless activities. He was building up an extensive knowledge of clinical leprosy, and developing a provocative lecture style that never failed to keep his audience alert. After another brief spell in India in 1933, he realized that he was going to achieve his growing ambition to make a significant contribution to leprosy as a world problem, and he needed further field experience. In 1935 he successfully applied for appointment as chief medical officer of the Lady Willingdon Leprosy Sanatorium at Chingleput, Madras. While there, he was asked to become adviser in leprosy to the State of Madras. He used the parent sulphone (dapsone, or DDS) orally to treat leprosy patients - being the first to do so. He had been in touch with the ICI chemists in Britain who were using the compound to treat streptococcal mastitis in cows. However, within a couple of years Cochrane left Chingleput to begin his association with the Vellore Christian Medical College and Hospital. He was responsible for the delicate negotiations which led to the upgrading and government recognition of the College, and the admission of men students. For his exemplary work in India he was awarded the Kaiser-i-Hind medal in gold, first class. He was a friend of Gandhi and other political leaders.

Bob Cochrane returned to England in 1951 and for the next three years he served, once again, as medical secretary of BELRA. Failing to enlist adequate financial sponsorship in Britain for his leprosy activities, he turned to the USA and was appointed technical medical adviser to the American Leprosy Missions Inc., having been assured that he could continue to make his base in Britain. With money from the USA and the Wellcome Trust, he founded the Leprosy Research Fund in London in 1951, as a coordinating and teaching centre, and a histopathological reference centre. This was renamed the Leprosy Study Centre in 1955.

By this time Bob Cochrane had gained a wide reputation as the leading leprologist in the English-speaking world. He was in great demand and travelled in many countries advising governments and missions on their leprosy programmes, and conducting seminars. He was always frank and outspoken in his opinions, qualities that did not endear him to everybody. When leprosy was made a notifiable disease in Britain, Bob Cochrane was the obvious person to be appointed consultant adviser to the Ministry of Health.

In the course of his travels he visited me in the Belgian Congo (now Zaire). In his best pontifical manner he had often declaimed that nowhere could leprosy attack more than 10% of the population. When I showed him some villages where over half the people had diagnosable leprosy lesions he could not believe the evidence of his eyes, but he did afterwards modify his somewhat sweeping generalizations. It was there that Bob advised me to concentrate on leprosy and abandon my interests in general tropical medicine. I next met him in 1960, when in England on my first leave from Nigeria where I had been appointed director of leprosy research. He discussed with me a riminophenazine compound known by its code number B663, which he said was 'tailor made for leprosy', being concentrated in macrophages in mice experimentally infected with *mycobacterium tuberculosis*. He persuaded me to try the compound in cooperative leprosy patients in Uzuakoli, Eastern Nigeria. Under its offical name clofazimine, and its proprietary name Lamprene (Ciba-Geigy), this drug is now generally recognized as an excellent anti-leprotic with anti-inflammatory properties.

Having asked me to succeed him as director of the Leprosy Study Centre, Bob felt free to return to India to resume clinical leprosy work. (He was president of the International Leprosy Association 1965-68, presiding over the Ninth International Leprosy Congress held in London in 1968. He was then made president emeritus.) Shortly after his arrival in India his wife was taken ill and died. Despite his loss, he went to Vadathorasalur in Madras State where he spent two useful years. Meanwhile, a lady doctor who had admired him from afar, and had written sympathizing with him in his loss, invited him to Tanzania. They were married in Kola Ndoto, and worked together under the auspices of the African Inland Mission from 1969-72. The couple then retired to Norristown in Pannsylvania, where Jeanie devotedly nursed her husband through many months of enforced inactivity and increasing weakness. The end came as he was approaching his 86th birthday.

No tribute to Bob Cochrane would be complete without mention of the driving force in his life - his Christian faith. He was a convinced, practising Christian and lost no opportunity of commending and sharing his faith. Some people might have thought him dogmatic and aggressive, but more admired him for his forthrightness and pertinacity. He was actively associated with the World Dominion Trust.

The Cochrane Annex at the Slade Hospital, Oxford, will remind dermatologists of the considerable influence that he wielded during his lifetime; as will the incomparable collection of stained series of histopathological slides - 16,000 of them - of sections of leprosy lesions and related conditions. These are housed at the Hospital for Tropical Diseases, London, and are available for study and research.

'This one thing I do' could be Bob's epitaph, whether the 'one thing' was campaigning on behalf of the victims of leprosy worldwide, bludgeoning people in governments or missions to abandon the stigmatizing epithet 'leper', or persuading clinicians and medical investigators to interest themselves in the disease called leprosy which to him was of overriding fascination and importance.

Bob Cochrane exercised a far greater influence in the cause of leprosy than any other individual of his generation. He made the study of leprosy 'respectable' and constantly advocated its inclusion in medical teaching and practice. He was a clinician to his finger-tips; an accurate observer of clinical signs and an excellent demonstrator of the bewildering variety of clinical manifestations of leprosy. His name will be perpetuated in the Robert Cochrane Fund for Leprosy, administered by the Royal Society of Tropical Medicine and Hygiene, which will provide bursaries to help young scientists in their studies in leprosy.

Bob is survived by his second wife and by three children of his first marriage: two doctor sons and a nurse daughter - all trained at St Bartholomew's Hospital.

SGB

[*Brit.med.J.*,1985,**291**,608; *Lancet*,1985,**2**,457; *The Times*,6 Aug 1985; Photo]

COLES, Handley Montague Townsend *b.*4 June 1917 *d.*6 Dec 1985 MRCS LRCP(1944) MB BS Lond(1945) DCH(1949) MD(1951) MRCP(1952) FRCP(1966)

Handley Montague Townsend Coles was a scion of the Coles family who came from Steeple Aston in Oxfordshire. He was born at Loughton in Essex and was the fifth son of William Francis Coles who, at an early age, had been obliged to work in the mines to support his widowed mother and sisters. In later life, Francis Coles became an evangelistic Plymouth Brother and lay preacher. Handley's mother, Rosalind Ethel Daisy Townsend, to whom he was very devoted, was the headmistress of a small preparatory school. Handley was named partly after his maternal grandfather and partly after a missionary whom the family admired. Early family life was governed by strict religious principles.

When Handley was twelve years old, his mother died, the school was closed and he was sent for further education to Wallingbrook School in Devon. On leaving school, he took a job with a shipping company in the City and remained there for two or three years until he was given the opportunity to study medicine. He had not found office work very congenial but after starting his medical studies at King's College, London, he made some good friends and life began to broaden.

When the 1939-45 war started the College was evacuated to Glasgow and Birmingham, where the preclinical courses were held, but most of the clinical teaching took place in the King's College Hospital sector at Horton and Leatherhead. It was at Leatherhead that Handley met Patricia Falle, his future wife and the younger daughter of Frances Huxley who was one of the early women consultants in London and a founder fellow of the Royal College of Obstetricians and Gynaecologists.

Handley qualified in 1944 and, after one house appointment and having gained his degree, he joined the Navy as a surgeon lieutenant RNVR. He served in destroyers in the North Atlantic, was transferred to the Naval base at Trincomalee in Ceylon and, after the end of hostilities, to Singapore.

After release from National Service, and marriage early in 1947, his career in paediatrics started at Southend General Hospital as house physician to Richard Dobbs whom he greatly admired. Two years were spent in general medicine before returning to paediatrics as an intern at Queen Elizabeth Hospital for Children, Hackney. His work and easy rapport with children impressed everyone there. Barnett Levin, the eminent biochemist, remarked that he knew Handley Coles would go far and described him as having 'the manner and manners of a Spanish Count'. Ian McAllister Anderson (q.v.) was also on the staff there and supported Handley in becoming paediatric registrar at Westminster and Westminster Childrens' Hospitals. Subsequently he was appointed senior registrar to Sir Charles Harris [*Munk's Roll*,Vol.VI,p.224] who allowed him a great deal of responsibility. The paediatric cardiac unit was being developed at Westminster Hospital and Handley trained and became proficient in the then new technique of cardiac catheterization, becoming a mainstay of the department in this respect. There was warm friendship and camaraderie between the registrars at that time and Handley, Kenneth Hugh-Jones and Seymour Mason, always looked back on those years with affection.

He was appointed to the staff of Queen Mary's Hospital for Children, Carshalton, in 1957. This had been the largest childrens' hospital in England, founded by the London County Council for the treatment of rheumatic fever and rheumatic cardiac disease. By the time he joined the staff, this was less of a problem and the hospital service was contracting. He saw through the difficult amalgamation of Queen Mary's with the Fountain Hospital and welcomed his new colleagues with his accustomed friendliness.

A year later he was appointed to the consultant staff at Westminster Childrens' Hospital, with sessions on the paediatric cardiology unit at Westminster Hospital. These were great days for both places; George McNab was undertaking pioneer work on hydrocephalus with the Spitz-Holter valve, and Cyril Scurr and Charles Drew, the cardiac surgeon, had developed the technique of profound hypothermia for intracardiac surgery in congenital heart disease. The role of the paediatric team was to investigate, discuss and propose treatment. The meetings did not always run smoothly but they were helped by

Handley's tact. To his colleagues, he seemed satisfied with things at the time though there were moments when he expressed anxiety about his position as a mere investigator, finding the role rather irksome. At this time, he wrote papers on selective angiocardiography of the infant heart, and a well remembered classic on endocardial cushion defects with Ian Anderson in 1961 [*Brit.med.J.*,11 March 1961,pp.696-705].

His principal interest remained general clinical paediatrics and he took the main burden of acute cases at the Childrens' Hospital, inspiring confidence in nursing staff, juniors, parents and children, by his kindly management and concern for their welfare. He was a most popular undergraduate teacher; his unorthodox approach and sense of fun made his rounds an entertainment. He thought it important to teach broad clinical principles and to stimulate students with varying subjects rather than try to instil the minutiae of academic paediatrics, which they could read for themselves. He served on the council of the section of paediatrics at the Royal Society of Medicine from 1970-78, and as its secretary in 1974 and 1975.

During the sixties and seventies he suffered two quite severe illnesses which he bore with characteristic stoicism but he was well enough to start the new department of paediatrics at Queen Mary's Hospital, Roehampton, in 1973, managing it and the neonatal service for five years as the sole consultant. In 1976 he largely wrote and edited an admirably lucid text book of paediatrics for students, *Paediatrics*, Tunbridge Wells, Pitman, 1976, and just before retiring received the high honour from the undergraduates of being elected chairman of the annual Shrove Tuesday dinner at Westminster Hospital Medical School.

He had an excellent relationship with all who worked with him and took a great interest in helping his juniors who benefited from his guidance and encouragement in their careers. Ward rounds at the Childrens' Hospital were often followed by drinks at the local hostelry. One remembers one such session after a clinical meeting with visiting consultants when an impecunious junior, waiting apprehensively for his turn to pay the next round, found a five-pound note quietly pressed into his hand by Handley.

Despite his heavy clinical commitments he made time for his family of two sons and two daughters, and his favourite relaxation of salmon fishing for which he would escape to Wales and, once a year, to the Spey, some times with his sons. He and Patricia travelled often in France and, in 1978, bought a ruined farmhouse there which gave Handley an opportunity to use his talents in designing and making the architectural drawings for its restoration. He had wide, cultured interests, and his wit, humorous view of life and disarming modesty made him a memorable companion.

In 1977 he had to undergo cardiac surgery and although he returned to work he felt, after two years, that he should retire from his Health Service commitments. After an enjoyable spell in Saudi Arabia, he retired finally in 1982. Sadly, his few years of retirement were clouded by the untimely death of his younger son following an accident in Italy, and by increasing ill health which was born with characteristic fortitude.

LS

[Brit.med.J.,1986,**292**,565]

COMPSTON, Nigel Dean *b*.21 Apr 1918 *d*.17 Oct 1986
CBE(1981) BA Cantab(1939) MRCS LRCP(1942) MB BChir(1942) MRCP(1942) MA MD(1947) FRCP(1957)

Nigel Dean Compston was a physician of exceptional talent who also had a gift for administration that would have ensured his success in many other walks of life. He will be particularly remembered for his work for the Royal Free Hospital and the College.

Nigel Compston was the son of a Leeds general practitioner who died when he was only six years old, leaving his family in some financial difficulty. Nigel and his brother Peter (Vice-Admiral Sir Peter Compston) were brought up and helped by various members of the family, and later Nigel was admitted to the Royal Masonic School which selectively was able to assist the sons of doctors. From there he won a scholarship to Trinity Hall, Cambridge, and went on to the Middlesex Hospital for his clinical training. He also had time for rowing and rugby. His childhood cannot have been easy but from it must have grown his determination, courage, dedication and capacity for hard work.

He took his qualifying examination and the MRCP in the same year, a remarkable achievement which would not now be possible. He took an equally short interval to marry Diana and then went straight into the Army. He landed with the British Forces in Normandy and at the end of the war, at the age of only 27 years, was a lieutenant colonel and responsible for the health of the British garrison in Berlin.

He left the Army in 1947 and in 1948 became a research fellow in his old hospital, the Middlesex, where his research with Douglas McAlpine [*Munk's Roll*,Vol.VII,p.349] was largely concerned with multiple sclerosis; he also assisted Alan Kekwick [*Munk's Roll*,Vol.VI, p.264] in teaching students. In 1951 he was E G Fearnsides Scholar at Cambridge and also Mackenzie-Mackinnon research fellow of the College. From 1952-54 he was assistant professor of medicine at the Middlesex.

Nigel Compston was appointed consultant physician at the Royal Free in 1954 and began his association with the hospital, which lasted until his retirement in 1983. He was a popular and painstaking consultant, always kind to patients, forbearing with his juniors and never ruffled in spite of the horrendous timetable he set himself. He was a popular teacher, and for some years was sub-dean of the medical school. His standards were set high and he expected his juniors to meet them; he was always available when needed and supportive when things went wrong. He was regarded by them as a good friend and indeed he was loved by them. For many years Nigel held an outpatients clinic on a Saturday morning at the old Hampstead General Hospital; the clinic always closed at 11.30am so that he and his juniors could gather with a set of Hampstead regulars at one of the local hostelries for a pre-lunch drink.

The hospital was able to make full use of his strength as an administrator, which led to his appointment as a leader of the medical planning team that so successfully transplanted the hospital and medical school from Gray's Inn Road to Pond Street, Hampstead. He kept the peace between ambitious clinicians, trying to maximise the facilities they would take over, and prudent administrators trying to minimise the costs. It was not an easy task, as extremists on both sides were able to blame him when their demands were not met. He withstood these criticisms with characteristic phlegm and fortitude and steered what he knew to be the correct course. As a tribute to his work in planning the new hospital, a ward at the Royal Free was named after him by the Duke of Kent in July 1988.

He was also consultant to the Royal Masonic Hospital, and to the King Edward VII Hospital for Officers. He was joint author of a work on multiple sclerosis, and co-editor of five volumes in the series *Recent Advances in Medicine*.

Election to the Fellowship of the College came in 1957 and with it the other focus of Nigel's life. The College recognized his academic talent and appointed him assistant registrar. In 1962 he introduced the annual conferences on advanced medicine and edited the first volume in the series, mentioned above, which still goes on. He became treasurer of the College in 1970 and held office for 15 years. During this time he put the College finances onto a secure basis and initiated two public appeals, while masterminding the purchase of the leases for the conversion of the 10 houses in St Andrews Place. The difficulties over the acquisition, planning and financial sides, were immense, but Nigel said that the opportunity would never recur and the difficulties just had to be overcome. He worried a great deal about the College and the Precinct project but was determined to follow it through; he even launched his own appeal to Fellows to invest some of the College's money so that it would be untouched for 100 years, and the Fellows would then be able to buy the next 99 year-leases of all the present College buildings from the Crown Estates Commissioners.

The project lasted the whole of his treasurership, and it was a great pleasure to all his College friends that Nigel and Diana were not only present at the opening of the Medical Precinct, the name he gave to the houses and area, by Her Majesty The Queen in June 1986, but that his grand-daughter Polly, daughter of his son Alastair and his daughter-in-law Juliet, both Fellows of the College, presented the

Queen with her bouquet on that occasion.

The College staff will always have cause to be grateful to Nigel's treasurership. He had their welfare at heart and put their pay and pension structure on a secure footing. He was ever ready to listen and advise, but any new project had to be properly presented and costed. He used to say that he could act as a minister or a civil servant, but his ojective advice was invaluable to both the senior staff and the College as a whole.

For relaxation he had now turned to skiing and golf, and occasionally during the week he would sneak up to Highgate for 18 holes. On these occasions his consulting rooms would be told he was at the College, and the College that he was consulting! Nigel loathed verbosity and could express profound disapproval through a series of grunts, accompanied by a scowl, at some of the more loquacious Councillors; sometimes the grunts occurred at lunches with those known as 'opinion formers' or guests from the media. Hypocrisy and humbug were always repugnant to him; anybody who was not on time, be it a patient or a luncheon guest, was told of their error in no uncertain terms - and some remembered it for years afterwards. Nigel trusted people, he would never let them down himself and felt it very deeply when his trust was misplaced. He was a good judge of character so this did not happen often despite the demanding standards that he applied.

In later years his health was not good but he surmounted even major surgery with equanimity. When the College telephoned, two hours after a hip replacement, to find out how he was, we were told he was awake and reading the newspaper.

Nigel's enthusiasms included domestic activities; he was very proud of Diana's painting talents and he had no mean brush himself. With their children, Fiona, Alastair and Robin, the Compstons were a close-knit family. He was particularly proud when Alastair became a Fellow of the College. Nigel was a man of many talents and diverse qualities, but one had to dig to find out about them as he seldom spoke about them himself. The digging was worthwhile; the deeper one got the greater the treasure.

It is only fitting that there is now a permanent memorial to him in St Andrews Place, echoing that of Sir Christopher Wren in St Paul's Cathedral: '*Si monumentum requiris, circumspice*'.

DAP

[*Brit.med.J.*,1986,**293**,1247; *Lancet*,1986,**2**,1048; *The Times*,18 Oct 1986; *Daily Telegraph*,20 Oct 1986; Memorial Service *Address*,All Souls,Langham Place,27 Jan 1987; Photo]

COMROE, Julius Hiram *b.*13 Mar 1911 *d.*31 July 1984
AB Pennsylvania(1931) MD(1934) *FRCP(1971) Hon DSc Chicago(1968) Hon MD Sweden(1968)

Julius Comroe was the son of a physician and was born in York,

Pennsylvania, USA. He did his early training in the University of Pennsylvania and became a faculty member in pharmacology. He was appointed chairman of the department of physiology and pharmacology in 1935.

After war service in the chemical warfare branch during the second world war, he collaborated with his lifelong friend, Robert D Dripps, professor of anaesthesia at the University of Pennsylvania. He carried out classic studies of the function of aortic chemo receptors. Dripps and Comroe developed sophisticated equipment to quantify pulmonary function. They studied artifical respiration, pulmonary gas exchange and the regulation of ventilation during exercise. Comroe's book *The Lung*, Chicago, Year Book Publishers, 1955, was a landmark which for many physicians was their introduction to pulmonary physiology and lung function testing.

In 1957 he moved to California to become the first director of the Cardiovascular Research Institute in the University of California. Here he developed a novel, multidisciplinary institution in which he mixed top flight basic and clinical scientists. Although he was extremely successful in attracting large amounts of funds to the Institute in the form of block grants from the National Institutes of Health, he also succeeded in the much more difficult task of persuading the University of California to grant tenure and full staff facilities to his recruits. He was far-sighted enough to link this research institute with the basic departments already existing in the hospital, in the form of joint appointments. This greatly strengthened the whole San Francisco medical campus. In all this he was staunchly supported by the vice-chancellor, Clark Kerr.

His other great gift was as a teacher. The CVRI became a major centre for training in research methods. He also did not neglect teaching the art of communication, and of critical evaluation of the work of others. In the early 1960s trainees were taught to improve their lecturing ability by the use of video tape recordings of their presentations. He was generous in his welcome to overseas visiting scientists, particularly young research fellows from Great Britain. These many disciples of Julius Comroe are scattered throughout the world.

Over 25 years the Institute has grown in stature and reputation. It has been responsible for many advances in biomedical research over a wide field, including the respiratory distress syndrome, neonatal intensive care, lipid metabolism, neonatal cardiology and many others. During this time Comroe played an unselfish role in furthering the careers of others.

Towards the end of his career he went back into the laboratory and carried out some extremely elegant experiments on his first love: separating the effects of the aortic and carotid chemo receptors by the introduction of a delay circuit into the cartoid circulation.

He had a fierce belief in the value of basic, nongoal oriented research, and when it became fashionable for politicians to interfere in medical research by demanding that this should be more clinically oriented, Julius Comroe characteristically countered this with painstaking investigation into the key steps which had led to the big advances in medical care over the last 20 years. He was able to show how major clinical advances, for example cardiopulmonary bypass, could not have taken place without basic research, which has as a side product produced Heparin and other anticoagulants. In the long run this might prove to be one of his greatest services to physiology. This collection of amusing and informative papers was published in the *American Review of Respiratory Diseases*, and was later collected into a fascinating book, *The Retrospectroscope: insight into medical discovery*, Menlo Park,California, Von Gehr Press, 1977.

All this may give the impression of a rather austere character, but Julius Comroe (or Uncle Julius, as he was universally known to his staff) had a tremendous wit and a wry, dry humour. One memorable lecture was on the subject of deans of American medical schools. These were illustrated by slides of dogs, from bulldogs through retrievers to pekingese and poodles.

He received honorary degrees which included an MD from the Karolinska Institute, Stockholm, and election to the National Academy of Sciences. He was awarded the Jesse Stephenson Kovalenko medal of the National Academy, the Eugenie Morelli International Award for pneumonology from the Accademia dei Lincei of Rome, the Ray C Daggs award of the American Physiological Society and the Golden Heart award of the American Heart Association, among many others. He was an honorary member of the Physiological Society of London.

He was survived by his wife Jeanette, the daughter of a pharmacist, whom he married in 1936, and one daughter.

PS?/GW

[*Brit.med.J.*,1984,**289**,842; *Lancet*,1984,**2**,532; *The Times* 31 August 1984; Photo]

COOK, John Bowen *b*.16 Apr 1921 *d*.18 Oct 1987
MRCS LRCP(1942) MB BS Lond(1947) MRCP(1949) MD(1950) FRCP(1969)

John Bowen Cook was born in London and educated at Hitchin Grammar School and St Bartholomew's Hospital, qualifying with the Conjoint in 1942 at the age of 21. The following year, after a single - and presumably eventful - house job in Mile End Hospital, he joined the Royal Naval Volunteer Reserve as a surgeon lieutenant. For the next three years, in HMS *Orwell*, he was involved in minesweeping and escort duties, taking part in the Russian convoys, the *Scharnhorst* incident and the D-day landings. After demobilization in 1946 he spent the next three years as a medical registrar at St James's Hospital, Balham, before making a definitive move to Yorkshire to work with the late Hugh Garland [*Munk's Roll*,Vol.VI,p.191] in the department of neurology at the General Infirmary at Leeds. Four years later he was appointed consultant neurologist and director of the spinal injuries unit at Pinderfields Hospital, Wakefield.

At Pinderfields his abilities as a clinician, a teacher, and a wise and perceptive administrator, soon became apparent and the hospital derived great benefit from his remarkable ability to anticipate the turn of events. Junior staffing, for example, was immeasurably improved by his insistence on what, at that time, appeared to be extravagant upgrading facilities for residents and the establishment of the postgraduate centre which now bears his name. His methods were sometimes unconventional and he was not above telling a thunderstruck regional committee that 'I meant to say that the Ministry had agreed to put a National Demonstration Centre at Wakefield.', when it was already half built. With the development of his unit, he soon acquired an international reputation in the sphere of rehabilitation, travelling widely to lecture and to attend meetings. Among many administrative duties he served as the United Kingdom's representative on the council of the Académie Médicale Européenne de Réadaptation, and as secretary to the research commission on neurological rehabilitation of the World Federation of Neurology. He was also regional adviser to the College.

Like his beloved bull-terrier, John Cook looked fit and fierce but was loyal and full of boisterous fun. He was by no means ostentatious but his appearance, his bearing, and the respect with which he was treated left no doubt that he was the director of his unit. To the timid he seemed rather alarming, for he had a deep forehead, rampant eyebrows, protuberant eyes and a penetrating stare. He spoke incisively, his head thrust forward, yet had a trick of suddenly dropping his gaze to search the floor as if he had lost something. With those he thought to be 'putting it on' he could be ruthless - which endeared him to the legal fraternity - and his attitude to paraplegics seemed at first to be unsympathetic. This was a deliberate ploy: as soon as they realized it was 'their problem not mine, so they'd better do something about it' his knowledge, experience, and deep personal concern for their well-being became apparent. His clear understanding of what was urgent, treatable or common, made him an excellent but demanding tutor; yet the same grasp of priorities would allow him to dismiss a DPM candidate's elemental ignorance of neurology with the comment 'don't worry - he's only going to do psychiatry.'

Most of his presentations were enlivened by this same malicious sense of humour. On one memorable occasion he announced that he would speak about HAA, which his audience took to be some sort of drug or enzyme. The first slide showed the subject - John Cook, his face distorted by a stocking mask, clutching his Hospital Activity Analysis number across his chest. There followed a devastating account of the way in which a reject from the typing pool had been left to code the work of the neurological department from an index which only recognized three tumours: Koch's peculiar, Pott's puffy and pearly of the ovary. Despite a protracted illness which, in his retirement, deprived him of the walking, fishing and other outdoor activities he loved, he never lost this sense of fun. Shortly before his death a colleague, who had tried to relieve his pain by altering his treatment, asked how he was. 'Fed up.' he replied, 'the place is full of flowers, everyone has been to see me, I've summoned the vicar and arranged my funeral and now I'm getting better.'

In many ways John Cook epitomized the great attributes, and relatively minor defects, of his generation. Starting during the horrors of the 'blitz' and the Arctic convoys, seeking a job in the chaos of the post-war years and appointed to an upgraded asylum, he knew nothing of postgraduate courses or career structures. Yet he trained himself, forged a fine unit and hospital, and established an international reputation. He demanded the right to control but accepted the conjoined obligations to serve and support his patients and subordinates. He attributed his early retirement to the fact that 'the scanner has taken all the fun out of neurology', but one suspects that with characteristic foresight he realized what would happen when administrative bumbledom took the control of hospitals out of the hands of practising clinicians. As usual, he was right.

He was survived by his wife Beryl, whom he married in 1943, a daughter and two sons.

MRP

[*Brit.med.J.*,1987,**296**,70-71; *Lancet*,1987,**2**,1285; Photo]

COOKE, William Trevor *b*.10 Jan 1911 *d*.12 Jan 1987
MRCS LRCP(1935) MB BChir(1936) MRCP(1936) MD(1940)
FRCP(1946) MD Birm(1955)

Trevor Cooke was born in Walsall, the son and grandson of doctors. He was educated at Uppingham School and Sidney Sussex College, Cambridge, graduating in the natural sciences tripos in 1932 and obtaining his MB BChir in 1936. After house posts he worked with Paul Dudley White [*Munk's Roll*,Vol.VI,p.457] at the Massachusetts General Hospital, Boston, USA, and obtained a research fellowship from Harvard University. Paul White was universally regarded as the outstanding cardiologist of his day and Trevor Cooke always considered this to be an important phase in his life, when his attitudes and approach to research were shaped.

Trevor was rejected for military service on three occasions because of a back injury, and in 1940 he was appointed medical registrar to the Queen Elizabeth Hospital, Birmingham. He became reader in medicine in 1946 and assistant physician to the united Birmingham hospitals in

1947. During this time he developed an interest in steatorrhoea, coeliac disease and gluten sensitivity. It was not long before he began to take a more general interest in disorders of the small intestine; steatorrhoea brought the 'new' condition of Crohn's disease his way and turned his attention to inflammatory bowel disease. There followed a productive period of nosological identification and therapeutic exploration; ulcerative colitis came into focus and was more easily identified, colitis due to Crohn's disease established for the first time, and right-sided colitis, which the Americans insisted on regarding as a form of ulcerative colitis, was shown to be a manifestation of Crohn's disease. Treatment with steroids, then new, and the recently introduced sulphasalazine, resulted in rational programmes, and the place and limitations of surgery were explored.

Trevor Cooke was intellectually honest and his scientific work bore testimony to this quality and secured his reputation. He abhorred dishonesty in public life just as he did in medical science, and he would not abandon his stance in his medicopolitical activities. When he felt that the prevailing climate made it impossible for him to pursue his researches, he relinquished his post as reader in the department of medicine at Birmingham, became a part-timer, and entered private practice. Many predicted the end of his academic career, but they were wrong. At the Birmingham General Hospital he soon established a gastroenterological unit of international repute. He established links with the biochemistry department of the new University of Aston and drew the staff into his researches.

He always kept scrupulous casenotes, maintained by a thorough and regular follow up which he conducted personally. He could claim one of the largest series of cases of steatorrhoea and Crohn's disease which he put to good purpose in what would now be known as a medical audit, checking results against treatment over many years and putting them to good use refining the management and control of treatment. He was a recognized authority on Crohn's disease, ulcerative colitis and coeliac disease. It was largely due to his efforts that the unit continued to flourish after his retirement, under the guidance of his trainees who have continued to build upon his data and methods.

Trevor Cooke was a sensitive and shy man, who seemed to cultivate anger and impatience to overcome what he thought were the shortsighted or unimaginative views of others. Although he was more than considerate to his patients and surgical colleagues, he was less so to those whom he regarded as being in competition. Yet the personal interest he took in those who worked with him enhanced the group ties of his unit. He was usually right in his opinions, but his anger did not endear him to some of his colleagues and it was not always easy to convert his ideas into practice. He was greatly respected by surgeons who worked closely with him, particularly in the management of inflammatory bowel disease, and he collaborated well with his colleagues in radiology and in the laboratories. Those who knew him well realized that he was an extremely kind man, who always had time and sympathy for people. Yet he received less recognition than was due to a man of his international standing, in particular from his medical school. It was partly the result of envy, but also of a reaction within the medical faculty to his irascibility and stubborn refusal to compromise.

In 1967 Trevor Cooke founded the Midland Gastroenterological Society and was its first president. He was a founder member of the Coeliac Trust in 1968 and served for over ten years as a much valued adviser and friend.

He wrote extensively on the management and prognosis of inflammatory bowel disease, on the neurological and malignant complications of coeliac disease, and a comprehensive monograph on this disorder, *Coeliac disease*, Edinburgh, Churchill Livingston, 1983.

He married Margaret Anne, née Foxell, and they had two daughters, Ann and Alison. His wife and daughters survived him. GW/VL

[*Brit.med.J.*,1987,**294**,517; *Lancet*,1987,1,285,397,639]

COOKSON, Samuel Harold *b.*4 July 1901 *d.*6 June 1986
BSc Birm(1920) MB ChB(1923) MD(1924) MRCP(1925)
FRCP(1945)

Samuel Harold Cookson gained a science degree in 1920 from the University of Birmingham and graduated in medicine in 1923. He obtained a post as house physician at the Brompton Chest Hospital, and then as medical registrar at the Middlesex Hospital. His primary interest was in cardiology and he was a consultant in this field to the Ministry of Pensions (now the DHSS). His other interests lay in thyroid diseases, and in diseases of the nervous system. He wrote several papers on these subjects.

He was appointed honorary consultant physician to Poole General Hospital, Hampshire. He was a very private, rather solitary man and after his retirement from medicine, around 1966, he had little contact with his old colleagues. He died in Poole General Hospital after a long illness. His wife survived him.

GW/VL

[Photo]

COOPER, Christine Elizabeth *b.*21 July 1918 *d.*1 Sept 1986
OBE(1967) MA MB BChir Cantab(1945) DCH(1948)
MRCP(1949) FRCP(1967)

Christine Elizabeth (Tina) Cooper was the eldest daughter of William Francis Cooper, an analytical chemist and physician in general practice in Watford. Her mother, Christine Hall-Jones, died when she was six months old. Her father remarried and Christine had a close relationship with all her siblings. Her early education was at St John's School, Bexhill-on-sea, during which she developed an interest in the welfare of children and later spent two years qualifying as a nursery nurse at Cromwell House in London. She soon realized that she should qualify in medicine if she were to use her gifts to the best advantage. She studied medicine at Girton College, Cambridge, followed by clinical work at the Royal Free Hospital, and graduated in 1945 after a distinguished undergraduate career.

After qualification Christine spent her early years in London as house physician, senior house officer in obstetrics, and first assistant in the children's department at the Royal Free Hospital. In 1949 she was invited by Elsie Wright to visit the children's department in Newcastle. She was captivated by the personality and philosophy of James Spence, later Sir James [*Munk's Roll*,Vol.V,p.386], and the following year became his senior registrar. In 1952 she succeeded Elsie Wright as consultant at the Newcastle General Hospital, where she built up a reputation as a gifted clinician and teacher.

Christine Cooper was always interested in the emotional as well as the physical needs of children, particularly those who were deprived. She therefore never worked full time in the National Health Service and devoted a few sessions a week to the wider aspects of paediatrics. Initially this led her to explore the problems of children who were fostered or adopted, and she became medical adviser to the Northern Counties Adoption Society. From 1960 she served on the council of the Association of British Adoption and Fostering Agencies. As a member of the Home Office departmental committee on adoption (the Houghton committee) she played an important part in producing the report which formed the basis of the 1975 Children's Act.

She was one of the first paediatricians in Britain to recognize the extent and nature of child abuse - physical, psychological and sexual. She had a remarkable gift for talking with families and gaining their trust. Her clear vision of the need to protect children was combined with an understanding and compassion for stressed and damaged parents. Her wisdom and experience, gained through much hardwork and pain, meant that she was consulted widely, whilst her example inspired many paediatricians in training. She was a member of the Tunbridge Wells study group on child abuse, which guided national policy in the 1970s, and she succeeded Alfred White Franklin (q.v.) as president of the British Association for the Study and Prevention of Child Abuse and Neglect. She had close friends who were also involved in this subject, such as Anna Freud, Dermod MacCarthy (q.v.),and Henry Kempe, and her library of relevant books - which she gave to the University of Newcastle-upon-Tyne - was unique.

Christine's interest in children extended beyond Britain. She travelled and lectured widely, and was an authority on cultural differences in child rearing. In 1964 she was invited to advise the government of Sierra Leone on child health services in that country and she worked there for two years. In the words of Davison Nichol, vice-chancellor of the University of Sierra Leone: 'She developed the paediatric services of the nation to a high level by her constant encouragement and advice to those who worked with her and for her. She travelled widely in the interior, and established a country-wide service of immunization against measles and other children's diseases. She saved many lives, and endeared herself to many by her directness, sympathy and warmth.' A friend who worked with her at that time described a characteristic visit to a health clinic in the 'bush'. When they arrived, dozens of laughing mothers with their babies on their backs were strolling in, and having walked long distances in the hot sun they made a rush for the standpipe. But Tina got there first and, to their amazement, no mother was allowed to drink until each had given her baby as much water as he or she could hold. Then the mothers drank, and this was followed by a talk on the dangers of dehydration in terms they could understand and in circumstances they would never forget. She was awarded the OBE for her work in Sierra Leone.

After she retired in 1983, Christine Cooper remained very active. She was invited to South Africa as a visiting professor, and it was a lasting sadness to her that she did not feel it right to accept an official invitation to a country whose children she longed to help. She continued to consult, advise and lecture virtually to the end of her life. She was thrilled that during her terminal illness she was able to treat a disturbed family which, as a result of her help and contrary to all predictions, was able to love and keep the baby.

Christine Cooper's work for children was untiring and she spent long hours seeing families at home and in hospital. Through her example and teaching, she opened the eyes of the toughest organic medicine enthusiasts to the need to consider all aspects of the child's environment. The majority of junior doctors who at first were taken aback by Tina's apparent lack of need for nourishment and sleep ended by doing without much of either themselves and, contrary to their first reaction, remained her faithful followers and friends.

Tina also managed to have wide interests. She was knowledgeable about many aspects of the arts, and in particular of opera. She loved travel in which work and play were combined, and in the ten years between 1969 and 1978 she visited Rotterdam, Uppsala, Athens, Jerusalem and Bangkok. She had an extensive library of books on art, travel, history and philosophy, most of which she had read. But above all she loved her family, her godchildren and her friends. She had a remarkable capacity for friendship and, once established, her friendships were maintained by planned meetings and faithful correspondence. It was always a delight to visit her.

Tina Cooper was a remarkable person. She was tall, imposing and good looking and her intelligence, honesty and strength of character meant that she was held in awe by those who met her only formally.

Yet she was essentially warm, feminine and a delightful, interesting companion who was dearly loved and respected by those who had the privilege to know her.

JMP

[*Brit.med.J.*,1986,**293**,1109; *Lancet*,1986,2,699; 1110; *The Times*,6 Sept 1986; Photo]

CRAIG, James Donaldson *b.*24 Apr 1917 *d.*10 May 1985
MRCS LRCP(1939) MB BS LOND(1939) MRCP(1945)
MD(1947) FRCP(1971)

James Donaldson Craig was born in West Lothian, the son of John Craig, an aeronautical engineer, and his wife Margaret Coutts Oliver. Educated at Bathgate Academy, Hampton School and London University, he qualified from St Mary's Hospital. In 1940 he joined the RAFVR medical branch as medical officer to 611 Squadron, subsequently becoming a squadron leader medical specialist. He obtained his MRCP in 1945. After the war he was medical registrar and research fellow at his teaching hospital, gaining his MD in 1947. After two years as senior medical registrar at St Stephen's Hospital he was appointed consultant physician to the Medway and Gravesend Group of Hospitals in 1950, where he remained until his retirement in 1983.

In 1940, 'JD', as he was commonly known, married Audrey Leonie Rowton who died in 1944 and by whom he had an only daughter, Phyllida. In 1947 he married Joan Lucia Dumaresq, daughter of Rear Admiral John Sumarez Dumaresq CBE CVO, who survived him.

Although his chief medical interest and research was in gastro-enterology, he remained a general physician who rapidly established an enviable reputation amongst his hospital colleagues and local family doctors and provided a much sought after opinion, especially for their own families - a considerable burden which he no doubt found flattering, but not without its problems.

He was a strange mixture of compassion and almost brutal honesty, because he never suffered fools gladly; his directness, bordering on rudeness, could occasionally be wounding and his own high level of intelligence inevitably ensured that the number of fools available was considerable. Nevertheless, even when his criticism appeared laceratingly harsh, his sense of humour was usually near at hand, as on the occasion of a long letter devoted to a particularly detailed and virulent description of the ineptitude of a colleague's management of his department to which he added the postscript: 'Rude letter to follow!'

He enjoyed a good discussion and was prepared to concede the occasional point if one worked at it, and he liked it even better if he had provoked the argument. JD was a brilliant teacher, and those juniors who were prepared to work hard and to learn were fortunate, for he had well deserved success in helping them with higher examinations and would go out of his way to encourage them, and help them up the medical ladder. He was never afraid of the young coming up - a trait not always to be found in more senior doctors. His whole team, who stood in considerable awe of him at first, soon learned to appreciate his sterling qualities of medical integrity and honesty. He discouraged needless investigation of 'blood onion' type and considered a normal ECG the equivalent of a good conduct medal in the Army catering corps.

On the social side, hospital lunch in his company was always stimulating with the cut and thrust that he generated, for he was a man of wide interests and very firmly held views. He was a patriot, who was not afraid to admit it, but if he had not been born a Scot he would have been delighted to have been born in France for he loved that country and its customs, particularly its art, cuisine and wine. He appreciated music, a great interest as his wife was a talented singer and his daughter a gold medallist in harpsichord and pianoforte at Lausanne Conservatoire. His main interests were his home, his family, his garden (he liked to watch his family weeding it), and his books. He was widely read outside medicine.

JD was a good raconteur, with a pithy wit tinged with asperity, and one of his colleagues meeting him after his retirement ruefully commented that JD was in his usual good form and had not lost his cutting edge! He had many friends in the Medway towns, being a 'regular' at the Castle Club in Rochester and enjoying their frequent convivial outings to international rugby matches (especially in Paris), horse race meetings and vineyard tours. He thought that the 'up and coming' business and professional men should know about claret and burgundy and was at pains to instruct them.

I suspect that JD would not appreciate any attempt at a 'potted biography' of him because he would know, and would not hesitate to say, that he could have done it much better himelf. It was typical of the man that he directed in his Will that there should be no memorial or similar service for him.

ODF

[Photo]

CROOM, ***Sir*** **John Halliday** *b.*2 July 1909 *d.*12 Apr 1986
Kt(1975) TD BA Cantab(1930) MB ChB Edin(1933)
FRCPE(1940) *FRCP(1972) FFCM(1974) Hon FRACP(1972) Hon FACP(1973)

John Halliday Croom had a lifelong association with Edinburgh. He was born in that city, coming from a distinguished medical lineage, his father being David Halliday Croom, a well known general practitioner, and his grandfather, Sir Halliday Croom, an eminent obstetrician and gynaecologist in Victorian Edinburgh.

John was educated at Glenalmond, Gonville and Caius College Cambridge, proceeding to the University of Edinburgh where he graduated in medicine in 1933. After early appointments in the Royal Infirmary, he decided on a career as a physician and obtained the important teaching post of clinical tutor (a term peculiar to the then Edinburgh teaching establishment) in the medical firm of A Fergus Hart and W Ritchie Russell. This post included inculcation of the elements of clinical examination to the large, but appreciative, classes of undergraduates in the years leading up to the second world war. Before the outbreak of war John had served in the Territorial Army for some years, in the 155 (Lowland) Field Ambulance. In 1939 he was appointed physician in the RAMC, with subsequent war service in hospitals in France, the Middle East, Malta and Italy. He was mentioned in despatches and by the cessation of hostilities had attained the rank of lieutenant colonel. He was awarded the Territorial Decoration in 1946.

On his return to the Royal Infirmary, that same year, he was appointed consultant physician and held this post until he retired in 1974. From 1960-66 he was attached as a NHS consultant to the University department of therapeutics where he and Sir Derrick Dunlop [*Munk's Roll*,Vol.VII,p.170] who were great friends, formed an impressive team of teachers. In 1963 he was appointed as a part-time senior lecturer in the University. During this period he was also a consultant physician at the nearby Chalmers Hospital.

John Croom was a general physician with a particular interest in diabetes mellitus, and for several years was closely associated with the diabetic and dietetic department of the Royal Infirmary. He was instrumental in setting up peripheral diabetic clinics in Fife, the Borders and East Lothian, which proved to be of considerable benefit to patients living in the more outlying areas of the south-eastern region of Scotland.

Amongst other important consultant and advisory appointments, he was principal medical officer to the Standard Life Assurance Company, medical adviser to the Royal Bank of Scotland and to the Northern Lighthouse Board, and from 1970-75 was honorary consultant physician to the Army in Scotland.

He gave long and distinguished service to the Royal College of Physicians of Edinburgh; from 1950-70 he held successively the offices of secretary, councillor, and vice-president. He was president from 1970-73. In 1968, in the light of the report of the Royal Commission on Medical Education, a committee was set up at the RCPE and chaired by John Croom. The subsequent 'Report of the Committee to consider the future of the College' was produced with remarkable expedition - by October 1969 - and is now customarily referred to as 'The Croom Report'. This was destined to have far reaching effects on post-graduate medical education and training in Scotland. The Committee also recommended the creation of the category of collegiate membership of the College, as a means of securing more active participation by the members. The move proved outstandingly successful and the collegiate members, now a flourishing body, have appropriately commemorated his name in the prestigious yearly Croom Lecture.

In his later years, following his retirement from the Royal Infirmary in 1974, he continued to devote much of his time to the development of postgraduate medical education both within the Royal College and elsewhere, with service on several committees. He was chairman of the Scottish Council of Postgraduate Medical Education from 1974-79; chairman of the Scottish Health Service Council 1972-75, and of the Scottish Committee of Action on Smoking and Health (ASH) from 1972-77. He was a member of the board of governors of St Columba's Hospice in Edinburgh. In 1972 he was made an honorary Fellow of the Royal Australasian College of Physicians, and of the American College of Physicians in 1973. He was knighted in 1975.

John Halliday Croom ('JHC') was an astute clinician who was held in the highest esteem by his colleagues, being one of the last of the old school of general physicians to serve the Royal Infirmary of Edinburgh, for which that hospital had become so justly famous over the years. His good counsel, sagacity, and sometimes downright blunt common sense, were much appreciated by many generations of his junior staff and students. His genuine interest and courtesy made him much beloved and long remembered by his grateful patients.

'JHC' was born at the end of the Edwardian area and perhaps in many respects he remained an Edwardian all his life. His very elegant attire, always with a buttonhole and bowler hat, made him a familiar figure and on retirement he became sadly missed from the corridors of the Royal Infirmary. There was also his other hat: at the weekend he might be seen, temporarily diverted to the Royal Infirmary, with cap, tweeds and binoculars, on his way to an enjoyable day's outing on the turf. As his health deteriorated he had to give up his outside activities, which included golf and fishing, but he pursued his love of racing almost to the end.

His marriage in 1940 to Valerie, daughter of David Samuel, a general practitioner, was a very happy one. His wife shared his interests fully and they were an extremely hospitable couple. During his final illness she gave him unstinting support and devoted care. There were two children of the marriage: a son, David Halliday, and a daughter, Elizabeth Anne.

During John's funeral service, in the Church of St John, Princes Street, a lone fiddler playing out the poignant strains of an early Scots lament was suddenly accompanied by the siren of a passing ambulance making its life saving dash to the modern day intensive care unit of the Royal Infirmary. A conjunction which, perhaps, symbolized the long continuity of tradition in Scottish medicine and John Croom's notable contribution to it.

BC/MFO/GW

[*Brit.med.J.*,1986,**292**,1146,1211; *Lancet*,1986,**1**,924; *The Times* 17 Apr 1986,Photo]

CRUICKSHANK, John Norman *b.*17 Nov 1891 *d.*9 Nov 1986
MC(1918) MB ChB Glasg(1916) FRFPS(1920) MD(1923) MRCP(1925) DSc(1931) FRCP(1935)

Norman Cruickshank was born towards the end of the nineteenth century, was educated at Glasgow Academy and the University of Glasgow and served this city all his life. Immediately after graduating in 1916 he joined the RAMC, serving as a regimental officer until 1919. He was awarded the military cross in the field.

On demobilization he took up house appointments at Glasgow Royal Infirmary, being appointed physician to the outpatient department in 1920. At the beginning of his career he developed an interest in glomerulonephritis, a subject said to have been the basis for his MD and DSc degrees. In 1921 he was appointed assistant to the Muirhead professor of medicine in the University of Glasgow, later becoming an honorary lecturer in clinical medicine. He became assistant physician to the Infirmary in 1926, and physician in charge of wards in 1937. He retired in 1957 having served the Infirmary for some 38 years, for around 20 of which he had been a chief.

Norman Cruickshank functioned very much as a general physician. He acted as representative on a number of medical societies, being a member of the representative body of the British Medical Association, secretary to the Glasgow Postgraduate Medical Association, member of Council and editorial secretary of the Medical Chirurgical Society of Glasgow and editor of the Glasgow Medical Journal. He was a Fellow of the Royal Society of Medicine. He wrote a number of articles for journals and a textbook, *Bright's disease: a clinical handbook for practitioners and senior students*, Baltimore, W Wood and Company, 1933.

After his retirement he went to live in St Andrews and lived there for some years. Shortly before his death he moved south to be nearer members of his family, and was planning to live to be 100.

ACK/GW/VL

[Glas.Roy.Infirmary *Report* 1957/58; Photo]

CURRAN, Richard Desmond *b.*14 Feb 1903 *d.*26 Sept 1985
CBE(1961) MRCS LRCP(1927) MB BChir Cantab(1928) MRCP(1928) DPM(1930) FRCP(1937) FRCPsych(1972)

Desmond Curran was one of the most distinguished clinical psychiatrists of his generation, and many of his pupils were to fill eminent positions in clinical practice and teaching in this country, and in many other parts of the world.

He was born in Devon, grew up near St Andrews, and was educated at Wellington and Trinity College, Cambridge, before entering St George's Hospital medical school where he graduated in the early twenties. Rare intellectual gifts, personal charm and a presence of distinction which was to change little with the advancing years, had equipped him well for an illustrious career in medicine. But he chose psychiatry, regarded in those days as a lowly walk of life in medicine, and was appointed house physician at the Bethlem Hospital in 1928. He received powerful encouragement from Edward Mapother [*Munk's Roll*,Vol.V,p.266] and, after a period in neurology at Maida Vale Hospital, he proceeded with a Rockefeller travelling fellowship to Johns Hopkins in Baltimore. He came under the influence of Adolf Meyer whose wide-spanning concept of psychiatry did a great deal to enrich and humanize the clinical practice of what had previously been a narrowly organic and descriptive specialty. His leading disciples in Great Britain included Desmond Pond (q.v.) later Sir Desmond; Aubrey Lewis, later Sir Aubrey [*Munk's Roll*,Vol.6,p.284]; R D Gillespie [*Munk's Roll.*, Vol.V,p.149] and D K Henderson, later Sir David [*Munk's Roll*,Vol.V, p.188]. Desmond was too level-headed to espouse any teaching in doctrinaire fashion and, although influenced by Meyer, developed his own pragmatic style of clinical exploration and practice in psychiatry. He achieved the feat of being appointed as consultant to St George's Hospital at the age of 31, and election to the Fellowship of the College at the age of 34.

With the outbreak of war in 1939 he joined the Royal Navy, and swiftly established within it a worldwide specialist service in psychiatry - becoming its youngest surgeon-captain. At the end of the war he returned to St George's and began work on what was to become the most important endeavour in the course of his long and distinguished career: the creation of the teaching hospital department of psychiatry at St George's Hospital. He had attracted to his side a number of men of clinical maturity and insight, including 'Bird' Partridge and Paul, later Sir Paul, Mallinson.

Desmond's exceptional clinical skills were the product of discriminating observation, an abundance of tolerance and compassion, and an understanding of human personality progressively enriched over years of painstaking first hand study of patients. The force of his example in practice was emulated by his colleagues and students, and St George's became pre-eminent among the medical schools for its department of psychiatry. Its national and international renown increased steadily during the three decades in which Desmond served it.

His textbook *Psychological Medicine*, Edinburgh, E&S Livingstone, 1943, was written in collaboration with Erich Guttmann, one of the small group of eminent psychiatrists who sought refuge in this country from persecution in Nazi Germany. Its lucidity, richness of clinical description and its practical wisdom, made it widely popular. It was to appear in numerous editions, including one with Maurice Partridge in 1955; recent ones having been edited by Curran's pupil, Peter Storey. In 1961 Desmond was appointed the first professor of psychiatry at St George's Hospital. He developed a many sided academic department, including a large unit devoted to psychotherapy, the latter notwithstanding his reservations about psychoanalysis.

Few senior psychiatrists burdened with manifold duties could have devoted themselves with such unswerving determination and energy to the professional advancement of those of his juniors whose clinical or scientific ability had earned his respect.

His clinical eminence, and the publications he devoted to the classification of personality disorders, sexual deviation and antisocial and violent conduct, generated wide demand for his services. He was a member of the Franklin Committee on punishment in prisons and borstals, the Wolfenden Commission on homosexuality, and he became adviser in psychiatry both to the Foreign Office and the Home Office. In 1948 he delivered the Croonian lecture at the College. He was president of the psychiatric section of the Royal Society of Medicine from 1951-52, and of the Royal Medico-Psychological Association in 1963-64. He was awarded the CBE in 1961.

Although Desmond did not originally favour the secession of psychiatrists from the Royal College of Physicians, he gave the new Royal College his full support once it received its Royal Charter in 1971. Many of its officers have reason to be grateful for his sagacious advice and support, and the exercise of his wide influence on its behalf during the early formative years.

Desmond Curran was a man who knew his own mind; integrity, personal simplicity, sharp wit, keen sense of humour and unfailing kindness rounded a personality of rare distinction. It explained, in part, the undivided loyalty and affection he inspired in his many pupils. The clinical excellence of those who trained under him, the scientific distinction of many, and the fine thriving department he created at St George's commanded respect for the subject he professed, as well as for himself, among many colleagues in many disciplines, and enabled him to leave a lasting imprint upon the psychiatry and medicine of his time both in his own country and wider afield.

He married Marguerite Claire Gothard in 1938, his spirited and delightful wife whose unflinching courage and understanding sustained him in triumph, tribulation, and during failing health in his last few years. She survived him, with their two sons.

Desmond is commemorated by the 'Curran Award' at the Atkinson

Morley Hospital and the Curran Memorial lecture delivered annually at St George's. MR

[*Lancet*,1985,2,903; *Daily Telegraph*,1 Oct 1985; *Brit.med.J.*,1961,1,p.838; *Lancet*,1961,1,p.569; Photo]

DAGNELIE, Jacques Georges Léon *b.*10 Dec 1906 *d.*6 Sept 1984
MD Brussels(1931) *FRCP(1972)

Born in Brussels, but spending some of his childhood in Paris, Jacques Dagnelie trained as a doctor at the Université Libre de Bruxelles and qualified with 'grande distinction' in 1931. Early in his career he demonstrated his talents for organization in the leadership of scientific societies, and later found fulfilment when he became the secretary and driving force of the European Association of Internal Medicine (AIME) which he served faithfully from its foundation in 1970. It was in this capacity that he first became associated with the College in 1971. He was dedicated to the protection of the quality of care provided by specialist general physicians, and what he perceived as the dangers of over specialization.

In his early career he served as Chef de Service and teacher in internal medicine at the Hospital of St Giles in Brussels before devoting himself to practice as an internist. His publications reflected his wide interests in medicine but his real love, apart from his devoted wife Maria and his three children, was AIME. Although some of us on the council were dignified by various titles, such as treasurer, Jacques Dagnelie actually did everything - and with painstaking care. The meetings of the council of AIME were often held in his house in Brussels, usually on a quiet Sunday when there was time to do justice to his wife's superb food and his own treasured wines.

When he first visited the College in 1971 he held all of the justified, and some of the unjustified, suspicions of a Francophile about the English but he soon recognized that the College not only shared his views on the importance of the specialist general physician but could, by its cohesive and efficient administration under the then president, Max Rosenheim, become the very citadel of European internal medicine. Dagnelie was deeply moved by his election to the Fellowship in 1972 but, alas, in the following years the College, for a variety of reasons, became more concerned with domestic problems and its earlier enthusiasm for a European initiative faded. His later years were troubled by prolonged illness. PAE

[Photo]

DATEY, Keshavrao Krishnarao *b.*7 Aug 1912 *d.*22 Apr 1983
BA MD Bombay MRCP(1949) FRCP(1967)

Keshavrao Datey was born in Jabalpur, India, the son of Krishnaji Hari Datey, an engineer, and his wife Anna Purnabai. He was educated at Model High School, Jabalpur, and the Ewing Christian College at Allahabad. He went on to study medicine at the G S Medical College, University of Bombay.

Datey was a pioneer in cardiology in India, when the specialty began to flower in the 1960s. He was an excellent clinician and teacher and was responsible for training many generations of young cardiologists. He had a large practice in Bombay and was known all over India. He pioneered the treatment of hypertension by Yoga and Shavasan and was largely instrumental in popularising this method of control by physiological means. In 1966 he became director of the department of cardiology at the K E M Hospital, Bombay, and honorary professor of medicine at the G S Medical College, Bombay. That same year he was appointed honorary surgeon-captain to the Indian Navy. He was president of the 5th World Congress of Cardiology, New Delhi, 1966, and also a JP. He was a founder director of the All India Heart Foundation, physician to the president of India and honorary physician to the governor of Maharashtra. He was a member of the WHO expert advisory panel on cardiovascular diseases, an honorary member of the British Cardiac Society, and member of honour of the Cardiological Society of Belgium.

Keshavrao Datey was editor in chief of the well known *Textbook of Medicine* by Indian authors. He wrote numerous original articles in scientific journals in India and abroad, and was the author of many chapters in textbooks. He also published three books for laymen: *Yoga and Heart, How to keep your heart young and healthy,* and *All you want to know about your heart.*

Always dapper and elegantly dressed, with impeccable manners, he was a familiar figure in important medical conferences in India and abroad. He was a genial host to his many friends and admirers, who will long remember his generous hospitality.

Outside medicine, his interests were largely sporting and he enjoyed cricket, tennis, hockey and athletics - winning a number of trophies.

He married Pramila, daughter of Shamrao Pathare, in 1956. There were no children of the marriage.

SP/GW

[Photo]

DAVIES, Iforwyn Glyndwr *b.*11 June 1901 *d.* - Jan 1984
MRCS LRCP(1923) MB BS Lond(1924) MRCP(1926) DPH(1930) MD(1944) FRCP(1954)

Ifor Davies, always known to friends and colleagues as 'IG', was one of the most important yet least histrionic members of a group of medical officers of health who contributed much to the development of the NHS, yet had little recogniton for so doing. He was born at Porth in South Wales and had his early education there before going on to his medical school at Bart's. After qualifying with a London MB and Conjoint in 1923-4 he held junior hospital posts in South Wales, and achieved the membership of the College in three years. He took the DPH and chose a career in public health, moving on from a series of junior posts to become Robert Parry's (q.v.) deputy in Bristol before transferring to Leeds to succeed Johnstone Jervis, one of the great figures between the wars, as MOH and professor of public health. He thus bridged the period between the rapid development of local authority health departments, and their concern with personal health and hospital services, and the first 25 years of the NHS. Some MOH's, and 'IG' among the foremost, did more for the development of community services and effective linkage between them and specialist and primary care than did any other element in the NHS.

'IG' was a considerable epidemiologist in the period when control of communicable diseases was still a major problem. When he joined the Ministry of Health staff as a principal medical officer in the early 1960s one of his first assignments was coordination of the containment of smallpox when it was imported, on no less than six occasions, from the Indian subcontinent in 1961-62. However, his most notable contribution in Leeds was the development of support services for home care. Much of his writing concerned that kind of service, including linkage with industrial health services. He was a member of both a hospital management committee and the local executive council. He organized, in Leeds, one of the best aftercare services for the mentally ill and handicapped in the country. He was perhaps the best equipped of his contemporaries to bring practical field experience into these areas when the Ministry was so fortunate as to recruit him. He helped to prepare Enoch Powell's health and welfare development plan of 1963 and the revision published by Kenneth Robinson, both of which had a seminal value too little recognized. He had become a Fellow in 1954 when there were few from the public health field, but he played little active part in the College.

'IG' was a quiet man, widely informed and a good negotiator. He became QHP and but for his slightly premature retirement would certainly have been further honoured. He was an invaluable ally and friend, and the kind of quiet, effective supporter for whom any chief medical officer must pray. He married Lilian James, from a background like his own, and they had one son who entered general practice. They were a very close family.

GEG

DAVIES, Leslie Graham Gwynfryn *b.*4 Aug 1922 *d.*3 Oct 1986
BSc Wales(1942) MB BCh(1945) MRCP(1950) MD(1956) FRCP(1969)

Leslie Graham Gwynfryn Davies, known to his friends and colleagues as 'LG', was born at Brighouse, Yorkshire, the son of Thomas Davies, a physical training organizer, and his wife Lilian, née Drake. Although he was born in Yorkshire, he spent most of his childhood and holidays in Wales. His early education was in local authority maintained schools and he went on to study medicine at the Welsh National Medical School, University of Cardiff, where he graduated after as short a time as a medical student as was then possible, achieving distinction not only in academic studies but also on the rugby field. He was a first class rugby player, known as a tenacious scrummager in the front row and a terrier in the loose.

After graduation he held house appointments at Cardiff Royal Infirmary and then enlisted in the RAF at St Athan as a medical officer, where he served from 1945-47, but listed his war record on his curriculum vitae as 'strictly none'. After the cessation of hostilities he continued his training in Cardiff, and then at the Hammersmith and National Heart hospitals. In 1960 he was appointed the first consultant cardiologist in Wales; his reference from Paul Wood [*Munk's Roll* ,Vol.V,p.456] described him as '...at least the second best cardiologist in Britain.' They were days when cardiologists and cardiac surgeons would come down to Sully from London on a Friday evening for clinical meetings that would continue well into the night. LG's time and energies, his care, and attention to detail were shared between the demands of his patients and the experimental transplantation programme. He contributed greatly to solving the problem of post-operative management in the early days of cardiac surgery. His diagnostic acumen was unsurpassed.

'LG' was a shy man, which belied the tenacity and the unselfishness which later became his hallmark as a doctor. He was a man of few words and had no small talk, and was diffident to the point of being mistaken for anybody but a consultant cardiologist. But he was totally committed to cardiology - to the benefit of thousands of grateful patients and their families; alone in Wales he looked after the children too, with a compassion sometimes hidden by his shyness. He was indeed a master clinical cardiologist. Discerning students idolized him, as did many who learnt their cardiology from him.

In 1945 he married Grace Price-Stephens, the daughter of a colliery official. and they had two daughters, Wendy and Linda.

Typically, he endured his long illness with dignity, continuing to work almost to the end. His wife and daughters survived him. GW/VL

[*Brit.med.J.*,1986,**293**,1380-81]

DAVIES, Patience Elizabeth (née Barclay)
b.28 July 1911 *d*.29 Mar 1985
MB BS Lond(1939) DCH(1943) MRCP(1945) FRCP(1970)

Patience Davies was born at Chippenham where her father, David Buxton Barclay, was a vicar. Her maternal grandfather was a bishop, and the discipline of service was a prevailing and pervading influence throughout her life. With this background, medicine was one obvious choice of career and, after preparatory school and public school in Essex, she entered the medical school of the Royal Free Hospital, taking her London degree in 1939.

After house appointments at the Three Counties Emergency Hospital at Arlesey, at Addenbrooke's, and at the Radcliffe Infirmary, Oxford, she became RMO and registrar at the National Temperance Hospital in 1942. In 1943 she was appointed medical registrar at Great Ormond Street Hospital where she laid the foundation of her wide knowledge and skills in paediatric medicine. During her two years in this post she obtained her DCH; her chiefs there spoke of her obvious affection for and rapport with children, her capacity for prolonged hard work, her meticulous notekeeping, and her wide reading of both general and paediatric literature, and they predicted a distinguished career for her.

In 1945 she gained her membership of the College and in the same year, on VE-day, was conscripted into the RAMC. She was posted with the rank of captain to India, to Army hospitals at Agra, Jhansi, and finally Darjeeling. In Darjeeling she met Arthur Davies of the Burma Frontier Service, who was then recovering at the Darjeeling (Lebong) Officer Convalescent Depot from his three and a half years as a Japanese prisoner of war. They were married in Darjeeling on her birthday, 28 July 1946, and after a honeymoon in Nepal they returned to England in September of that year. The following year she joined F L King-Lewis, a children's consultant physician, for three years in his general and specialist practice in London.

In December 1950, Patience and her husband moved to Kenya where she joined the Nairobi Clinic, a group of consultants, and it was in that country that the definitive part of her professional life was spent. She quickly established herself as the acknowledged consultant in her own field and she retained that position throughout her 26 years in Africa. The centre of her hospital work was Gertrude's Garden Children's Hospital, but she also held appointments at the Kenyatta

Memorial, the Aga Khan, and the Mater Misericordiae Hospitals. She gave regular series of lectures and contributed several articles to the *East Africa Medical Journal.* She organized and dominated the practice of exchange transfusion for haemolytic disease of the newborn and her skill in intravenous therapy was legendary. She was responsible for founding the East African Paediatric Association and was its president for the first two years.

Patience was held in affection and immense respect by every member of the medical and nursing professions in East Africa. Some mothers were said to be in holy fear of her, for there was no sentimentality about her approach to her work or her patients and any criticism she might make of a parent's inadequacies of care was direct, even forceful. But once the air was cleared and the way forward established she enjoyed their full cooperation and admiration. She gave selfless and devoted service to parents and children of all races and her positive manner, together with clear and concise appraisal and instruction, quickly established a feeling of great comfort and confidence that in her hands everything possible would be done. And, of course, it was done. In difficult and critical situations she had the gift of being both incisive and imperturbable.

Apart from her private work and hospital clinics, she had a large practice among the poor and needy whom she treated with equal devotion, in love, and for no reward save their recovery.

She and her husband left Kenya in 1976 and returned to England, where they settled in a charming house overlooking Dartmoor. For the next six years she did sessional work in the children's clinics in Plymouth. She finally retired in 1982.

Off duty, when she could relax, Patience showed the warm, generous, amusing person she was, and she was a marvellous friend. She and her husband enjoyed music and, for outdoor recreation, sailing. In Kenya they had kept a boat on the Nairobi dam. They had two children of their own: their son John is a chartered accountant and a successful business man in Kenya and Tess, their daughter, is married and lives in London. They also unofficially adopted, from early childhood, a brother and sister whom they saw through schooling and training, the girl becoming a nurse. Patience was also survived by a brother, an ENT consultant.

Her life was one of great value to many people, who will remember her with enormous gratitude.

RRH

[*East Afr.med.J.*,July 1985,531-32,Photo]

DAVIS, Harold Sidney *b.*6 Aug 1908 *d.*6 Nov 1988
ERD BA Cantab(1930) MRCS LRCP(1933) MA(1935) MB BChir(1935) MRCP(1936) FRCP(1956)

Harold Davis was born in Kingston, Jamaica, where his father was an architect and his mother Edith May, née Goodrich, was the daughter of a planter. He attended Jamaica College, West Indies, and Dulwich College, London. He started his medical career at Gonville and Caius College, Cambridge, and carried out his clinical duties at Charing Cross medical school, where he obtained a university scholarship in pathology and a prize in clinical medicine.

His first house appointment at Charing Cross Hospital was to Sir Gordon Holmes [*Munk's Roll*,Vol.V,p.195]. He was later house surgeon to J Bright Bannister. From 1936-37 he was house physician and then medical registrar at the London Chest Hospital, and later became clinical assistant to Sir Arthur Hurst [*Munk's Roll*,Vol.IV,p.509] at Guy's Hospital, following this with postgraduate work at the Hôpital Vaugirard in Paris. In 1938 he returned to Charing Cross Hospital as medical registrar. His registrarship was interrupted by the outbreak of war in 1939, when he was commissioned in the RAMC as a first lieutenant and seconded to Ashridge Hospital (EMS) in 1940. In 1941 he was discharged from the RAMC owing to ill health, with the rank of captain. He resumed his post as medical registrar at Charing Cross and remained there until 1945, when his persistent efforts to rejoin the RAMC were rewarded and he joined the Army Reserve. By 1955 he had been promoted to lieutenant colonel and was in charge of the medical division of the 51st General Hospital (AER). In 1964 he became officer commanding 308 (Co London) General Hospital RAMC/AER with the rank of colonel.

Harold Davis had been appointed consultant physician to King Edward VII Hospital, Windsor, in 1945, and to the Hampstead General Hospital in 1946, where he became chairman of the medical advisory committee. With the integration of the Hampstead General into the Royal Free Hospital group he was appointed physician to the Royal Free in 1954. He also served at the Florence Nightingale Hospital, and the New Victoria Hospital at Kingston. In 1973 he retired from all these appointments, but undertook the local consultant position at Highlands Hospital from 1973-78. He was honorary physician to the Queen 1967-68.

During his time at the Royal Free Hospital he was examiner for London University and served on the board of governors from 1963-66. He was very much a general physician but his enthusiasm was for gastroenterology and it was in this area that he published. He was not research orientated.

In 1973 he retired to Skirmett, near Henley on Thames, which had been his home since 1957. Being a former gamekeeper's cottage this encouraged him to develop his hobbies of riding, fishing, shooting in the local woods and boating on the Thames. He was an active member of the Hambleden British Legion. His major sporting interest was in skiing and he became a gold medallist. In 1964 he was vice-president of the Ski Club of Great Britain. He was also president of the Eagle Ski Club from 1960-63, and a member of the Kandahar Ski Club. He was indeed a polished performer on the slopes, particularly at his favourite resorts at Grindelwald and Murren. Apart from sport, he was a collector of old silver and English porcelain.

Although a generous host and a good raconteur, not least at the Bath Club, he tended to confine his social gathering to small groups of friends. He was noted for his considerate attitude to colleagues. He had a quiet dignity, with the sort of bedside manner that provides comfort and reassurance to patients. His manner was enhanced by his being very particular about the way he dressed. His usual attire was striped trousers and black coat, the latter with a buttonhole flower, a starched collar, and a watch chain displayed across his waistcoat.

In 1940 he married Molly, née Stimson, who survived him together with their only child, Caroline Penelope. During his final and frustrating illness Molly's support for him was unfailing. She was the first to acknowledge the tender, loving care he also received during his frequent periods as an inpatient in Watlington Hospital.

ATR

[*Henley Standard*,Dec 1988; *Hambledene Valley Group Mag.*,Jan 1988; Photo]

de SILVA, Percival Cholmondeley Chalmers *b.*25 Feb 1904 *d.*20 May 1987

MB BS Lond(1930) MRCP(1930) MD(1938) FRCP(1956)

Percival Cholmondeley Chalmers de Silva was born in Colombo, Sri Lanka. He was the son of William Henry de Silva, a pioneer eye surgeon who was responsible for establishing ophthalmology as a specialty in Sri Lanka. 'CC de Silva', as he later came to be known, lost his father at the early age of four years and was therefore deprived of paternal guidance in his pursuit of a medical career. However, his mother, who was the daughter of Sir Henry de Soysa, one of Sri Lanka's best known philanthropists, discharged her responsibilities with much credit to herself.

CC de Silva had his early education in a fashionable girls' school in Colombo, which refused to have him any longer when he reached the objectionable age of 10 years. He moved on to St Thomas' College, Mount Lavinia. Having decided on a medical career, he entered Ceylon Medical College in 1923 but left soon afterwards to pursue his studies in England. He first entered University College London and then Kings College Hospital.

After qualifying, gaining his MD and his membership of the College, he returned to Sri Lanka in 1931 and set up in general practice, an unusual step for a young physician to take at that time. He was attracted to paediatrics early in his career. When the first chair of paediatrics was created at the University of Ceylon, they acted wisely in selecting him for it; CC de Silva set about upgrading the teaching of paediatrics both at undergraduate and postgraduate level. The status that this specialty enjoys in Sri Lanka today is in no small measure due to his efforts.

His post at the Lady Ridgeway Hospital for Children gave him much scope to conduct original research on many aspects of paediatrics in Sri Lanka. His work stamped him as an authority on tropical paediatrics, especially kwashiorkor and avitaminosis. He was invited to many conferences abroad, some of which he chaired.

CC de Silva was awarded the highest honours that any medical man or scientist could aspire to. In 1949 he was elected president of the Ceylon branch of the British Medical Association, now the Sri Lanka Medical Association. He was elected president of the Ceylon Association for the Advancement of Science in 1961, and in the following year he became chairman of the Ceylon Institute for Scientific and Industrial Research.

In 1931 he had married Irene, daughter of Sir Henry de Mel and sister of the Rev Lakdasa de Mel, Bishop of Calcutta and Metropolitan of India, Burma, Pakistan and Sri Lanka. They had a son and a daughter.

De Silva had an inspiring personality which was evident in any gathering of people. He spoke English with a distinctive drawl, which in later years became more pronounced. He was a hard task master to his students and to colleagues on his staff, but even his rebukes were draped in such charming language that no one could possibly take offence. His concern for his small patients was always uppermost in his mind. It was a well known practice of his to telephone his house physicians in the dead of night to enquire about the condition of a patient - somewhat to their irritation. When he retired in 1966 his staff demonstrated their regard for him by presenting him with a congratulatory volume to which many eminent paediatricians had contributed. The University recognized his service by appointing him emeritus professor and conferring on him an honorary DSc.

For CC de Silva retirement did not spell retreat into inactivity. He was involved more than ever in various social organizations devoted to family planning, tuberculosis, nutrition and the deaf and the blind, in all of which he played a leading role.

In the evening of his life, when ill health had impaired his physical faculties, he continued to attend important academic meetings. Much of his later years were devoted to writing his memoirs. It must have given him the utmost joy when the first copy of his *Life as I lived it* was shown to him just before his death.

CGU

[Photo]

DEAN, John *b.*5 Sept 1917 *d.*23 July 1983
MA MB BChir Cantab(1942) MRCP(1947) FRCP(1973)

John Dean was born at Walsall, Staffordshire, the eldest son of a tool manufacturer Herbert Dean and his wife, Edith Elsie Yates, daughter of Simeon Yates, a collier. He was educated at Wednesbury High School, obtaining an exhibition in natural sciences to St Catherine's College, Cambridge, where he was awarded first class honours in Part I of the natural sciences tripos in 1938. He received his education in clinical medicine at the Westminster Hospital medical school.

After holding one house physicianship at his teaching hospital, he was commissioned as temporary surgeon lieutenant RNVR and served in the Royal Navy, in HMS Ready, for three years. After demobilization he held house physician posts at the Hospital for Sick Children, Great Ormond Street, and the Westminster Hospital, before being appointed senior registrar in paediatrics at the Westminster and the Westminster Children's Hospitals from 1947-50, when he became assistant to the professor of child health in London University, A A Moncrieff [*Munk's Roll*,Vol.VI,p.343], for four years; working mainly at Hammersmith Hospital in the care of newborn babies.

At that time vacancies for or new appointments of consultant paediatricians in the NHS fell far short of the numbers of suitably qualified applicants. Moreover, Dean - although greatly liked and respected by his contemporaries in paediatrics - was not always popular with his seniors for understandable, if unworthy, reasons. For example, during ward rounds he would present all the relevant facts in a case history succinctly and fairly, but did not deem it one of his duties to assist his chiefs to arrive at the correct conclusions. Thus it was to the regret, but not altogether to the surprise of his friends when in 1955 he felt compelled to seek a livelihood overseas. He emigrated, with his young family, to Canada and until his death, only three months after retirement, he lived and worked in Vancouver.

For one year he held the post of senior resident in paediatrics at the General Hospital, and on obtaining certification in paediatrics from the Royal College of Physicians and Surgeons of Canada was appointed to the paediatric department of the medical faculty of the University of British Columbia, where he rose steadily in academic rank from teaching fellow in 1956 to associate professor in 1975. he had clinical responsibilities at the Children's and at the General Hospital.

John Dean was too shy and self-effacing, and too indifferent to the promotion of his own career, ever to acquire a national, let alone an international reputation in paediatrics. He did little research worth the name and published little, other than contributions to paediatric textbooks. He had a cynical attitude concerning the evanescent value of most medical publications, and of the motives for publishing. But, if Dean's reputation was not worldwide, he was greatly admired and valued in British Columbia for his teaching, his clinical excellence, and for his willingness to shoulder important but personally unrewarding planning and administrative responsibilities. In the words of his friend and ex-colleague Robert Hill, professor and head of the department of paediatrics at the University: 'Dean was a valued consultant whose written opinions were models of brevity and precision, as were his verbal contributions to rounds, committee meetings and conversation.'

Dean had a profound knowledge of child health and ill health, which he kept up to date throughout his professional life. In respect of recent advances, however, he once said of himself that he '...would not feel too desperate if separated from a laboratory.' His greatest strength, and the quality for which he will be remembered, lay in his gift for teaching. He had the ability to dissect a problem and to get at the essentials in simple terms, and to express himself clearly. He was deeply and retentively read in 'worthwhile' medical literature, and possessed an inexhaustible fund of apposite and often amusing anecdotes. Thus his ward rounds were memorable.

John Dean was short and spare. He spoke slowly and deliberately, with a distinct but unaffected drawl. To his friends he was the best of company: an ever interesting conversationalist and possessor of a ready wit. 'Sardonic' is the adjective most appropriately applied to his sense of humour but not to his character; since he was uncensorious of individual human failings and a most kind-hearted man. Contrasting, but by no means incompatible with a sense of humour, was his melancholic temperament - not depressive, and never remote. His life was indeed marred by tragedy: the deaths from malignant disease first of his only daughter when aged 26, and then of his wife in her 61st year; each was a beautiful and vivacious woman and together they had supplied a perfect foil to Dean's habitual gloom. His family life, which had been a particularly happy one, was redeemed from utter tragedy by his affection for and satisfaction in the successful career of his son, John Melville Dean MB London, FRCPC, now an allergist and clinical immunologist in Vancouver.

Dean's interests, apart from his work, were intellectual rather than athletic, although he enjoyed watching, in England, first class cricket and, in Canada, American football, a game which he claimed - and half-persuaded the writer - to comprise numerous esoteric subtleties. He was a talented amateur mathematician and had represented Cambridge University at chess for four years. A more practical interest was that of cooking; with the assistance of his son he prepared delicious meat and fish dishes, allowing - nay encouraging - his womenfolk to cook the vegetables and do the washing-up.

He was a keen and learned gardener: and he died as a consequence of head injuries sustained on falling from a ladder while pruning a tall vine that had infiltrated the eaves of his home. PT

[*Brit.Col.med.J.*,1983,**25**,489; Photo]

DICKSON, Robert Clark *b.*24 Sept 1908 *d.*19 Feb 1984
CM(1967) CD(1952) OC OBE(1945) MD Toronto(1934)
FRCPC(1945) MACP(1976) *FRCP(1972) Hon LLD Dalhousie
and Toronto

Born in the small Ontario town of St Mary's, and having received his primary education there and his secondary education in Ottawa, the capital of Canada, Robert Dickson was destined to become a truly outstanding physician, teacher, administrator and medical statesman; a man who combined a strong character with personal warmth and good humour. He was the son of William Murray Dickson, a Federal civil servant, and his wife Mabel Earl Clark.

Such is the magnitude of this medical giant's contributions that they are best considered in relation to three separate categories, namely academic medicine, military medicine and medical statesmanship.

Early during his undergraduate medical education in the University of Toronto, Robert Dickson demonstrated his superb scholastic ability and was made a member of the Alpha Omega Honor Medical Society. In 1934, two years later, he graduated with honours and was awarded the Cody silver medal. He had been greatly influenced by Duncan Graham [*Munk's Roll*,Vol.VI,p.204] who was at that time professor of medicine of the University of Toronto, and also serving as the third president of the Royal College of Physicians and Surgeons of Canada. Indeed, his powerful influence was reflected throughout Dickson's illustrious career.

Motivated to become an academic physician, Robert Dickson spent five years' postgraduate training in internal medicine at the Toronto General Hospital, during which time his record was exemplary. After six years of service in the Royal Canadian Army Medical Corps, he was invited to join the teaching staff of the faculty of medicine in the university, where he proved to be a magnificent teacher not only in the lecture hall but also at the bedside - in the tradition of another great Canadian physician, Sir William Osler [*Munk's Roll*,Vol.IV,p.295]. Many medical students vividly recall three important lessons they learnt from Robert Dickson: first, 'come on time or not at all'; second, 'knowledge is useless unless it is organized', and third, 'you learn medicine from patients and not from books'.

Having served on the teaching staff of the Toronto General Hospital for four years, Robert Dickson was appointed physician in chief of the Wellesley Hospital (at that time a newly created division of the Toronto General Hospital) and during the ensuing seven years he rapidly achieved an enviable reputation as an outstanding physician and teacher, especially in his field of gastroenterology within the specialty of internal medicine.

Robert Dickson was already recognized as an academic leader when he was appointed professor and head of the department of medicine in Dalhousie University, Halifax, a position which he filled with great distinction until his retirement in 1974. During these 18 years Dickson's brilliance as a postgraduate teacher was demonstrated by the fact that the specialist residents he trained had the highest pass rate in Canada in the examinations of the Royal College of Physicians and Surgeons of Canada - a most impressive record. At least four of his postgraduate students eventually became professors of medicine in other Canadian universities. Robert Dickson also contributed some 40 articles to scientific literature. After his appointment as professor emeritus, he continued to teach postgraduate students in Camp Hill Hospital, the Department of Veteran Affairs Hospital in Halifax, and continued to exemplify both the art and the science of medicine. Both nationally and internationally he was considered to be one of Canada's most distinguished physician-teachers and in 1974 his contribution was recognized by the Royal College of Physicians and Surgeons of Canada when he was given the Duncan Graham Award - the College's highest honour in the field of medical education.

In 1975, Dalhousie University established the Robert Clark Dickson Lecture in his honour. These lectures, which Robert Dickson faithfully attended, have been delivered by many American and Canadian colleagues including Franz J Ingelfinger [*Munk's Roll*,Vol.VII,p.296], Robert G Petersdorf, John C Laidlaw, Samuel O Freedman, Douglas G Cameron, Robert B Salter and John R Evans. His contribution to academic medicine was also recognized, in 1982, by the Victoria General Hospital of Halifax when it named the Robert Clark Dickson Centre for Ambulatory Care and Oncology Research.

Before the outbreak of the second world war, Robert Dickson served for two years as a medical officer in the Militia with No 16 Field Ambulance of the Royal Canadian Army Medical Corps. In September 1939 he was among the first Canadian physicians to volunteer for active service. He went overseas with No 15 Canadian General Hospital as officer in charge of medicine. He rose to the rank of lieutenant colonel and served with distinction in North Africa and Italy. His contributions to military medicine were recognized in 1945 when he received his OBE. At the end of hostilities, Robert Dickson returned to the Militia, in which he served from 1946-52 as RMO to the 48th Highlanders of Canada. He was awarded the CD in 1952 and appointed Queen's Honorary Physician in 1973. He was a member of the Canadian Forces Medical Council from 1966-74 and subsequently its chairman.

Throughout his medical career, Robert Dickson accepted increasing responsibilities as a national leader. He served as a member of council of the RCPSC for a record period of 12 years, including two years as one of the College's most distinguished presidents, 1970-72.

During his presidency of the Royal College, in a particularly difficult period in the history of the College, Dickson made a personal tour of every major centre across Canada in a successful effort to bring about important changes in Royal College policy. This gruelling tour he undertook despite a serious and painful physical handicap; he had arthritis of both hips, the surgical treatment of which he deferred until he had completed his two-year term as president of the college.

It was an example of the dedication he exemplified throughout his professional career. He also served Canadian medicine as chairman of the board of governors of the R S McLaughlin Examination and Research Centre of the Royal College.

Robert Dickson received a number of well merited honours: he was appointed an Officer of the Order of Canada, and received the honorary degree of Doctor of Laws not only from Dalhousie University but also from his *alma mater*, the University of Toronto. The American College of Physicians appointed him a governor for the Atlantic Provinces, 1962-68, and he later became a member of their board of regents, 1968-74. More recently, he was elected a Master. In the United Kingdom he was elected to the fellowship of the College.

Robert Dickson was a man of vision, combined with common sense and practical application. At Dalhousie he founded the first Canadian School of Outpost Nursing in 1961. From 1970 he served as a member of the board of directors of the Grenfell Labrador Medical Mission, and from 1974 as a member of the board of directors of the International Grenfell Association.

He was a physician in the best Oslerian tradition: a physician, teacher, administrator and medical statesman, and kind and gracious to both patients and colleagues. Those who were fortunate enough to be his friends will always cherish his memory.

He married Constance Fraser Grant, daughter of a civil engineer, in 1906 and they had two daughters, Shelagh and Jane. His main interest, apart from medicine and his family, was in sailing.

RBS

[Photo]

DONAGHY, John Patrick *b.*4 Mar 1919 *d.*2 Oct 1987
MB BCh BAO Belf(1943) MRCP(1947) FRCP(1970) FRCPI(1979)

John Donaghy was born in Holywood, Co Down,and died while on holiday in Spain. His father, Matthew George Donaghy, was a police officer and his mother, Julia Brady, came of farming stock on both sides of her family.

John's first appointment, in 1944, was to the Mater Infirmorum Hospital, where he had been a student. He was rejected for the medical services of the armed forces on health grounds and worked for a short time in general practice. From 1945-47 he served in Crumpsall

Hospital, Manchester, and then went to the Postgraduate Medical School at Hammersmith. He subsequently became chief assistant in medicine at the Sheffield Royal Infirmary, returning to the Mater Infirmorum Hospital in 1948 as consultant physician.

He brought with him an enthusiasm for investigative clinical medicine which had a strong impact on the hospital, and also on doctors who sought his advice in consultation. There are no erudite publications to prolong his memory, but there are many who will always be grateful for his insistence on the teaching of basic, fundamental medicine at the bedside. He was a devoted teacher who regarded his association with students as a privilege.

Extensive and successful as was John's professional career, it was only one strand in a life into which was woven a deep love for his family, his country and his faith. His horizon was wide enough; his father was an officer of the old Royal Irish Constabulary and John was born into an Ireland still united under the Crown. But although his early appointments were in English hospitals, it was to the Mater Infirmorum that he devoted himself with a fierce loyalty that was characteristic of him.

As a university representative in the old Northern Ireland Senate he seized every opportunity to draw attention to the difficulties experienced by the hospital which was in those days still excluded from the National Health Service. But he had the joy, before he died, of seeing new buildings well on the way to completion.

Over and above his work as clinician and teacher, John had a passionate concern for a philosophy of medicine so much wider than mere bioethics. For generations of students he interpreted medical ethics with admirable adherence both to theological orthodoxy and scientific truth, between which there was, for him, no hint of conflict. Some students may have found his approach too scholastic for their youthful emotions, but they never had to complain of ambiguity.

Neither medicine nor politics absorbed him totally. He was a considerable Irish scholar, speaking the language with enthusiasm and elegance. He also had an informed interest in the field of Celtic art.

John was never careless in his speech. Even in heated disagreement he spoke slowly, with a clarity and formality which sometimes disconcerted his close colleagues, but neither friend or foe ever heard him use an unconsidered or discourteous word.

Loving loyalty to his religious faith was the mainspring of his personal and professional life. His devotion to his hospital, his work for the Guild of Catholic Doctors and his ethical stance, all stemmed from that strong root. It was typical of the man that immediately after retiring he embarked on a serious course of Biblical theology and had just completed a thesis before leaving for the holiday from which he never returned.

For many years John suffered from major illnesses, including a need for cardiac surgery, which would have defeated a lesser man. He showed great courage in facing them, helped by the devoted attention of his wife Una whom he had married in 1947. They had seven children: one is an orthopaedic surgeon in Canada, two are in general practice, one is a nurse, and three became teachers.

On May 12, 1986, John was invested with the insignia of Pro Ecclesia et Pontifice, a papal honour awarded *egregia opera studioque conspicuis.* We had no premonition at that time, but there could have been no better epitaph.

BM/GW

[*Brit.med.J.*,1987,**295**,1215; Photo]

DONALD, Ian *b.*27 Dec 1910 *d.*19 June 1987
CBE 1973 BA Cape Town(1930) MB BS Lond(1937) MD(1937) FRFPS Glasg(1958) FRCSG(1962) FRCOG(1965) FCOG S.Afr(1967) FACOG(1976) *FRCP(1987) Hon DSC Lond(1981) Hon DSc Glasg(1983) Hon FRCR(1983)

Ian Donald, emeritus regius professor of midwifery at the University of Glasgow, was one of the most eminent obstetricians of his time. He pioneered the diagnostic use of ultrasound in obstetrics and gynaecology which made possible the detection of abnormalities in the earliest stages of pregnancy, a technique which is now used routinely not only in obstetrics but also in many other specialties.

Ian was born in Liskeard, Cornwall, the son of John Donald, a general practitioner, and his wife Helen Barrow, née Wilson. His father's practice was in Paisley, Scotland, and Ian's early education was first at Warriston School, Moffat, and then at Fettes College, Edinburgh. When he was 14 years old his father moved to Cape Town, South Africa, for health reasons, and Ian continued his education at the Diocesan College in Rondebosch, subsequently obtaining a BA from the University of Cape Town. After the death of his parents he returned to the United Kingdom and in 1937 graduated in medicine from the University of London. He served in the RAF as a medical officer from 1942-46, being awarded the MBE and mentioned in despatches for his bravery in saving several airmen from a crashed blazing aircraft.

After the war he entered St Thomas's Hospital. It was Joe Wrigley who inspired him to specialize in obstetrics and gynaecology and Ian became a reader in the subject at St Thomas's. In 1952 he moved to take up a similar position at the Hammersmith Hospital, now the Royal Postgraduate, where he first showed his talent for applying his electronic expertise to medicine when he developed a respirator triggered by the feeble respiratory efforts of the infant. His early researches were concerned with respiratory disorders in the newborn child, and his Blair Bell lecture at the Royal College of Obstetricians and Gynaecologists in 1954 was on this subject.

Donald's interest in ultrasound diagnosis began when he was at the Hammersmith but it was his appointment, in 1954, to the Regius chair of midwifery at Glasgow which enabled him to develop his ideas. Glasgow had the engineering expertise he needed and he had the help of a brilliant engineer, Tom Brown, and the Scottish electronics company Kelvin Hughes. But there can be no doubt that it was Ian's own inspired approach to research that provided the driving force behind the creation of the first contact compound sector scanner. Donald's celebrated paper on the subject, with T G Brown and J MacVicar, appeared in the *Lancet*, 1958,1,p.1188. The early images were of very poor quality and at that time many saw no future in the new technique. Only a man of Ian's obsessive drive and vision could have seen the project to its successful conclusion. For the diagnoses did come: first pelvic tumours, next hydatid moles and, following the development of the full-bladder technique, the growth of the foetus in the first trimester of pregnancy. This was soon followed by early diagnosis of multiple pregnancy, measurement of the foetal biparietal diameter, and placental location.

Donald's initial work on ultrasound was at the Royal Maternity Hospital in Glasgow, but within 10 years he had campaigned for and helped to design the splendid new Queen Mother's Hospital at Yorkhill. He moved there in 1964 and the hospital became a centre for visitors from all over the world to study ultrasound diagnosis. Donald built up a strong team of many young researchers and allowed them great freedom to pursue their ideas, which resulted in making many advances. Ian Donald was a man without a trace of envy and he was always generous to his American rivals who had by this time built their own copy of his contact scanner. But until the development of the real-time scanner in the 1970s, Donald's 'Diasonograph' was the world's best and his team in Glasgow were always one step ahead of the competition.

Ian was a great teacher and a forceful lecturer. He always said that the basis of good teaching was enthusiasm, of which he himself possessed an abundance. His book *Practical obstetric problems*, London, Lloyd-Luke, 1955, was a bestseller and gives an idea of his unique didactic style. It was often recommended as bedtime reading, and his entertaining style made even the dull aspects of his subject interesting and absorbing.

Ian Donald held very strong views on the sanctity of life and vigorously opposed the 1967 Abortion Act, which earned him almost more media publicity than his scientific contributions had done. One might disagree with him, but none challenged the integrity of his views. He was a passionate believer in life and was horrified that obstetricians should be called on to destroy it, the very antithesis of their true calling.

He received many honours, including the Blair gold medal, the Eardley Holland gold medal, the Victor Bonney prize and the Maternity prize of the European Association of Perinatal Medicine. He was elected a Fellow of the College in 1987.

Shortly after receiving his fellowship of the RCP, he attended the fellowship admission ceremony of the RCOG to see his friend and colleague, Eric Saling, receive the fellowship *ad eundem.* It was fitting that his last public appearance should be with his wife at his own College.

Ian had married Alix Richards of Bloemfontein in 1937. It was a very happy marriage and they had four daughters of whom he was very proud. His happy family life was a valuable source of strength because for some 30 years he had to fight a continuing battle with ill health, undergoing major cardiac surgery on three occasions.

Ian was not only a man of indomitable spirit, he also had many talents: an accomplished pianist, a competent watercolour artist, and an enthusiastic yachtsman, as well as being a pioneer in his own field.

He was survived by his wife and their four daughters. GW/VL

[*Brit.med.J.*,1987,**295**,126; *Lancet*,1987,**2**,168; *The Times*,20 June 1987; *Trans.Coll.Med.(SA)*,July-Dec 1987,**31**,No.2,146; Photo]

DOWNIE, Allan Watt *b.*5 Sept 1901 *d.*26 Jan 1988
MB ChB Aberd(1923) MD(1929) DSc(1937) FRS(1955)
FRCPath(1964) *FRCP(1982) Hon LLD Aberd

Allan Downie and his identical twin, Richard, were born in the Mid Street cottage in Rosehearty, Aberdeenshire, where William and Margaret (née Watt) Downie had raised their other six children. Three generations of Downies had lived by sailing deep-sea drifters from the tiny harbour, and for 14 years the twins seemed set to follow. However, their outstanding intellect had been spotted by their village school headmaster who, in 1915, coached them into Fraserburgh Academy.

In 1918 they entered Aberdeen medical school, leaving in July 1923 with first class honours, every academic prize in each year being shared between them, and as the professor of medicine, Sir Ashley Mackintosh, disclosed on graduation day, each with equal marks in the four professional examinations. As their 'year' celebrated in the Kirkgate Bar, Allan was nearer to the door than Richard; thus it was he who received a job offered by the astute John Cruickshank who wanted one of the twins as his assistant, starting in the following October. Three months in general practice in Sheffield helped Allan to repay some of the debt he owed his parents - and he was set on his remarkable career.

For four years he worked in Aberdeen as Cruickshank's assistant on diagnostic research and public health bacteriology; he taught himself German in order to profit from Cruickshank's German-speaking evenings; he spent four months in Breslau with Karl Prausnitz, of Prausnitz-Kustner fame, and published in German, with Meissner, a paper which finally disposed of the heresy that the tubercle bacillus had a filterable form. On his return he prepared his MD thesis; in 1929 this was awarded with the highest honours and received the Thursfield prize.

Then a move to Manchester's pathology department, followed by six years in bacteriology with H B Maitland, introduced Downie to pox viruses. He was later to become recognized as a world authority on smallpox, its epidemiology and pathogenesis, and the virus that caused it.

In 1934, Allan Downie was appointed to the prestigious Senior Freedom research fellowship at the London Hospital. The appointment was conditional on first spending an academic year in New York at the Rockefeller Institute. He found himself working with O T Avery at a time which we now see as a turning point in medical science. Downie himself recognized that year as the most formative and stimulating year of his scientific life. Fred Griffith, in London, had demonstrated capsular transformation of pneumococci; Avery's group was close to success in demonstrating that DNA was the 'transforming principle'; leaving Watson and Crick in Cambridge to ice the cake - with the aid of Rosalind Franklin's crystallographic findings - and to reveal the beauty of the double helix. It was the start of molecular genetics.

Downie already had a major interest in pneumococcal infection and immunity and, on his return to London, he wrote his DSc thesis. Sulphapyridine had however taken the grave import out of pneumococcal infection, making immunotherapy redundant, and Allan Downie returned to his pox viruses; making the surprising discovery that vaccinia virus as used for smallpox vaccination was not after all the cause of naturally occurring cowpox.

In 1939 Downie left the London Hospital for the National Institute of Medical Research at Hampstead - but never arrived! Instead, the MRC drafted him to take charge of the laboratory in Cambridge set up by the newly formed Emergency Public Health Laboratory Service. East Anglia was thought to be a prime target for German invasion, and the Cambridge laboratory had to be prepared for any infective catastrophe which enemy action might bring. In the event it was largely civil public health problems rather than military needs which formed the bulk of the workload, although the latter did include making, on the open bench, batches of rabies vaccine for the troops. There was no Howie Safety Code in those days.

Downie's extensive experience in Manchester and London with pox viruses led to his being given responsibility for laboratory diagnosis throughout the UK but, curiously, this task never burdened him until 1944 after he had arrived in Liverpool to fill the chair of bacteriology at the university.

The mode of Downie's appointment bears telling: Hedley Wright had died and Liverpool University was desperate to find a worthy successor. Downie was invited to apply but, for his own reasons, declined. He agreed however to visit his good friend Tom Davie [*Munk's Roll*,Vol.V.,p.95], who was professor of pathology and dean, to look around. Davie and the vice-chancellor had recently laid on a selection committee and Downie, without warning, was thrust in and to his shocked surprise appointed.

Downie held the chair for 23 years, serving for periods also as dean and pro-vice-chancellor. His 110 published papers show the breadth of his interests, but smallpox proved to be the most fascinating.

Allan Downie's role in the WHO eradication campaign has been told elsewhere; suffice it to say that without his initial planning, his expertise and practical cooperation, and particularly his generous advice given throughout without thought of personal recognition, the campaign might not have succeeded. When eradication was complete the RCP honoured Allan Downie, in a way which pleased him as no other honour could have done, by making him a Fellow of the College.

Allan married Nancy McHardy in 1936 and they had three children, a son and two daughters. Honours, distinctions and fame never changed him. He played football in his youth, golf and table tennis well into his 70's, gardened, cooked fudge or oatcakes and was a dreadful nuisance in his Nancy's kitchen. Each summer he sailed his old fishing boat out of Rosehearty's harbour, first with his three children and later with his grandchildren, or with any friends who had the courage to face those treacherous waters. He was a man of incomparable charm and total honesty, loved by all and living on in the work of those whom he trained so well.

KMcC

[*The Times*,28 Jan 1988; *The Independent*,1 Feb 1988; *Lancet*,1988,1,312,426; Photo]

DREADON, Ross Graham *b.*11 Oct 1918 *d.*9 Sept 1987
MB ChB NZ(1941) MRCP Lond(1948) DCH(1949)
MRACP(1952) FRACP(1964) FRCP(1974)

Ross Graham Dreadon was born in Auckland, New Zealand, the son of Joseph Dreadon, a farmer, and his wife Anne Bull, whose father owned a gold mining battery. He was educated at the Papatoetoe Primary School and Auckland Grammar School, and then entered the University of Otago to study medicine. His clinical studies were undertaken at Dunedin Hospital and Auckland Hospital, and on graduation he took up a post as house physician in Auckland Hospital, from 1941-42.

In 1942, during the second world war, he enlisted as a medical officer in the Royal New Zealand Air Force and saw service in the Pacific Zone, until 1945. On demobilization, he became medical registrar at the Auckland Hospital until 1948, when he decided to come to the UK for postgraduate training. He was appointed medical registrar at Newcastle General Hospital, Newcastle upon Tyne, in 1948, and during this period obtained his membership of the College and his diploma in child health. On return to New Zealand and Auckland, he became visiting physician to the Middlemore Hospital, 1950-62, and then physician to the Greenlane Hospital which he served until his retirement.

In 1973 he was president of the Auckland branch of the Medical Association of New Zealand.

He married Mary, daughter of John Berry, in 1955 and they had two children, a son and a daughter. He had few interests outside medicine and his family, but he was a keen collector of antiques, with a special interest in English porcelain.
GW/VL

[Photo]

EDHOLM, Otto Gustaf *b.*14 June 1909 *d.*18 Jan 1985
BSc Lond(1930) MRCS LRCP(1933) MB BS(1934) *FRCP(1975)
Hon DSc Surrey (1979)

Otto Gustaf Edholm was born in London, of Swedish parentage. From Tonbridge School he went to King's College, London, where in 1930 he took a BSc with honours in physiology before graduating in medicine from St George's Hospital in 1933. From 1934-35 he combined hospital appointments with general practice but his interest in physiology led him back to King's; first as a research fellow and then as an assistant lecturer.

In 1938 he was appointed lecturer in physiology at Queen's University, Belfast, where Henry Barcroft and he carried out pioneering work on human peripheral circulation. During the war, under the aegis of the shock committee of the Medical Research Council, Barcroft and Edholm, using their plethysmographic methods, collaborated with McMichael and Sharpey-Shafer [*Munk's Roll*, Vol.V,p.372] who had introduced cardiac catheterization in the UK, in studies of circulatory changes induced in man by severe haemorrhage; this work elucidated the mechanism of the post-haemorrhagic fainting reaction.

In 1944 Edholm returned to London as professor of physiology at the Royal Veterinary College where he and his associates worked on the venous valves in the giraffe, the okapi, the camel and the ostrich. Otto's dramatic description of his ride from the Zoo to the College, with the head of his dismembered giraffe staring at unbelieving passers-by through the side window of his taxi, once heard was never forgotten. He also resumed his work at Hammersmith, applying plethysmographic techniques to clinical problems.

In 1947 Edholm was persuaded to take the chair of physiology in the University of Western Ontario, and it was during his stay in Canada that he became interested in the effects of extreme climatic conditions on man. His book *Man in a cold environment*, London, E Arnold, 1955, written with A C Burton, remains a classic work.

After the war, when it was clear that the Services would have to adapt to extremes of temperature and climatic conditions, the Medical Research Council - which was already involved in research on heat adaptation - decided to set up a division of human physiology at the National Institute for Medical Research with a primary remit to study adaptation to cold. Edholm was invited to head the division, and he returned to the UK in 1949 to face the formidable task of supervising the building of the unique climatic chambers at the Institute. Inevitably, and with a rapidly increasing momentum, the division became the hub of environmental studies of a highly diverse nature, as requests for help poured in from various sources. Edholm remained at the centre with his fingers in innumerable pies, for the baking of which he was not of course entirely responsible but which, without his inspired advice, direction and coordination, both at Hampstead and on the relevant committees and working parties, might never have reached the oven. But he was not office-bound. He was seen - a lean, smiling, Parka-clad figure striding through a snowstorm over the cluttered decks of a minesweeper, to view at first hand the testing of the latest inflatable life-craft designed to increase the rate of survival at sea. He collaborated with nutritionists in energy expenditure studies on officer cadets at Sandhurst, after a general had expressed concern that in their supposedly arduous life the cadets were not gaining weight for lack of good red meat. He participated in studies of the Bedouin in the Negev. The work of his division included high altitude studies which contributed to the conquest of Everest by the Hunt expedition, the unique investigations made on the Tristan da Cunha islanders during their stay in the UK, and advice to athletes due to compete at high altitudes or in the heat. Edholm took increasing responsibility for the planning and evaluation of polar physiological research. The Commonwealth Trans-Antarctic expedition leaned heavily on him for advice. His division provided a base for medical officers and physiologists participating in the British Antarctic Survey; there they prepared their research projects, often using modest unsophisticated equipment,and they returned there to evaluate and write up their results under the guidance of Edholm and his team. It was fitting that just before his retirement Edholm went to the Antarctic in the Royal Research Ship *Bransfield*, calling at the Falklands and South Georgia to see for himself those stations from which so much hard won and interesting data had been obtained. Perhaps the largest and most elaborate physiological experiments ever mounted in the UK were carried out after the divisional climatic chambers, built originally for work on cold, were modified for heat acclimatization and men acclimatized in those chambers were flown to Aden and elsewhere for testing. A practical system for artificial acclimatization resulted from this work.

To his staff Edholm's enthusiasm and interests seemed to have no bounds. He was an excellent teacher and it is fortunate that some of those he worked with have continued to practice and teach climatic physiology. Although immensely hard-working himself his door was always open to anyone seeking discussion, advice or encouragement on matters either professional or personal. One of his former staff remarked that he set an example of what a good boss should be like.

Edholm's philosophy as a human physiologist was that the proper study of man is through experiments on man himself. He expressed this forcibly at one International Physiological Congress when he wrote on the blackboard in capital letters 'MAN IS NOT A RAT'. This firmly held conviction radiated from him and energized others into human studies.

After his retirement from the Institute in 1974 Edholm became visiting professor at the School of Environmental Studies at University College London, where he applied himself to the difficult problems of the built-up environment. He continued to write; indeed only two days before his death he saw advance copies of his last book, written with R P Clark, *Man in his thermal environment*, London, E Arnold 1985. His publications - scientific papers, reports and books - were numerous. He was a member of many societies, and a founder member of the Ergonomics Society. Many honours came his way: the College made him a Fellow, the University of Surrey gave him an honorary DSc, he was consultant to the Army in applied physiology, and he was proud to be Bellinghausen Medallist of the Antarctic Committee of the USSR Academy of Sciences.

In spite of these honours and all his outstanding achievements, Edholm remained an essentially modest man. He was both an enthusiast and an incurable optimist. Although severely crippled in his last years by painful disorders he retained until the end that cheerfulness which had always characterized him. He was a great conversationalist and his alert and questioning mind could move at ease in the realms of music and the arts, and also into philosophy, politics and international problems. He called forth a loving loyalty from his staff, and he was a marvellous companion and friend.

He married Elizabeth Frances Margaret, daughter of William Bevan OBE, and they had three daughters.

HB'/SH

[*Brit.med.J.*,1985,**290**,569; *Lancet*,1985,1,650; *The Times*,16 Feb 1985; Photo]

EDWARDS, Thomas Archibald Watkin *b.*5 Oct 1912 *d.*27 June 1986
BA Cantab(1935) MB BChir(1938) MRCP(1944) FRCP(1968)

Thomas Archibald Watkin Edwards (Wattie) was born in Middlesborough, Yorkshire, where his father John Watkin Edwards was a general practitioner. He was educated at Dulwich College and Christ's College, Cambridge. He carried out his clinical duties at St Bartholomew's Hospital, where he was house physician to the medical professorial unit from 1938-39, just before the second world war. When war was declared he immediately joined the Royal Air Force as a medical officer, serving throughout the 1939-45 war and being mentioned in despatches. He was proud of being one of the very few British doctors to have passed the MRCP examination, held in Cairo in 1944, at the first attempt. He reached the rank of wing commander in command of the medical division, and later as commanding officer of No 1 General Hospital, RAF Algiers, 1944-45. After the war he returned to Bart's in 1946, as supernumerary chief assistant, before entering chest medicine as registrar at Clare Hall Hospital, South Mimms. In 1948, shortly before the start of the NHS, he was appointed consultant chest physician to St Albans City Hospital and Chest Clinic, and to Clare Hospital. From 1964 until his retirement in 1980, he was also on the consultant staff of Queen Elizabeth II Hospital, Welwyn Garden City. His tact, humour and urbane approach made him a most successful chairman of meeting in St Albans, and as vice-president of the British Thoracic Association and chairman of its Joint Tuberculosis Committee.

Wattie was of short, stocky build and exuded goodwill and genial helpfulness. It was always a pleasure to meet him and work with him in the care of patients, or in hospital development. He was quietly determined, however, in building up the chest services. In the early days of the St Albans Clinic, for instance, the old laundry was converted not into a casualty department, as originally planned, but into a chest clinic. This was his empire, with a family atmosphere of devoted loyalty from secretaries, nurses and social workers. All his working life, records and x-rays of his patients were filed there separately, and in the new outpatient department an office was planned for the chest physician and for no other consultant. Wattie's career spanned the development of chemotherapy and his research was the integration of chemotherapy with other treatments for tuberculosis. He had great regard and concern for his patients, for their jobs and their families, and in return he was held in high affection by them.

A quiet, scholarly man, he started life as a classicist and one of his retirement ambitions was to translate his 16th-century copy of Swammerdam's *De respiratione*. He had a warm sense of humour and enjoyment in simple games and television programmes. He was also of an enquiring mind; climbing to the rim of Vesuvius soon after its 1944 eruption. Gardens and golf, numismatics and music, were his recreations; he was proud of being invited as a student to play the toccata from Widor's 5th Symphony on the organ of the Royal Albert Hall.

Medicine was very much his life and he never accepted retirement. His idea of relaxing bedtime reading was a large textbook of medicine. He was unfulfilled after leaving his hospitals, despite the affectionate care of his wife, a Bart's ward sister, and his pride in the families and achievements of his daughter, who was a nurse, and his son, professor of clinical medicine at Edinburgh.

DNP/GW

[*Brit.med.J.*,1986,**293**,1035-36; Photo]

EVANS, Philip Raymond Charles *b.*26 Sept 1917 *d.*26 Apr 1984
GM (1943) MRCS LRCP(1942) MB BS(1942) MRCP(1946)
MD(1963) FRCP(1966)

Philip Evans was born in Edinburgh, the elder son of a Welsh general practitioner and his French wife. He was educated at the Nautical College, Pangbourne, and at Guy's Hospital. He qualified in 1942, having won the year before the Golding-Bird gold medal in obstetrics and gynaecology. In the same year he joined the Royal Navy as a temporary surgeon lieutenant RNVR. In 1943, when on active service in a destroyer, he was badly injured in both legs when the ship hit a mine. Despite his injuries he continued to look after other wounded men until help arrived, and for his courage and devotion to duty was awarded the George Medal. But the damage done led to his being invalided out of the Service and the after effects (recurrent osteomyelitis) were to plague him all his life and lead eventually to the amputation of one leg below the knee.

After leaving the Navy, Evans worked for the next three years as a medical registrar at Guy's, and at some of the other hospitals in the Guy's sector, and while doing so qualified MD and MRCP. In 1947 he was appointed first assistant consultant and then, in 1948, consultant physician to the Wrexham Group of Hospitals and he continued to work there until his retirement in 1979. He was for some years a member of the hospital management committee and was a founder member of the Society of Physicians in Wales. He published a number of papers, mostly on the treatment of peptic ulcer. In 1963 he was appointed a Justice of the Peace in Denbigh, and was elected a Fellow of the College in 1966.

Evans' temperament was mercurial, which is perhaps not surprising in view of his mixed French and Welsh blood. He could be devastatingly comic on occasion and full of gaiety. But, at times, he suffered periods of depression, though this was known only to his close friends. He bore the burden of recurrent foot infection for many years with little complaint, even though it meant that nearly every form of sport and active recreation was denied him. At one time he was a keen yachtsman; and he always enjoyed the gastronomic delights of France on his regular visits to the family house in the Haute Loire.

Kindly, shrewd and wholly lacking in conceit, Evans was much loved as a physician and well deserved the devotion and regard he received from a wide circle of friends.

In 1942 he married Margarita (Peggy), daughter of John Wheelwright Barnett, an exporter and British Consul in Chile, and they had three sons. She nursed him devotedly in his many bouts of illness and survived him, together with their sons.

JRF

[Photo]

EVANS, William *b.*2 Nov 1895 *d.*20 Sept 1988
MRCS LRCP(1924) MB BS Lond(1925) MD(1927) MRCP(1928) FRCP(1935) DSc(1944) Hon DSc Wales(1961)

William Evans was born in an isolated farmhouse near Tregaron, Cardiganshire, where he died, aged 92. His father farmed surrounding land and his mother was the daughter of a neighbouring farmer. He grew up in a strictly religious atmosphere, characteristic of the time and place, which together with the serenity of the environment had a profound influence on his attitude to the coming war, to his career in medicine and his later interpersonal relations. After attending the local county school, where he matriculated, he worked in a bank, whilst still considering a possible career in the church. As for countless other young men of his time, the advent of the first world war changed everything. He left the bank to join the Army, and was trained in the Buffs before being commissioned in the Lancashire Fusiliers. Although he served in the Ypres salient and at Passchendaele, he survived unscathed - unlike the majority of his battalion. He was demobilized in September 1919. Attracted by Army discipline, he thought of making a career in the regular Army but eventually opted for medicine, being greatly influenced by the contact he had had with human trauma and by a near relative in the profession.

A brief period of premedical studies was followed by admission to the London Hospital medical college in September 1920, where he graduated five years later. After an appointment in the pathology department he became house physician to John Parkinson, later Sir John [*Munk's Roll*,Vol.VII,p.443], to whose inspiration he attributed his future career as a clinician and researcher. A short period as assistant director of the medical unit was followed by his appointment as assistant physician to the cardiac department in 1936. Thereafter his fruitful collaboration with his mentor and revered senior colleague led to the pre-eminence of the cardiac department in the London. He became physician in charge of the department on Parkinson's retirement in 1948. Four years before this, during the war, he had been appointed to the staff of the National Heart Hospital.

Affectionately known as 'Willie' to all and sundry because he was always pleasant, affable, rarely ruffled and rather portly, he retained throughout his long life a special liking for rice pudding. He was, however, something of an enigma; the real man was difficult to know, he was close hauled and few if any of his closest colleagues knew him well.

Willie was fond of the pen and he used it very well. His writings included nigh on 100 medical papers, five books on cardiology, an autobiography with the unattractive and uncharacteristic title *Journey to Harley Street,* London, Rendel, 1968, and the *Diary of a Welsh Swagman 1869-1894,* by Joseph Jenkins, abridged and notated by William Evans, London, Macmillan, 1975, which was appreciated by Australian historians, together with several poems.

William Evans was a memorable teacher, almost a preacher, dogmatic, lyrical at times, but there was always much pithy common sense. Thus his lectures had a great appeal for the majority but the absence of uncertainty, the tendency to reject the rational and the dilution of basic science led to some scepticism among the more academic of his listeners. His investigations in cardiology were numerous and recorded in many papers. His description of the anatomy of coarctation of the aorta, his pioneering contribution to the use of controls in drug trials, and some of his early studies on electrocardiography, together with the first sound description of 'familial cardiomegaly' were the most important. Although the possessor of a DSc - rather rare among clinicians - he was not the most scientific of men, for his thinking was more intuitive, instinctive and empirical rather than deductive and rational.

He was a great advocate of specialization in medicine and his chosen subject gained thereby. Not surprisingly therefore he was one of the architects of the British Heart Foundation, and his efforts also led to the foundation of the Society of Cardiological Technicians, of which he became the first president. He was perhaps at his best as a doctor for he was caring, conscientious and straightforward. Patients loved him, and all his juniors liked him. He had a sense of humour which was never misplaced.

On retirement he returned to the Wales he loved. The Principality reciprocated by making him High Sheriff of Cardigan and an Honorary Druid. He placed both these honours above the legion of professional ones which came his way. He was not the most eminent cardiologist of his time but was one of the most unusual, well known and well liked. In retrospect, medicine in general and cardiology in particular seemed to be only an interlude, albeit an absorbing and exciting one, which occurred between childhood in the Teify Valley, overlooked by the Cambrian hills, and his return to the same unchanged scene to pass the 25 years of his retirement.

He had married Christina Downie, a nursing sister at the London Hospital, and daughter of John Downie, an engineer, in 1936. There were no children of the marriage and Christina died in 1964. William retired to a little house he had built in a corner of a field of his old home, continuing his interest in farming by helping his niece, Frances, to run the home farm. He was also a keen fisherman, and a conscientious gardener.

WB

[*Brit.med.J.*,1988,**297**,913-14,1040; *Lancet*,1988,**2**,859,1376; *The Times* 22 Sept 1988; *The Independent*,24 Sept 1988; *The Guardian*,28 Sept 1988; *The Times*,15 May 1963; *The Lond.Hosp.Gaz.*,Mar 1961,**64** (1),6;May 1961,**64** (2),39-40; Photo]

FAIRBURN, Anthony Clare *b.*22 Sept 1927 *d.*3 Oct 1988
MB BS Lond(1951) MRCP(1953) DCH(1957) DPM(1960) MRCPsych(1972) FRCP(1973) FRCPsych(1981)

Tony Fairburn was born in Singapore, where his father was Inspector General of Police. The only child of parents who married late, he passed his early years in the home of his maternal uncle and aunt in Beckenham, and his five cousins were like brothers and sisters. His mother's family were Buckleys, Catholics from Lancashire, and Tony went to Stoneyhurst; cross-country running took him into the countryside, and he excelled in the school's OTC.

His clinicals at St Mary's, and house appointments in London, were followed by nearly three years in the RAMC, mainly in Malaya, where he produced *A Handbook of medical Gurkhali* and wrote on the antibiotic treatment of leptospirosis in British troops. In Kuala Lumpur he married Margaret Dingwall, in 1954; they had met in Chester and she had been posted to Japan in the QARNS. Their marriage had solid foundations and their five children, and the coming of grandchildren, brought them great joy.

Back in London, Fairburn trained in paediatrics at Great Ormond Street, Queen Charlotte's, and the Westminster Children's Hospitals; and in psychiatry at the Maudsley. In 1960 he was appointed senior registrar in paediatrics and child psychiatry at the Bristol Royal Hospital for Sick Children, reading the first UK paper on child abuse in 1963. He took up his consultant appointment in 1965.

His duties included child and family psychiatry in Bath, and mental handicap in both Bristol and Bath. In mental handicap, he was initially responsible for over 300 patients in Hortham Hospital, Bristol, and for hostels in Bath, as well as a consulting and assessment network in neighbouring parts of Wiltshire, Somerset and Gloucestershire. In time he was joined by other teams; he opened the first residential home outside hospital for Hortham patients and latterly guided their move into the community. He was best known for working closely with his paediatric colleagues at the Royal United Hospital, Bath, where he built up his own department from scratch, and for his innovations in Bath city. He was a pioneer in the West Country of specialized day units for children with communication disorders, the maladjusted, and disturbed and inadequate parents. From 1970 he chaired one of the earliest child abuse monitoring panels. His emphasis was on ease of referral, early intervention, and preventive work with families.

Tony Fairburn was the kindest and gentlest of men, invariably understanding of parents and superb in his manner with children and adolescents. He was also courageous in forming his opinions and tenacious in acting on them. He incurred odium and condemnation by arguing that exceptional circumstances could justify the statutory removal of babies from their mothers at birth; a national survey some years later showed that his arguments had become accepted by many. His research on removals at birth led to recommendations to the DHSS on the review of the law on child care. He also spoke and wrote on alternative early interventions and preventive measures, and campaigned for the inclusion of preparation for parenthood in the school curriculum for teenagers. He sat on working parties in the two Royal Colleges of which he was a Fellow, as well as the DHSS/DES and had a hand in numerous reports. During 1985-86 he visited Pakistan, through the British Council, and advised on the development of child mental health services there, recommending exchange placements and training, of which he saw the first fruits.

Slight of figure, Tony Fairburn loved the countryside, mountains and rock climbing. At home he was a musician, gardened with pleasure, and did his own tree surgery - duly roped. He went ice-skating several times a week. On holidays he took his family far afield on his narrow-boat, north to Ripon and, on one occasion, on to the Thames tideway, with inches of freeboard to spare. In middle life he left the church of his upbringing and joined no other.

Everyone recognized his influence but few realized its extent. Those in Bristol who valued his qualities knew too little of his achievements in Bath; if he were chairing a group of Wessex child psychiatrists, as like as not he would be convening a similar group in the south western region. He trained doctors in psychiatry, paediatrics and general practice but also taught students of many other professions. He supported voluntary organizations and had a regular 'spot' on local radio. People could remark on his absences, being unaware that he was up in London or Pakistan.

Tony Fairburn remained modest, enthusiastic and alert, welcoming newcomers, hopeful amidst change, and always moving forward. He never ceased to enjoy his work and looked forward to a third visit to Pakistan. He had a warning attack at Easter, 1988, but resumed his usual activities and appeared to be fully recovered. He died suddenly in Bath.

IDC

[*Psych.Bulletin*,1989,**13**,104]

FANTHORPE, Roland Henry John *b*.4 Oct 1918 *d*.26 Dec 1984
MB BS Lond(1943) MRCP(1946) MD(1948) FRCP(1969)

Roland Fanthorpe was born in Torquay, the son of a production engineer. He received his medical education at the London Hospital where, being a keen sportsman, he also obtained his colours for cricket and rugby football. After qualifying he held various junior hospital appointments and then did his National Service in the RAMC, 1945-49, becoming a medical specialist with the rank of major. After release he had further hospital appointments, including house physician at the Brompton Hospital, medical registrar at Bristol Royal Infirmary and assistant chest physician at the Edmonton Chest Clinic.

In 1952 he was appointed consultant chest physician in Croydon. To start with he had well over 100 beds, chiefly at the Mayday Hospital, and most of his patients had pulmonary tuberculosis. Later, fortunately, this disease was brought under control and in 1960 he became a general physician but continued his interest in his chest clinic work. He played a full part in all hospital activities, serving as chairman of the medical advisory committee and later as chairman of the medical division. In 1965 he became the first clinical tutor in Croydon and laid the foundation for the present thriving medical centre and library. He was an active member of the Croydon Medical Society and was president from 1980-81.

Fanthorpe was an excellent physician, with good judgement, and was popular with his hospital colleagues and with the local general practitioners. He never did any private practice and his hospital and chest clinic were the centre of his professional life. In spite of his general ability he never published any papers.

He married in 1947 and had two daughters. Unfortunately, he was not happily married and the stresses at home may have contributed to his coronary artery disease, from which he suffered during the last two or three years of his life. This did not interrupt his hospital work and he did not retire until he was 65. Immediately after his retirement he left his home and went to live in Somerset, where he died suddenly only fifteen months later.

His great popularity was demonstrated at his Memorial Service when the Mayday Hospital's chapel was full, with many people standing at the back, on a cold and snowy day in January 1985. I felt honoured by being asked to give the address.

GPB

[*Brit.med.J.*,1985,**290**,863; Photo]

FELIX-DAVIES, Derrick David *b.*30 June 1925 *d.*1 May 1988
MB BS Lond(1948) MD(1952) MRCP(1952) FRCP(1974) FLEX Lic USA 1976

Derrick Felix-Davies was born in High Wycombe of South Wales parentage. His mother, Mary Elizabeth Felix, was the daughter of a draper. His father, John Llewellyn Davies, was a dental surgeon who worked in the school dental service, which resulted in several changes of home for the family. They finally settled in Swansea where young Derrick attended the Grammar School. He gained a junior scholarship to St Bartholomew's Hospital and graduated in 1948. After house officer posts at Bart's, and at the Brompton Hospital, he held appointments as junior registrar in Swansea and Bart's and as medical registrar and tutor at the Postgraduate Medical School in Hammersmith. In 1956 he was appointed senior registrar at the Queen Elizabeth Hospital in Birmingham where he developed his special interest in rheumatology and published original work on rheumatoid factor. In 1960 he was appointed consultant physician to the East Birmingham Hospital, but before taking up this post he spent a year as a Sir Henry Wellcome travelling fellow at the Harvard University arthritis unit in Massachusetts General Hospital, and also a short spell in the New York State University medical school in Buffalo.

In January 1962 he embarked on his life's work as the first full-time consultant physician in the new East Birmingham Hospital, which was being developed on the site of the Little Bromwich Fever Hospital. With boyish enthusiasm and boundless energy Derrick threw himself into the task of establishing a major department of medicine in a thousand-bedded hospital. Overriding numerous administrative and financial difficulties, not to mention the scepticism of some of his colleagues, he largely achieved his objectives within some 15 years. In addition to developing his primary interest in rheumatology, and its associated physiotherapy and rehabilitation, he was instrumental in setting up a cardiology department with ancillary services, as well as postgraduate and undergraduate teaching programmes; and he played a large part in creating one of the first postgraduate centres in the Midlands. In 1966 he extended his consultant commitments to Solihull Hospital where he was responsible for developing a coronary care unit and a new rheumatology clinic.

He was a tall man of handsome appearance, with charm and courtesy, but he was also inherently restless and a compulsive achiever, relishing every challenge in life whether it be intellectual or physical. Having been rejected as medically unfit for military service, he thereafter showed an abiding passion for outdoor sport. His prowess at dinghy sailing and skiing, tennis and squash conceded little to the passing years. There was nothing he enjoyed better than leading hospital parties on *macho* camping expeditions in Snowdonia. He acquired an old cottage in the Mumbles, and personally excavated a mountainside to build a bathroom. He was a connoisseur of art and antiques, with specialist knowledge of Victorian watercolours.

His professional interests were also extensive. Although not a prolific writer, he continued to pursue clinical research, particularly in the development of second-line drugs for arthritis. He was a member of the Heberden Society and the British Society of Immunology, and an honorary senior clinical lecturer in Birmingham University. He was an enthusiastic lecturer and teacher, usually preparing his lectures in the small hours. Widely acknowledged as an astute and caring clinician, he had a busy private practice and also acted as physician for the National Exhibition Centre in Birmingham. He was quick to perceive the usefulness of computers in medicine and set out to master the subject at a mature age. In his later years he also devoted much time to the problems of pain relief and terminal care in Solihull. In 1975-76 he spent a sabbatical year as clinical associate professor of medicine in Buffalo. During a number of subsequent visits to the United States he worked for short spells as a federal medical officer in the Sioux Indian reserves of South Dakota, and became much involved in their social problems.

Derrick was a shy and complex man, who did not readily show his deep emotions. His impulsiveness and eccentricity were exciting, but could sometimes be exasperating. He had few really close friends, but many admirers and no enemies, and he never spoke ill of anybody. For all his competitive drive he was never ambitious for power or influence.

As a family man he was very domesticated, with a flair for cooking. There were two sons by his first marriage to Joan Kellett. In 1970 he married Joan (Anne) Sly, only daughter of the vicar of Hampton-in-Arden, who was a modern language teacher. They lived in Solihull and had three daughters.

Derrick was well known for his fast driving; his end came swiftly in a road accident and deprived him of an old age which he probably would not have enjoyed.

RDM

[*Brit.med.J.*,1988,**297**,418; *The Times* 3 May 1988; Photo]

FITTS, *Sir* Clive Hamilton *b.*14 July 1900 *d.*7 Feb 1984
Kt(1963) MB BS Melb(1926) MD(1929) DTM Syd(1930)
MRCP(1933) FRACP(1938) FRCP(1948)

Clive Fitts, consultant physician, cardiologist, educator, orator, man of affairs and tennis player, was born in Melbourne. His schooling was at Scotch College, Melbourne, and Melbourne Grammar School. He studied medicine at the University of Melbourne during which he was resident at Trinity College. In one appreciation of Clive Fitts, his course through medicine was described a 'blithe and leisurely with billiards, football and tennis at inter-state level taking their proper place... he liked to think that his example inspired those who decided that an extra year should be added to the medical course'. Fitts' final year marks were insufficient to secure him a resident post at the Melbourne Hospital after he graduated in 1926 and he was, as he said, 'banished to the Alfred Hospital'. He was influenced there by a charismatic surgeon, Fay McClure, and become attracted to surgery as a profession. His training in this art was not entirely wasted - T H Hurley, in an appreciation, records that Clive Fitts read a paper 'High Tracheotomy'. It seems that after his return home from London he did locum work and performed an emergency life-saving tracheotomy in the kitchen of a patient suffocating with laryngitis. He accompanied the patient to hospital, keeping the airway open, only to hear a disdainful admitting officer remark that he had made the cut too high in the neck!

The glamour of surgery notwithstanding, Fitts was destined for medicine. In his early postgraduate years he obtained his doctorate of medicine from the University of Melbourne, and the diploma of tropical medicine from the University of Sydney. He travelled to London, as was done in those days, as a surgeon on a merchant steamer by way of Cape Horn, and was detained a month in Holland waiting for cargo before sailing on to England. He enjoyed the museums and galleries in Holland, an experience which inculcated a life long interest in art. In London, he first obtained a hospital position at the Battersea Anti-Vivisection Hospital. Other possible more formative appointments were at the National Heart Hospital, where he was influenced by Sir John Parkinson [*Munk's Roll*,Vol.VII,p.443], and at the Brompton Hospital where he developed his skills as a cardiothoracic physician, and established many lasting friendships. In London he furthered his interest in art and, without much bother, he glided through the examination for membership of the College. He worked also at a sanatorium in Switzerland, so adding expertise in the management of tuberculosis to his other talents. He developed a passion for mountaineering, and gained the satisfaction of contemplating the world from the summit of the Matterhorn.

Back in Melbourne in the 1930s, Clive Fitts was initially appointed to St Vincent's Hospital and subsequently to the Royal Melbourne Hospital, which he served for 20 years. He practised cardiac and thoracic medicine, and in time became Melbourne's most eminent consultant cardiologist.

Clive Fitts would not have regarded himself as a research physician, but he had a keen and influential interest in the development of research in cardiology in Melbourne. It was his concept that a full-time department of cardiology be established at the Royal Melbourne Hospital and, as a result of his influence among private donors, sufficient funds were raised for a department to be created. In 1956 an appointment was made of the first director, with resources for an associated cardiac catheterization laboratory. He also had a substantial influence in the setting up of the National Heart Foundation in Australia in 1961, and this organization has grown to be a highly important source for promoting advances in cardiology in Australia, and for developing research money for investigation into various aspects of heart disease and hypertension. He also sponsored and supported the concept of cardiac rehabilitation units to act as centres of 'resurrection from the scrap heap'. He was knighted in 1963 for service to medicine.

Clive was a noted, albeit somewhat controversial, teacher. His teaching style was impeccable but his reputation among students tended to be mixed, because in his bedside teaching he concentrated as much or more on the patient as on the disease. Theorists and students with an eye to examinations found this not always to their liking. Foremost in his interest in patients was their adaptation to their disease and disability and how this could best be modified or augmented - he enjoyed their foibles. He picked up from some untutored patient the term 'legarthic', which he enjoyed applying to himself when he appeared sluggish on the tennis court! When asked the secret of success as a respected physician, he said that one needed 'to create the illusion of timelessness'.

Clive Fitts became very prominent in community affairs in Melbourne. He was for many years chairman of the Felton Bequest Committee and the National Gallery Society of Victoria. He was a member of the council of the University of Melbourne,1951-67; a Councillor of the Royal Australasian College of Physicians, and vice-president 1956-58. He was a president of the Medico-Legal Society of Victoria. He was also a member of council of Melbourne Church of England Grammar School from 1960-69, and was a well known member of the Melbourne Club and president of the Club in 1965.

Many will remember Clive Fitts' prowess at tennis from the time of his University days when he was a Victorian representative in inter-state matches. He maintained a very active interest in tennis through-out his life and his tennis parties at home in Melbourne, or at his seaside place, were a scene of many tense and hard fought struggles. Clive was quite a figure on the court as he often appeared in the dress of yesteryear - clean white flannels - and never failed to follow his serve quickly to the net, where his volleying was faultless. He was heard to remark that when he died he would like to have crossed tennis racquets over his bier. He also had a great fondness for trout fishing, which he enjoyed with his friend John Halliday.

Clive Fitts' scholarship and his interest in literature were profound. His learning, together with his eloquence and easy wit, led him to be one of Melbourne's most popular after dinner speakers, his address being marked by pertinence to the topic, enriched by many literary allusions, with a few gentle barbs directed towards aspects of the establishment which had happened to displease him. He was a gregarious, amiable, and very friendly man. But it was said by one of his protégés - and many might agree - that he was just slightly 'encapsulated', so that even close associates might wonder how close they had really come to the inner man.

Clive Fitts was married in 1939 to Dr Yrsa Osborne, a daughter of a professor of physiology at the Univerity of Melbourne, and they had two sons and three daughters.

IRM

[*Lancet*,1984,1,522; Photo]

FITZGERALD, Oliver Patrick Noél Alexis *b.*23 Dec 1910 *d.*1 Dec 1987
BSc(1933) MB BCh BAO NUI(1936) MSc(1937) MD(1941)
MRCP(1967) FRCP(1972) FRCPE(1972) FRCPI(1973)

Oliver Fitzgerald was born in Waterford where his father was medical superintendent of the local mental hospital. His mother came of the O'Halloran family of east Clare, one of whom, Sylvester O'Halloran, was one of the great Irish surgeons of the 18th century who hastened the establishment of the Royal College of Surgeons of Ireland.

Oliver was one of six siblings. Of the five brothers, three of whom were medical, all achieved intellectual distinction in their various careers. Being one of a brood of brilliant sibs must have inherent disadvantages but Oliver, being the eldest, quietly established his own independent identity. He entered University College, Dublin, in 1929, and already had an outstanding record in Clongowes Wood College. His student years were decorated with honours and distinctions but he had not yet acquired any terpsichorean skills so that he moved around a ballroom floor with a rugged independence, to the dismay of his partner since it bore little relation to the music.

After he qualified he veered towards the basic sciences (biochemistry and physiology) and in 1937 won the travelling studentship of the National University of Ireland. This academic vehicle, which has unfortunately fallen into disuse, enabled young graduates to travel around Europe in a manner reminiscent of the peripatetic scholars of the Renaissance. He spent a year with Verzar in Basle and so arose a lifelong productive friendship founded on more than gastroenterology. He spent almost a year with Adrian in Cambridge, later Lord Adrian [*Munk's Roll*,Vol.VII,p.3] but, though he contributed to some esoteric papers on cat's whiskers, neurology remained outside his interests.

Fitzgerald was appointed to the staff of St Vincent's Hospital in 1940 and began a productive career in clinical research. In 1958 he was appointed to the chair of therapeutics at University College. He was the obvious choice as chairman of the new National Drugs Advisory Board in 1966, and remained in this post until he retired in 1985. He was also a conscientious member of the Medical Research Council of Ireland (now defunct) from 1957-76.

In manner he was gentle but positive in his views. At controversial meetings, if a harassed member looked to him for support it would be tendered in quiet tones. One could not conceive of him banging the table. When one of his sons was chosen as Lord Mayor of Dublin he was described as being positively coy about it. He was devoid of malice, but he would vividly recall the peccadillos of colleagues with dry humour and a twinkling smile punctuated by subdued chuckles.

Some men's habits or enthusiasms may exceed the bounds of common acceptance, whether it be in smoking, drinking, golfing or gabbling, but Oliver's passion was an admirable love of books - not just reading them, he had to possess them. His library was one of his major achievements. It was catholic in diversity and readily explains his erudition in so many unexpected subjects.

He enjoyed travelling to Great Britain; London held the attraction of friends and better bookshops. He attended the routine meetings of the Medical Defence Union with religious fervour. He welcomed the posts of external examiner in many centres. He was a triple Fellow of the Royal Colleges in London, Dublin and Edinburgh.

The guiding force in Oliver Fitzgerald's life was his family. They were very united, yet each had an independence of mind which both he and his wife, Cliodna, cultivated. Fortunately, the medical tradition continued; their second son is an established rheumatologist and one daughter has also joined the profession.

DKO'D

[Photo]

FORD, Edmund Briscoe *b.*23 Apr 1901 *d.*22 Jan 1986
MA DSc Oxon(1943) FRS(1946) Hon FRCP(1974) Hon DSc Liverp(1964)

Edmund Briscoe ('Henry') Ford died at Oxford where he had worked all his life. He was born at Papcastle, near Cockermouth, Cumbria, the son of Harold Dodsworth Ford and his wife Gertrude Emma Bennett. He was an undergraduate at Wadham College, then a lecturer in zoology and comparative anatomy, and later became an honorary fellow. He was elected a fellow of the Royal Society in 1946, being awarded the Darwin medal in 1954 and the medal of Helsinki in 1967. He was also one of the first scientists to be elected a fellow of All Souls College since the 17th century, where he served two terms as senior dean. From 1946-49 he was president of the Genetical Society of Great Britain.

Henry Ford's interest in genetics led to his appointment as director of the genetics laboratory of the university, 1952-69, and as professor of ecological genetics, 1963-69. His work on natural selection, mainly carried out on butterflies and moths, was outstanding. He demonstrated that what had been purely theoretical predictions actually worked in the woods and the fields, and thus he was the originator of ecological genetics. He was the first to describe and define genetic polymorphism, that delicate balance between conformity and diversity, and he prophesied, correctly, that susceptibility to particular diseases was a factor in maintaining the human blood group polymorphisms, The associations of cancer of the stomach and group A, and of duodenal ulcer and group O, bore this out. In the sickle cell haemoglobinopathy the advantage of the heterozygote was excellently demonstrated as it protected children against malaria. In addition, his investigation of the pigments of *Lepidoptera* was one of the most successful early attempts at relating chemistry to classification.

He had a knack of selecting the right material for a particular investigation and a gift for picking good research workers and then giving them their heads: H B D Kettlewell throve on the peppered moth and industrial melanism; Arthur Cain excelled on the snail *Cepaea nemoralis*, demonstrating that natural selection by predators acted on a colour polymorphism; and Philip Sheppard [*Munk's Roll*,Vol.VII, p.532] concentrated on the evolution of mimicry, particularly in the swallow-tailed butterfly *Papilio dardanus*, on which Ford's monograph was the Bible. But there was more to come: Sheppard applied Ford's suggestion about human blood groups to the Rh (rhesus) system, and with researches at the department of medicine at Liverpool University found a method of preventing Rh haemolytic disease of the newborn. It was for this type of work that the Nuffield Foundation, of which Henry was a trustee, set up the unit of medical genetics there, and it was a nice quirk that in the Rh polymorphism the heterozygote does not obey the rules, for it is always at a disadvantage.

Butterfly collecting hints at eccentricity, but in E B Ford this trait was combined with an extensive knowledge of history, literature, archaeology and religion - as well as genetics - which resulted in writings of great educational value. He also took an enthusiastic interest in heraldry. Further, he contributed much to the Prehistoric Society.

In 1984, with J S Haywood, he produced *Church treasures in the Oxford district*, Gloucester, Alan Sutton. The titles of his other books demonstrate his versatility: *The Study of heredity*, London, Thornton Butterworth, 1938; *Mendelism and evolution*, 6th ed. London, Methuen, 1967; *Ecological genetics*, London,Methuen, 1964, 4th ed. London, Chapman & Hall, 1975; Genetic polymorphism, London, Faber and Faber, 1965; *Genetics for medical students*, 6th ed. London, Methuen, 1967, 7th ed. London, Chapman & Hall, c.1973; *Moths*, London, Collins, 1955; *Butterflies*, London, Fontana, 1975, *Genetics and adaptation*, London, Edward Arnold, 1976, *Understanding genetics*, London, Faber, 1979, and *Taking Genetics into the countryside*, London, Weidenfeld and Nicolson, c.1981. He was still writing at the time of his death. Some of these books went into many editions and every one is characterized by his lucid prose, so different from much modern scientific gobbledygook - he regarded molecular geneticists as incomprehensible interlopers. *Butterflies* proved to be a bestseller and it is not surprising that the ordinary reader devoured it with enthusiasm, for it opened up entirely new aims in butterfly observation and led the amateur painlessly into the mysteries of genetics and ecology.

Oxford has produced many eccentrics, but few in the same class as Henry. Although an endearing character in many ways he had a feline capacity to make his disapproval felt. He was a faithful and generous friend, but he froze those of whom he disapproved, and his attitude to women was exemplified when lecturing to undergraduates: when his audience was of mixed sex he always began 'Gentlemen', and on one occasion when only girls were present he is alleged to have walked out - '... since there is no one here, there will be no lecture.' He was unmarried.

Henry was a stimulating teacher and his breadth of knowledge profoundly impressed the young - who were also fascinated by his unpredictable reactions. He delighted to present himself to students as a terror. When apprehension faded, the imprint which endured was of a passion for poise, for meticulous thoroughness and, above all, for style. We shall never see his like again, and how far evolution can account for such a unique individual is a matter for wonder. CAC/GW

[*The Times*, 23 and 30 Jan 1988; *The Independent* 23 Jan 1988; *Nature*,Vol.**332**,3 Mar 1988,*Memorial Address* All Souls Chapel,12 Mar 1988]

FORD, *Sir* Edward *b.*15 Apr 1902 *d.*27 Aug 1986
Kt(1960) OBE(1945) MB BS Melb(1932) DTM Syd(1938) MD Melb(1946) FRACP(1946) DPH Lond(1947) *FRCP(1958) FACMA FRSH FZS Hon LittD Syd(1971) FRAHS Hon FRCPA

Edward Ford was born in the small country town of Bethanga in north eastern Victoria, Australia, the son of Edward John Knight Ford and Mary Doxford, née Armstrong, and in 1908 came to Melbourne with his family to commence his education. In 1918 he obtained high marks in the public service examination and entered the postal service where he worked until 1928. He wanted to study medicine at the University, but found it necessary to obtain the required certificates for entry. These he obtained by attendance at night classes. He entered the medical school in 1926, still working in the postal service and supporting himself by coaching students at home. He graduated with distinction in 1932. Twelve months were spent in clinical work at the Melbourne Hospital and at the end of 1933 he was appointed Stewart Lecturer in anatomy at Melbourne University, becoming senior lecturer in anatomy and histology from late 1934 until December 1936.

At the anatomy department he came under the influence of Frederic Wood Jones whom he greatly admired, and with whom he shared an interest in physical and cultural anthropology. It was Wood Jones who fostered Ford's interest in book collecting and the history of medicine. Their friendship continued until the death of Wood Jones in 1954.

In 1937 Ford left Melbourne to become lecturer in the School of Public Health and Tropical Medicine at Sydney University, and having obtained his DTM early in 1938 was asked by the Papuan Administration to undertake a survey of native health, particularly malaria and venereal disease, in New Guinea, the Trobriand Islands, Goodenough Island and the D'Entrecasteaux Group. The survey extended from early 1938 to the middle of 1939 and he found this field work absorbing and demanding, but richly rewarding - specially when the area later become involved in war. At the same time he made a study of the cultural heritage of the Papuan people.

On his return to Sydney, he next went to Darwin as medical officer in charge of the Commonwealth Health Laboratory until 1940, when he was released so that he could enlist in the Army, taking command of the 1st Australian Mobile Bacteriological Laboratory, with the rank of major, and being posted to the Middle East.

During the Syrian campaign he worked first at Nazareth investigating cases of malaria, sandfly fever and dysentery. Later, with the occupation of Syria, he moved to Tripoli with a casualty clearing station to provide all laboratory investigations.

Japan entered the war in December 1941 and by January 1942 had captured Rabaul. This direct threat to the safety of Australia resulted in troops being returned from the Middle East for service in Malaya and New Guinea. Ford was appointed assistant director of pathology to the New Guinea Force with the rank of lieutenant colonel, but unofficially acted as its malariologist. Here he made his most important contribution to the problems of malaria in the field among troops at first untrained and ill-equipped to cope with the situation. It was at Milne Bay that the problems became critical. In September 1942 the malaria rate among the troops there was steady at 33 per thousand per week, but later this rose to 82 per thousand, and in the week ending 25 December 1,083 men were admitted to hospital out of a force of 12,000. Had this alarming increase in rate continued the whole force would have been lost in less than three months.

The medical services were unable to obtain adequate supplies as priority was given to combatant stores, and attempts to tighten up poor malarial discipline produced only a weak response. An appeal was therefore made to the commander in chief and, through the DDMS, Brigadier W W S Johnston, Ford obtained an interview with General Sir Thomas Blamey in December 1942. Quite bluntly he pointed out that malarial discipline was bad and should be an integral part of unit discipline, that personal protection must be enforced and breaches of discipline visited by punishment, and that there must be a steady and unfailing supply of all anti-malarial material, especially quinine and atebrin. He pointed out that the medical units had been evacuating 300 men a week for weeks past and that this could not be allowed to continue. He asked for higher priorities for all malarial work, particularly increased quantities of equipment and stores for mosquito control, and for 1000 men for additional labour. Ford received an immediate response from Blamey, and an intense anti-malarial campaign began at once. Discipline was tightened, the wearing of protective clothing and the taking of atebrin in suppressive doses made compulsory. Within a short time the malaria rate started to fall.

In March 1943 Ford was appointed senior malariologist to the Army and, because one was not available, prepared a memorandum on malaria, illustrated by maps of the hyperendemic areas, which was widely circulated to all Australian and American units and to the British Forces in Burma: *Malaria in the south-west Pacific*, Australian Military Forces Publication, 1943. This short, but comprehensive and masterly acccount, was written over the course of three days while he was on leave in Sydney, but his name does not appear as the author. At the request of the British Army he visited Burma to discuss anti-malarial control; his recommendations being accepted without modification.

The director of medicine, Brigadier (later Sir Neil) N H Fairley [*Munk's Roll* Vol.VI p.171] recommended the establishment of a malarial research unit at Cairns and this became a most important centre for all aspects of investigation into the problems of malaria. In June 1944 Fairley convinced a conference of the Australian General Staff of the effective use of atebrin and that unit commanders should be made responsible for the proper dosage to be regularly taken. Orders were immediately issued to this effect, with the result that suppressive measures were taken out of the hands of the medical services and made the responsibility of combatant commanders.

By the end of 1944 malaria was largely under control. This followed dedicated work by all sections of the medical services, but there can be no doubt of the great importance which must be given to the outstanding contribution made by Ford in the field. In the latter stages of the war Ford was director of hygiene, pathology and entomology for the Army, with the rank of colonel. For his work he received the OBE (Mil.) in 1945 and had been mentioned in despatches in 1943.

After the cessation of hostilities Ford returned to Sydney University and, in 1946, was awarded a Rockefeller Fellowship to study at the London School of Hygiene and Tropical Medicine, and in the same year obtained the MD (Melb) with a thesis on malaria and anti-mosquito methods. In 1947 he received the DPH (Lond) and returned to Sydney to become director of the School of Public Health and Tropical Medicine and professor of preventive medicine at the University. These positions he held until 1968, when he retired. He was responsible for many changes in the School; the building was enlarged, the staff increased and all activities expanded. His skill as an administrator was seen both inside and outside the University. From 1953-57 he was dean of the faculty of medicine and a fellow of the senate. From November 1960 to March 1961 he was acting vice-chancellor. He had received a knighthood in 1960.

He took part in the planning of the medical school of the University of Western Australia and was a member of the first council of Macquarie University, as well as being a member of the committee on the future of tertiary education in Australia. He served on the council of the National Health and Medical Research Council, being chairman of two of its committees; on the World Health Organization in the field of medical education and research, and as a member of the New South Wales State Cancer Council. He was a member of the board of the Sydney University Press and, with his friend Sir John Ferguson, was active in the formation of the Friends of the Library of the University of which he acted as president from 1962-72, when he became patron. Sir Edward was active in the affairs of the Royal Australasian College of Physicians to which he had been elected a Fellow in 1946. He served on the editorial committee of the *Australasian Annals of Medicine* (now the *Australian and New Zealand Journal of Medicine*) from 1951-65; on the research advisory committee 1952-63; on the committee on occupational health, and was vice-president of the College 1970-72. In 1958 he became curator of the historical library of the Australasian College and chairman of the library committee, both of which he held until his death. During his curatorship the library grew significantly in stature as an important reference collection in the history of medicine, but especially in relation to Australian medicine, for he had presented to the College his collection of 19th-century medical works by Australian authors which formed the basis for his *Bibliography of Australian Medicine 1790-1900*, published by the Sydney University Press in 1976, a magnificent reference work on which he had worked for many years. He bequeathed his large collection of early printed books on medicine to the College.

Throughout his life he was always interested in the history of medicine and this is reflected in his publications, and in the orations which he was frequently called upon to give. At first his papers dealt with physical anthropology, but after the war they became more biographical and bibliographical, always with the evidence of careful research. In 1950 he wrote 'The malarial problem in Australia and the Australian Pacific territories', *Med.J.Aust.*,1,749-760. This is a classic account of the problem, but his own very significant contributions are not mentioned.

In 1969 he was the first recipient of the Neil Hamilton Fairley medal, awarded by the Royal Australasian College of Physicians and the Royal College of Physicians of London in alternate years, which was presented on the occasion of the Arthur E Mills Memorial Oration, when he spoke of the life of his great friend Sir Neil, with whom he had worked so hard during the war. His last few years were marred by illness which curtailed his activities but not his spirit.

In appearance Ford was tall, with a large frame, and one was surprised at the quiet gentleness which was present. He was a modest, retiring, almost shy person with great common sense and wisdom, but his gentleness masked a firmness and strength of character which only occasionally needed to be expressed. When he spoke in his characteristic, rather hesitant way he commanded instant attention, and this was very evident when on committees or other administrative occasions, for he was fearless and firm in debate. Always delighted to meet his friends, the greeting given to those who were close to him was often boisterous. To all with whom he came in contact he was most courteous, and those who came to him seeking advice or help seldom left without the assistance they sought. He had a keen sense of humour, often shown by a chuckle or a twinkle in the eye, and a sharp wit with an apt turn of phrase or well chosen anecdote. He was a staunch friend, who always showed his never failing kindness and help, particularly in sickness. Very fond of children, his nieces, nephews and godchildren returned his affection to the full, being always delighted to see him and be with him. He had not married.

Generous almost to a fault, he was a great letter writer to his family and many friends; letters which were informative, giving freely of his knowledge, particularly to fellow book collectors, or just giving details of his day to day events. Letters to children were illustrated with sketches, often of flowers for he was keen on botany, never letting an opportunity go by without a visit to the botanical gardens in the cities he lived in or visited. As a form of relaxation, whenever possible he walked to his destination with a slow, deliberate step, never in a hurry.

He enjoyed listening to good music and good conversation, being an excellent host with a keen knowledge of food and wine as well as being a competent cook. He collected the works of many artists whom he knew personally and was a very discriminating bibliophile. He possessed documents or letters signed by Samuel Pepys, Captain James Cook, Charles Darwin, Florence Nightingale and others. At one time he became interested in scrimshaw and taught himself the art of engraving the teeth of whales. Indeed, there were few aspects of art on which he did not have knowledge.

In all he was one of the most remarkable medical men that Australia has produced.

KFR

[RACP *College Newsletter* March 1978,10,1,p.1; April 1979,11,1; Photo]

FORMAN, Louis *b.*15 June 1901 *d.*13 Nov 1988
MRCS LRCP(1923) MB BS Lond(1924) MRCP(1925) MD(1926) FRCP(1939)

Louis Forman was born at Portsmouth, the eldest son of Itzhak and Annie Sarah Fogelman. Brought up in an orthodox Jewish family, Louis was proud of being a Jew whilst regarding religion as a superstition; he was an agnostic with little time for orthodoxy.

He went to Guy's as a medical student in 1918 and did very well, winning the Golding-Bird gold medal and scholarship in 1922. After graduation he was appointed assistant house surgeon to Sir Arthur Fripp from February to April 1923, outpatients officer May-July, house physician to John Fawcett [*Munk's Roll*,Vol.IV,p.437], and was appointed medical registrar 1924-25. He obtained his membership of the College in 1925 and his MD the following year, and was appointed as general physician to St Olave's, and in the fever service.

After an argument with Sir Arthur Hurst [*Munk's Roll*,Vol.IV,p.509], in which Louis accused the great man of tampering with experimental results, he was not appointed to a medical post - and medicine's loss was dermatology's gain. He was appointed chief clinical assistant and registrar to the skin department in 1928, and chief clinical assistant to the actinotherapeutic department 1930-32.

On Geoffrey Dowling's appointment to St Thomas's [*Munk's Roll*, Vol.VII,p.163], Louis Forman was appointed to the staff of Guy's as consultant physician in diseases of the skin and remained in this post until September 1967. After his retirement he continued to come to weekly histology meetings and ward rounds, and was a respected and constant attender at meetings where his encyclopaedic knowledge and wise counsel were much in demand. From his gastroenterological experience he retained a vivid interest in the relationship of the gut and skin, and it remained one of his research interests.

During the war Louis Forman was responsible for dermatological patients at over a dozen hospitals in the Guy's sector; he was sub-dean for students at Farnborough Hospital, and active in the Home Guard with the rank of acting major.

He was president of the British Association of Dermatology and of the St John's Dermatological Society, as well as the section of dermatology at the Royal Society of Medicine. He was also an honorary fellow of the RSM.

A small man with thick glasses, he was a constant figure at meetings, carefully peering at patients, taking exact histories and examining every patient minutely, even when others had given up. He was an excellent teacher and encouraged the young; he was a good listener and made even inexperienced doctors feel they had much to contribute. For years he had an interest in *Demodex* and made an early film of this inhabitant of the human hair folicle. This film is now in the collection of the Natural History Museum, at their request.

Louis Forman had great charm and good humour. He was completely without malice and, although a shrewd observer, always thought the best of everyone. A frugal man, he always went by tube or bus and rarely spent money on himself, except on books and his glass collection; he was a knowledgeable collector of glasses and glass paper weights.

Lolly, as he was known to generations of dermatologists, enjoyed tennis, wrestling, and playing the piano and although in his later years failing eyesight prevented all these he still enjoyed music and concerts. His main 'hobby' however was dermatology and in the last few years a devoted friend read to him for three or four hours daily. Until shortly before his death he was more 'up to date' than anyone else, thanks to constant reading and a remarkable memory, complementing his own experience. Lolly enjoyed holidays - as long as they were short, and he particularly liked coming back from them. As a young man he had enjoyed fast cars - and had owned some.

SA

[*Brit.med.J.*,1988,**297**,1604; *Guy's Hosp.Gaz.*,Feb 1989; Photo]

FRANCIS, Reginald Sterling *b.*21 Jan 1921 *d.*5 Sept 1984
MRCS LRCP(1943) MB BS Lond(1943) MRCP(1949) MD(1951) FRCP(1971)

Reg Francis died of a cancer borne privately and with great fortitude for a number of years. Informed of the diagnosis quite early on, he and his wife Pamela reached a brave decision not to let this distort or upset relationships among family and friends, who were accordingly protected from the knowledge until fairly late on. Two decades earlier he had borne the protracted ill effects of a parathyroid tumour, undiagnosed for years, with a similar unassuming patience. He was in this an exception to the rule that doctors make bad patients.

Reg was the youngest child of a Merthyr Tydfil fishmonger and greengrocer who took over the premises of a jeweller called 'Sterling'; hence Reg's middle name. He won a place at University College Hospital at a very young age and then set off on a bicycle to France and Germany to spend a year learning the languages. He qualified at the age of 22 and then served in the Navy in Normandy and the Far East as surgeon-lieutenant. After holding junior hospital posts he was appointed consultant physician at Whipps Cross Hospital in 1962. In 1967 he added the duties of director of the North East Thames Region mass X-ray service.

He taught himself the discipline of research. Throughout his professional career he continued to produce original work. He wrote a series of papers on surgery for pulmonary tuberculosis, antibiotics for chronic bronchitis and corticosteroid aerosols for asthma, and he was still at work on this last subject at the time of his death.

Reg was a man of great resource and energy. He brewed his own beer, made his own wine and baked his own bread. A government minister on a consumer affairs 'phone-in' was startled by a doctor ringing to ask what he was going to do about grain wholesalers who refused to sell bulk flour to a private citizen. He undertook feats of carpentry and plumbing which were, at times, of heroic Heath Robinson ingenuity. He fitted central heating in his large three-storey house with radiators from scrap yards, ably assisted by his five year old son, putting pipes in place beneath the floor boards, and later built his own swimming pool - during the only April in memory to suffer from an appreciable fall of snow. In an age when the privilege of being a junior hospital doctor did not extend to a living wage he became an adept with a cobbler's last and a chimneysweep's brushes. His originality was tempered with economy, but never with parsimony and usually with realism. On one occasion he decided on a plan for savings on dog food for his beloved Norfolk terrier, Packly. He visited the local Home of Rest for Horses and enquired of a military-looking gentleman whether there was ever any horsemeat to be bought cheaply: the reply was in the negative and unprintable. Characteristically he told the story against himself, but it emphasized a principle he lived by and also taught to his family: 'Always ask questions, people can only say "No".' He was never afraid of appearing ingenuous in enquiry if something useful could be gained or learned.

He often said that had he not been a doctor he might have been a musician. Much to his wife's embarrassment he took his trumpet to dances and joined in, playing a fine 'Stormy Weather'. On the piano he played everything in F sharp, explaining that when he was learning he had found the black notes easiest to pick out. He could whistle a remarkable trill and no one ever worked out how it was done. But his lasting interest lay in the music of Mozart and Schubert.

Reg had an admirable disrespect for authority. It was a point of honour with him always to wear an outrageous red bowtie on his first day in a new job. He was an adept mimic, particularly of his chiefs. And it was entirely in character that when confronted with the College's rather presumptuous biographical details form (now reworded) he struck out the words 'Preparatory' and 'Public' referring to schooling and replaced them with 'Elementary' and 'Secondary'.

He was always a fervent advocate of the National Health System, although in later years he was never sentimental about it, recognizing its shortcomings and its problems. He served on many committees with much forebearance, and was particularly active in the planned development of Whipps Cross. He maintained an unflagging interest in the evolution of such hospitals and in the medicine practised in them. The Wellcome Museum of Medical Science contains many of the medical antiquities from his large and quixotic collection.

Reg married Pamela Valerie Spratt, daughter of a Lloyds broker, in 1943 and they had five children, two sons and three daughters. They were a very happy family and their hospitality will long be remembered with pleasure by all who were fortunate enough to enjoy it.

His wife and children survived him.

PDBD/DF

[*Brit.med.J.*,1984,**289**,1011]

FRANKLIN, Alfred White *b.*2 June 1905 *d.*20 Sept 1984
MRCS LRCP(1929) MRCP(1932) MA(1933) MB BChir Cantab(1933) FRCP(1942) Hon FCST

Alfred White Franklin, nicknamed 'the bishop' by those who knew him well, was born in London, the son of an otorhinolaryngologist, Philip Franklin and his wife Ethel Julia. He was educated at The Hall, Hampstead, and Epsom College. He then gained a scholarship to study medicine at Clare College, Cambridge, pursuing his clinical studies at St Bartholomew's Hospital, London, where he was Lawrence scholar and a gold medallist. From 1934-35 he was Temple Cross research fellow at Johns Hopkins Hospital, returning to Bart's as assistant physician to the children's department.

In 1937 he was appointed assistant to Alan Moncrieff, later Sir Alan, [*Munk's Roll*,Vol.VI,p.343] at Queen Charlotte's Maternity Hospital, and became one of the earlier neonatologists working in this country. In a sense he was a progenitor of what has now become a major sub-specialty. During the war years he was paediatrician to Sector 3 of the Emergency Medical Service, working at Hill End and St Alban's Hospitals. In 1942 he was elected a Fellow of the College, and in 1945 he was appointed physician to the children's department at Bart's, becoming physician in charge of the department in 1965 under its new title, the department of child health.

Alfred White Franklin had a natural affinity with children. He had a quiet, reassuring approach which gained their trust and affection. As a teacher he was sympathetic and stimulating. His teaching sessions were held in the playroom of the Kenton Ward at Bart's, where students were offered an object lesson in his concern for the total needs of sick and handicapped children. Some twenty years or more ago, students and their child patients were more apprehensive in each others' presence than they are today and Alfred's style of teaching put both sides at their ease. His friendliness and gentle voice made it possible to conduct clinical consultation and demonstration without distress to the young patient. He also had an intuitive understanding of the problems and distress, both emotional and practical, suffered by the family of a sick - perhaps mortally - or handicapped child. To such families, and particularly to the parents, he gave not only his skill and wisdom but also the sympathy, counselling and understanding which they needed so badly. He felt deeply for mothers who had lost a baby, and for the parents of a newborn child with a congenital defect, and would spend long hours of a busy and tiring day consoling and advising those who needed his help.

White Franklin was on the Council of the College from 1966-69, president of the British Paediatric Association from 1968-69, president of the British Society for Medical History from 1974-76, and president of the International Society for the Prevention of Child Abuse and Neglect from 1981-82. He guided the Invalid Children's Aid Society through difficult years and during its development of the dyslexia centre and schools for boys with asthma. When the full extent of child abuse became apparent in the late '60s, White Franklin became the acknowledged doyen of those working in that field in this country. He took a prominent part in the Tunbridge Wells symposium and was instrumental in holding together the various professions - doctors, social workers and lawyers, whose collaboration was essential if problems were to be resolved in the best interests of the child, the family and the State. Alfred recognized that it was necessary for paediatricians to become knowledgeable and active in their diagnosis of children suffering from abuse and neglect and to seek ways of preventing it. His capacity for organization, his personal authority and gift for writing, were all widely used to promote meetings for postgraduate education in this field.

Alfred White Franklin was also the founder, with W R Bett, of The Osler Club of London, and was Osler orator at the College in 1971. He was a friend of the Osler Library of the History of Medicine at McGill University, Montreal, from its beginning. Over the years his interest in Osler, in the Osler Club and the Osler Library, never flagged.

After his retirement Alfred edited several books on child abuse, a subject which particularly interested him, and also on medical history, including a biography of William Osler. His own writings included *Widening Horizons of child health...*, Lancaster, MTP Press, 1976 and *Pastoral Paediatrics*, Tunbridge wells, Pitman Medical, 1976.

Alfred White Franklin was a short man of portly build, always well dressed in the traditional style of the teaching hospital consultant. His appearance, his mannerisms and even his sly sense of humour had their roots in a past age, but he was a shrewd judge of men and affairs and was always receptive and productive of new ideas, even into old age. His intellect was outstanding, linked to kindliness and humility, and a fierce determination to do what he considered to be right.

He had married Ann Vaisey in 1943, daughter of Francis Vaisey, a clerk in Holy Orders, and they had four children - two daughters and two sons. He delighted in his family and considered them the crowning adornments of his career. His wife and family survived him. GW/VL

[*Brit.med.J.*,1984,**289**,1082,1703; *Lancet*,1984,**2**,880,994; *The Times*,4 Oct 1984; *St Bart's Hosp.J.*,1970,**74**,325-26; 1967,**71**,242; Photo]

FRANKS, Raymond Basil *b.*28 July 1917 *d.*19 Oct 1981
MRCS LRCP(1941) MA MB BChir Cantab(1941) MRCP(1947) FRCP

Raymond Franks was a very astute and competent physician whose talents were never fully recognized. Soon after qualifying he joined the RAMC, serving from 1942-46, first as a junior company commander with a field ambulance and later as a regimental medical officer with infantry and armoured battalions. He also saw brief service in India and Egypt. On demobilization he was caught up in the senior registrar bottleneck of the 1950s and went to St Helier Hospital, Carshalton. It was about that time that he developed diabetes, the complications of which were to dog him for the rest of his life. He decided that he would be wise to opt for geriatrics, rather than the physically arduous life of a general physician, and in this capacity he went to Worthing in 1961 as a consultant physician in geriatric medicine to Worthing Health District and West Sussex County Council.

Raymond was a very good organizer and a very valuable member of hospital committees. He was adept at making trenchant criticism without giving offence. He also lent his skill to various charitable organizations, particularly to St Barnabas Terminal Home, which was one of the earliest institutions of its kind.

In 1952 he married Ulla Margareta Douhan, daughter of Emil Douhan, and they had four children: two sons and two daughters. His family occupied much of his attention. During the winter he enjoyed shooting and he was a very keen and accurate shot. He also enjoyed fly-fishing and foreign travel. Sadly, he gave up playing the piano - which he did by ear, with great skill and entertainment. In spite of retinopathy, a stroke, and increasing arterial disease in his legs which led to double amputation, he remained remarkably resilient and continued to maintain his interests.

Raymond was a man of many acquaintances but few friends. Nevertheless, he was always generous in his advice, time and expertise, when it was required.

JFPS/GW

[Photo]

FRASER, Peter Kennedy *b.*25 Mar 1909 *d.*4 Nov 1984
MB ChB Glasg(1931) MD(1940) MRCP(1964) FRCPath(1964) FRCP(1972)

Peter Fraser was born in Glasgow, the son of Alexander Fraser, a merchant, and his wife Mary Stewart. He was educated at Allen Glen's School and Glasgow University, where he graduated in medicine in 1931. He had house appointments at Bradford Eye and Ear Hospital, Victoria Infirmary, and the Royal Hospital for Sick Children, Glasgow, and then joined the RNMS in September 1934 - being a lively and amusing member of a batch of ten new surgeon lieutenants under training at Haslar. Details of Peter Fraser's early appointments are no longer obtainable from the Ministry of Defence, but it is probable that his first ship was *HMS Hood* and he subsequently served on the China station, where his ultimate interest in the microscope and the laboratory was foreshadowed by an informative article on the diagnosis of tinea infections. In 1940 he was awarded the Gilbert Blane medal for his work.

Fraser was promoted to surgeon lieutenant commander in 1939 and appointed to the Royal Naval Hospital, Bermuda, where he remained for several years. On completion of the normal spell of duty he was reappointed and promoted to surgeon commander in 1945. In 1947 he returned to Britain for a course at HM experimental station at Porton. Following a brief spell at sea in *HMS Victorious* he returned to Haslar in 1950 as senior specialist in bacteriology. He was appointed senior medical officer to HMS Collingwood in 1953. He had always been an avid reader and an enquiring epidemiologist and this appointment gave him ample demographic material, much of which he correlated and printed.

Peter then spent a brief period in *HMS Triumph* and later returned to Haslar as senior specialist in pathology, and medical officer in charge of the zymotic section in 1955. During this time he had articles accepted and published in both the *BMJ* and the *Lancet* on influenza, adenovirus infections and Q fever. In 1958 he was again promoted, this time to surgeon captain. He gained the Parkes Memorial prize and the Erroll-Eldridge prize in 1960. He then served for three years at RNH Malta before being appointed to RNH Haslar as professor of medicine and consultant in pathology. He was made an honorary physician to HM The Queen in 1965.

In March 1966 he retired from the Navy, but not from medicine as he continued to work in Ipswich as director of the public health laboratories and consulting bacteriologist to the Regional Hospital Board. Unfortunately, a cardiac degeneration began to impose restrictions on his activities and after treatment in intensive care units he was advised to retire from official appointments. This was much against his will, but he continued with part-time medical work in general practice. He enjoyed doing this, in Felixstowe, until two years before his death when he became physically unable to continue working and reluctantly accepted retirement. During his last years he continued to read voraciously.

Many Naval and civilian colleagues will have happy memories of 'PK' during his time in the Portsmouth area, and of his wife Doris, née Haigh, and the cheerful hospitality at their house in Lee-on-Solent. They had two daughters, Rosemary and Susan. Peter was a sagacious

Scot with a strong character, a lively sense of humour and a friendly personality.

EDC/GW/VL

[*J.roy.nav.med.Serv*,1985,**71**; *Lancet*,1965,1,1286; *Glasgow Herald* 12 June 1985; *Medical News* 25 June 1965; Photo]

FREW, *Sir* John Lewtas *b.*10 Sept 1912 *d.*8 May 1985
Kt(1981) OBE(1976) MB BS Melb(1935) MD(1938) MRACP(1938) MRCP(1938) FRACP(1951) FRCP(1960) PRACP(1972-74) Hon AM Singapore(1971) Hon FACP(1973) Hon FAMA(1980)

John (Jock) Lewtas Frew, consultant physician to the Royal Melbourne Hospital, Australia, was born in Melbourne, the son of Joseph Davidson Frew, master mariner, and his wife Charlotte Lewtas Neale. He was educated at Camberwell Grammar School and Scotch College, then at Melbourne University and Ormond College, graduating in 1935. He pursued his clinical studies at the Royal Melbourne Hospital and began an association which was to last until his death 53 years later. It was only to be interrupted by the second world war when he served as a captain in the AAMC 1941-45, being assigned to the 13th Army General Hospital, and by the short period he spent working with the late Lord Rosenheim [*Munk's Roll*,Vol.VI,p.394] at University College London. He saw war service in Malaysia, Burma and Thailand, spending a period as a prisoner of the Japanese after the fall of Singapore. In collaboration with Lord Rosenheim, he published an important paper on the labile neurogenic component in hypertension, *Clinical Science* 1949,7, 217-229. From 1948 Jock Frew served the Royal Melbourne Hospital, and its clinical school of the University of Melbourne, without further interruption.

With the hospital and its clinical school he held successively the posts of RMO, medical superintendent, honorary physician to outpatients and then to inpatients, secretary and chairman of the medical staff, sub-dean of the clinical school, member of the faculty of medicine, lecturer in therapeutics, member and vice-president of the board of management, and finally president for six years.

He was an active member of many organizations, notably the Royal Australasian College of Physicians. His association with the College began when he was appointed to the board of censors in 1954. In 1961 he was elected to the Council, and he became censor-in-chief of the College in 1966. During his four years in this office he introduced the first multiple choice question papers which have now become an established part of the examination. In 1970 he was elected vice-president. He had a deep interest in the teaching of medicine in South-East Asia. In 1972 he was elected president of the College and served for two years.

Jock Frew also served the Repatriation Commission for many years as visiting physician to Heidelberg Hospital and was an active contributor to its postgraduate teaching programme. As a member of the government of Victoria's medical salaries (Dillon) committee his work led to significant developments in hospital services in the State. He was a member of the Australian drug evaluation committee, and for many years its chairman. He was a councillor of the Victorian branch of the Australian Medical Association and honoured by this body, by election to the Fellowship in 1980.

For his work under the Colombo Plan, as a postgraduate teacher in medicine, he was elected an honorary member of the Academy of Medicine in Singapore in 1971. He was made an honorary Fellow of the American College of Physicians in 1973 in recognition of 'his outstanding contribution in the health service of his country'. In 1976 he was awarded the OBE, and in 1981 he was knighted for his services to medicine.

Jock would have admitted that he was a workaholic. After Friday meetings in Sydney or Canberra he would fly back to Melbourne, visit his patients in various private hospitals, and chair one or other of his many committees before going home. He remained thoroughly up to date in his knowledge of medicine despite the size of his practice, his advancing years, and his many professional and community activities, and was always very carefully prepared for committee meetings. He was a thoroughly competent professional, who saw his own knighthood as a recognition of the role and status of the consultant physician. Believing that a physician is called to serve, he willingly gave his time and energies to many professional and community bodies and activities.

In these days of highly specialized medical practice, Jock Frew may have been almost the last of a generation of great Australian consultant general physicians. His wisdom in clinical practice, his humour and humanity, the extent of his knowledge together with his quickness and penetration of intellect, and his slightly unusual cast of mind, are quite irreplaceable. Even when mingling easily in a group of men there was something in his quality of mind which set him a little apart from his fellows. He was a gifted clinical teacher and spent much time in undergraduate and postgraduate teaching, almost always at the bedside. In later years he concentrated especially on preparing young postgraduates for the examination of the RACP. His skill as a physician was evinced by the large number of doctors, and members of doctor's families, who were his patients. His understanding of people resulted in many seeking his advice, including members of the nursing and administrative staff at the hospital. He was one of a group of outstanding Australian doctors whose entire hospital service was without any financial reward, and he firmly believed in the value of honorary service.

Jock enjoyed rugby football, being a Melbourne and Australian University blue, but it was of cricket that he talked. Much of his leisure was spent watching first class matches at the Melbourne and Sydney cricket grounds. He was also a keen amateur photographer. But the great concern of his life was medicine and all that went with it.

In 1940 Jock married Joy Bell, who had graduated in medicine two years before him. They were mutually supportive and their marriage brought peace and happiness to them both, and to their son John. Joy's invaluable contribution to Jock's welfare and achievements was fully understood and appreciated by him. To be a privileged guest at the Frew's home was to be among friends, to be wined and dined royally, and to enjoy the relaxed atmosphere and spirited conversation. SG?/GW/VL

[*Med.J.of Aust.*,1985,**143**,467-68; *Minute of Board of Management,Royal Melb.Hosp.*,1985; *Convocation Program of Amer.Coll.Phys.*,9 April 1973; *College Newsletter* RACP,**4**,no 9; AMA Gaz.,Feb 1981,**34**; Photo]

FRIEDMAN, Max *b.*17 Dec 1931 *d.*8 Dec 1987
MB BCh Wits(1958) MD(1966) MRCPE(1961) PhD(1968) MRCP(1971) FRCPE(1971) FRCP(1977)

Max Friedman's career covered a remarkable range of medical practice and research. He was born and brought up in South Africa and studied medicine at Witwatersrand University. After his first house posts he came to London in 1960 where he quickly established himself on the academic side of adult clinical medicine, initially at the Hammersmith with Cuthbert Cope [*Munk's Roll*,Vol.VI,p.118], then at University College Hospital with Charles Dent [*Munk's Roll*,Vol.VII,p.148]. It was at this time that his interest in paediatrics was first stimulated. In 1966 he was appointed as a senior registrar in the paediatric department at UCH. Having made the switch from adult medicine, he never looked back on this decision and quickly grew into an enthusiastic and highly skilled paediatrician. From 1968-69 he was an MRC travelling fellow in paediatric endocrinology at Johns Hopkins. On his return to the UK he was appointed consultant at the Whittington Hospital and set about creating an extremely effective paediatric centre which became renowned all over the world for the quality of the medical care provided, and for its excellence as a centre of undergraduate and postgraduate teaching. Generations of medical students at University College Hospital owe a substantial part of their paediatric teaching to Max Friedman, and many became imbued with his infectious enthusiasm for the subject. Indeed enthusiasm, vitality, and an incredible capacity to see things through whatever the effort required were the hallmarks of all Max Friedman's activities, whether professional or personal. In his early life he faced many difficulties; he had to work nights as a plumber to finance his medical studies. On qualification as a doctor he is quoted in a Johannesburg newpaper interview as saying: 'It is surprising what one can do in 18 hours a day if one organizes and wants something badly enough.' He never stopped working those 18 hours.

Max made a number of important contributions to metabolic medicine and endocrinology and, in particular, observed that the use of ACTH as a therapeutic agent in children caused much less retardation of growth than equivalent doses of corticosteroids. But above all, it is as a master of clinical medicine that he will be remembered. He cared very deeply, indeed passionately, for the well being of his child patients and fought vigorously on their behalf against the restrictions in services which darkened his last years in the NHS.

Faced with mounting evidence of coronary insufficiency, he retired from his hospital appointment in 1985 while continuing with a diminishing commitment to part-time private practice. Those who knew him understood very well that in his short life Max lived twice as intensely, and accomplished at least twice as much, as any ordinary man. He married Clarice Ann in 1958 and they had three sons. They all survived him.

LBS

[*Brit.med.J.*,1988,**296**,798; *Lancet*,1988,1,252; *The Times*,30 Jan 1988]

GAISFORD, Wilfrid Fletcher *b.*6 Apr 1902 *d.*24 Mar 1988
MRCS LRCP(1925) MB BS Lond(1926) MD(1928) MRCP(1928) FRCP(1940) MSc Manch(1951)

Wilfrid Gaisford was born in Somerset into a seafaring family. His father was a captain in the Royal Navy, who served through the first world war, and love of the sea, of sailing and of the west country remained with Wilfrid throughout his life.

After early education at Bristol Grammar School, Wilfrid qualified in medicine from St Bartholomew's Hospital, London. During these years he became a distinguished rugby player; starting as a schoolboy, playing for Somerset over 20 years, for Bart's, the British Lions - touring South Africa in 1924 - and the Harlequins.

It was during resident posts at Bart's that his interest in paediatrics was aroused, and he became resident medical officer to the East London Children's Hospital from 1927-31. A year of this time was spent in the USA as assistant resident at the Children's Hospital, St Louis, and he was then appointed deputy medical superintendent at the Alder Hey Children's Hospital, Liverpool, where he broadened his paediatric experience before his appointment as physician and paediatrician to Dudley Road Hospital, Birmingham, in 1934. His research interests were nurtured and encouraged by Sir Leonard Parsons [*Munk's Roll*,Vol.IV,p.588], one of the doyens of paediatrics in that era, and he was able to maintain them when he moved to Warwick in 1941 to the new post of consultant paediatrician and physician to Warwickshire County Council. This new appointment gave him scope to place the interests of the sick child before all other considerations. It also brought him into contact with the Czechoslovak hospital sector at Warwick. Applications to follow his father into the Royal Navy had been rejected and he was placed in charge of the resuscitation units at Dudley Road Hospital, Birmingham, Warwick and Stratford, where the Czechoslovak Forces had their headquarters. For his interest and encouragement to the Czechoslovak refugees he was awarded the military Medal of Merit by the Czechoslovak government, and he also gained a long and fruitful association with Aron Holzel [*Munk's Roll*,Vol.VII, p.274] whom he appointed to his staff in Warwick and took with him as his first assistant to Manchester. Holzel later became a professor.

It was with some reluctance that in 1947 Wilfrid Gaisford accepted the invitation to take up the foundation chair of child health in Manchester. He held strongly to the view that children should not be admitted to adult wards but that appropriate nursing and technical facilities should be made available to them in paediatric wards. Having recognized the important contribution of a friendly ambience to the recovery of the sick child, he elected to establish his main professorial paediatric unit at the Royal Manchester Children's Hospital, Pendlebury, with an academic neonatal unit at St Mary's Hospital, Whitworth Park, adjacent to the Manchester Royal Infirmary. He encouraged the development of paediatric surgery and neurosurgery, and the appointment of a full-time paediatric surgeon at Pendlebury instead of a series of part-time surgeons. These ideas are now accepted as common practice but were achieved by the considerable efforts of the early postwar paediatricians and foundation professors, such as Gaisford.

With the interest and comfort of the child ever in mind, Gaisford insisted on warm hands for patient examination (a hot water bottle was carried on his own ward rounds) and never more than two attempts at venepuncture were permitted. He was technically highly skilled and would himself undertake procedures after one unsuccessful attempt by a member of his staff. Before oral preparations of penicillin were available, 'Stoss' therapy (a single or two high dose injections) was used whenever appropriate, and painful antibacterial injections were prescribed dose by dose. Intrathecal treatment, so dreaded by children with tuberculous meningitis, was abandoned immediately effective oral alternatives became available, and this course was also followed for non-tuberculous meningitis. The strict and limited visiting hours of that time were relaxed on his wards. His frequent ward rounds, much enjoyed for their fun and kindness by his patients, and his 'green-fingered' medicine, rendered his wards happy ones in spite of the many longstay conditions with which they were filled.

Although Gaisford's principle commitment was to the sick child and the practice of paediatrics, the preventive aspects of child health were of considerable concern to him. He played a major role in association with Wallgren [*Munk's Roll*,Vol.VI,p.447] in the introduction of BCG (bacillus-Calmette-Guérin) vaccination in infancy, and with Fanconi and others in infant health and nutrition. He initiated one of the earliest paediatric tumour registries in this country, was actively involved in promoting immunization programmes, and fostered research in the newborn into the problems of immaturity, including neonatal hyperbilirubinaemia and retrolental fibroplasia. He was author, editor and adviser on a number of books and paediatric journals.

One of his greatest contributions was in teaching both undergraduate and postgraduate students, where his wit and enthusiasm, and emphasis on the prime importance of the child's care and comfort, were an inspiration to many who later followed a paediatric career.

Apart from his prowess on the rugby field, Wilfrid Gaisford's interests were far ranging and included sailing, which was accessible off the Lleyn peninsula in North Wales and in Cornwall, and gardening - which he continued in the beautiful Cornish home to which he retired, although sadly hampered in his later years by osteoarthritis.

In 1935 he had married Mary Guppy and the hospitality of their family home in Bowden, Cheshire, was legendary among friends, colleagues and paediatricians from many parts of the world. This continued in retirement when former staff and colleagues were frequent visitors. They had one son and four daughters, including a pair of twins, and in spite of having heavy commitments to the University and writing, and being available to visit or advise on sick patients day or night, Wilfrid Gaisford entered fully into his family's life and interests. Sadly one of his daughters, who became a nurse at Great Ormond Street, died eight years before him.

JMS

[*Brit.med.J.*,1988,**296**,1338; *The Guardian*,21 Apr 1988; Photo]

GANADO, Walter *b*.21 Apr 1907 *d*.[?] 1987
BSc Malta(1927) MD(1931) MRCP(1945) BSc Lond(1947)
FRCP(1964)

Walter Ganado was born in Malta. His father, William Ganado, was a captain in the Royal Malta Artillery and his mother, Bianca, was the daughter of Sir Filippo Sciberras, one of the foremost political leaders of Malta at that time. In 1935, Walter married Carmelina, née Pace, whose father had been Treasurer to the government of Malta, and they had two daughters, Maria and Anna.

Walter was educated at the Lyceum and the Royal University of Malta, where he obtained his BSc in 1927, followed by his MD in 1931. Soon after qualifying he spent a year in London as a postgraduate student at the London Hospital and the National Hospital for Nervous Diseases in Queen Square. On returning to Malta he undertook his clinical duties at the Central Hospital, where he was assistant to the professor of medicine. During the war years, 1939-45, he was assistant physician to the Emergency Medical Services and later acting professor of physiology; receiving honourable mention for war service to the Crown.

In 1945 he spent another two years in London, obtaining his membership of the College at his first attempt, and undergoing postgraduate training at University College London where he obtained a BSc in physiology. In 1947 he was appointed professor of physiology and biochemistry at the Royal University of Malta and physician to St Luke's Hospital, the teaching hospital of the Island. In 1963 he became professor of medicine and the senior physician in Malta. A year later he was elected to the fellowship of the College.

Walter Ganado was a stimulating teacher and physician. He was remarkably good company and very popular with students and post-graduates. He was an irrepressible conversationalist, which on occasions was exhausting. He established the specialty of neurology in Malta, and it was in this area that he had his main research interest. He was a man of wide culture, with a profound knowledge of the history of Malta, the arts and classical music. He was also very much involved in medical politics, holding leading positions in the Medical Officers Union, which later became the Medical Association of Malta. Walter had a complex and controversial personality. He played an important role in both the medical and academic fields in Malta for over 30 years, and remained active throughout his retirement until a few years before he died.

FFF

[Photo]

GAULD, Edgar Norman *b.*21 Jan 1915 *d.*19 Sept 1986
MB ChB Aber(1937) MRCP(1950) FCCP(1955) FRCPC(1972) FRCP(1984)

Norman Gauld was born in Aberdeen, the second son of William Alexander and Elizabeth Paterson Robertson Gauld, and died suddenly at his home in Hamilton, Ontario.

He was educated at Aberdeen Grammar School and the University of Aberdeen at a time when the chairs of medicine and surgery were held by Stanley Davidson [*Munk's Roll*,Vol.VII,p.136], later Sir Stanley, and James Learmonth, later Sir James, both dedicated and inspiring teachers and both to become doyens in the Scottish medical scene. He graduated in 1937, and after a year of resident posts in Aberdeen Royal Infirmary, which had just moved to its prestigious site at Foresterhill, he went to the Bolingbroke first as house physician and later as resident in medicine. Here he came under the influence of Ernest Lloyd, Purdon Martin [q.v.] and Zachary Cope (later Sir Zachary) all of whom were to have an enormous influence on his career and on his life. He stayed at the Bolingbroke throughout the 1940 air raids on London, and thereafter joined the RAFVR, reaching the rank of squadron leader and serving in this country with a night fighter squadron, and overseas for a time in an Air Force hospital in the Bahamas.

Returning to the Aberdeen Royal Infirmary, following demobilization, as a post-Services registrar for further training in general medicine, he took the London membership and like so many of his contemporaries emigrated in 1952 to Canada to a post, obtained for him by Ernest Lloyd, at the Mountain Sanatorium on the outskirts of Hamilton. From there he sat the Ontario State qualifying examination, and after postgraduate courses at the Bellevue Hospital in New York, and at Duke University in North Carolina, he became a fellow of the College of Chest Physicians in 1955. In the following year he was appointed consulting physician to the Hamilton Civic Hospitals, and in 1960 he became head of department of medicine in the Chedoke General Hospital, with an attachment to McMaster and a teaching commitment there.

In his early fifties he suffered a relatively minor myocardial infarction but was left with angina which greatly restricted his activities, especially during the winter months. He was advised to have coronary bypass surgery, at that time in its infancy and regarded as a hazard, and this was carried out successfully in the Cleveland Clinic in Ohio in 1968.

His interests were in cardiology, but he was a very sound general physician whose opinion was highly valued. He was a man of integrity and a man who brought a sense of calm to the emergency situation. His standards never varied and he earned the respect of all with whom he worked. Alan McNabb, chief of medicine at the Hamilton Civic Hospitals, wrote: 'Norman was not only a professional colleague but a close personal friend, and gave me good advice and strong support over the years. He was a staunch supporter of all hospital and educational programmes, and was considered an exceptionally good clinical teacher at the bedside. He was an active consultant who in later years did more private practice, looking after a large number of personal friends and patients of long standing.' He had a large measure of common sense, and his warm human understanding ensured that those who came under his care were treated kindly and sympathetically: his rapport with patients was excellent, later enhanced no doubt by his own health problems.

He enjoyed work and he enjoyed life. He had a wide circle of friends both inside and outside the profession, and nothing gave him greater pleasure than dining and wining with them and their wives. He was a good conversationalist and a gifted raconteur: a story lost nothing in the telling when he was in full flight, and he seemed to have the right riposte for most occasions.

He took a great interest in all sporting activities, and by nature he was a participant and not a spectator. He was blessed with a good eye for ball games, and at school and university he excelled at cricket and tennis. Subsequently he took to golf and, without passing through the happy hacker stages, quickly reduced his handicap to single figures. When stationed in the Bahamas he partnered the Duke of Windsor on the golf course on several occasions, and he played in the RAF golf team just before he was demobilized. He maintained his low handicap into his late sixties - a very good advertisement for cardiac surgery.

In 1956 he married Frances Jefferson Scarlett, the daughter of Emmet Scarlett, MD, an eminent family physician in Hamilton. They had no family but were very happy and there is no doubt that but for her staunch support and devoted care in the late 60s, and afterwards, he would not have survived to celebrate his 71st birthday. She, too, died suddenly some two weeks after his cremation service. Perhaps she wished it that way.

WRG

[*Brit.med.J.*,1987,**294**,381; *Lancet*,1987,**1**,285; Photo]

GAUNT, Raymond Tom *b.*17 May 1912 *d.*9 Oct 1985
MSc Sheff(1934) MB ChB(1936) MRCP(1947) MD(1948) FRCP(1968)

Raymond Tom Gaunt was born in Sheffield where his father, Tom Gaunt, was managing director of a well known firm which manufactured small tools. He was educated at King Edward VII School Sheffield, going on to Sheffield University and Medical School where he graduated, gaining the Holdroyd gold medal. He held house appointments at the Sheffield Royal Infirmary, and was clinical physiologist from 1937-39. Prior to the war, Tom (as he was known) joined the RNVR. In August 1939 he was posted to Malta, where he remained until 1943. He was then posted to Stornoway, in the Hebrides, and later spent two years in Melbourne, Australia. He had married Ruby Anne Phelan in 1939 and she went to Malta with him, and worked there throughout the siege.

After the war, having attained the rank of lieutenant commander, he returned to Sheffield, where he was appointed medical registrar at the Royal Hospital in 1946. He remained there until 1949 when he took up an appointment as consultant physician to the Chesterfield and North Derbyshire Royal Hospital, where he was the first physician to devote himself entirely to consultant work. Naturally, he had to deal with general medical work although his special interests were diabetes and cardiology. It was not until 1965 that he was joined by a second consultant. Tom rapidly built up a first rate department in the hospital, and a good private practice. His opinion was widely sought in the district and he had an excellent relationship with the general practitioners.

Tom Gaunt was a fine physician, possessed of exceptionally sound judgement. A good, kind and generous man, he was always concerned with the best interests of his patients rather than merely seeing them as objects of scientific interest. He had a rare ability to communicate with them and they always understood what he was talking about. He was a born teacher and there is little doubt that he could have been appointed consultant to his old teaching hospital, but he preferred to do pioneer work in a fresh field. He enjoyed administrative work and for many years was a member of the Chesterfield Hospital management committee, and of the regional hospital board.

Outside his work, Tom led a full life. He and his wife had a charming house where they entertained their friends in a delightful way. He was good at games and, in earlier life, played tennis and squash racquets for the Sheffield Hallamshire Club. Later on he took up fishing and, after his retirement, painting in water colours. Bridge and caravanning were also among his hobbies. He enjoyed listening to music and was a regular attendant at concerts. But perhaps outside working hours his greatest interest was his family; he and his wife Anne had a son and a daughter, and five grandchildren.

FJM/GW

[*Brit.med.J.*,1985,**291**,1729; 1986,**292**,210; Photo]

GELFAND, Michael *b.*26 Dec 1912 *d.*9 July 1985
CBE(1959) OBE(1956) MB ChB Cape Town(1936) MRCP(1938)
DMR Eng(1939) MD Cape Town(1948) DPH Lond(1950)
FRCP(1952) Hon LLD Rhodesia(1978) Hon DLitt Cape
Town(1978) Hon LLD Birm(1981)

Michael Gelfand was born in Wynberg, Cape Province. After education at Wynberg High School, he entered Cape Town University and graduated in medicine in 1936. During his undergraduate career he was noted more for his skill on the rugby field than for outstanding academic achievement. In 1937 he came to England where, among other appointments, he was house surgeon at the North Devon Royal Infirmary, Barnstaple, and then house physician at Salisbury Infirmary. He obtained his membership of the College in 1938, followed by the diploma in medical radiology.

The critical point in his career came in 1940 when he was appointed to the Federal Health Service of Rhodesia (now Zimbabwe) and Nyasaland and stationed at Salisbury, Rhodesia. The establishment was small and for some years Michael had to be his own radiologist and pathologist, as well as physician. He led a strenuous life; being in constant demand as a physician yet also finding time to play rugby for the Police Club and for the Mashonaland team. On Sundays he would frequently visit the Catholic missions, and it was in these early years that he developed his deep interest in the African and his life style; particularly that of the Shona people who form the majority of the population of north-east Rhodesia (Zimbabwe). He slept little and much of the night was spent writing. *The Sick African*, Capetown, Juta, 1943, ran through many editions and was a landmark in clinical tropical medicine, while his monograph on *Schistosomiasis...*,Capetown, Postgraduate Press,1950, had gained him in 1948 the MD of Capetown University, with first class honours. Over the years he was the author of some 45 monographs and 200 articles, and these provided not only much new knowledge of medicine and pathology in Central Africa but also added a new insight into the history and sociology of that area. Notable among them are: *Medicine and magic of the Mashona*, Capetown, Juta,1956; *Shona ritual...*,Capetown, Juta, 1959; *Tropical victory...*,Capetown,Juta, 1953; *Medicine in tropical Africa...*, Edinburgh, E & S Livingstone, 1961; *Northern Rhodesia in the days of the Charter...* ,Oxford, Blackwell, 1961, and *Livingstone, the doctor...* Oxford, Blackwell, 1957. In 1955 he founded *The Central African Journal of Medicine* and was its editor up to the time of his death.

When, in the early 1960s, the proposed medical school in the University College of Rhodesia and Nyasaland (later the University of Rhodesia) was incorporated into the University of Birmingham, England, as an affiliated college, Michael Gelfand was appointed professor of African medicine. His was the main driving force in the development of the medical school and its outstanding success was due more to him than to anyone else. He was an extremely able teacher and his ward rounds were fascinating, emphasizing the Oslerian dictum: '...as is our pathology so is our practice.'

As a man Mike was kind and gentle; with love for all and malice for none. He played no part in politics, and his wide popularity as a physician and man was shown by the fact that he was almost automatically the physician of governors, prime ministers and leading citizens. This eminence in no way impaired his devotion to the ordinary sick African. Prime Minister Robert Mugabe gave an address at his funeral. Although a devout member of the synagogue he was completely ecumenical - evidenced by the award to him of the Roman Catholic knighthood of St Sylvester, and the Order of Distinguished Auxiliary Service of the Salvation Army.

In 1937 he married Esther Kallenberg, daughter of a company director. They had three daughters, two of whom married doctors. His wife Esther survived him. Over many years together they gave warm hospitality to visitors, particularly to those from Birmingham.

MA

[*Brit.med.J.*,1985,**291**,1057; *Lancet*,1985,**2**,458,512,568; *S.Afr.med.J.*,1985,**68**,432; *Central Afr.J.Med.*,1978,**24**,5-11; Photo]

GEMMELL, John Patmore *b.*15 Sept 1917 *d.*25 June 1987
MD Manitoba(1941) MRCP(1946) FRCPC(1947) FRCP(1970)

John Patmore Gemmell was born in Winnipeg but grew up in Assiniboia, Saskatchewan, where his father was a general practitioner. His mother was Janet Victoria Patmore, daughter of a farmer. Like many prairie sons he loved the West, and naturally entered his father's profession and pursued it with devotion and distinction throughout his life.

He took premedical studies at the University of Manitoba and entered the faculty of medicine in 1937 where he excelled as a student, winning many academic awards throughout the course. He was interested in debating and was an articulate speaker. After graduating MD he spent a year as resident in medicine at the Winnipeg General Hospital, where he demonstrated exceptional clinical skills and a capacity for bedside teaching which was to flower in later years. On completion of his residency he joined the Royal Canadian Army Medical Corps where he was soon appointed a medical specialist with the rank of major - at the time he was the youngest major in the RCAMC. He served with the 5th Canadian General Hospital in England, Sicily, Italy and northwest Europe. Despite the strenuous demands of a busy hospital, in constant proximity to active warfare, he maintained an avid interest in medicine - and could always quote from recent journals to the delight and edification of his colleagues. He was the authoritative source of information about unusual diseases with which his fellow medical officers were unfamiliar. With dedication and compassion, he would stay up all night with critically ill soldiers when nurses were scarce or unavailable. Because of his fat, round face and clear complexion he was often known as 'Baby-faced Johnnie'.

Following discharge from the Army he joined the medical department of the Winnipeg Clinic, but shortly afterwards left to take up a fellowship in the department of pharmacology at Yale University. On his return to Winnipeg in 1948 he became a research fellow of the Medical Council of Canada and pursued his interests in endocrinology and metabolism. The following year he was appointed one of the first full-time members of the department of medicine. In 1964, on the retirement of his friend and preceptor Lennox Gordon Bell [*Munk's Roll*,Vol.VI,p.39], Gemmell became professor and head of the department of medicine, University of Manitoba, and physician in chief of the Winnipeg General Hospital (which became the Health Sciences Centre in 1974). During these years the department experienced rapid growth and he was active in recruiting new faculty members. He had an unusual talent for keeping in touch with young physicians while they were away doing postgraduate studies, and he served as mentor and friend to them during their years of training.

With his research interests, John Gemmell early became a member of the Canadian Society for Clinical Investigation and served as a member of council and as secretary for several years. He also served as a member of council of the Royal College of Physicians and Surgeons of Canada, and for many years was a member of the committee on examinations and the committee on scientific meetings. He was an examiner for the College, and took great interest in the development and progress of young physicians. He was a fellow of the American College of Physicians, and was governor for the Prairie Provinces, 1976-79.

Always a prolific reader, he was not a productive writer although he was author of 11 papers.

John married Nancy Mathers, daughter of a lawyer and herself a doctor, in 1948, and they had two daughters and a son who became a lawyer. Rather late in life this marriage ended in divorce. He later married Patricia Beaufoy who survived him.

John Gemmell lived a good and fruitful life and will long be remembered by his friends and colleagues, by grateful patients, and perhaps particularly as an outstanding teacher by his students - who are now to be found all over the world.

REB/CBS

GHALIOUNGUI, Paul *b.*18 Sept 1908 *d.*2 Jan 1987
MB BCh Cairo(1929) MD(1933) MRCP(1934) FRCP(1964)

Paul Ghalioungui was born in Lower Egypt, in a city called Mansurah where young Paul received his preparatory education in the French school Collége des Fréres. When his family moved to Cairo he continued his education at the Collége des Fréres, Khoronfish, Cairo. At that time there was only one university in Egypt, first known as the Egyptian University, then as Fouad el-awal, and finally as Cairo University. Paul entered the medical faculty and graduated in 1929, pursuing his clinical studies at the Kasr-el-aini Hospital. He spent four years as house officer and registrar to A G Biggam, later Sir Alexander, [*Munk's Roll*,Vol.V,p.37] during which time he obtained his MD. This was followed by his membership and fellowship of the College. The years between 1933 and 1964 were passed in teaching, examining, and conducting research in endocrine and metabolic diseases; a time when diagnoses were made on purely clinical grounds sometimes supported by biological or biochemical assays. In 1949 he was appointed professor of medicine to the new Ain Shams University and later became chairman of the department of general medicine.

Paul was a highly cultured man and a brilliant linguist. He was interested in the history of medicine, both Arabic and Pharaonic. His English and French culture enabled him to acquire six languages and he could both read and write hieroglyphs. His publications included some 20 books, both in Arabic and English. As the first historian of

Egyptian medicine he contributed a great deal to this field. He was an honorary member of the Pan-American Society of the History of Medicine and of the Ateneo de Historia de la Medicina , Buenos Aires; correspondent member of the National Academy of Medicine of Buenos Aires, and an honorary member of the Argentinian Medical Association. He was also president of the Egyptian Society of Endocrinology and Metabolism and vice-president of the Egyptian Society of Rheumatic Diseases, and a member of the board of the Egyptian Medical Association.

Outside medicine his interests lay in photography and archaeology. He was president of the Cairo Rotary Club in 1956 and secretary-general of the Institut d'Egypte in 1963.

Paul was instrumental in organizing the very successful 29th International Congress for the History of Medicine which was held in Cairo at the Marriott Hotel. Only one week before his death he was busy lecturing at Mansora University and the Suez Canal University.

He married Huguette Aimee, daughter of Aziz Mancy, a lawyer, in 1934 and they had two sons. His wife and family survived him. AG/GW

[Photo]

GIBSON, Ronald Valentine *b.*14 Feb 1920 *d.*13 Jan 1986
MRCS LRCP(1943) MA MB BChir Cantab(1947) MRCP(1949) FRCP(1965)

Ronald Valentine Gibson was born a Guernseyman, his father Ronald Gibson OBE, a medical practitioner, having settled there before the first world war. He was educated at Marlborough College, Pembroke College, Cambridge, and the Westminster Hospital. After the usual house jobs, he served as a captain in the RAMC with the South Staffordshire Regiment until the Battle of the Orme, and subsequently crossed the Rhine with a battalion of the Seaforth Highlanders. He was among the first to enter Buchenwald concentration camp after its inmates were released, and later helped the civilian population in Kiel and Hamburg at the end of the war, taking part in the denazification programme.

After demobilization he specialized in the newly developing field of cardiology, and in 1953 was appointed senior registrar to the cardiac department at the Brompton Hospital, joining the staff two years later. At that time both cardiology and cardiac surgery were making rapid strides forward with the introduction of cardiac catheterization and open heart surgery. Ronald was intimately involved in all the new developments and in particular was closely associated with Russell Brock, later Lord Brock [*Munk's Roll*.Vol.VII,p.62], and other surgeons in the new approach to congenital and valvular heart disease. In 1962, following the death of Paul Wood [*Munk's Roll*,Vol.V,p.456] he became physician in charge of the department. He was elected a Fellow of the College in 1965.

Ronald was first and foremost a clinical cardiologist. He consulted in Bromley, Cornwall and Wessex, and by his efforts a large surgical practice was established at the Brompton Hospital. Like his predecessor, Paul Wood, he correlated the results of invasive examination with clinical signs, continuing a tradition which still distinguishes British cardiology. His skill in assessing patients for operation and in detecting potentially remediable complications of surgery were widely appreciated. In addition to running an active clinical department at the Brompton, he had a large international practice and was among the first to deal with the huge influx of patients with heart disease from the Middle East in the 1960s. It was from him that many of his younger colleagues learned the art of treating patients whose cultural and religious backgrounds were so profoundly different from their own. In spite of this workload, he punctiliously maintained his clinical work and teaching at the Brompton.

Ronald was no academic and by today's inflated standards his output of published work was small. In his early years he wrote seminal papers on tricuspid regurgitation and constrictive pericarditis, followed later by a number of case reports. He continued to maintain an interest in new technology and was rapidly able to assimilate echocardiography into his practice. Those who never came directly into contact with him were unaware of his flair as a clinical teacher, for which he was widely and affectionately remembered by many who later specialized in cardiology, both in this country and in other parts of the world.

His style of leadership of the cardiac department was never forceful and in these days of strategic planning might have been considered indecisive. But when judged by results it is clear that under his management the department prospered and, by the time of his retirement was well placed to merge with the National Heart Hospital in the new cardiothoracic centre.

Ronald was not an easy person to know and many meeting him for the first time would find his manner withdrawn. He did indeed find it difficult to relate to many professional colleagues, but to his junior staff he was a friend, teacher and loyal supporter in their later careers. He had few interests outside his profession and his family, and his final illness darkened the end of his career, denying him the opportunity to widen his activities after retirement.

He married Thelma Pauline Gibson in 1945 and they had two sons and three daughters. He had a son and a daughter from his second marriage.

DGG

[*Brit.med.J.*,1986,**292**,773; *Lancet*,1986,1,812; *The Times*,20 Jan 1986]

GILL, Alan Morton *b.*11 Jan 1909 *d.*25 May 1985
CBE(1970) MRCS LRCP(1932) MB BS Lond(1932) MD(1934)
MRCP(1936) FRCP(1946)

Alan Morton Gill was born in Gibralter. His father, a Gibralterian, was a doctor and his mother was a Scot. Alan was bilingual in English and Spanish. He was educated at Epsom College and subsequently entered the Middlesex Hospital medical school with an entrance scholarship. After a period as research assistant in the department of physiology, he was appointed house physician and later medical registrar. In 1939 he was appointed physician to the West London Hospital.

At the beginning of the second world war he attempted unsuccessfully to join the Royal Navy, being rejected on account of a family history of tuberculosis; indeed he himself had been infected as a child. Instead, he joined the Emergency Medical Service and was posted to the Old Windsor and Stoke Mandeville Hospitals. Later, he opened the EMS gastric unit at the White Lodge Hospital, Newmarket. In the meanwhile, in 1940, he had married Mary Hammond, a qualified dispenser and also a nurse, and a member of a remarkable family of four generations of churchmen. There were three children of their marriage, two sons and a daughter.

After the war he returned to the West London Hospital, where he built up a department of gastroenterology. He taught general medicine in the medical school and played a prominent part in its eventual fusion with the Charing Cross Hospital medical school. He was a pioneer in the field of gastroscopy; today his personal gastroscope is on display in the London Science Museum. He published some 50 papers, mainly on gastroenterology. He was elected a Fellow of the College in 1946, was an examiner from 1966-74, and an appointee to the DHSS Medical Appeal Tribunal in 1976 from which he was forced to retire three years later after suffering his first coronary infarct.

Alan was a quiet, shy man who dedicated himself to the care of others, indeed only a few knew of his many and varied achievements. A keen horseman, he rode very well and had played polo in his younger days in Spain. He played rugby for the Middlesex Hospital, was a good swimmer and enjoyed skiing. He and his wife travelled widely in Europe, North Africa and through the Americas. They were both invited to attend the Independence Celebrations in Barbados at the personal wish of the prime minister. In 1970 he was created CBE for his services to the Barbados and Guyana High Commission in the United Kingdom. For 35 years he also served as senior medical consultant to the Canada Life Insurance Company and was later elected president of the Assurance Medical Society.

For six years after his retirement he enjoyed his newfound leisure in the country with his family and his dogs; gardening, bird watching, and assisting his daughter with her stallions and the stud. His son, Robin, a priest and lecturer in the divinity faculty of Edinburgh University, wrote eight books on religion. Alan read all the proofs and made useful amendments to sentence structure, vocabulary, and sometimes to logic. An inveterate pipe smoker, Alan developed another extraordinary addiction: he not only enjoyed but became an authority on all the Western cowboy films broadcast on television. He never missed one, and during each showing no one was allowed to talk or in any way to disturb his concentration. He died at his country home in Charlwood, Sussex.

BGP-S

[*Brit,med.J.*,1985,**291**,62; *Lancet*,1985,2,227; *The Times*,6 June 1985; Photo]

GILLIE, *Dame* Katherine Annis Calder *b.*8 Aug 1900 *d.*10 Apr 1985
DBE(1968) MB BS Lond(1925) MRCP(1927) FRCP(1964)
FRCGP(1967) Hon MD Edin(1968)

Dame Annis Gillie was one of the great ladies of medicine, and by that I mean that she had a most unusual dignity and charm. She attracted attention from the moment she entered a room. When young she was tall and very dignified, but as she got older she became very bowed.

She died, aged 84, very much retired and out of the round of hospital services and general practitioner services to which she had contributed so much. She was devoted to her family, especially to her husband Peter Smith who was ill for many years with multiple sclerosis. She nursed and cared for him unsparingly. They had two children; a son and a daughter. The latter was a medical practitioner and died young.

Annis was born at Eastbourne, a daughter of the Manse. She was educated at Wycombe Abbey School and University College Hospital, qualifying in 1925. She entered general practice in West London and remained active until 1963, when she retired to Gloucestershire to look after her husband. She was a Fellow of the College and of the Royal College of General Practitioners, and was given an honorary MD by Edinburgh University. After the College of General Practitioners received its Royal Charter she became its first president; being the only woman to become president.

She was a member of the Central Health Services Council from its inception, and from 1956-70 she chaired the Council's committee which produced in 1963 the Gillie report on the fieldwork of the family doctor. She was a great champion of women in medicine, and in 1954 was president of the Medical Women's Federation. From 1950-64 she was a member of the Council of the British Medical Association. She served on the North West Metropolitan Regional Hospital Board, and later on the Oxford Regional Hospital Board for ten years. She received the OBE in 1961, and the DBE in 1968 during her presidency of the Royal College of General Practitioners.

Annis Gillie was quite the best general practitioner in London during the late 1930s and early 1940s, and had practically all the professional women in London on her list. She also had a considerable number of very distinguished men, including Sir Hugh Casson, president of the Royal Academy, and Lord Clark. She knew just how to run this rather difficult and articulate group of patients and had just the right light hand on the reins, knowing exactly when to pull them up sharply and when to let them run. Altogether, her contribution to medicine and general practice was very great, and her influence on women doctors was considerable.

ALW

[*Brit.med.J.*,1985,**290**,1360; *Lancet*,1985,1,996; *The Times*,16 Apr 1985; Photo]

GOLDING, Frederick Campbell *b.*4 June 1901 *d.*17 July 1984
MB ChM Sydney(1926) MRCP(1929) DMRE Cantab(1933) FRCP(1949) FFR(1949)

Cam Golding died at the age of 83. He was an outstanding and much appreciated specialist in the discipline of diagnostic radiology and first joined the Middlesex Hospital on the invitation of Sir Harold Graham-Hodgson [*Munk's Roll*,Vol.V,p.162] to assist him in the development of the new X-ray diagnostic department in 1933.

His knowledge and teaching ability immediately made a great impression and he was largely responsible for the important reputation developed by this department. In 1956 he succeeded to the post of director of diagnostic radiology and held this until his retirement in 1967.

Cam Golding was born and educated in Australia, and his schooling was firstly in Sydney and then at Scots College in Melbourne. He went to St Andrew's College of the University of Sydney to study medicine and qualified in 1926. During his university career he showed great keenness in various sports. Taking on rowing, he became a member of the successful College crew, and then became stroke of the University eight. He began to play golf, and got down to a low single figure handicap, but soon gave that up and played tennis instead. He did not play for the University but was in the running, practising with University and New South Wales representatives.

A few years after qualification he came to England and obtained a post as registrar at the Royal Free Hospital. He then began to study diagnostic radiology and gained the Cambridge DMRE in 1933. After various appointments, including the directorship of the x-ray diagnostic department at the Royal Marsden Hospital, he came to the Middlesex Hospital which, as well as private practice, remained his principal place of activity until his retirement.

Golding had wide general interests, lectured in the medical school and served as chairman of the hospital medical committee, and was particularly involved in orthopaedic and chest diseases. He was consultant radiologist at the Royal National Orthopaedic Hospital, at the Arthur Stanley Institute for Rheumatic Diseases, and the Chelsea Hospital for Women. He was also civilian consultant in radiology to both the Royal Navy and the Royal Air Force.

He became a member of the College in 1929 and a Fellow in 1948, and was one of its examiners in radiology for the three years 1954-57. He gave the Mackenzie Davidson lecture for the British Institute of Radiology in 1961, and the Watson Jones lecture for the Royal College of Surgeons in 1964. Among his various medical publications were contributions to *A Textbook of x-ray diagnosis*, by British Authors, ed.S C Shanks, Philadelphia, Saunders, 1957(3rd ed); W S Copeman's *Textbook of rheumatic diseases* Edinburgh, E & S Livingstone, 1948, and 'Survey of radiology of the chest' in the *British Encyclopaedia of Medical Practice.*

Cam, as he was known to his friends, was a modest member of the staff, much appreciated by his many friends and colleagues. He continued his great interest in sport throughout his life. He was an expert fisherman and, after retirement, each summer he had a weekly appointment for trout fishing in the south and a month in Scotland for salmon. In the winter he was an excellent shot. He continued to play tennis until the last year or two, and he and his wife - who gave him great support in all that he did - used to have regular daily tennis games together. She was Barbara Hubbard and they married in 1942. She survived him, with two sons and six grand-daughters.

BW

[*Brit.med.J.*,1984,**289**,1464; *The Times*,19 Sept 1984; *J.roy.Nav.med.Serv.*,1985,**71**,63; *Middx.Hosp.J.*,Oct 1966,**66**(5); Photo]

GOLDING, John Raymond *b.*30 July 1926 *d.*16 Sept 1988
BA Oxon(1947) MRCP LRCP(1950) BM BCh(1950) MRCP(1956) MA(1958) DM(1962) FRCP(1975)

John Golding was the son of Herbert Melhuish Golding DFC, a general practitioner in Bristol, where John was born. He was educated at Bristol Grammar School and St John Baptist College, Oxford, where he was a Sir Thomas White scholar. He gained a BA with honours in physiology and won an entrance scholarship to Westminster Hospital medical school. After qualifying he worked as a house physician at Westminster Hospital, Westminster Childrens' Hospital, the Brompton Hospital and Addenbrooke's Hospital, Cambridge. From 1952-54 he served with the RAF in Iraq, before becoming medical registrar and later senior registrar at Westminster Hospital to W E Lloyd [*Munk's* Roll,Vol.VI,p.297] and Frank Dudley Hart.

John Golding became particularly interested in rheumatic disorders and in 1960 published a paper on the neuropathic and vascular complications of rheumatoid arthritis, and the early loss of vibration sense in some patients with this and other disorders; a subject which later formed the basis of his DM thesis. Corticosteroids had recently been introduced as a form of therapy for rheumatoid arthritis, and John had been involved in a number of clinical trials. In 1961 he was appointed consultant physician to the Royal Bath Hospital in Harrogate and to the rheumatology department of St James's University Hospital in Leeds; he was also physician in charge of the chronic sick unit at Chapel Allerton Hospital, Leeds. He was clinical tutor in postgraduate medicine in the Harrogate area in 1970, and became examiner in clinical medicine and senior honorary lecturer in the faculty of medicine at Leeds University in 1972.

Golding took an active part in the first controlled double-blind multicentre clinical trial of D-penicillamine in the treatment of rheumatoid arthritis, which proved so worthwhile that it was introduced into orthodox anti-rheumatic practice. He was an excellent teacher and popular with the undergraduate and postgraduate medical students at Westminster, and later at Leeds. He was a good all-round clinician, interested in all fields of medicine and particulary in rheumatic disorders. The bulk of his work latterly was at the regional rheumatology centre in Harrogate, where he had lived for many years. Although severely incapacitated by a number of serious illnesses over a period of many years, he worked as a consultant physician until shortly before his death.

A devout, serious-minded Christian, he was an honorary medical officer to the general synod of the Church of England and a member of the Anglican Hospital Chaplaincy Council, and also a church warden of St Wilfred's Church, Harrogate. He had a delightful sense of humour and a shy, gentle manner; an excellent speaker, with an occasional slight and attractive stammer, he did not push himself forward in any conversation.

John had married Myfanwy Miles, daughter of a farmer, in 1961 and they had two children, Joanna and James. The marriage was a very happy one. John's brother, Philip, also qualified in medicine at Westminster Hospital medical school.

FDH

[*Brit.med.J*,1988,**297**,1123; *Lancet*,1988,2,859]

GOODWIN, Frank John *b.*26 Feb 1938 *d.*29 Sept 1987
BA Cantab(1959) MB BChir(1962) MA(1963) MRCP(1964)
FRCP(1977) MD(1980)

Frank John Goodwin was born in Stratford, London, and died at the tragically early age of 49 years after a long and debilitating illness. He was married to Mary Rose Goodwin, ophthalmologist, and was the only son of John Leonard Goodwin, a cost accountant, and his wife Edith Maud.

Frank Goodwin was educated at Brentwood School, Essex, from 1946-56. He matriculated at Gonville and Caius College, Cambridge, with a BA in the natural sciences tripos. His clinical studies were undertaken at the London Hospital medical college where he graduated MB BChir, gaining a distinction in medicine and the college prize. Following this auspicious start he was appointed house officer to the medical unit and then to the surgical unit at the London. House physician posts at the Hammersmith Hospital and the National Hospital for Nervous Diseases, Queen Square, followed and in 1965 he returned to the London as lecturer in medicine. Having been awarded the Lewis Smith postgraduate travelling scholarship in clinical medicine and a Lilly international research fellowship, he went to work with John Laragh at the Columbia University Presbyterian Medical Center, New York, in 1967. The research work he carried out there formed the basis of his MD degree, and resulted in several original publications. He returned to England two years later and was appointed lecturer in the medical unit at the London Hospital medical college. In 1971, at the unusally early age of 33, he was appointed consultant nephrologist to the London Hospital and senior lecturer to the medical college.

The passion Frank had for clinical medicine and research was inculcated from his undergraduate days by such eminent nephrologists and teachers as Clifford Wilson, John M Ledingham, Michael Floyer, and later John Laragh. It is not surprising therefore that he developed an abiding interest in hypertension, the role and control of renin, the workings and diseases of the kidney and, in particular, his patients and students.

In 1965, under the guidance of Clifford Wilson and John Ledingham, he first undertook regular haemodialysis for the long-term treatment of patients with terminal renal failure. It was as a result of this early experience that the first Ministry of Health dialysis unit was set up in 1968 at the London Hospital, and a renal transplant programme was started the same year. Frank's return from the USA, and his later appointment as consultant nephrologist, enabled him to maintain his interest in renal replacement therapy. The number of patients treated by dialysis and transplantation increased rapidly, so that 20 years later the London Hospital dialysis unit had become the largest in the United Kingdom. Although the pressure of work was very heavy he always refused to restrict his interests to nephrology, believing that the illnesses of patients were not confined to one organ and neither should be the sympathies of their physicians. He was proud of the fact that he was senior lecturer in medicine and honorary consultant physician, and in both capacities he practised general medicine throughout his professional life.

Frank remained active in research which, from the time he returned from New York, was almost entirely patient orientated. He gained a considerable reputation for his work, publishing over 40 articles and reviews, which had a special bias towards the renin-angiotensin system and its role in hypertension.

In his dealings with patients he was meticulous. His assessments took into account every aspect of their illness: medical, social and occupational factors were recorded with great care, together with his interpretation and decisions on management. Woe betide the notes which were in poor condition. Out would come the scissors, scalpel, sticky paste and sellotape, and tatty notes were soon restored to pristine condition. This punctilious, even obsessive, approach characterized everything he did, whether clinical practice, research, teaching, or wallpapering a room in his beautiful old house in Kent.

Frank Goodwin loved teaching and was good at it. His formal lectures were held in high esteem, perhaps because they were deliberately simple in presentation, his having considered carefully what his audience needed to know and how best this could be conveyed. His informal teaching was equally popular. He was a tutor to students at the medical college for 20 years, and one of only three senior clinical tutors for some 10 years. As a member of the committee which selected students for entry to the medical college he met them again during their clinical studies and took very seriously his responsibility for monitoring their progress, advising on their electives, helping them when they were experiencing difficulties and stimulating those likely to achieve particular success.

As if this were not enough, he actively encouraged a heavy administrative load. His meticulous habits helped him greatly in coping with this, for all arrangements and results of discussions and meetings were documented precisely and filed for future use. This provided the basis for his considerable administrative ability, together with a critical faculty which enabled him to separate what was important from the unimportant and the accurate from the imprecise. He was a member of the executive committee of the Renal Assocation from 1974-78 and its secretary from 1978-81. He was also the local organizing secretary of the 1983 congress of the European Dialysis and Transplantation Association held in London. Sadly, his illness prevented him from completing the work of the same post in connexion with the 1987 congress of the International Society of Nephrology, also held in London.

A taught, smart, spare figure, with a very direct gaze, he could be both forbidding and authoritarian. He tolerated no nonsense, and his responses could be brusque, even cutting, particularly for those whose abilities or dedication were of doubtful calibre. His severest rebukes were reserved for people who were less than totally honest, or would not try. He himself was intensely honest and strove to maintain the highest of clinical and academic standards. Privately he would sometimes admit to a fear that he could not live up to the standards he set for himself. His incisiveness made his judgements much valued by colleagues, and his presence on a committee ensured that its deliberations were not merely well-intentioned but also properly thought through.

Despite his apparent austerity, he could be very kind and he would go to much trouble to ensure the health and happiness of his patients, and also that his students were given every opportunity to succeed.

Frank's main relaxation was music. Having learned to play the clarinet at school, apparently without achieving greatness, he took piano lessons in his late thirties and became an enthusiastic and proficient player. He loved opera, particularly Wagner, sang madrigals at Cambridge, and was president of the London Hospital music society, in whose choir he sang.

He also painted watercolours, enjoyed art exhibitions, walked the fells and dales of northern England, and went birdwatching on the Norfolk Broads. Although not a great natural sportsman, he was a competent skier, and took much pride in having coxed for Caius in the Bumps and at Henley. He abhored television and for many years refused to have a television set. What time others spent watching television found him improving the flowers and foliage of his country garden, double-glazing the windows of his house or preparing for the next day's work.

Frank met his wife when she became one of his tutorial students at the London Hospital medical college. Mary Rose, an ophthalmologist, was the daughter of the late Frank Burnet Byrom, an experimental pathologist with an international reputation for his work on the pathogenesis and pathophysiology of hypertension and hypertensive encephalopathy. She survived Frank, with their three children, Harriet, studying at Balliol College, Oxford; Nicholas, studying at King's College, Canterbury, and Katherine, at Bromley High School and already following her father's musical interest.

The memorial Frank John Goodwin would most have liked for himself would surely have been the simple statment: 'I tried my hardest.'

FPM

[*Brit.med.J.*,1987,**295**,1149; Photo]

GORDON, Hugh Walker *b*.5 Aug 1897 *d*.11 Dec 1987
MC BA Cantab(1922) MB BChir(1926) MRCP(1928)
FRCP(1940)

Hugh Gordon was born in Maxwellton, Kirkcudbrightshire, the son of H Sharpe Gordon, a solicitor in Dumfries. His forbears for many generations had been lawyers in southern Scotland. After his education at Marlborough he immediately went into the Army, being commissioned in the RFA, and was wounded and invalided out in 1918. He won the Military Cross. During his Army period in France thoughts of a career in medicine gained strength but after the war he entered Pembroke College, Cambridge, with a history exhibition. Once there he changed and read natural sciences.

He had no particular reason for choosing to go to St George's Hospital but he won a scholarship to the medical school there and that determined the site of his clinical studies. After qualification and the usual house jobs, he thought of specializing in medicine. In fact he obtained the coveted post of resident medical officer for one year, but by then it had been suggested to him that there was a future in dermatology and he was partly influenced by Wilfrid Fox [*Munk's Roll*, Vol.V,p.140].

He married Jean Robertson at St George's, Hanover Square, and decided to go abroad for postgraduate studies. At this time the great medical centres of Europe were an essential part of dermatological training and he spent six months at the St Louis Hospital, Paris, living with his new bride in a little apartment not far from the hospital. There was only one other English speaking student in his group. He came under the spell of the great French morphologists, and indeed consulted Sabouraud personally as a patient, who proved successful where other previous consultants had failed. Hugh Gordon became greatly interested in radiotherapy, then a relatively novel procedure, and he studied this technique, acquiring a skill which was later to prove useful in consultant practice. Three months in Vienna followed the Paris visit, and then he returned to London where he quickly obtained three hospital appointments, at St John's Hospital, Lewisham, the Shadwell Children's Hospital and the East Ham Memorial Hospital. There was soon to be a vacancy at St George's and it was assumed that Hugh was the likely appointee. He joined its staff in 1933 as assistant skin physician, relinquishing his three previous appointments but adding two others - the West London Hospital and the Cancer (now the Royal Marsden) Hospital in Fulham Road, both prestigious and central appointments.

In 1939 Gordon was appointed sector dermatologist to the EMS Sector Seven which entailed weekly visits to a number of hospitals, but he also became the lynch pin at St George's at Hyde Park Corner, not only as physician but also as administrator and acting dean of the medical school. The inpatients had mostly been evacuated westwards, so the empty wards were available to receive air-raid casualties. It was not an easy time for teaching but it had to continue; clinical material was short and improvisation was the key. Without doubt, Hugh enjoyed this period and was very effective in wearing his various hats. After the war he continued for a time as acting dean. In fact it was he who was responsible for the introduction of female students from Cambridge to St George's, and a very talented collection they proved to be.

At this time Hugh was living with his family at Wimbledon, in a house with a fairly large garden which allowed him to indulge in his agricultural interests; his smallholding of cows, pigs and poultry was a prolific producer and many of his friends benefited from the welcome home grown produce which supplemented the meagre rations which still prevailed.

A busy consulting practice which included three hospitals was characteristic of the peripatetic life of a dermatologist at this period and Gordon much valued his association with the Cancer Hospital, where he became expert at treating the various skin malignancies, and it was here that his previous interest in radiotherapy proved valuable. He contributed the relevant section in Mitchell-Hegg's [*Munk's Roll*, Vol.VI,p.341] *Modern Practice of dermatology*, London, Butterworth, 1950.

He soon became intrigued by the possible role of hypnosis in the treatment of various skin conditions and with the assistance of Kathryn Cohen and S Mason, he undertook a study and presented the results to the 1952 International Congress in London, a paper which attracted considerable attention.

Gordon and his family had now moved to Queen Anne Street and his private practice was flourishing. In 1963 he was president of the British Association of Dermatologists, and also president of the dermatological section of the Royal Society of Medicine. He was chairman of the medical staff committee at St George's for three years and was also on the board of governors, at a particularly critical time for the hospital if it was to survive and evolve in the new London medical scene.

His father had retired from legal practice at the age of 55, moving south and acquiring a large house at Yateley in Surrey. On his father's death, Gordon moved to Yateley and was again able to indulge his agricultural interests. He cultivated asparagus, but his attempts to produce eggs with a low cholesterol content proved unsuccessful! Once his children had married and left the family home, he retired from active practice and decided to return to his old haunts and lived happily with Jean in the High Street, Kirkcudbright. He was not far from his son, Graham, who ran a successful fish farm.

Hugh Gordon was a remarkable and apparently ageless figure, notable for his slim and upright physique, handsome coutenance and somewhat authoritarian voice. He was one of a close knit group of post war (1914-1918) Cambridge men who entered St George's Hospital, most of whom were appointed to the consultant staff, and throughout his time there he was a devoted and dedicated servant of the hospital and medical school. ÓÇ'

[*Lancet*,1989,1,56; Photo]

GREAVES, John Ley *b*.26 July 1918 *d*.15 May 1987
MRCS LRCP(1941) MB BS Lond(1941) MRCP(1947) MD(1948) FRCP(1971)

John Greaves was consultant physician at the Royal Hampshire County Hospital, Winchester, and the Southampton General Hospital. He came from a distinguished medical family in Derby, where his father had been a surgeon. He graduated from St Thomas's Hospital and for the next five years was commissioned in the Royal Air Force, serving in Britain and the Far East. He returned to St Thomas's to develop his career in medicine and quickly gained his membership of the College and his MD. He became increasingly attracted to children's medicine and in 1948 was appointed to Sir James Spence's unit in Newcastle [*Munk's Roll*,Vol.V,p.386]. After a spell at Great Ormond Street as resident assistant physician he became consultant paediatrician at Middlesborough in 1952.

After 11 years of devoted single-handed service on Tees-side, John returned south with consultant appointments at the Winchester and Southampton hospitals, where he remained until his retirement in 1978. He was a paediatrician with an encyclopaedic knowledge of his subject, meticulous in his clinical standards and unfailing in his courtesy and sensitivity to his patients and their families. Nor was it ever long before those around him became aware of his gentle and ready sense of humour.

Despite his busy professional life, John took an active interest in local affairs; in the musical life of the neighbourhood and in service to Winchester Cathedral. But perhaps he was never happier than with his wife Elizabeth, sharing their beautiful and varied garden with a few friends and regaling them with accounts of some of his more recent exotic acquisitions. He could also turn his hand to skilled cabinet making and his much used music cabinet was a piece of beauty and ingenuity.

In retirement, near Alton, although dogged by ill health he was never far from the centre of the village activities, and he once more developed another lovely garden from which his many friends were able to enjoy the wide vistas of the North Hampshire downs, of which he was so fond.

His career spanned and to a great extent reflected the great post war development of paediatrics in this country, and it was fitting that in the last few years of his practice he was able to convey something of this experience to the students at the new medical school in Southampton.

He met his wife Elizabeth Fiennes when she was nursing at St Thomas's. They married in 1950 and they had four chldren. It was a marriage of great happiness and John was a devoted father and grandfather. He was survived by his wife and children.

ICSN

[*Brit.med.J.*,1985,**295**,613; Photo]

GREENBURGH, Harry *b.*28 June 1917 *d.*28 Mar 1985
MRCS LRCP(1939) MB BS Lond(1939) MRCP(1952) MD(1952) MRCPath(1963) FRCPath(1966) FRCP(1973)

Harry Greenburgh was a Devonshire man, being born at Devonport where he was also educated. He qualified in medicine from Guy's Hospital, London, at the early age of 21. After house appointments at Guy's, the East Sussex Hospital, Ipswich, and the Royal United Hospital, Bath, he joined the Royal Air Force Volunteer Reserve and was wounded on active service. He spent the remainder of the war in South Africa, with the rank of squadron leader. Demobilized in 1946, Harry elected to specialize in pathology. After a brief spell in Plymouth as clinical assistant at the South Devon Hospital, he returned to Guy's as chief assistant in clinical pathology, taking his London MD and MRCP. In 1954 he retraced his steps to Plymouth to become consultant haematologist to the district, a post he held until his enforced retirement due to a massive coronary occlusion in 1981.

Harry had a deep-seated concern for his fellows and was always willing to give unobtrusive service to others. Blessed with a deliciously robust sense of humour, and a capacity to see and to absolve all traces of human frailty, he was rarely at a loss to cap a story and join in the appreciative chuckles. Fundamentally his was a retiring mien and to many of his colleagues he remained a stranger; partly this could be traced to his choice of the burgeoning specialty of haematology, of which he was obliged to remain a lonely exponent in Plymouth for almost two decades. At last, in the '70s, he was joined by a colleague of comparable experience, and his detailed grasp of the intricacies of his subject blossomed as he was able to share the demands of departmental administration: now he could enjoy his gifts for clinical observation and diagnosis and shed the sense of isolation that could so easily have marred his professional contributions.

Outside medicine, Harry's strong sense of family made him a tower of strength to the many who sought his advice and help; for some of whom he went to quite extraordinary lengths to ameliorate quite dauntingly intractable family complications. In 1971 he married Helen Rindl, daughter of a South African professor of chemistry, whom he had known from his earlier visits to South Africa. This truly devoted partnership gave to Harry, until then a confirmed bachelor, the care and affection that were to be so important during his last clouded years, when both health and resilience tended to desert him.

Apart from his congregational commitments, Harry was a keen fisherman, a shrewd and knowledgeable philatelist and a connoisseur of fine things, including carpentry and furniture making. He had a gourmet's discernment of the more important things in life. His scientific motivation may occasionally have bordered on the obsessional, but this reflected the integrity that he valued above all else. He never consciously sought to give offence or to cause humiliation; he did his best for others and was unfailingly loyal to his own staff. Until his ultimate myocardial ischaemia he enjoyed good health, taking part in both district and regional advisory functions in his specialty, and for more than a decade acting as indefatigable secretary to the division of pathology. Were he to catch a glimpse of this description of him as a slightly skew cross between the Archangel Gabriel and St Francis of Assisi, Harry Greenburgh would disappear behind a cloud of mischievous laughter, perhaps proferring a mild extenuation of the solecisms but, ever the gentleman, wishing not to criticize. That is how his friends will remember him: in the old-fashioned sense of the word, a gentleman.

GRS

[*Brit.med.J.*,1985,**290**,1436; Photo]

GRIEVE, Sydney *b.*1 July 1917 *d.* - Aug 1984
MB BCh Wits(1939) MRCP(1946) FRCP(1973)

Sydney (Bobby) Grieve was born in Johannesburg, South Africa, the son of John Miller Grieve, accountant, and his wife Mary Florence Tucker. He was educated at Pretoria Boys' High School and the Univerity of Witwatersrand, qualifying in medicine in 1939, after clinical studies at the General Hospital Johannesburg.

After the usual house appointments, Grieve joined the South African Medical Corps in 1941 and served with them in South Africa, India and Egypt until 1946, being demobilized in London, England; the year when he also obtained his membership of the College. During this period, as a junior doctor at the King Edward Hospital, Durban, he met Jane (Jean) Morrow Cook, a nursing sister, whom he married in 1946. They were a devoted couple and their marriage was an exceedingly happy one, but they had no children.

Grieve was an outstanding practitioner of clinical medicine: first physician in charge at the Coronation Hospital, University of Witwatersrand, then successively senior physician, principle physician, associate professor and, finally, professor and head of the department of medicine. He cherished the long traditions of the medical profession and will be remembered by generations of students, doctors and patients, for his kindness and sincerity.

Bobby was a man of shy charm, hidden beneath a protective brusque exterior. He was an enthusiastic follower of sports of all kinds. In the end he faced a long and exhausting illness with great courage. His wife survived him.

GW/VL

GRIFFITHS, Robert Amos *b.*11 July 1922 *d.*14 Sept 1988
BA Oxon(1944) BM BCh(1947) MA(1958) MRCGP(1958)
MRCP(1964) FRCP(1976)

Amos Griffiths, consultant physician in geriatrics at Oxford, had been retired only three years before his untimely death. He was born in a small mining village, Royston in Yorkshire, and educated at Normanton Grammar School where he won a scholarship to read medicine at University College, Oxford; for his clinical training he went to Guy's Hospital medical school. Shortly after qualifying he joined the RAMC as a regimental medical officer in the Grenadier Guards. After the war he held several appointments in the Guy's Hospital sector before entering general practice. He was an outstanding general practitioner in Derby for 12 years, but he always wanted to increase his knowledge and clinical skills and he decided to enter hospital medicine. He first had to obtain his membership of the College, in which he was successful from a busy general practice, a rare achievement in those days. He then became the first senior registrar in geriatric medicine at Cowley Road Hospital, Oxford, a leading teaching hospital in the care of the elderly.

In 1965 he was appointed consultant physician in geriatric medicine at the Horton Hospital, Banbury, which he developed into a modern district general hospital. He established a post of consultant physician in general medicine with a special interest in the elderly, and obtained the appointment by open competition. He became chairman of the medical executive committee and played a major role in the development of the hospital services in Banbury.

In 1973 he moved to Oxford as consultant physician in geriatric medicine to the United Oxford Hospitals, and clinical lecturer in the University of Oxford. With the opening of the John Radcliffe Hospital in 1979 he joined a general medical firm where he developed a geriatric service linked to the Radcliffe infirmary and Abingdon Community Hospital. His research interests were in degenerative neurological disorders of the elderly and he organized an international conference at Oxford which formed the basis of a book. He continued with research and private practice after his retirement until his final illness; he never lost his interest in medicine and continued to read the medical journals assiduously.

Amos was a kindly, conscientious and caring clinician with high clinical standards; his patients had great confidence in him and he was held in high regard by his medical colleagues. He played an active part in the development of the hospital and community services for the elderly and was on the Health Advisory Service, the Alzheimer's Disease Society, and chairman of Age Concern, Oxfordshire.

Apart from medicine, Amos was a committed freemason and in his later years he held high rank both locally and nationally. In this sphere, as in medicine, he was able to fulfil in a practical way his deep felt wish to help others.

To his many friends he was a genial host at his lovely home in Shotover, but above all he was a family man. In 1948 he married Grace Margaret ('Patsy'), a Radcliffe trained nurse and the daughter of the Reverend Robert Henry Wilson. They were devoted to each other and to their family of two sons, twin daughters, and a growing number of grandchildren.

LW

[*Brit.med.J.*,1988,**297**,1124; Photo]

GRIMBLE, Allan George Seaverns *b.*8 Mar 1920 *d.*9 Mar 1984
MB BS Lond(1945) MD(1950) MRCP(1953) FRCP(1972)

Allan Grimble was born in Hong Kong, where his father Eric Norton Grimble was a broker. He was educated at Cottesmore Preparatory School and at Shrewsbury, before entering the medical school at St Mary's Hospital. Shortly after beginning his medical studies he took six months off work to attend art school. While at St Mary's he became intensely interested in the sea and sailing, and later in his undergraduate career he took another six months off work to go to sea with the Scottish fishing fleet.

He was an impressive figure: tall, upright, with a weatherbeaten complexion and blue eyes, and even at the age of 64 his dark brown hair was scarcely touched with grey.

In the obituary notices in the *BMJ* and *Guy's Gazette* he was described as 'a very private man'. And in a list he made of his interests and hobbies he mentions offshore sailing, painting and portraiture, and 'natural quiet'; perhaps this last may give some hint of his desire for privacy. Once, when asked for the name of his yacht by one of his staff, who was also a sailor, he smiled - but made no reply.

In his own field of genital medicine he wrote some excellent papers on the prognosis in cardiovascular syphilis, and mycoplasma urethitis, with others on immunology and epidemiology. Just before his death he had completed a very scholarly work entitled *Tradition, science and medicine: towards a metaphysical paradigm for medicine*, and the breadth of knowledge displayed is most impressive. In addition to a consideration of homoeopathy, faith healing and nutrition, he dealt with the traditional methods of medicine in Tibetan, Hindu Vedic, Chinese, Egyptian, Islamic, Assyrian, Babylonian, Persian, Polynesian and American Indian cultures. This helps to explain some of his professional quirks. The term venereology was anathema to him: 'Any specialty centered on an infection is doomed to early extinction; I wish I could convince my colleagues of this fact.'

In the Lloyd clinic every new patient had to have a complete physical examination. No patient could be referred to another consultant without all reasonable investigations being performed and a careful study of the results. Another of his foibles was that he insisted that no member of staff must make an internal examination of a woman while seated, an edict that some of the older assistants who were of less athletic build than their chief found somewhat trying. He also refused to employ any professional contact tracers, holding that this was not only to abdicate the duty of a doctor to his patients and to the public health but also that it subjected the patients to an unjustifiable invasion of their privacy.

As a chief he was inspiring and helpful, but sharply critical of any slackness. It is, I think, a pity that no publisher has yet seen fit to produce his last work.

He married Phyllis Muriel Bournes in 1946 and there was one child of the marriage.

FJA/GW

[*Brit.med.J.*,1984,**289**,563; *Guy's Hosp.Gaz.*,May 1984,145]

GRUSIN, Hyman *b*.30 Sept 1913 *d*.18 Sept 1987
MB BCh Wits(1937) MRCP(1948) FRCP(1977)

Hyman (Hymie) Grusin died suddenly and unexpectedly in Johannesburg. He had previously sustained an acute myocardial infarction in 1966.

Educated in Johannesburg at Jeppe High School, he qualified at Witwatersrand University. He joined the South African Medical Corps shortly after the outbreak of the second world war and with the fall of Tobruk in mid-1942 became a prisoner of war in Italy. When Italy capitulated in September 1943, Grusin elected to remain with his patients in a military hospital rather than escape, as was possible for many South African and other prisoners of war at that time since the German Army had not yet occupied the whole of Italy. He served the remainder of the war in prisoner of war camps in Germany.

After the war Hymie completed his postgraduate studies in London and obtained his membership of the College. He worked at the Royal Free Hospital and also spent several months at the Hammersmith Hospital and Postgraduate Medical School with the renowned cardiologist, Paul Wood [*Munk's Roll*,Vol.V,p.456]. When he returned to South Africa he was appointed a full time physician at the Coronation

Hospital but soon transferred to Baragwanath Hospital where he remained until he took up private practice in 1955. He was elected a Fellow of the College in 1977.

Hymie married Rhona Berman, daughter of an attorney, in 1955. They had no children, but his wife had a son by a previous marriage and they were a close knit family.

Grusin did not enjoy didactic teaching and preferred discussion at the bedside, and the examination of the patient with students or junior staff. He challenged dogma and made noteworthy original observations during his several years at Baragwanath Hospital. These included the recognition, with Priscilla Kincaid-Smith, of the prevalence of scurvy, its clinical manifestations and its association with haemosiderosis and osteoporosis. His paper relating to electrocardiographic alterations in the black population of South Africa, remains a classic and is still quoted in the pertinent literature [*Circulation*,June 1954,9,860-67]. He was interested in the congestive cardiomyopathy of black patients, and wrote meaningfully of beri-beri heart disease. His enthusiasm and enquiring mind were an inspiration to junior staff at Baragwanath.

Hymie was intolerant of insincere or superficial behaviour and seldom failed to express criticism of those who were guilty of it, in somewhat vehement terms. Despite his outspoken disposition he was humble about his own professional ability and achievements. Because of this humility, his astute clinical acumen and discerning assessments were sometimes underestimated by colleagues. His patients were his principal interest and he left no stone unturned in order to clarify a diagnosis and formulate optimal management. Many of his patients were also his friends.

At one time he was an enthusiastic and accomplished golfer, but he turned to bowls after his myocardial infarction. His joviality, amusing comments, and overall enthusiasm on the bowling green were sorely missed by his fellow players.

JBB

[*South Afr.med.J.*,1988,**73**,198-99]

HADDOCK, David Robert Wigston *b.*[?] *d.*10 June 1985
MD Liverp(1950) MB ChB(1952) MRCP(1954) DTM&H(1959) FRCP(1972)

David Haddock was born in Southport and graduated from Liverpool University. He had an outstanding undergraduate career, gaining distinctions in physics, chemistry and pharmacology, medals in pathology and surgery, and prizes in pharmacology and obstetrics. He also won the Milne medal for tropical medicine. After completing his postgraduate training in various Liverpool hospitals he joined Her Majesty's Overseas Colonial Service in 1956, at first as a special grade medical officer, being promoted later to specialist and consultant physician. He spent nine years in Tanzania where he gained a reputation for being a shrewd physician who would champion his patient's cause in a manner that earned the gratitude of many from all walks of life.

In 1965 he returned to England and joined the Liverpool School of Tropical Medicine as a senior lecturer (Technical Assistance) and in 1966 he was seconded to the Ghana Medical School where he spent six very happy and productive years, helping to educate a generation of Ghanian doctors who at all times held him in the highest esteem. He was appointed professor of medicine and head of the department of medicine at the University of Benin, Nigeria, in 1973, where he was responsible for the clinical training of the first medical students ever to graduate from that university. After a short spell back in Liverpool he was offered the chair of medicine at King Abdul Aziz University, Jeddah, where once again his dedication, sympathy, clinical skills and teaching ability endeared him to both students and colleagues.

His marriage was a very happy one and throughout his many years overseas his wife, May, gave him loyal and valiant support. They had two sons, Keith and Clive.

David's clinical interests were wide, but one of his special subjects was tropical neurology, on which he wrote several excellent papers which gained him an international reputation as an acknowledged expert. In 1980 he returned to Liverpool, taking an active part in the clinical and teaching activities at the School. He was always a glutton for work and undertook extra medical duties at Whiston Hospital, and towards the end of his life acted as consultant in tropical medicine at Monsall Hospital in Manchester, where he earned the respect and affection of his colleagues, as he also did at the School.

He had several interests outside medicine and was a keen and talented photographer.

David's loyalty was exemplary; his readiness to help at all times, no matter how inconvenient, became almost legendary. His quiet and unassuming personality masked a very strong character, which became evident during his short and painful terminal illness. He knew that he had malignant disease but accepted it with fortitude, concerned about some literary commitments which he felt he was not going to be able to meet. This was typical of him. His last days were made easier by his wife's devotion and skilled nursing.

HMJG/GW

[*Lancet*, 1985,2,226; *Liverpool Echo*, 7 Mar 1969; Photo]

HADLEY, George Dickinson *b.*30 June 1908 *d.*14 Aug 1984
MRCS LRCP(1934) MB BChir Cantab(1934) MRCP(1936)
MD(1939) FRCP(1947)

George Hadley (Dickin, at home) was the son of Laurence Hadley, editor of the *Birmingham Post.* To him he owed the nucleus of his collection of books on angling and his bibliophily, as well as his civic sense and a close circle of friends from his youth in Birmingham. He was educated at King Edward VI School, Birmingham, and Clare College, Cambridge, where he obtained first class honours in natural sciences. He qualified at the Middlesex Hospital, and was house physician to Charles Lakin [*Munk's Roll*,Vol.VI,p.272] and house surgeon to Gordon Gordon-Taylor, whose clinical style he had clearly admired. From 1936-38 he became an Elmore research student at Cambridge and developed his lifelong interest in peptic ulcer. He was then registrar to Donald Hunter [*Munk's Roll*,Vol.VII,p.288] at the London Hospital. He joined the RAMC in 1939 and was captured at Dunkirk. While a prisoner he helped to start a camp orchestra, with instruments provided by the Red Cross. He was himself a 'cellist of professional standard. The same camp achieved fame from some of the earliest continuous field observations on nesting brids. The corporate spirit which directed such activities is well caught in John Buxton's monograph *The Redstart* in the New Naturalist series.

Hadley became resident medical officer at the Middlesex Hospital directly after demobilization and built into the post an authority and breadth of responsibility for which his successors were grateful. After a year he was appointed assistant physician, and was later also physician, at the Canadian Red Cross Hospital, Taplow. He was not a prolific writer and he is best remembered for his teaching, which had the same concise quality as that of his contemporaries Richard Asher [*Munk's Roll*,Vol.VI,p.16] and Horace Joules [*Munk's Roll*,Vol.VII, p.307]. His special interest was in endoscopy: he used the Hermon Taylor and Schindler instruments from 1949; himself brought back from Japan the first gastrocamera in 1963, and went on to be a pioneer of fibreoptic endoscopy. He was a firm treasurer of the British Society of Gastroenterology in its early and more intimate days.

The man was much more complex than this professional biography indicates. He returned from the war intensely shy and often monosyllabic, and seemed at ease only with patients (unfailingly), and with small groups of people whom he knew well. His capacity to make a few very close friendships was characteristic of him. After about 1960 he mellowed, only to be struck by serious illness in the last five years of his professional life, which renewed his isolation. His reputation for epigrammatic description, verbal and written, was never diluted by professional small talk. He was distinctly impressionistic about dates and commitments. Perhaps as a consequence of his experience in the war, his clear political vision was not matched by any public commitment to change, at a time (the years immediately before the first reorganization of the National Health Service) when such vision was at a premium.

His private life was very happy and full. His wife Jean, née Stewart, was a professional viola player, and his three daughters all played instruments at home and abroad. The Hadleys were the friends of many of the musicians of their time: Vaughan Williams was his patient and confidant. They lived in George Street, in a tall house full of books, with two Volkswagen Beetles at the door.

PAJB

[*Brit.med.J.*,1984,**289**,1082; *Lancet*,1984,**2**,587; *The Times*,1 Sept 1984; *Middx.Hosp.J.*,1974; Photo]

HALONEN, Pentti Ilmari *b.*27 Apr 1914 *d.*[?] 1983
MD Helsinki(1941) *FRCP(1978)

Pentti Halonen was born in Kuopio, in the Finnish lake district province of Savo, and he spoke the distinctive Savo dialect all his life, even accenting his spoken English with it. His father was a master builder. From school in Kuopio, Pentti went to university in Helsinki, where the war delayed his graduation until 1941. He had then become attracted to cardiology and wrote his MD thesis on the venous circulation of the heart.

Halonen's scientific bent in internal medicine led to his appointment as professor of medicine at the University of Helsinki at the age of 39. In 1947, some six years before his appointment as professor, he had been appointed chief physician to the hospital of the privately funded Wihuri Research Institute, where he founded the first cardiac catheterization laboratory in Finland. Experience in this new technique was gained at Hammersmith Hospital where he was made warmly welcome by John McMichael (later Sir John) in spite of the fact that Halonen then spoke very poor English and never became really fluent in it.

In 1959 he invited the Hammersmith heart team to Finland to initiate open heart surgery there. The team was led by W P Cleland, a Fellow of the College. Pentti also had close contacts with the leaders of cardiology in Europe, USA and Scandinavia, and these international connexions helped him in his pioneer work of training the first generation of modern cardiologists in Finland.

However, Halonen was not a narrow specialist. He took an active part in hospital administration and helped to plan the new Meilahti Hospital in 1965, before becoming chief physician at the Helsinki University Central Hospital from 1970-74. He personally cared for a large number of patients until the end of his life, and his great clinical ability led him to become personal physician to two Finnish Presidents - Juho Passikivi and Urho Kekkonen.

Halonen was actively engaged in the work of national and international societies, being founder of the Finnish Cardiac Society, a member of the Finnish Acdemy of Sciences and Letters, and a corresponding member of the British and other cardiac societies. He was also vice-chairman of a bank and a pharmaceutical company.

Pentti was a man of great personal charm, with no hint of stuffiness or pretension despite his high position. He kept his family life entirely private and one never met his wife or daughter. His hobby was salmon and trout fishing, and when this became difficult in Norway he went regularly to Iceland for it. A heavy cigarette smoker - usually lighting one from the butt of another - he paid the price he must have expected. He died from cancer of the lung, actually an adenocarcinoma, at the age of 69.

ÁÈ[1]

[Photo]

HAMILTON, Max *b*.9 Feb 1912 *d*.6 Aug 1988
MRCS LRCP(1934)MB BS Lond(1937) DPM(1946) MD(1950)
MRCP(1963) FRCP(1970) FRCPsych(1971)

Max Hamilton was born near Frankfurt, the son of Henry Himmelschein and his wife Sarah Kleinberg. His family emigrated to England when he was three years old, settling in the east end of London. He was educated at the Central Foundation school, Cowper Street, London, before undertaking medical training at University College Hospital medical school, University of London. After qualification he returned to the east end, working as a physician in one of the most deprived areas of the country. During the second world war he joined the Royal Air Force as a medical officer and saw service from 1939-46. Towards the end of hostilities he took Part 1 of the DPM and after demobilization began his clinical psychiatric training at the Maudsley Hospital, London. There he had sharp differences of opinion with some of his teachers, memories of which remained with him for the rest of his life, and he failed to be appointed to a post which would allow him to continue his studies at the hospital. He continued his psychiatric training at UCH for the next four years, where he came under the influence of Sir Cyril Burt, one of the first clinical psychologists, and became particularly interested in psychology and statistics.

From 1950-51 he worked at King's College Hospital as senior registrar to Denis Hill, later Sir Denis [*Munk's Roll*.Vol.VII.p.264], a man two years his junior. While at King's he submitted his MD thesis on the personality of dyspeptics. He subsequently worked at Tooting Bec Hospital in the senior hospital medical officer grade; an unusual career step for a future academic but one which gave him experience of mental hospital psychiatry, and of responsiblity for the care of a large number of patients. He remained proud of his unique feat of progressing from SHMO to professor of psychiatry.

He moved to Leeds as a senior lecturer in the department of psychiatry in 1953. After a few years he again took an unusual step in resigning his post at the age of 45 years, to take up a research post locally which would allow him to spend more time in research with fewer demands from routine clinical work. He was visiting scientist at the National Institute of Mental Health, Bethesda, Maryland, USA, in the year 1959-60, and returned to Leeds to work as a member of the external staff of the MRC. Following the unexpected death of G R Hargreaves in 1963, he was appointed to the Nuffield chair of psychiatry at the age of 51.

During Max Hamilton's tenure of the Nuffield chair the activities of the department were considerably expanded: the diploma in psychological medicine, which had been established shortly after the war, was developed to include a research dissertation; a diploma course in social work was established jointly with the Department of Adult Education and Extra Mural Studies, and later a course in mental health social work leading to an MSc. Max Hamilton believed that social workers made a vital contribution to the care of psychiatric patients and he sought to increase the academic component in their training. Regrettably, both these courses were closed as a consequence of the university economies of 1981-82. In 1965 a course in clinical psychology, leading to an MSc, was established jointly with the department of psychology. Happily, this course is both thriving and expanding and at the time of its 20th anniversary in 1985 it had been responsible for the training of 10% of the clinical psychologists in Britain. All these developments stemmed from Max Hamilton's belief that the contributions of many disciplines were essential to advances in psychiatry. At the same time the Leeds School of Medicine was expanding, and there was an increasing demand from undergraduates for instruction in psychiatry.

But it was in the fields of teaching and research that Max Hamilton made his most distinctive contributions. He was an excellent lecturer, giving a clear, concise account of many psychiatric topics with a degree of critical appraisal which made them memorable. His Hargreaves lecture to the Yorkshire Regional Psychiatric Association, in which he reviewed established information on the nature of depressive illness and offered new interpretations of the data, was typical. He was in demand as a lecturer, not only in the UK but throughout the English-speaking world. He was also in demand in Europe and frequently delivered lectures in French and German. And he was a frequent visitor to the United States, where he spent long periods after his retirement Many a speaker at conferences had good reason to remember him; he had an ability to ask the most penetrating questions and if he felt he was right, and he usually was, he was prepared to persist with his opinion in the face of overwhelming odds - and even hostility.

As a teacher Max is remembered as being somewhat severe, but he produced a lasting impression on his many pupils. He always stressed the central position of the patient in both clinical work and in research. He had no love for fools but was prepared to argue a point even with the most junior trainee if he or she could hold their ground. In spite of his fearsome reputation as a teacher there are many accounts of his kindness to individual students, and of the help given to people privately. He often held parties at home for a wide range of students both from Britain and abroad. He was particularly concerned to ensure that foreign students should feel at home in Leeds, and in the department.

Max Hamilton's major achievements in research are likely to form his most enduring memorial. He was a pioneer in the development of rating scales designed for measuring the severity and extent of psychopathology. His first published scale, known as the Hamilton Depression Rating Scale, published in 1960, is now the most widely used of all such scales; so much so that the paper in which it was described has been among the most cited of medical papers on two occasions. Hamilton was one of the first to appreciate the need for objective rating scales of psychiatric psychopathology, and for these to be well designed and critically evaluated. He subsequently developed other rating scales, including one for anxiety. He also had a long standing interest in the methodology of research. A series of lectures on this topic were published in a book, *Lectures on the Methodology of clinical research,* Edinburgh, Livingstone, 1961. His other books included *Psychosomatics,* New York, Wiley, 1955, which stemmed from a study of the influence of personality in peptic ulcer sufferers and was the basis of his MD thesis.

Max Hamilton's lifelong interest in research was marked by the University of Leeds establishing a prize in 1983 for the best dissertation submitted for the M Med Sc degree in clinical psychiatry in each year. The award was named the Max Hamilton Prize, the first recipient receiving it from Max Hamilton himself.

Among the many honours bestowed upon Max Hamilton were the presidency of the British Psychological Society in 1972. He was only the second psychiatrist to be appointed to this position. His remarks as president were regarded as controversial by a section of the membership and a protest meeting was arranged to demand a retraction - which Max Hamilton himself attended. He was foundation president of the British Association for Psychopharmacology, and a managing editor of the journal *Psychopharmacology.* In 1980 he received the Paul Roch Award for contributions to psychiatric research from the American Psychopathological Association.

Outside his professional life one of his great interests was politics, where he always favoured the advancement of the under-privileged and oppressed. He never hesitated to hold an unpopular or minority opinion, indeed he rather relished it. He was also a keen photographer and his subjects included many professional colleagues. He had an interest in gardening, particularly in narcissi and orchids, which he tended in his own garden at his home in Leeds.

After moving north in 1953, he strongly identified with Leeds and was active in local affairs. He was married twice. There were two sons of his first marriage in 1933, and two daughters and a son from his marriage to Doreen Margaret Moody in 1947. He found a central interest in his family and maintained regular contact with all its members throughout changing circumstances. He was survived by his wife Doreen, who was for many years a city councillor, and all five of his children.

RHSM

[*Brit.med.J.*,1988,**297**,914-5; *Lancet*,1988,**2**,582; Photo]

HAMILTON, Patrick John Sinclair *b.*28 July 1934 *d.*12 June 1988
BA Cantab(1955) MB ChB Edin(1958) DTM&H Lond(1963)
DPH(1966) FFCM(1980) FRCP(1986)

Patrick Hamilton was born at Edinburgh, the son of a Presbyterian minister, John Edmund, and his wife Elizabeth Lilias, a medical practitioner. He was educated at Winchester College and Christ's College, Cambridge, and his life in medicine and public health spanned four continents. His interest in tropical diseases developed in Nepal with the Royal Army Medical Corps, but his work as a teacher and research worker really began in Uganda in 1963 at the Makerere medical school, with studies of splenomegaly, sickle cell disease, tetanus, and a growing interest in epidemiology. From there he moved to the London School of Hygiene and Tropical Medicine under Donald Reid [*Munk's Roll*, Vol.VII,p.493] and applied himself to the organization of an epidemiological study of heart disease involving 20,000 civil servants, and to an examination of the risks of smoking and the effects of anti-smoking programmes.

His abilities proven, Hamilton was the natural choice to head a newly created tropical epidemiology unit within the School. Then, after establishing the unit in research on onchocerciasis, schistosomiasis and African arvoviruses, he went to Trinidad as director of the Caribbean Epidemiology Centre. Again he started virtually from scratch, building up a large and effective international organization engaged in the control of communicable diseases in the area.

After seven years in Port of Spain he came back to the London School of Hygiene and Tropical Medicine to take up the chair of community health. During recent years the specialty has had to struggle against outsiders casting doubts as to its worth, while many of its own practitioners have sought to defend an increasingly limited territory. In contrast, Patrick saw this as a time to push the case for community medicine all the harder, both in academic and health service spheres.

His vision was characterized by the range of potential contributors - clinicians, epidemiologists, economists, statisticians, sociologists - and the breadth of its applications both in industrialized and third world countries. And he was forever seeking to extend these activities still further. One of his last major initiatives was the establishment, with the help of the Nuffield Provincial Hospitals Trust, of a research group to study the social impact of AIDS, a development that included the appointment of a historian to the School's staff. This reflected his personal delight in history, which he saw as crucial to our understanding of contemporary medicine.

At a time of the increasing emphasis on the scientific and technical aspects of health and health services research, Patrick continued to recognize the importance of the managerial and personal contribution for achieving change. There was probably no country in the world in which he lacked a personal contact - an ex-colleague, a past student, an acquaintance from a previous visit. This unrivalled international network meant that regardless of topic or location he could, by picking up the telephone, help students and colleagues with even their most obscure problems. It was in his capacity as adviser to the WHO programme on onchocerciasis, a disease causing widespread blindness in some tropical areas, that he was visiting Ouagadougou, Burkina Faso, in West Africa, at the time of his death. He had characteristically refused to allow his very active life to be restricted by his diabetes.

Patrick married Fiona Hunter, a financial journalist, in 1972 and they had two children, a son and a daughter. He and his wife filled their London house with unpredictable combinations of overseas visitors, colleagues, students, friends and neighbours. As well as being a fine ambassador for this country, Patrick was an enthusiastic host. He also found time to collect a variety of things. Besides the tribal spears and weavings that decorated his office were antique mahogany medicine chests, packages of historic books on public health, and old medical instruments of obscure purpose.

Patrick's sudden death came as a great shock to all who studied under him or worked with him.

NAB/CS

[*The Times*,17 June 1988; *The Guardian*,17 June 1988; *The Independent*, 20 June 1988; *Brit.med.J.*,1988,**297**,550; *Lancet*,1988,1,1469; Photo]

HARRIS, John Stuart *b.*13 Oct 1900 *d.*2 May 1986
MB ChB Edin(1922) MD(1924) DPM(1924) MRCP(1929)
FRCP(1953) FRCPsych (1971)

Stuart Harris, a Scot, was physician superintendent of Claybury Hospital, Woodford Green, Essex. from 1938 until his retirement in 1961. He was among the best of this country's physician superintendents; his outstanding contribution to psychiatry being adminstrative rather than clinical. As a director he was patient and benign; a man who chose his staff carefully and supported them fully. As a man, he was extremely reticent and rarely talked about himself, but he was well known for his unflappability - a quality which contributed greatly to his outstanding administrative ability.

Stuart was educated at George Watson Academy and Edinburgh University where he obtained his MD and DPM in the same year. Before his appointment to Claybury he had served on the staff of the Maudsley Hospital, where he was a great admirer of his chief, Edward Mapother [*Munk's Roll*,Vol.V,p.266] and as deputy superintendent of West Park Hospital. After the introduction of the NHS, Stuart Harris and George Graham, the regional medical officer, contributed greatly to the development of psychiatry in the area covered by the NE Metropolitan Regional Hospital Board.

Claybury Hospital became well known in the 1950s and '60s for its development of the therapeutic community under the supervision of Stuart's successor, Denis Martin, who had been carefully selected by Stuart. There is no doubt that Claybury would not have been ready for these developments but for the patient work Stuart had put in during his time as superintendent, particularly during the difficult war years, and his calm and accepting support and encouragement were essential ingredients in an exciting experiment, which is unlikely to be repeated because of the changing psychiatric scene. After his retirement from Claybury, Stuart became one of the Lord Chancellor's Visitors and served for some four years in a post which had been formerly occupied by a distinguished predecessor at Claybury, the first physician superintendent, Sir Robert Armstrong-Jones [*Munk's Roll*, Vol.IV,p.480].

He met his wife, Kathleen, when he was on the staff at the Maudsley where she was a ward sister. They had two children; his son Robert is a senior probation officer in Newham, and his daughter Eileen, a consultant psychiatrist, married Graham Evans who was a psychiatric registrar at Claybury. They are now practising as consultant psychiatrists in New Zealand.

For many years they has a house at Sheringham, Norfolk. Stuart and his wife were both enthusiastic golfers; Kathleen being the better player. After his retirement they moved house several times; first to Effingham in Surrey and then to Chigwell, Essex, finally settling down in Sheringham. For many years they spent their holidays in New Zealand where Stuart would spend some time putting case notes in good order. He remained alert and interested in psychiatry to the end. Unfortunately, during the last three or four years his sight began to fail so that he was unable to play golf. He eventually died of carcinoma of the prostate.

JSP

[*Brit.med.J.*,1987,**294**,64]

HARRIS, Ronald *b.*31 Dec 1918 *d.*4 Apr 1987
MB ChB Leeds(1943) DPhys Med(1949) MD(1951) MRCP(1967) FRCP(1974)

Ronald Harris was born in Leeds, where his father, Alec Ellis Harris, was a tailoring manufacturer. His mother, Dorothy, née Pliener, was the daughter of a haberdasher. He was educated at the City of Leeds School and the University of Leeds, where he graduated in medicine, undertaking his clinical studies at Leeds General Hospital.

He held postgraduate house appointments at Tynemouth Infirmary, Doncaster Royal Infirmary, the Radcliffe Infirmary, Oxford, the Royal Infirmary, Leeds, and the Royal Free Hospital, London. He soon developed an interest in physical medicine and moved to Buxton in 1952, where he set up a comprehensive rehabilitation service for a wide range of locomotor disorders, in addition to complementing existing rheumatology practice. He wrote many articles which contributed to a better understanding of physical methods of treatment and was the author of several chapters in a series of books edited by Sidney Licht, *Physical Medicine Library*, New Haven, USA, E Licht.

During the second world war he served as a surgeon lieutenant RNVR, from 1945-47.

He married Natalie Denise Sacof, daughter of a company director, in 1949 and they had three children, a son and two daughters. Outside medicine and his family, his main interest lay in horticulture and rowing; he coached for Bristol University Boat Club and also for Leeds University and Northampton College, London.

Ronnie was greatly admired at Buxton for his drive and enthusiasm for rehabilitation and also because he recognized that all the members of his team - whether medical, paramedical or nursing - should be regarded as being on an equal footing. His final illness forced him to retire at 55, and he spent many years in hospital. His wife and family survived him.

GW/VL

[*Brit.med.J.*,1987,294,1360; Photo]

HARRIS-JONES, John Norman *b.*8 Nov 1919 *d.*17 Sept 1987
MRCS LRCP(1943) MB BS Lond(1945) MRCP(1946) MD(1949) FRCP(1970)

John Harris-Jones was born at Skewen in South Wales, the son of David Lewis Jones, a general practitioner, and his wife Amy Hopkins, daughter of a colliery manager. He was educated at Neath Grammar School and studied medicine at St Bartholomew's Hospital, London, qualifying with the conjoint in 1943. After house appointments at Bart's and the London Chest Hospital he became medical registrar and resident medical officer at Southend General Hospital. In 1947 he joined the RAMC and served with the rank of major in various military hospitals. He left the Army in 1949 and was appointed senior medical registrar to the Royal Hospital, Sheffield, where he served until 1957. In 1958 he returned to his home territory as consultant physician to the Llanelli Hospital, Swansea.

John's skills as an excellent and dedicated physician were soon recognized throughout South Wales. In 1966 he fell seriously ill but on recovery, instead of reducing his workload he continued to work in a way which was typical of him; he renewed a longstanding interest in endocrine disease and attended a refresher course in endocrinology at the Hammersmith Hospital, London. This influenced his career as a general physician with a special interest in endocrine medicine. He built up a first class endocrine service in West Glamorgan, and established a thyroid service for the whole of west Wales. In his latter years he had a close liaison with the academic department of medicine at the University Hospital in Cardiff.

John was a very loyal colleague and a delightful companion. He loved his work but had many outside interests, including golf, rugby, music and bridge. He was also a member of the Bristol Channel Yacht Club.

In his final illness, which he characteristically diagnosed himself, he showed courage and determination to live life to the full.

In 1946 he married Caroline (Carole) Agnes Mackenzie and was blessed with a very happy family life. They had two children, a son and a daughter. His son also entered the medical profession, representing an unbroken tradition for three generations. His children and three grandchildren brought great joy to the last years of his life.

DD/GW

[Photo]

HARRISON, Robert John *b.*15 Oct 1905 *d.*27 May 1988
MRCS LRCP(1937) MRCP(1943) FRCP(1961)

Robert Harrison was born in London, the son of Charles Harrison, a builder. He was educated at Sir George Monoux Grammar School and studied medicine at St Bartholomew's Hospital, qualifying with the joint diploma in 1937. Six years later he obtained his membership of the College.

Ill health, due to surgery for renal calculi, caused his rejection for military service during the war, but he was active in the Emergency Medical Service. Junior hospital appointments included that of lecturer in the department of medicine at the Postgraduate Medical School, Hammersmith, and with the inception of the NHS he was appointed consultant physician at Fulham Hospital in 1948. In 1951 he was appointed to the staff of St James Hospital, Balham, where he served until his retirement in 1970. After retirement he undertook several locum appointments and in 1975 he was instrumental in helping to establish the Worthing Postgraduate Medical Centre, organizing meetings and inviting speakers from his wide circle of distinguished colleagues.

Widely known as 'Happy' Harrison, he was a committed and supremely well informed clinician with a 'nose' for the unusual. His junior staff were devoted to him and under his teaching, at which he excelled, several obtained their membership of the College and became physicians themselves. 'Happy' greatly enjoyed serving as examiner for the College, and he kept in contact with the Royal Postgraduate Medical School where he was regularly to be seen at the Wednesday morning staff rounds. He had worked as a lecturer in Sir John McMichael's department and enjoyed the cut and thrust of these meetings throughout his professional career. Near retirement he worked for the Hospital Advisory Service and greatly enjoyed the visits to other units.

As a fellow of the Royal Society of Medicine he was exceptionally active, being a member of council, a president of the clinical section, and vice-president of the section of medicine. In 1969 he was elected an honorary member of the clinical section, the first fellow to have been so honoured. He published articles on a variety of topics, including the first described case of Q-fever endocarditis.

Harrison was a quiet, rather shy and private person. He coped with increasing deafness, and then his prolonged final illness, with characteristic fortitude and good humour. His work, clinical medicine, was also his hobby, which he discharged with great enthusiasm and he contributed much to the profession. He married late and was survived by his widow.

MD

[*Brit.med.J*,1989,**298**,520; Photo]

HART, Edward Watson *b.*[?] 1911 *d.*23 May 1986
MBE MRCS LRCP(1936) MB BChir Cantab(1937) MRCP(1938) DCH(1946) FRCP (1949)

Edward Hart was born in Glasgow and went to the University there for a year before going to St John's College, Cambridge, where he gained first class honours in the Tripos. He won an entrance scholarship to the Middlesex Hospital in 1933. After qualification he held house physician posts at the Middlesex and at the Hospital for Sick Children, Great Ormond Street, where he developed his interest and love for paediatrics. During the war he joined Sir Lionel Whitby's [*Munk's Roll*,Vol.V,p.444] blood transfusion service and ultimately became responsible for blood transfusion in the whole of India. He was rightly awarded an MBE for his valuable work.

After demobilization he returned to paediatrics and in 1947 was appointed paediatrician to the Middlesex Hospital where he worked single-handed for 20 years, building and developing the department to the highest standard. He was also paediatrician to the Hampstead General Hospital (now the Royal Free). During this time he was the sole teacher of paediatrics to students. He was superb; always clear and decisive. Both his students and colleagues held him in great respect, and not a little awe, for he did not hesitate to criticize their work and efforts. There are many ex-students who are most grateful for his teaching. In a tribute paid to him on his retirement from the Middlesex, he was described as 'A strict disciplinarian and woe betide any colleagues making loose statements which could not be supported by facts.' He was a keen and active member of the British Paediatric Association, being honorary secretary in 1959-68. He contributed greatly, and was made an honorary member in 1974.

Edward Hart was a true Scot, with an attractive accent, who always went straight to the point. He had a typical Scottish sense of humour - a wee bit cynical but on the nail. Outside medicine his great pleasure was golf, in which he was no mean performer. In his younger days he had a handicap of four. As in paediatrics, he played with accuracy and scrupulous attention to the task he performed. His other hobbies were woodwork and clock making, at both of which he became a real expert, and his services were always available should they be required in his village.

All who knew Edward Hart were his devoted companions and friends. He retired to the Cotswolds with his wife Peggy, and two daughters, one of whom trained as a radiographer and the other as a nurse.

AGW

[*Brit.med.J.*,1986,**292**,1742; *Middx.Hosp.J.*,Nov 1973,**73**(3); Photo]

HASSANEIN, Mahmoud Amed *b.*29 Nov 1919 *d.*28 Oct 1988
MB BCh Cairo(1942) DM(1947) MRCP(1947) FRCP(1970)

Mahmoud Hassanein was born in Sharka Province, lower Egypt, where his father was mayor. His mother Eisha Abdalla el-Attar was the daughter of a landowner. His early education was at Zagzig School and he went on to Cairo University to study medicine, his clinical studies being carried out at Kasr el-Ainy Hospital, Cairo. After graduation he held house appointments from 1942-44, when he became demonstrator in medicine and later lecturer at the University. He was appointed assistant professor 1958-63, and physician in charge of the occupational diseases unit. In 1963 he was appointed professor in the faculty of medicine.

He married Amira Mostafa el-Sharawy, daughter of a general and chief of the Air Force, and they had two children; their son Khaled followed his father into the medical profession and was appointed assistant professor of chest diseases at Cairo University, and his daughter was known as 'Hala'.

Mahmoud Hassanein's great interest lay in aviation medicine, which was affiliated to the department of industrial medicine. He died suddenly, at Cairo airport, from massive myocardial infarction while he was returning from a congress on aviation medicine in Australia.

He was an excellent teacher, a sympathetic man and a good friend. Outside medicine his interests lay in football, music, reading and fishing. He was survived by his wife and children. AG/GW

[Photo]

HAWKING, Frank *b*.b.18 May 1905 *d*.d.4 Mar 1986
BA Oxon(1927) MRCS LRCP(1929) BM ChB(1930) DM(1933) MRCP(1952) FRCP(1965)

Frank Hawking was born in Beverley, Yorkshire, the son of Robert Hawking, a farmer, and his wife Mary Atkinson who was the daughter of a cabinet maker. Frank Hawking's intellectual abilities were early recognized and he was educated at Ley's School, Cambridge, and the University of Oxford, where he obtained his BA with first class honours in 1927. He carried out his clinical studies at Bart's, later obtaining his DM in 1933. He was elected a Fellow of the College in 1965.

After house appointments in Oxford and East Ham, his first contact with tropical medicine was in 1930 when he became the Caton Memorial research fellow at the Liverpool School of Tropical Medicine, working on trypanosomes under the guidance of Warrington Yorke [*Munk's Roll*, Vol.V,p.466]. From 1933-35 he held a Radcliffe travelling fellowship at Oxford, and then spent four years as a lecturer in pharmacology at the Welsh National School of Medicine. In 1939 came his most important appointment: to the scientific staff of the National Institute of Medical Research, then at Hampstead. When the Institute moved to Mill Hill, Hawking became head of the division of chemotherapy, later renamed 'parasitology', a position he held until his first retirement in 1970.

His major scientific interests in the early part of this period related to chemotherapy. In 1950, with J S Lawrence, he published an important book *The Sulphonamides*, London, H K Lewis, jointly edited with R J Schnitzer over the period 1963-65, and made a number of valuable contributions to a series entitled *Experimental Chemotherapy*, New York, Academic Press. This interest in how to kill parasites led him directly to studies on their biology and biochemistry, and to around 150 publications on protozoan and helminth parasites. Some of his most interesting work related to the periodicity shown by blood microfilaria, and he spent a great deal of effort in determining what cues from the host were telling the parasites when best to enter the peripheral circulation in order to meet their vectors.

His big disappointment was being beaten to the discovery of the liver stages of malaria by Shortt and Garnham.

Following his retirement from Mill Hill, Frank spent a year at a primate centre in the United States, three years at the Clinical Research Centre at Harrow, three more as a senior research fellow at Brunel University, and then worked as editor of *Protozoological Abstracts* at the Commonwealth Bureau at St Albans until July 1982. Work meant a great deal to him and he was active to the end.

Frank Hawking was not a 'hail fellow well met' person; a basic shyness was often misjudged as being compounded by a mixture of self-interest and disinterest in others, but this was wrong. He could be, at times, very kind but his shyness prevented him from offering his help in special circumstances.

Frank would have never thought of himself as having gifts as a teacher but throughout the years the Mill Hill Laboratory attracted a large number of workers, many of whom now occupy university chairs or are highly placed in the pharamceutical industry. They all learned something from him.

He left a wife and four children; one being Stephen Hawking FRS, his brilliant but sadly handicapped son who was, at the same time, both a great joy and a great sorrow to his father.

SRS/RJT

[*Brit.med.J.*,1986,293,455; Photo]

HAWKINS, Leslie Alan *b.*13 July 1909 *d.*4 Aug 1985
BChir Cantab(1934) MRCS LRCP(1934) DCH(1935) MB(1935)
MRCP(1943) FRCP(1970)

Leslie Alan Hawkins died in the Doncaster Royal Infirmary, the hospital where he had been a consultant paediatrician for 27 years. At the time of his birth the family lived in Calcutta, India, but they returned to England when Leslie was five years old. The new family home was at Cowslip Green, some 10 miles from Bristol, and Leslie went to a preparatory school at Weston-super-Mare. From there he went to Clifton College where he became head of his house. He received his medical education at Cambridge, and his clinical studies were undertaken at the London Hospital. Whilst at Cambridge he distinguished himself by obtaining first class honours in both Parts I and II of the natural sciences tripos (physiology). At that time Leslie came under the influence of 'Joe' Barcroft, and he frequently spoke in later years of his teaching and how it had shaped and sharpened his own approach to medicine. On qualification he joined the house staff at the London Hospital. At the outbreak of war he was asked, in view of his organizing experience as receiving room officer at the London Hospital, to join the Emergency Medical Service. He served as a sector hospital officer until 1946, being responsible for the allocation and utilization of hospital beds for a large area of southern England.

In 1940 he married Phyllis, the daughter of Harry Roberts, a general practitioner and industrial medical officer who practised in Ayrshire. They had met at the London Hospital when Phyllis was a medical secretary.

After a short spell at the Northern Hospital, Winchmore Hill, Leslie was appointed as a consultant physician and paediatrician to the Doncaster hospital group in 1947. From 1964, until his retirement ten years later, he practised solely as a paediatrician. During his time in Doncaster he saw, and was involved in, a great number of changes culminating in the hospital paediatric care being based in one unit at the Doncaster Royal Infirmary. Quite early in his time in Doncaster, Leslie appreciated that full paediatric care meant more than the confines of a hospital. He initiated close liaison with the schools' medical service, and was a frequent and welcome visitor to the local special schools.

Leslie had obviously enjoyed a happy childhood and he frequently spoke with great fondness of the family home in Somerset. It was there that he developed his initial interest in natural history, and this was to remain with him for the rest of his life.

He was a slim man of short stature, with alert eyes that betrayed his versatile and active mind. He was modest to the point of being self-effacing, and would express genuine surprise at his own achievements and successes. He had a gentle, yet robust sense of humour. He was never malicious, but enjoyed recounting anecdotes, particularly those directed against himself. His prime interest was as a clinician and he begrudged time spent on committee work. Despite this, his contributions at such meetings would be accurate, scholarly and logical - if at times lengthy enough to erode any opposition to his argument. He was loyal to his colleagues, and displayed a lasting pride and affection for his school, his university and his hospital.

Unfortunately Leslie's retirement was marred by increasing ill health. He developed polymyalgia rheumatica and increasing respiratory difficulties due to chronic bronchitis. As a consequence he became increasingly house-bound. But his mind remained active, and he maintained his wide medical and scientific interests. During this period he exercised his knowledge by considering such diverse problems as the resilience of woodpeckers' brains, and the care of mentally handicapped young adults.

Leslie's life was devoted to his work, and consequently he enjoyed a limited social round. He had, however, a loving family. His wife and three sons survived him.

AFC

[Photo]

HAYTER, Clive James *b.*b.3 Sept 1922 *d.*d.2 Mar 1986
BSc Lond(1950) MB BS(1953) MRCP(1956) FRCP(1971)

Clive Hayter, senior clinical lecturer in medicine, head of the department of nuclear medicine and consultant physician to the General Infirmary at Leeds from 1960-85, was born in Portsmouth without any strong family background in medicine. Before beginning his medical training he served as a pilot with the RAF from 1941-46, in both night fighter and low level bomber squadrons, in the UK, India and Burma, and attained the rank of flight lieutenant. He retained his interest in flying, both as a civil pilot and flying instructor, for most of his life. He was also a keen motorcyclist.

In 1946 he married Iris Hedges, daughter of a civil servant, and entered St Thomas's medical school. He had a brilliant undergraduate career, qualifying BSc with first class honours in physiology, and later MB BS with a distinction in medicine. After junior appointments at St Thomas's Hospital he joined their department of medicine as a lecturer, becoming involved in research led by Sharpey-Schafer [*Munk's Roll*,Vol.V,p.372] and was closely involved with the experimental work that Mills and dc Wardener carried out related to the effects of blood volume expansion in animals. It was at this time that he developed his lifelong interest in the diagnostic and therapeutic uses of radioisotopes in human pathophysiology. Contemporaries from this time remember his rather 'Punch-like' figure, often squatting cross-legged on a bench, smoking his pipe and airing distinctly radical political views, scoring intellectual points off his fellows. He was known to his small circle of intimates as 'Grumps', but no one doubted his enthusiasm, reliability and superb technical capabilities particularly in the field of radioisotopes.

In 1960, the General Infirmary at Leeds decided to establish a department of nuclear medicine and Clive Hayter was appointed as its head; the first such independent department in the NHS. With enthusiasm and skill he built up his department over the following years, maintaining a blend of clinicians, biochemists, physicists and technical staff, and explored almost every aspect of this new and exciting field but in particular radioisotope imaging and radioimmunoassay. He was involved in the fight for recognition of nuclear medicine as an independent discipline in the NHS, and was a founder member and first president of the British Nuclear Medicine Society in 1968. He also took on more clinical duties, becoming a consultant physician to the Infirmary. He was elected to the fellowship of the College in 1971.

In later years he assumed a greater role in management in the NHS as consultant representative on the district management team of the Leeds Western District Health Authority, and was its chairman in 1981. As may be imagined, this led to a number of conflicts with colleagues who could not (or did not wish to) appreciate that there were changes afoot in the NHS, but Clive Hayter always attempted to defend any decisions made and to warn his colleagues of the dangers of hospital administrators managing without good medical advice.

To the outsider his private life seemed tough and uncompromising but he did not complain. His wife Iris had developed chronic renal failure at a relatively early age and was on regular haemodialysis for a good part of her life. Clive took on the responsibility of her haemodialysis three times a week, year after year, including the practical aspects of cannula insertion etc. When she died in 1983 it seemed that much of his inner drive left him. He suffered from long-standing asthma which forced him to give up his beloved pipe and in his last year laid him low in hospital. Indeed it was a cruel irony that, after retiring due to ill health in 1985, he spent the rest of his remaining months in hospital. The state of his asthma often gave his colleagues in a tense committee meeting an indication of his involvement in the discussion, for as the argument increased so did Clive's wheeze.

Coupled with these formidable personal problems was the clear feeling that he was at heart an academic who was forced to take on a good deal of humdrum service work. Together this may perhaps explain the view which many of his colleagues had of him: a rather melancholic and unsympathetic figure. But he had a fierce loyalty to his department and its staff and this was equally reciprocated. He had a clear and active mind with a wide-ranging intellect and considerable foresight. At lunch he could be an interesting raconteur, with an occasional devastating side comment, at times directed against his full-time university colleagues.

He was survived by two sons and a daughter. One of his sons is a consultant geriatrician in Stoke and a member of the College. Few who knew Clive Hayter will easily forget him.

JKW²/GW

HELLIER, Francis Findlay *b.*b.30 June 1904 *d.*d.19 Nov 1986
OBE 1944 BA Cantab(1929) MRCS LRCP(1929) MBBS Lond(1929) MD(1931) MA Cantab(1931) MRCP(1931) FRCP(1941)

Francis Findlay Hellier, who was always known as Frank, was born in Leeds. He was proud of his city; his father had been professor of obstetrics and gynaecology in the University of Leeds and his grandfather, the Revd Benjamin Hellier, was at one time principal of Wesley College, Leeds. His father was born in 1853 and was 51 years old when Frank was born, so the lives of these two men extended over 133 years. Frank was educated at Oundle, where he was head boy, and then went on to Caius College, Cambridge, for his preclinical training.

He obtained firsts in both parts of the natural sciences tripos. He was taught by several famous physiologists, including Sir Frederick Gowland Hopkins [*Munk's Roll*,Vol.IV,p.535] and E D Adrian (later Lord Adrian) [*Munk's Roll*,Vol.VII,p.3]. He returned to Leeds for his clinical undergraduate training and after qualification went on to the Brompton Hospital, intending to become a general physician. He subsequently returned to Leeds General Infirmary as RMO , and later medical tutor, where he was soon attracted into dermatology by J T Ingram [*Munk's Roll*,Vol.VI, p.252] who had become the first full-time dermatologist in the city in 1927. Frank used to help him in the skin department for two sessions a week when working as medical tutor. Ingram was a powerful character who considered dermatology to be a branch of general medicine of equal status to that of other specialties. This concept appealed to Frank and he decided to specialize in dermatology. He spent six months with H W Barber [*Munk's Roll*,Vol.V,p.23] at Guy's Hospital and later studied with Pautrier, one of France's outstanding dermatologists. He often referred to this period of training in France, but later admitted that the French were ignorant of dermatology in countries other than their own. He was awarded the Chesterfield medal of St John's Hospital in 1936. When he returned to Leeds he was appointed assistant physician to the skin department and helped Ingram to build up an outstanding clinical department of dermatology on the basis of good history taking, including social and psychological aspects, accurate clinical description of the skin lesions and complete general medical examination.

Ingram went off to France soon after the outbreak of war, leaving Frank to open a new, purpose-built skin department in 1940. Ingram returned to Leeds after Dunkirk, and Frank was released to join the Army and was appointed as dermatologist to Western Command. He believed in the rehabilitation of patients with skin disorders and encouraged graded physical exercise in the soldiers in his care. When he went to Normandy with the 21st Army Group he was one of the first to treat skin infections with topical penicillin, using tablets dissolved in sterile water. Purpuric eruptions were also common in the troops and he, with Geoffrey Hodgson, was instrumental in showing that this was due to the finish in khaki uniforms and not to the khaki itself.

He returned to Leeds after the war and when Ingram went to Newcastle in 1959 Frank became head of the department, and was given a personal chair in the University in 1968. His main interests were in histopathology and in industrial dermatitis. He had trained in histopathology when he worked as a demonstrator in the University of Leeds, and later with Woringer in France and Freudenthal in London. This interest continued throughout his life and after retirement he continued to teach histopathology to the junior staff of the department. Because of his expertise in industrial dermatitis, he did a considerable amount of legal work and for many years served as a member of the Medical Appeals Tribunal. He was widely read, had an excellent memory for the literature and for cases, but he was not a scientific laboratory worker. He edited, with W N Goldsmith [*Munk's Roll*,Vol.VI,p.201] the second edition of *Recent advances in Dermatology*, London, Churchill, in 1954, and gave the Watson Smith lecture at the College in 1969 on the 'Application of dermatological concepts to general medicine'. He was honoured by being made president of the British Association of Dermatologists in 1964, and of the dermatological section of the Royal Society of Medicine in 1967. His father had been dean of medicine in 1917-18, but Frank failed to be elected dean in the early 1960s. He was chairman of St James Hospital management board, a member of the Leeds regional hospital board, and chairman and later president of Leeds Marriage Guidance Council.

Frank was an active man and a good sportsman. He just failed to get a blue at rugby, and continued to play golf until the end. Everyone remembered him as a friendly person and an excellent conversationalist - although sometimes one felt he was happier to talk than to listen. He was devoted to his wife Doreen, who had been a leading singer with the D'Oyly Carte Opera Company, and it was a great tragedy for him when she died suddenly in 1974. He was sustained by his deep religious faith. He served the Methodist Church for many years.

Fortunately, he was able to enjoy his garden, music, reading, travelling and his family, until the very end. He died suddenly 'in harness', as he would have wished, while attending a consultation with lawyers in Manchester. He was survived by his son, a consultant physician in Swindon, and his daughter Sally. His eldest son had been killed in a road accident when a schoolboy.

NRR

[*Brit.med.J.*,1987,**294**,130; *Lancet*,1987,**1**,55; *The Times* 24 Nov 1986; *J.Am Acad Dermatol*, 1981,**4**,747-752; Photo]

HICKLING, Richard Anderson *b.*6 Dec 1900 *d.*30 Oct 1984
BA Cantab(1922)MRCS LRCP(1924) MB BChir(1925)
MRCP(1925) MD(1929) FRCP(1932)

Richard Anderson (Andy) Hickling was the son of a clergyman of the same name and was born in the mission field at Bangaloor in India. This close association with the church in his early years clearly played a large part in forming his character.

He was at Taunton School during the first world war and played for the first rugby and cricket teams. He also gained an open scholarship at Sydney Sussex College in 1919. He played rugby, soccer and cricket for his College, and obtained his BA with first class honours in the natural sciences tripos. For 20 years he was secretary and later president of the Cambridge Graduates Medical Club. A university scholarship took him from Cambridge to the Westminster Hospital medical school in 1922.

In August 1923 the Westminster Hospital medical school was closed for renovation and all its students were transferred to Charing Cross Hospital. He would never have imagined that 60 years later the two schools would amalgamate, nor would he have approved of the changes that have taken place in the London teaching hospitals.

Hickling played cricket and rugby for his new hospital and in his last year as a student he was asked to captain the First XV. It was typical of his modesty, a characteristic that he never lost, to decline the honour - saying that it should go to a Charing Cross student. He played regularly for Rosslyn Park, which was a leading London rugby club at that time. He qualified in 1924 and was house physician to W J Fenton [*Munk's Roll*,Vol.V,p.127] and W J Adie [*Munk's Roll*,Vol.IV,p.596]. He then became resident obstetric officer to C Lockyer [*Munk's Roll*,Vol.V,p.245] and J B Banister [*Munk's Roll*,Vol.IV p.21] and, later, a demonstrator of pathology in the medical school. He was elected to the fellowship of the College in 1932. In later years he served on the Council of the College, and he was a member of the Association of Physicians for many years.

Hickling worked for a short time with Salzmann at Graz in Austria, before spending two years in the USA. The first year, in New York, was devoted to pathology and in the second year he was resident physician in the Cornell division of the Bellevue Hospital, where he worked with Eugene Dubois in the metabolic ward.

On his return from the USA he became a medical registrar at the Westminster Hospital and on Fenton's retirement in 1929 he was appointed assistant physician at Charing Cross Hospital; he was only 28 years old. He made it clear for the rest of his life that he considered it a great honour to be appointed to the teaching staff of the hospital, and it was his great pleasure and duty to devote himself to the hospital and medical school. If there was a clash between private and hospital work, it was always the hospital work that came first. He did an enormous amount of clinical work and was equally active in all aspects of hospital and medical school life. When he was senior physician he performed the hat trick at cricket in the annual match against the students. He was appointed dean of the medical school during the 1939-45 war and with his characteristic enthusiasm undertook the combined duties of warden, tutor and dean. He was intolerant of the endless discussions which bedevilled proceedings in hospital committees, and was equally outspoken in his views on the needless investigations that he felt were done to patients in hospital.

Much of my knowledge of Hickling's earlier career stems from N S Plummer [*Munk's Roll*,Vol.VII,p.474], who knew him in his younger years and who, like Hickling, was a great general physician and teacher with a special research interest. It was difficult to find out much of Hickling's attainments from himself because he was modest almost to a fault. This modesty and his forthright manner probably prevented him being appreciated fully outside Charing Cross Hospital itself. His modesty was such that each year he always kept it a secret when he went as external examiner to Singapore University, in case he was thought boastful.

In his later years he was probably less popular with some of his students because of his attitude with patients; he was always extremely polite and kind to all his patients, but if he found one exasperating he might make a caustic comment after the patient had departed. This was sometimes interpreted as hypocrisy by some students when, in fact, it was the reverse. He felt it was his professional duty to be kind to the patient, but made no pretence about his true feelings for the sake of the students.

He was a great proponent of the apprenticeship system of teaching. He viewed with abhorrence the German model of so-called 'scientific' medicine with its emphasis on the laboratory. Before Peter's principle was propounded about promotion to levels of incompetence, he believed that the pyramid of the professorial unit would result in research being done to obtain promotion, whereas with the plateau system of the part-time London teaching hospital consultants of his day research was done for its own sake. He would have been delighted to learn of the recent appointment of the chairman of Daimler Benz as a part-time professor at the local university, and the most popular teacher in his department. It would have seemed to Hickling that the world had moved full circle.

Hickling did research work throughout his busy life as a practising consultant. He had a small room at Charing Cross, with a bench and a microscope, where he worked without the aid of a technician. For nearly 40 years he himself examined the blood and histological sections of spleen and marrow of the cases referred to him. He would travel any distance to follow up these patients, and must have had the longest follow-up of the largest number of cases of splenomegaly in this country if not in the world.

Hickling will always be remembered for his original observations on the life history, diagnosis and treatment of myelosclerosis. His paper on chronic non-leukaemic myelosis in the *Quarterly Journal of Medicine* in 1937 was a landmark. He first reported the association, in the *Lancet* in 1953, of gout with leukaemia and polycythaemia. He showed that it was: 'much more common in patients with myeloid metaplasia involving erythroblasts, leucoblasts and megakaryocytes than in patients with typical myeloid leukaemia or with typical primary polycythaemia.' He continued his work for years after he left Charing Cross Hospital, as a result of facilities offered him by wise and courteous colleagues at the Royal Free Hospital. Again, his extraordinary modesty about his research meant that it was not fully appreciated by his colleagues at Charing Cross Hospital nor, one suspects, by the haematologists who met together in diverse places, as is now the rule in most specialties. He rarely spoke at, or even attended, meetings since his time was so fully occupied in Charing Cross Hospital treating, teaching and doing research.

He was a man of great courage. Two years before he died, he collapsed at a convivial Charing Cross Hospital dinner, with severe chest pain. It was almost certainly a heart attack but he would not allow a cardiogram to be done. He was taken home, and on his own insistence left to fend for himself. Two days later he wrote an uncharacteristically testy letter complaining of the unnecessary fuss that had been made.

He was a man who inspired many young men, students, housemen and registrars, and he helped them greatly in their careers. He was also a very warm family man. He and his wife were a devoted couple and he never completely recovered from her death. They used to take their three sons to North Wales each year, where they were keen sailors. I suspect that they excelled in this sport but, again, they would give no hint of their own ability.

Hickling would be considered by some to be a reactionary influence in teaching hospital medicine. There are those who have seen the change from apprenticeship teaching to so-called 'academic' teaching, and who realize that we shall not see again such physicians as Andy Hickling who, through a combination of humanity, honesty and wisdom, made British medicine famous.

PBSF

[*Lancet*,1984,**2**,1287; *Charing Cross Hosp.Gaz.*,Winter 1966-67,**64**,(3),137-139; Photo]

HIGGINS, Francis Edmund *b.*3 Nov 1924 *d.*26 Dec 1985
BA Loyola College,Canada(1944) MRCS LRCP(1951)MB ChB Birm(1951) MRCP(1953) FRCP(1973)

Francis (Frank) Higgins was born in Derby, into a medical family. Both his parents were doctors, and his mother Mary Alice, née Gallagher, had been the first woman medical graduate of Queen's University Belfast. His early education was spent at Stoneyhurst College, but at the outbreak of war - when he was 15 - he was evacuated to Canada. At first he lived in Newfoundland but later moved to Montreal, where he studied at Loyola College. He graduated in 1944 in economics and philosophy and, being a staunch Catholic, considered life in the priesthood. However, he was drafted into the armed forces and spent the next two years in the Royal Canadian Infantry Corps, rising to the rank of second lieutenant. On demobilization he took up teaching posts, and did some broadcasting, but on his return to England he decided to take up the study of medicine at Birmingham University.

Despite his lack of a scientific background, he had a brilliant undergraduate medical career, being awarded prizes throughout the course, the culmination being the Arthur Foxwell memorial prize in medicine in 1951. At this time he developed his interest in amateur dramatics and light opera, being a founder member of the Gilbert and Sullivan Society. He continued to have a pleasing tenor voice until an exploratory neck operation left him with a vocal cord palsy. His sister had qualified as a doctor, at Birmingham, a year before Frank and she worked for some time as an anaesthetist.

After graduation in 1951 he was house physician to Sir Arthur Peregrine Thomson [*Munk's Roll*, Vol.VII,p.575] at Birmingham General Hospital and went on to hold other junior medical posts in Birmingham. He obtained his membership of the College in 1953, and in September 1955 he was appointed senior medical registrar to the professorial medical unit at the General Infirmary at Leeds.

Frank spent 1958 as a research fellow at the Mayo Clinic, Minnesota, USA, in endocrinology and gastroenterology, and then returned to Leeds to spend a further three years as the tutor in medicine. Although his commitment to research was not great, he was an excellent teacher and clinician, able to put complex matters in a simple way, with an encyclopaedic memory and always very well read. His interest in diabetes had begun with his association with John Malins in Birmingham, and in Leeds was fostered by Ronald Tunbridge, later Sir Ronald (q.v.).

Frank was appointed to the Huddersfield group of hospitals in 1961, having married one of his students, Elizabeth (Liz) Rowe, six months previously. He was one of two physicians in Huddersfield and by hard work, good medicine and excellent teaching, his impact on Huddersfield medicine was extremely beneficial and widely recognized by colleagues everywhere. The hospitals in Huddersfield were old and he soon became involved in the design and planning of the new Huddersfield Royal Infirmary which later became the District General Hospital.

Frank built up a vigorous and renowned diabetic clinic with strong community nursing links - a relatively new idea at the time. He served on the children's and medical advisory committees of the British Diabetic Association as well as the hospital management committee locally, and many regional committees; the last being as chairman of the regional working party on endocrinology and diabetes.

Frank Higgins was the epitome of what is best in a district general hospital physician. He was extremely hard working, conscientious, fiercely independent, knowledgeable and well read, an excellent bedside teacher and staunch ally. He attracted clinical students from all over England and Ireland, and his junior medical staff normally were successful in examinations. He did little research himself and did not publish widely once he was in Huddersfield, donating all his energy to building up the standards of medical care. He occasionally lost his temper, which could be impressive, but it was short lived. His square frame, plethoric face and friendly, jocular greeting were always welcome at meetings which he often attended with his wife, who helped him in the diabetic clinic. Of his four children, only Sarah, his youngest daughter, followed him into medicine, becoming an undergraduate at the Sheffield medical school.

His later years were dogged by ill health, brought on by his incessant hard work, and he took early retirement but died within a few months. There is little doubt that the high standard of medical care in Huddersfield was due, in no small way, to his clinical expertise, good management and teaching.

JKW2/GW

[*Brit.med.J.*,1986,**292**,704; *Lancet*,1986,**1**,569]

HILL, James Maurice *b*.25 Sept 1920 *d*.1 Sept 1986
MRCS LRCP(1944) MB BS Lond(1944) MD(1049) MRCP(1949) FRCP(1974)

James Maurice (Mick) Hill was born in Glasgow, the son of a regular serving officer in the RAMC, whose wife, Mabel Hoy, was the daughter of a bank manager. He was educated at Campbell College, Belfast, and graduated in medicine from the University of London, at St Thomas's Hospital. After a house appointment at St James Hospital, Balham, he enlisted in the RAMC, serving most of his time in India. After demobilization he was appointed medical registrar at St Stephen's Hospital where he came under the influence of the formidable Sir Adolphe Abrahams [*Munk's Roll*,Vol.VI,p.1], and in chest diseases, Alec Wingfield [*Munk's Roll*, Vol.VI,p.469]. Spurred on by this experience, he obtained both his London doctorate and his membership of the College in 1949, and was appointed senior registrar in thoracic medicine to the St Helier Hospital and Mitcham Chest Clinic, later being upgraded to consultant, a post he held until his retirement.

At the St Helier Hospital he joined an active and progressive group of young chest physicians in a thoracic unit where the surgical staff was led by William Cleland, and later by Angus MacArthur. With enthusiastic clinical judgement he overcame the problems of the long waiting list for the treatment of pulmonary tuberculosis by short stay admission for the induction of pneumothorax with domiciliary maintenance, and later by the extended use of domiliciary-based chemotherapy. As clinical and epidemiological control of tuberculosis was achieved, he realized the changing nature of thoracic medicine and took six months' study leave, which he spent at the Royal Postgraduate Medical School and Leeds General Hospital. He returned to St Helier and was appointed to general medical sessions in the group, reshaping his career in general medicine. It served to underline his considerable ability as a clinician with excellent judgement, and a deep concern for the welfare of his patients. He never came to terms with what he considered to be the mystique of the pulmonary function laboratory but, always practical, he set up a flourishing clinic for fibreoptic bronchoscopy.

Mick Hill was an enthusiastic teacher, and as secretary of the postgraduate medical centre he tackled the task of raising money to secure the rebuilding of the centre and the medical library with great success.

In 1975, following the reorganization of the NHS, he was tempted into medical administration. He had not by nature, nor inclination, the patience to reach consensus and realized that consultation with colleagues at the bedside was vastly different to persuading them in committee to support the principle of the greater good. To his credit he always acted with integrity, although he had sometimes to recommend policies with which he was not in sympathy.

He married Audrey Gallimore, a graduate nurse and the daughter of an artist, and enjoyed great happiness at home with her and their two children, tending his garden and especially his trees for which he had a natural interest and considerable knowledge. With his wife, he pursued his interest in painting in water colours.

Mick was a well set up, quite handsome man, with an engaging, rather 'laid back' personality, and with much to look forward to on his retirement. Sadly, this was not to be. Unknown to him, what seemed to be a trivial illness turned out to be a carcinoma of the stomach. He faced his last illness, with the devoted help of his wife and family, with uncomplaining fortitude.

HFH

[*Brit.med.J.*,1986,**293**,1512; Photo]

HODSON, Cecil John Charles Guy *b.*20 Dec 1915 *d.*1 Dec 1985
MRCS LRCP(1938) MB BS Lond(1939) MRCP(1941) DMRE Cantab(1942) FFR(1954) FRCP(1958) Hon FRCR(1981)

John Hodson - known to all as 'Hoddie' - was one of the outstanding original thinkers in British diagnostic radiology. He will be remembered by those who trained under him as a fine clinical radiologist, and as a stimulating and exciting teacher whose highly original mind was a constant source of inspiration. Even early in his career he would turn down requests for examinations if he thought they would not affect patient management. As a chief he was encouraging and supportive and those who worked with him could not fail to recognize his keen, inquiring mind and his readiness to welcome any intellectual challenge, but as an inveterate researcher he had little patience with clinicians who tended to overlook the research implications of the cases they dealt with routinely. Yet, as they came to know him, they found that his personal charm would temper his somewhat scathing comments, which could be infuriating, although some time might pass before they realized that they had worked for one of the few really original thinkers in radiology.

'Hoddie' was born at Blackheath, where his father was a general practitioner, and educated at Eastbourne College and St Mary's Hospital. After house appointments at St Mary's, St Giles Hospital Camberwell, and the Brompton Hospital, and a brief period as medical registrar at Harefield Hospital, he gained his DMRE in 1941 and joined the RAMC the following year as a specialist radiologist. He served in Algeria, Tunisia, Sicily, Italy and Greece. Once, on arrival at Bougie with the invasion forces, his ship was sunk but he was able, after some fine emergency work, to set up his X-ray plant in the grounds of a vineyard. He was mentioned in despatches in 1944.

After demobilization in 1947 he was appointed radiologist to the Queen Elizabeth Hospital for Children, Hackney. The following year he became deputy director of the department of radiology at University College Hospital and in 1960 succeeded Cochrane Shanks [*Munk's Roll*, Vol.VII,p.527] as director. He was also consulting radiologist to the Queen Alexandra Military Hospital from 1948-70.

After 22 years at UCH, personal and professional frustrations made the offer of a professorship in Newfoundland attractive, and in 1970 he was appointed first professor of radiology to the Memorial University of Newfoundland, where he worked until 1975. He then became professor of uroradiology at the school of medicine of Yale University and held this post until his retirement in 1984. He was secretary and vice-president of the Faculty of Radiologists, Baker travelling professor of the Royal Australasian College of Radiology, William Julius Mickie fellow of the University of London, fellow of the Royal Society of Medicine and a member of the Renal Association, the Thoracic Society, the British Paediatric Association, the Harveian Society and the Medical Society of London. He contributed many original publications and chapters in books, and participated in numerous national and international meetings.

Few radiologists have had so great an influence both within and outside their specialty. The terms 'chronic pyelonephritis', 'pyelonephritic scar' and, later, 'reflux nephropathy' and 'postobstructive atrophy' were all either coined or brought into everyday use by him. He made radiologists aware of the importance of measuring renal lengths and of serial measurements in detecting children whose kidneys were not responding to treatment for infection. Quite apart from his impressive research achievements, he was often consulted by eminent colleagues when they had a difficult or abstruse clinical problem that included possible kidney involvement. Although he is especially remembered for his renal work, he made contributions in other fields: he was one of the first radiologists to stress the blurring of the cardiac and diaphragmatic contours in detecting and localizing chest consolidation - which later became known as the silhouette sign. He also did early work on the significance of bronchial wall thickening, particularly in asthmatics. In the field of urinary tract radiology his dominating interest was in the process of scarring in kidneys. By drawing attention to the fact that the relationship between the renal outline and the calices was the key to interpreting the urogram he widened its scope to permit the detection of a whole range of parenchymal disease processes. In so doing he laid the foundations of modern radiology of the urinary tract and became one of its foremost figures. His work with David Edwards on the relation between vesicoureteric reflux and renal scarring fundamentally altered the approach to the investigation of children with urinary tract infection. It also led to his work with Rolleston, Maling and others, which established the importance of intrarenal reflux as the underlying mechanism for localized renal scarring in vesicoureteric reflux. His strong conviction that reflux could produce this damage in the absence of infection led to his continued research interest in this field throughout his life.

In Newfoundland, 'Hoddie' quickly developed a superb research programme, with major help from the Canadian Research Council. He concentrated at first on the study of urinary reflux as a cause of pyelonephritis in pigs while simultaneously working with paediatric nephrologists and urologists to confirm his suspicion that a similar condition existed in children. His subsequent work at Yale, where he had increased research opportunities, was to be recognized by a lecture named after him, and by a memorial plaque in the library.

'Hoddie' always loved challenges - physical or intellectual. He loved rugger and overland skiing. For many years he owned a boat, *The Galloping Bully,* sailing off the coasts of Britain, Newfoundland and the USA; his last trip being to the Virgin Islands in 1985. In Newfoundland the boat's temperamental engine became part of the local lore. He was a born mimic and an entertaining raconteur. A man of many talents, he could turn his hand to anything and do it well. He was an amateur singer of some repute, and a skilled painter. Although he had many friends he was at heart a loner; in a rugged exterior there lived a romantic whose dreams found some satisfaction in his watercolour paintings of sea and hills, and in a profound love of the countryside. He was married for a few years and his only child became the centre of his life.

He was looking forward to retiring and returning to England, planning to stimulate others to continue research work on kidney function, when in April 1985 he diagnosed his own serious condition. Despite rapidly progressing ill health he continued to work, while his son Ian helped him to settle his affairs in the USA and return to England. He spent a peaceful last month at home below his beloved Sussex Downs.

RL/IM/HS

[*Brit.med.J.*,1986,**292**,419; Photo]

HUMPHREY, John Herbert *b.*6 Dec 1916 *d.*25 Dec 1987
CBE(1970) BA Cantab(1931) MB BChir(1940) MD(1947)
FRS(1963) FRCP(1971)

John Humphrey was born in West Byfleet. His father was one of seven children of the town clerk of Finsbury, London, who attended the Finsbury Technical College and became the chief engineer and later the consulting chemical engineer to ICI. One of his aunts was also an industrial chemist. John himself was the senior scholar (*aulae praefectus*) at Winchester and subsequently a governor. His interest in science, his desire to study medicine - and not engineering - as a humanitarian activity, and his concern for the welfare of others, were established there and were much influenced by R L M (Dick) Synge. He went to Trinity College, Cambridge, and obtained his BA in 1937. Like many of his generation, he was moved by the desire for a better society to join the Communist party, but left at the time of the German-Russian pact in 1939. He met his wife, who later became a child psychiatrist, at University College Hospital, her father was the Foulerton research professor of physiology at University College and a Nobel Laureate. Her maternal grandfather, a Nobel prize winner, was J N Keynes; she was a niece of Sir Geoffrey Keynes [*Munk's Roll*,Vol.VII,p.319], and a great niece of Sir Walter Langdon-Brown [*Munk's Roll*,Vol.IV,p.491]. John Humphrey became a close friend of his father-in-law, A V Hill, who was president of the Academic Assistance Council which helped Jewish academic refugees from the Nazi persecution, and in due course John became president of its successor, the Society for the Protection of Sciences and Learning, from 1978-87.

John worked with Charles R Harington [*Munk's Roll*, Vol.VI,p.222], as a medical student on the immunogenicity of tyrosyl-proteins, and showed an interest in chemistry unusual in British biologists. He graduated from University College Hospital medical school in 1940 and was so fascinated by research that he boasted that he did the minimal work necessary to pass. He was a houseman at the Hammersmith Hospital, and was the Jenner research student at the Lister Institute, 1941-42. His work at the Institute may explain his lasting interest in polysaccharides which led to his discovery of the thymus independent antibody response. He was assistant pathologist at the Central Middlesex Hospital from 1943-46, covering all branches of pathology, and was in sympathy with the medical director, Horace Joules [*Munk's Roll*,Vol.VII,p.307]. John showed that mixing procaine with penicillin slowed absorption and his notes printed for the nurses subsequently prevented an American patent. His MD thesis, 1946, was on 'The aetiology of pneumonia in North West London 1942-44'. He joined the external staff of the Medical Research Council as a bacteriologist at University College Hospital in 1946. He remembered the greater cooperation that existed between the clinic and the laboratory at the Central Middlesex.

John then joined the National Institute of Medical Research at Hampstead (division of biological standards) where he established, with J W Lightbown, many international standards for antibiotics. He then moved to the Mill Hill site and founded the division of immunology in 1957. He regarded a division which started with two scientists, (Dr B A (Eta) Askonas and Dr Brigid Balfour, who had also been at Cambridge) two technicians and space for 1-2 visiting workers as large enough to be viable. There was no need to submit a detailed scientific programme to the MRC as the philosophy was to support good scientists. He was deputy director of the NIMR, 1961-76, and acting director following Sir Peter Medawar's (q.v.) stroke in 1969. The substantive post was given to a 'tougher' man. In 1975, John was professor of immunology at the Royal Postgraduate Medical School, Hammersmith, and emeritus professor in 1981. He studied follicular dendritic cells and antigen presentation and also supervised the department of haematology.

The postwar era was an outstanding time at the NIMR. There was a renaissance of immunology and John embodied the pioneering feeling of the period. He liked to work at the interface between immunology and other sciences. By way of example, studies with K F (Frank) Austen showed that crosslinking of antibody on the surface of the mast cell caused histamine release; work with Stephen Black showed that the erythema of the tuberculin skin reaction could be inhibited by hypnotic suggestion. His key approaches were the importance of quantitation, the use of radioactive and later fluorescent labels for determining the distribution of antigens, and a liking for microscopy. His interest in the localization of antigen in germinal centres led to collaboration with Michael Sela of the Weizmann Institute, Israel. He found that the antibody response to a random amino acid copolymer differed in Sandylops and Himalayan rabbits. A visiting worker interested in genetics, Hugh O McDevitt, followed up this observation which led to his pioneering work on immune response genes and their relation to transplantation antigens. Overall, John Humphrey was intellectually generous, did not seek credit for himself and encouraged openness in those around him. He liked medically trained workers and had a flair for collaboration.

As head of division, he helped members of the department to look for important questions and elegant approaches. The number of papers was entirely unimportant. He helped scientists find and develop their own areas of interest, and their own approaches, rather than learn a set of techniques or contribute to the long term themes of the department. He enjoyed and elicited lively discussion, often over coffee at the bench, but had little liking for regular departmental presentations. His experience with biological standards caused him to produce, and persuade members of his department to produce antibodies and to undertake quantitative precipitation curves. The sera were freeze dried and were valuable 'scientific presents'. He preferred to make reagents rather than purchase them, and this reflected the limited range and quality of biological reagents commercially available.

John's advice to members of the department who wanted to know what to do next was - to experiment and find out. For him science was fun, both intellectually in designing elegant experiments and discussing exciting ideas, and practically as he enjoyed working with his hands and was a gifted experimentalist. His approach was not primarily theoretical and he emphasized the role of serendipity in biology, i.e. observations made casually at the bench. He had an enormous breadth of knowledge. John was one of the last of the 'general' immunologists able to work usefully in many areas of immunology and proud of his range. He was an excellent discussant at meetings and went out of his way to talk to and advise young scientists. The financial stringency and the changing political scene of the '70s made this general approach less popular, but it is not clear that the more directed 'value for money' research of the '80s provides a better method for studying natural phenomena. His scientific standing was recognized by the Royal Society in 1963, by the College in 1971, and by the US National Academy of Science in 1986. He was an honorary member of several societies including the South African Society of Pathology, 1975; the Purkinje Medical Society of Czechoslovakia, 1977, and the Hungarian Society for Immunology, 1981.

John Humphrey regarded the advancement of immunology as very important. Together with Robin A Coombs, he started the British Society for Immunology as a splinter group from the Pathological Society. He was coeditor and assistant editor of *Immunology* for 10 years. His interest in Eastern Europe facilitated the meeting on the Mechanisms of Antibody Formation in Prague, 1959. His international interests also led him to serve on the WHO expert committee on biological standardization and immunology. He was president of the International Union of Immunological Societies, and chairman of the organizing committee for the Second International Congress of Immunology in Great Britain, 1974. He was also chairman of the advisory board of the Basle Institute for Immunology. He served as secretary of the standing committee on safeguards of the pursuit of science of the International Council of Scientific Unions. John placed great value on his work for peace, but he was not a pacifist. He took part in the second Aldermaston March, and other activities of the Campaign for Nuclear Disarmament (CND). He was also president of the Medical Campaign against Nuclear Weapons, and wrote against bacteriological and chemical warfare. He repeatedly wrote and acted on behalf of Russian, South American, and other scientists and doctors who had been badly treated.

John had developed pulmonary tuberculosis during the war and initially attributed severe deafness due to bilateral acoustic neuroma to a delayed effect of streptomycin. Fortunately he had a robust constitution and would work indefatigably even when he had a fever. He was at peace with himself, and had time for many things and for many people. Some time after his brother's death he became a ward of his nephew and niece, and later brought up his nephew with his own five children. He remained active until two days before his death on Christmas Day, continuing to work with a technician and to make observations on the localization of acidic and neutral polysaccharides in the antigen presenting cells of the spleen. John's strong feeling for the world in which we live, his effective desire to help other people, and his sophisticated straightforwardness, provided a model of proper behaviour.

[*The Times*,31 Dec 1987; *The Independent*,5 Jan 1988; *Brit.med.J.*,1988,**296**,796; *Lancet*,1988,**1**,193-4; *Doctor*,28 Jan 1988,p.32; *MRC News*, No.**39**,June 1988,37-8; *Medicine and War*,**4**,71-73,1988; *Lancet*,1976,1,652; *Brit.med.J.*,1976,1,721; *Lancet* 1989,1,115-6; Photo]

HUNT, Jack Naylor *b.*29 Apr 1917 *d.*14 Apr 1986
LMSSA(1940) MB BS Lond(1945) PhD(1949) DSc(1954) MD(1956) MRCP(1966) FRCP(1973)

Jack Naylor Hunt, known as 'Michael' to his intimates and affectionately as 'Hunt' by the rest of his colleagues, was born in London. His father, Charles Frank Hunt, was an engineer and his mother, Mary, was the daughter of James Moss, a builder. He was educated at elementary school and the Royal Masonic School before entering Guy's as a pre-clinical student. He gained his LMSSA in 1940, spent the war years in the Royal Navy, and graduated MB BS in 1945. He returned to Guy's as demonstrator in the physiology department, became reader in 1955, and university professor and head of the department of physiology in 1962, where he remained until 1977 when he left England to take up a similar post at Baylor College of Medicine, Houston, USA. Arriving in the United States he took - and passed first time - all the preclinical and clinical examinations necessary for medical practice in Texas, at the age of 60.

During his years at Guy's his standards of teaching and research were exceptionally high, and he expected similar standards both from his colleagues and from those whom he taught. He introduced new methods in teaching: multiple choice questions, interpretive questions based on clinical papers, and practical classes where students ran a research project for a full term in conjunction with a member of the staff. He could be critical if anyone was found wanting and former students will remember him as the professor who insisted that they arrive for classes and lectures on time; lecturers will remember him as the person who insisted that the students be respected, and physiologists will remember him as the man who would ask a penetrating question at the end of a paper read to the Physiological Society. Yet he was always ready and willing to help anyone trying to get a research project started, or someone with a new idea for a practical class. He left the department at Guy's when the UGC was seriously considering severe financial cut-backs, and unfortunately Hunt's predictions proved correct.

An extremely able man, with an original and inquiring mind, his thirst for knowledge was insatiable - whether it was physiology, Limoges enamel, or the price of a vintage Rolls. He was a fine exponent of Claude Bernard's premise that physiology represents the greater part of experimental medicine. His primary interest was the upper alimentary tract, and in particular the stomach. He did much to increase understanding of the control mechanisms governing its emptying. His view was always clinically oriented; a view which received recognition when he was awarded membership, and later fellowship of the College. He fully deserved to be counted among the great international figures of the 1960s, such as M I Grossman [*Munk's Roll*, Vol.VII,p.231] C F Code and R A Gregory. His intellectual ability made him impatient with fools, but he was always ready to change his mind if there were sufficient scientific evidence to prove him incorrect.

'Michael' Hunt was by nature shy and extremely reserved; some people considered him difficult and even rather abrasive, but on close acquaintance he proved to be warm, charming and witty, with a keen and uninhibited sense of humour. Whatever his current interest it became a passion, whether it was breeding sheep at his island home in Nevis, sailing a catamaran, restoring old Georgian houses, or extolling the hypothesis that drinking water with a meal delayed gastric emptying and so aided body weight loss.

In 1948 he married Anne-Claire Haley, daughter of William John Haley, an editor. There were no children of the marriage. GW/VL

[*Brit.med.J.*,1986,**292**,1535; *Lancet*,1986,**1**,1337; *Guy's Hosp.Gaz.*,1977,**91**,50-51; 1986,**100**,147; Photo]

HUNT, John Henderson Lord Hunt of Fawley
*b.*3 July 1905 *d.*28 Dec 1987
CBE(1970) MRCS LRCP(1931) BM BCh(1931) MRCP(1934) MA MD(1935) FRCS(1966) FRCGP(1969) FRCP(1974)

John Hunt was born in Secunderabad, where his father was a surgeon to the Hyderabad State Railways and to the Ruler and his family. He was educated at Charterhouse School, of which he was later to become a governor, at Balliol College, Oxford, where he had a distinguished career, and at St Bartholomew's Hospital. At Bart's he was house surgeon to George Gask before a period at the National Hospitals for Nervous Diseases, Queen Square, as house physician to

Sir Charles Symonds [*Munk's Roll*,Vol.VII,p.563]. He returned to Bart's as second and then first assistant on medical firms, and developed a particular interest in peripheral vascular disease, being awarded his doctorate in medicine for a thesis on Raynaud's Disease. He gained his membership of the College in 1934 and was elected a Fellow in 1974, the first from the ranks of general practice to be so honoured.

It was already becoming clear before the second world war that Hunt's broad interests were leading him towards general practice, and a year or two before the outbreak of war he went into practice in Chelsea. During the war he served in the RAF, where he became a wing commander medical specialist. After the war he returned to general practice and started on his own at 54 Sloane Street, which later became the second headquarters of the College of General Practitioners. Here he installed a small pathological laboratory, staffed by a technician with whom he had worked in the RAF, and in the mews house behind he organized the setting up of a diagnostic X-ray unit. As the practice grew he had several assistants and formed a partnership that continued excitingly until his retirement.

His practice was one of the largest in London and attracted patients from all walks of life and from all over the world, for the depth and breadth of his knowledge of medicine and of people, together with his infinite capacity for pursuing any turn of thought that would benefit his patients, marked him as an exceptional doctor by any standards. His patients relied on his wise and kindly counsel and friendship, and his colleagues recognized his clinical skill and judgement which helped specialists and other family doctors alike. Those who were fortunate enough to work with him in his practice drew continually on his strength of character and immense experience; no problem was too small for him to discuss and give advice, all leavened by well remembered gusts of laughter which relieved moments of tension and anxiety. Above all, he had the rare gift of inspiring excitement and anticipation.

But also in his character was a single-minded determination that brooked no obstruction. After the 1939-45 war, when the National Health Service was being planned, the position of general practitioners was uncertain and unsatisfactory. The notion of an academic body to promote the efficiency of general practice had been put forward as long ago as 1844, but widespread efforts came to nought. With the advent of the NHS a new purpose was seen in uniting the widely felt aspirations of many general practitioners to bring order out of disorder through a college of their own. Following a crucial meeting at the BMA in October 1951, John Hunt and Fraser Rose wrote a letter which was published in the *British Medical Journal* and *The Lancet* proposing '...a possible College of General Practice.' Memoranda from them both were published two weeks later and much comment was provoked, both favourable and unfavourable. In particular, the presidents of the Royal

Colleges expressed their opposition to the whole project, but the newly formed steering committee, chaired by Sir Henry Willink, with Hunt as secretary, pressed on with its plans, undeterred. After the most careful consultations, a memorandum and articles of association of the College of General Practitioners were signed by the members of the steering committee on 19 November 1952 and published, with the committee's report, a month later.

There was widespread support for the initiative from both medical and lay sources and within six months of its foundation over 2000 doctors had joined the college. Of the many men and women who had worked so hard to form the college, none had worked harder or displayed such determined leadership as John Hunt, and his work in developing the role of the college in this country, and extending its influence across the world, was to continue for the rest of his professional life, most notably as honorary secretary of council from 1953-66, and later as president from 1967-70.

It was during his presidency that HM The Queen graciously granted the Royal prefix, and the college became the Royal College of General Practitioners. The Royal Charter followed in 1972 when HRH Prince Philip, Duke of Edinburgh, was appointed an honorary fellow and elected as president for the following year, later becoming patron of the college.

John Hunt was honoured many times by medical and lay organizations throughout the world, and the award of the CBE as he finished his three years as president of the college was followed, three years later, by his elevation to the House of Lords. In the upper house he spoke with the wisdom of wide experience in many debates on medical affairs, notably in the debate on the 1978 Medical Act which materially revised the status of the General Medical Council.

Of all the eponymous orations he was asked to give, his Lloyd Roberts lecture 'The renaissance of general practice', delivered in 1957, was in many ways the keynote speech of his life, for it was uncannily prescient in its proposals for the future work of the college and of general practitioners. And in that lecture, discussing and gently deflating the popular notion of 'the bedside manner', he epitomized his own approach to his patients as he said: '...its components, I think most people will agree, are personal interest, kindness, sympathy, friendliness, understanding, cheerfulness and humour.' Perhaps unconsciously he was describing his own personality, ethos and example.

In all his busy life John was singularly fortunate in the help and support of his wife Elizabeth, whom he married in 1941. She bore even more than the usual burden of a doctor's wife, for not only his large practice but his work for the college, and his writing and travelling, made immense inroads on his time. In his years of retirement, when he was increasingly incapacitated by illness, she cared for him with quite remarkable devotion. They had five children, two daughters, a son who sadly died in early childhood, and twin sons. Both the latter are general practitioners.

ML

[*Brit.med.J.*,1988,**296**,218,300; *Lancet*,1988,1,194; *The Times*,30 Dec 1987; *News Journal of RCPG*,Feb 1988,p.87-88; Photo]

HURFORD, John Vernon *b.*30 Mar 1906 *d.*3 Mar 1988
MB BCh BAO(1929) MD(1935) DPH(1935) MRCP(1936) FRCP(1950)

John Hurford was born in Dublin, of English parents. His father, Arthur Lionel Hurford, was a civil servant, and had served in the Indian Army, and his mother was the daughter of Ivan Henry Herford, an Army officer. John was educated at Campbell College and Queen's University, Belfast, from where he graduated with honours, obtaining the gold medal in surgery. From 1930-31 he served as extern surgeon at the Royal Victoria Hospital, Belfast, and for the next three years he worked in general practice in order to support his widowed mother. After her death, he moved to Winsley Sanatorium from where he obtained the diploma in public health, and in 1935 he was awarded his doctorate with commendation. Membership of the College followed in 1936 and he was elected a Fellow in 1949.

To John's great disappointment, his ardent desire to do military service during the second world war was officially frustrated by his reservation. In 1939 he became assistant physician to William Snell at Colindale Hospital, and for a short period was seconded to Highwood Children's Hospital. He was subsequently appointed consultant physician to Ealing Chest Clinic, and in 1949 he was appointed physician superintendent at King George V Hospital. This was the large sanatorium near Godalming, administered by the London County Council (later the Greater London Council) as a main treatment centre for pulmonary tuberculosis. He served in that post until the closure of the hospital in 1971. During the period 1949-71 the chest clinics serving the Farnham, Camberley and Aldershot areas were amalgamated and successfully administered under his able supervision.

John Hurford was closely involved in research and administration as well as clinical practice, and was honorary secretary to the British Tuberculosis Association from 1943-47. In 1946 he was the prime mover in the formation of the Association's research committee, to which he was honorary secretary. With his staunch support, this committee played a leading role in the rapid diagnostic and therapeutic advances taking place. For some years he was a member of the joint tuberculosis council which ably advised the Ministry of Health on the development of the chest diseases services. In all these offices his outstanding ability in committee was widely recognized and respected.

His methods in the treatment of chest diseases were all

encompassing, extending beyond purely medical management to include developments in occupational and art therapy. He always had a full appreciation of his patients' personal problems which inevitably arose in the long term care of tuberculosis. Being an enthusiastic amateur painter himself, King George V became one of the first hospitals to adopt the principles of art therapy, originated in the early 1940s by Adrian Hill at Midhurst. He also supported the latest developments in occupational therapy. He was well aware of the importance of respiratory physiology in the diagnosis and management of chest diseases and he pressed for and built one of the first laboratories for respiratory function studies in the hospital.

Hurford was a kind disciplinarian, who ensured that his hospital was a happy place for both his patients and his staff; his concern was not only for the care of his patients but also for the welfare of all the hospital staff, and he took part in many hospital social activities. He captained the hospital's tennis team, and was no mean Thespian in the annual pantomimes of which he was usually part author.

As well as his talent for painting in oils, his leisure interests were wide, embracing literature, music and drama. He had a particular interest in the works of Jane Austen and Anthony Trollope, a special love for the music of Bach and Mozart, and he greatly enjoyed the classical theatre.

At the outbreak of war, when they were both medical officers at Highwood Hospital in Essex, he met his beloved wife Olive, the daughter of Francis James Browne, professor in obstetrics, and sister of J C McClure Browne. John himself was the great-grandson of Anthony Todd Thomson [*Munk's Roll*,Vol.IV,p.29]. Olive and John were married in November 1939. She was able to continue her career, and she faithfully encouraged her husband in his dedication to his vocation. Latterly she devotedly sustained and cared for him during a prolonged period of failing health. They had two children, a son and a daughter, and they have respectively pursued successful academic careers in science and letters.

One of John Hurford's colleagues writes '...he was an example of honesty, uprightness, open-mindedness and helpfulness. His knowledge was encyclopaedic though in his modesty he never paraded it. My admiration is truly boundless.'

John gave his advice generously and freely and he surely left his mark as a physician, widely regarded as having made a significant contribution not only in the battle against tuberculosis but also in the building of the modern approach to the diagnosis and treatment of thoracic disease.

GML/LJR

[*The Times*,17 Mar 1988; *Brit.med.J.*,1988,**296**,1204; Photo]

HUTCHISON, James Holmes *b.*16 Apr 1912 *d.*27 Dec 1987
CBE(1971) OBE(1945) MB ChB Glasg(1934) FRFPS(1937) MRCP(1938) MD(1939) FRCP(1947) MRCPE(1957) FRCPE(1960) FRCPG(1962) FRSE(1965) Hon FACP(1968)

James Hutchison was born in Rangoon. He came to Glasgow with his parents when he was eight years old, received his schooling at Glasgow High School, and proceeded to Glasgow University where he graduated with commendation at the age of 22. After residencies in the Royal Hospital for Sick Children and Western Infirmary in Glasgow, and St Luke's Hospital in Bradford, he was awarded a McCunn research scholarship in 1936. His work on tuberculosis in children during the three year tenure of that scholarship formed the basis of an MD thesis for which he was awarded honours and a Bellahouston gold medal.

From 1939 he served for seven years in the RAMC, in France with the British Expeditionary Force, in the United Kingdom, Algeria, Italy, and Austria, reaching the rank of lieutenant colonel and latterly commanding a medical division. He was awarded the OBE (military).

On demobilization he was appointed assistant physician to the Royal Hospital for Sick Children and the Royal Maternity Hospital in Glasgow, and in 1947 was appointed physician in charge of wards at the RHSC - a prestigious and accelerated appointment in the Glasgow medical structure - at the early age of 35. This appointment he combined with an academic appointment to the Gow lectureship in the University of Glasgow and private paediatric practice. In 1961 he was appointed to the Samson Gemmell chair of child health as professor and head of department, and also became consultant to the Queen Mother's Hospital. During his 16 year occupancy of the Glasgow chair his department was productive in research and in influencing the development of improved standards of child care despite the fact that during much of that time the Glasgow RHSC was being rebuilt and the temporary accommodation in which it was housed was far from ideal. He also made an early and significant contribution to the developing branch of paediatrics, community paediatrics, by establishing within his department a social paediatric unit.

In research James Hutchison maintained his interest in childhood tuberculosis while this disease was still a major cause of childhood morbidity and mortality, and his bronchoscopic studies did much to elucidate primary tuberculous disease of the lungs. He later turned his attention to hyperthyroidism, which was endemic in certain tinker families in the West of Scotland, and his studies with E M McGirr demonstrated the metabolic basis of a number of these inherited disorders.

He was an excellent undergraduate and postgraduate teacher in the lucid didactic mould. His single author textbook *Practical paediatric problems*, London, Lloyd-Luke, first published in 1964,became a standard text in many departments. It had reached its sixth edition, now co-edited with his successor F Cockburn, by the time of his death.

Retiring from the Glasgow chair in 1977, Hutchison immediately accepted an invitation to become professor of paediatrics in Hong Kong. He served there for three years with great distinction. His academic ability, his qualities of leadership and the loyalty which he attracted, did much to upgrade the Hong Kong department and at the same time to improve the links between Hong Kong and Britain.

Of Jim Hutchison's many outstanding qualities perhaps the most impressive was his competence as a committee chairman. His effectiveness was based on an excellent memory, a firm grip of essentials, a willingness to listen to reasoned and logical argument but not verbosity, a natural air of authority and an acute sense of time. These qualities put him in great demand as a medical statesman. In Glasgow he was president of the Royal Medico-Chirurgical Society, president of the Royal College of Physicians and Surgeons, and dean of the faculty of medicine. Nationally, he was president of the British Paediatric Association, president of the Association of Physicians of Great Britain and Northern Ireland, chairman of the MRC working party on childhood leukaemia, chairman of the Scottish Health Services Council and chairman of the Royal Scottish Society for the Prevention of Cruelty to Children. He was awarded the CBE in 1971, but many felt that that did not adequately recognize his contribution to medicine and society.

Personally, Jim Hutchison showed unshakable loyalty to his family, friends and colleagues. His friendship was deep and compounded of trust, responsiveness, a keen appreciation of the qualities and contributions of others, and a great sense of humour. He was devoted to his wife Ness, and to his family, and in his dealings with them a very gentle man indeed. He and Ness were delightful hosts. In his department he was something of a disciplinarian but he applied discipline more strictly to himself than to anyone else. Despite the enormous amount of professional and administrative work he undertook, he never seemed in a hurry and always had time to take on any new commitment which he considered his duty. His weekly round of golf and Saturday evening dinner with his wife in the Royal Automobile Club were important relaxations.

I last saw him on 14 December 1987 when he knew his life was ebbing fast. Characteristically he had worked out the final programme. He wished to die at home, he wished to see Christmas with his family, but he did not wish to burden his family much beyond that. He died on 27 December.

JOF

[*Brit.med.J.*,1988,**296**,365,1476; *Lancet*,1988,**1**,132; *The Times*,1 Jan 1988; Photo]

IBRAHIM, Muhammad (Mohamed) Ahmed *b.*[?] 1899 *d.*[?] 1981
MB Cairo(1921) MRCP(1930) MD(1931) FRCP(1951)

For the best part of a century Egyptian medicine was administered by Europeans - the British and the French. Muhammad Ibrahim belonged to the generation of Egyptian physicians who helped found the new national administration when the independent Arab Republic of Egypt was born in 1922. He was the son of a mayor of Sherif Pasha, a village in middle Egypt, and was born at a time when education was generally limited to instruction in the Koran. Ibrahim was the first person in his own village to receive a modern education on present day lines; at first in his home town and then in Cairo at Khedival Secondary School. In 1921 he graduated in medicine from Kasr-El-Aini School of Medicine, which later became the medical faculty of the University of Cairo.

After graduation Ibrahim joined government service and worked in provincial hospitals for a few years. His ability was very soon recognized and the Egyptian government sent him to England to sit for the MRCP examination. This he passed successfully, becoming the third Egyptian physician to obtain this distinction. On returning to Egypt, where the University of Cairo had just inaugurated an MD degree, he obtained his doctorate in 1931, having the honour to be the first physician to be awarded this degree.

From 1931 to 1956 he was successively tutor at the University; assistant professor to A G Biggam (later Sir Alexander) who was at that time serving as lieutenant general in HM Forces [*Munk's Roll*, Vol.V,p.37] and who later became dean of postgraduate medicine at the University of Edinburgh; then professor of medicine at his *alma mater* and subsequently dean of the faculty. After his retirement he continued to be very active as emeritus professor until his death, and was a much sought after external examiner in practically all Egyptian schools of medicine.

Muhammad Ibrahim instituted many innovations in Egyptian medicine, among them the creation of laboratories attached to clinical departments, blood tests and test meals, electrocardiography, bronchoscopy (for which he had been trained in Vienna), intrathecal and intratracheal lipiodol injections, and therapeutic pneumothorax.

In 1932 he founded the Kasr-El-Aini Clinical Society and was instrumental in launching its first journal, which is still very much alive today. He also founded the Egyptian Cardiological Society in 1951, and at the time of his death he was president of the Egyptian Medical Association.

Ibrahim's reputation as a clinical cardiologist extended throughout the Middle East. He became firmly established when, at the beginning of his career, he correctly diagnosed the cardiac infarction of the Crown Prince - much to the surprise and derision of senior consultants who had never heard of such a diagnosis in their day. It was the first ever such diagnosis to be made in Egypt and was confirmed by Sir John Parkinson [*Munk's Roll*,Vol.VII,p.443] who had been called in as a consultant. Muhammad Ibrahim went on to show that this condition was fairly common in the Middle East but that it usually went undiagnosed.

His many publications covered all aspects of cardiology, including statistical studies of cardiovascular disease. He also wrote papers on the diagnosis of spinal and pulmonary tumours. He received many honours, among them: Bey First Class under the monarchy; the Republican Order of Merit First Class, and the National Gold Medal of Science. He was a member of the board of the International Cardiological Society and of the Institut d'Egypte, and was elected a Fellow of the College in 1950.

In his dealings with his superiors, Ibrahim was courteous and cooperative but also firm and decisive. With his colleagues and juniors, he was always fair and understanding. In the generations of students - and their students - many of whom head the medical departments of Egyptian universities, Ibrahim left a lasting legacy to his country and a personal testimonial.

Muhammad Ibrahim had four children: two sons and two daughters. One son followed his father into the medical profession and also became a professor of medicine.

PG/GW/VL

[Photo]

JACOBSON, Solomon *b.*23 Apr 1918 *d.*[?1975]
MB BCh Wits(1942) MSc(1946) MRCP(1948) FRCPE(1948) DPM Lond(1950) FRCP(1973)

Solomon Jacobson was born in South Africa, the son of Maximilian Jacobson and his wife Bertha, née Stein. He was educated at Doornfontein preparatory school, Athlone High School, and the University of Witwatersrand, pursuing his clinical studies at Johannesburg General Hospital. After the usual house appointments at the hospital, he joined the South African Medical Corps in 1944, as a temporary captain, and served until 1946. On demobilization he was appointed a part-time lecturer in anatomy at the Johannesburg General Hospital and then came to Britain where he was a registrar at the Maudsley from 1948-50. In 1950 he returned to South Africa, being appointed as neurologist and psychiatrist to the Johannesburg General and the University of Witwatersrand, a post he held until 1961 when he returned once more to England. He settled down in England, being first appointed as consultant psychiatrist to the St Francis, Brighton and Lewes Group of hospitals, and subsequently clinical tutor in psychiatry in the University of London and visiting professor to the University of Sussex. He was a member of the medical consultative committee at the latter university, and from 1971-72 held a grant for research in suicide in Sussex from the South East Metropolitan Hospital Board.

Outside medicine, Solomon Jacobson enjoyed writing poetry and short stories and collecting rare books. He was chairman of the Poetry Society of Brighton. As well as literature, he was actively interested in art and music.

He married Shelley, daughter of Herman Lowenstein, an engraver, in 1943, and they had two children; a son and a daughter. VL/GW

[Photo]

JAIN, Saroop Chand *b.*11 May 1926 *d.*23 Oct 1986
MB BS Luck(1950) MD(1953) MRCPE(1955) MRCP(1955)
FRCP(1983)

Saroop Chand Jain was born in Allahabad, India, where his father Phul Chand Jain was a businessman. After early education in Allahabad he entered the King George Medical College at Lucknow, where he graduated in 1950. His particular interest was cardiology and in 1954 he came to the United Kingdom for postgraduate education, obtaining his MRCP in 1955 and returning to India in 1956, when he was appointed as lecturer in medicine at King George Medical College.

In 1958 he joined the GSVM Medical College, Kanpur, as reader in medicine, being appointed professor in 1965. During this period his interest in cardiology increased and he spent over two years in the United States at the Buffalo School of Medicine and Childrens' Hospital, Medical Center, Boston, for further training in his specialty. In 1971 he was appointed head of the newly formed department of cardiology at the GSVM Medical College. His outstanding leadership and untiring efforts resulted in the department becoming the LPS Institute of Cardiology, Kanpur, and Jain became its professor and director in 1975, a post he held until his retirement in 1986.

Saroop Jain had a brilliant academic record and a high degree of clinical expertise. He was an excellent teacher and greatly respected by his students and colleagues. Perhaps his major contribution to medicine in India was the establishment of the Institute, which confers both degrees and diplomas. This Institute is devoted to teaching, research, and the medical care of cardiac patients. It is the only one of its kind in Uttar Pradesh, the largest state in India. It was here that Jain carried out his research in the field of cardiology. He wrote some 35 papers and pioneered cardiac electrophysiology in India.

He married Shakuntala, daughter of Sumeru Chand Jain, in 1952, and they had two sons and a daughter. One son, Anoop Jain, is also a doctor. SSS/GW/VL

[Photo]

JARVIS, Stuart Edmund Mullenger *b.*1 Oct 1916 *d.*15 Jan 1982
MB ChB Cape Town(1940) DCH(1949) MRCP(1949) FRCP(1969)

Stuart Jarvis was born at Umtali, an important town in what was then Southern Rhodesia, on the Mozambique border, where his father Edmund Mullenger Jarvis was a veterinary surgeon. His mother was Constance, daughter of James Allen Methven, a business man. Stuart was first educated at Grey College, Port Elizabeth, and then at Umtali School before studying medicine at Cape Town University. He held a Beit Memorial fellowship during clinical training at Groote Schuur Hospital, where he became house physician in the paediatric department. After holding a similar position in general medicine at Wynberg General Hospital, Cape Town, he volunteered for service in the South African Medical Corps and from 1942-47 was a regimental medical officer, and a medical officer in a field ambulance in Egypt and Italy.

After demobilization he came to London for postgraduate study at the Hospital for Sick Children and other London hospitals, and attended a course in Edinburgh. Thus well grounded, he passed the examinations for the DCH and MRCP. Having decided on a career in paediatrics he undertook a year's rotating internship at Queen Elizabeth Hospital for Children in Hackney Road, London, where he remained as registrar until 1952 when he was appointed first assistant in the Royal Free Hospital group. While working there he was awarded a travelling fellowship to visit paediatric centres in Scandinavia and Finland.

At that time there was no paediatrician in the Army although the RAMC was expected to serve the children of soldiers, sailors and airmen overseas, as well as the children of embassies and English schools abroad. On the representation of the civilian paediatric adviser to the Army, Bernard Schlesinger FRCP (q.v.) it was decided to appoint a paediatric consultant and Jarvis was encouraged to apply. He became consultant paediatrician with the rank of major in 1957.

Stuart Jarvis immediately set about improving the medical care of children in hospital and in general practice abroad, particularly in maternity units. Within a few years he had two consultant colleagues. They rarely met for each moved around every three years; one based in England, one in Germany, and the third in Singapore. Successively, Jarvis was responsible for paediatrics in six hospitals in Germany, four in the Far East from Nepal to Hong Kong, and the Cambridge Military Hospital at Aldershot, as well as being adviser to the Ministry of Defence. He established an organization for which he received well deserved credit. He retired in 1974 with the rank of colonel. He was then 58 and worked for a while as a medical assessor for the Ministry of Health. Two years later he became paediatric consultant at the Corniche Hospital in Abu Dhabi. Here he was in his element and happy, living life to the full. His hobbies were botany and horticulture, and when he retired to England in 1981 he went to live quietly in Worcestershire. Soon afterwards he had a fatal heart attack.

Jarvis was tall and slim, with a soldierly open-air appearance. His movements were quick, he talked well and was mannerly. A colleague wrote of him: 'I am sorry his effervescence was extinguished so prematurely... he impressed new colleagues with his boundless enthusiasm, wit and charm... what never failed to impress me was his apparent mastery of all the latest advances in paediatrics while giving the impression of a rather old-fashioned type of character. A great entertainer, a great teacher, and an excellent children's doctor of the old school.'

Jarvis married Annice Beeley of Worcester, a state registered nurse, and they had two daughters and two sons, one of whom followed his father into the Army, but not into the RAMC. PRE

JOHNSTONE, Rupert Desmond Charles

*b.*5 June 1907 *d.*14 Mar 1987
MRCS LRCP(1939) MB BS Lond(1939) MRCP(1940) MD(1942) FRCP(1966)

Rupert Desmond Charles (Hec) Johnstone's first choice of career was the Army. He was educated at Wells House, Malvern Wells, and Malvern College, and entered the Royal Military Academy, Woolwich, as a gunner. He was duly commissioned and had the distinction of being selected for the equitation course at Weedon before being posted to his regiment. But Army service in peace time did not suit him; he resigned his commission in 1930, learnt Arabic, and went to the Sudan to grow cotton. There he indulged what were to become his lifelong interests in natural history, fishing, shooting and riding. However, this life gave him no sense of fulfilment and he returned to the United Kingdom to study medicine at St Thomas's Hospital, London, and so found his true vocation.

'Hec' was the second son of John Eccles St George Johnstone, a Gloucestershire general practitioner, and his wife Geraldine, daughter of a barrister. He graduated from St Thomas's in 1939, gaining his MRCP and MD while holding house appointments at the hospital. With the outbreak of the second world war, he rejoined the Army as a medical specialist RAMC, serving in a field hospital in Nigeria and at the 37th General Hospital, Accra, where he met his future wife, Sheila, who was herself then serving as a physiotherapist. On returning to the United Kingdom he was posted to Colchester Hospital and later worked on malarial research in London until demobilization.

In 1947 Hec was appointed Burton on Trent's first consultant physician and in 1948, with the coming of the National Health Service, he laid the foundations of what is now a comprehensive medical unit. Until he retired in 1972 he served the hospital with distinction and devotion. He was meticulous in pursuit of diagnostic accuracy and all that benefited his patients and, although at times his manner may have seemed militarily brusque and forbidding, he had a caring and sensitive nature and a discerning intellect. The hospital and his patients profited much from his foresight, energy and drive, and many house physicians had cause to remember him with gratitude and affection.

His heavy clinical commitments did not prevent him from serving for many years on the hospital management committee, before the days of the district health authority, and he was for some time chairman of the consultants' advisory committee. In the early days of the NHS his common sense and wise counsel proved invaluable. Subsequently, with others, he made a notable contribution to the creation of the graduate medical centre and after retirement continued to be concerned with its activities as chairman of its committee.

His marriage to Sheila, in 1946, was a happy one and he was devoted to his family. They had two children, a son and a daughter. He was a keen gardener, a fine shot, and had a full and productive retirement until advancing emphysema limited his activities. He was survived by his wife and children; his son is also a doctor.

GW/VL

[*Brit.med.J.*,1987,**294**,981; Photo]

JOLLY, Hugh Reginald *b.*5 May 1918 *d.*4 Mar 1986
MRCS LRCP(1942) MA MB BChir Cantab(1942) MRCP(1948) DCH(1949) FRCP(1965)

Hugh Jolly was probably better known to the general public than any other living doctor.

He was born in the Isle of Man, where his father was a Reverend Canon and reputedly an excellent speaker, an ability that Hugh inherited and utilized to the full. Hugh was educated at Marlborough College and Sydney Sussex College,Cambridge, where he was in the water polo team, before proceeding to the London Hospital. He used to recall that even as a student he was fascinated by paediatrics, so much so that he would run the two miles from his teaching hospital to the Queen Elizabeth Hospital for Children, Hackney, in order to arrive for the rounds on time - chewing sandwiches on the way.

He qualified with the Conjoint and MB BChir in 1942. This was during the 1939-45 war and post-qualification jobs had to be few before joining the Services. Hugh's were at the London, and included that of children's house physician to Maitland Jones [*Munk's Roll*, Vol.V,p.263] and Doyne Bell [*Munk's Roll*, Vol.VI,p.37]. He then joined the RAMC, serving in the United Kingdom, Normandy and North West Europe; ending his service with the rank of captain and dermatologist to the Allied Forces in the Netherland East Indies, a specialty he quoted as being invaluable in his subsequent career.

After the war he renewed his acquaintance with paediatrics as house physician to Wilfrid Sheldon, later Sir Wilfrid, [*Munk's Roll*,Vol.VII, p.531], the first Royal paediatrician, at the Hospital for Sick Children, Great Ormond Street. There was immediate empathy between them and they shared an enthusiasm for a clearer understanding of children's illnesses, as well as a sense of humour and a lifelong love of gardening.

Hugh Jolly realized that to be a 'whole' paediatrician he needed a knowledge of paediatric surgery, and at considerable personal and financial sacrifice (he was poor in those days) he opted to postpone promotion and become a house surgeon at the same hospital before climbing the usual medical ladder to senior registrar. In November 1951 he was appointed consultant in Plymouth; the first full-time paediatrician in an area which included large parts of Devon and Somerset as well as the whole of Cornwall. There were only part-time paediatricians in Taunton and Truro, between Bristol and Land's End. For the first three months he elected to become a 'resident' again, at the Children's Hospital, Cincinatti. Back in Plymouth, only a man of his enormous energy could have achieved so much. His new 'cubicalized' ward attracted innumerable admiring visitors, even from abroad. He broke the isolation of children in hospital by encouraging parents to live in with their children, as well as by daily visiting. He also established combined clinics, then a novel concept, with orthopaedic and ear, nose and throat colleagues, which were of mutual benefit and helped his adult specialist colleagues to a better understanding of children.

Hugh made over 400 domiciliary visits annually, always with the general practitioner (the limit for payment was 200 visits), and he found time to visit prisoners in Princetown gaol to talk to the pre-release groups about the problems they would have to face on return to family life, and also to understand better the criminal mind.

During his first 10 years as a consultant he attended a vast number of medical meetings, where he always made penetrating and usually apposite remarks. In later years they were often more controversial. At this time he also managed to be honorary secretary of the paediatric section of the Royal Society of Medicine. There were no motorways to London then, so he travelled by car in the small hours. He infuriated his junior staff by doing a full ward round at about 4 a.m, with only the frightened night sister, before leaving for London. When the juniors came on duty at the normal time they had to sort out the myriad instructions left by 'the boss'. They got their own back: Jolly crept in for his usual morning round, but this time he found his whole staff standing strictly to attention in starched white coats, or nurse's uniforms.

Small wonder that in 1960 he was invited to join Doyne Bell at Charing Cross Hospital, London. In 1965 he became paediatrician in charge at Charing Cross, a post he retained until he retired in 1983. He turned his energies to the handicapped, and the child development centre at Charing Cross is a fitting memorial. He was also medical adviser to the Chailey Heritage Hospital. Generations of medical students flocked to his ward rounds and outpatients, as indeed did swarms of visitors - including many from overseas.

From 1961-62 he was seconded as professor of child health to University College, Ibadan, Nigeria. This fired his interest in the child health problems of the developing world, which continued until his death.

Jolly was probably the most widely travelled paediatrician. He would often suggest alterations to the running of a service which were clearly common sense, but sometimes his enthusiasm to improve a situation blinded him to local difficulties so that his efforts were counterproductive. Not surprisingly, his ideas would either be so welcome that he was invited to return or the reaction would be 'Never again!'

Hugh was particularly welcome in Australia where he was a member of the Panel of Assessors, National Health and Medical Research Council, as well as adviser to 'Parents' Centres' and the Nursing Mothers' Association. His views on bereavement were movingly set out in a broadcast in Sydney in 1977 and later published: 'Loss of a baby', *Aust.Paediatr.J.*,14(1):3-5 March 1978.

He examined abroad at the Universities of Benin and Nsukka, Nigeria; at Singapore and Riyadh, and at home for the Universities of Glasgow, Liverpool and London, as well as for the MRCP and DCH. As always, he demanded high standards which earned him a hawkish reputation, Other appointments included the Tropical Medicine Research Board, Medical Research Council, the advisory council of the British Society of Music Therapy, the editorial board of *Excerpta Medica*, the Bureau of Overseas Medical Service, the Down's Children Association, the National Association for Hospital Play Staff, Nuclear Weapons FREEZE, Family Planning Association, Exploring Parenthood, and the Institute for Social Inventions - all examples of his catholic interest in wide areas of child care which was reflected in his departments both in Plymouth and at Charing Cross. Psychiatrists, psychologists and play therapists (Hugh was one of the first to introduce the latter) cooperated with him to establish a free and happy atmosphere where previously the rigidity of departments geared to adults had been the rule.

His first book, *Sexual precocity*, Springfield,Ill.,Thomas,1955, gave the clearest understanding of the problem at that time and was a culmination of his thesis for MD Cantab, which won him the coveted Raymond Horton Smith prize. There followed a succession of books and articles, nearly all written in the small hours - sleep hardly seemed necessary. His student textbook *Diseases of children*, Oxford, Blackwell, first appeared in 1964; the fifth edition, in collaboration with

M Levene, appeared in 1985 only shortly before his death. Its universal popularity speaks for its clarity and explains why it was translated into other languages. Even more widely successful was his *Book of child care,* London, Allen and Unwin, 1975, which ran to four editions and appeared in six other languages. His regular column in *The Times* during the 1970s culminated in collected articles (c.1973) entitled *Common sense about babies and children,* London, Unwin Paperbacks, 1983 and *More common sense about babies,* London, Pelham (also Sphere),1978. Two new books appeared the year before he died: *The First five years* and *The Grandparents' handbook,* London, Pagoda Books, 1985. In his later years he concentrated on bereavement and greatly influenced obstetricians, paediaticians and general practitioners to react more sympathetically to parents of stillbirths and miscarriages. He also drew attention to the problem of the dying child. It was not surprising that the British Paediatric Association elected him an honorary member in 1985. He was also an honorary member of the section of paediatrics of the Royal Society of Medicine.

Through his numerous writings in medical and lay journals, and in newpapers, and by his appearances on television and his talks on radio, Hugh sought to establish a wider understanding of his patients: of their health, their behaviour, their illnesses and anxieties. This naturally involved parents and he was never happier than when talking through their difficulties. As a parent, it is unpleasant to hear oneself upbraided, but Jolly never spared the feelings of others if he believed he was acting in the best interests of his patient, whether the latter be newly born or a tyrannical teenager. He recognized that this upset parents, even to anger, but he would argue passionately in defence of his stance.

Hugh Jolly was tall, with a slightly stooping, lolloping gait, a shock of hair and usually a large bow tie. He drove himself hard and fast, and all who worked with him; this included his cars. He was indomitable and indefatigable. Even when ill, and in a wheelchair, he flew to Berlin to fulfil lectures he had promised. He was equally energetic in his leisure pursuits whether gardening, breeding budgerigars and peacocks, or water skiing and, later, gliding. Even when on holiday his thirst for knowledge continued, especially in regard to historic buildings and their treasures, which just had to be visited no matter how long the drive. Passengers were allowed no 'stops', not for food or even calls of nature - at least, not until pleas were desperate.

Professor Sir Peter Tizard allows me to quote from his address at the Service of Thanksgiving at All Souls, Langham Place, London on 14 May 1986: 'And what of Hugh as a person? First, he was a man of high principles, but he did not wear his conscience on his sleeve. Next, he was a man of boundless energy and boundless vitality. In his dealings with children and their parents he was a model of patience and the unhurried approach. As a friend he was always good company and always good humoured, and he never took offence even when deliberately baited. I generally attacked him on the elegance of his clothes - large bow ties, white dinner jackets and so on - or on the successful opening on TV, or in the Press, of his spring or autumn advertising campaign. All this needling Hugh accepted with amusement and good-humoured tolerance.'

In everything he did his wife Geraldine supported him devotedly. Equally, he was devoted to her, his three children and latterly, while he was so sadly ill, his grand-daughter.

BML

[*Brit.med.J.*,1986,**292**,835; *Lancet*,1986,**1**,629; *The Times*,5 Mar 1986; *London Standard*,5 Mar 1986; *Guardian*,5 June 1978; *Charing X Hosp.Gaz.*,Feb 1961,**59**,(1),p.4; Photo]

JOULES, Francis Edwin *b.*16 Mar 1916 *d.*31 Dec 1986
MRCS LRCP(1939) MB BS Lond(1940) MD(1946) MRCP(1947) FRCP(1970)

Francis Edwin Joules was born at Woodseaves, near Stafford, where his father, Richard Edgar Joules, was in business. His mother Emily Annie, née Hyatt, was the daughter of William Hyatt, a gentleman of independent means. His brother Horace was also a physician and a Fellow of the College [*Munk's Roll*,Vol.VII,p.307].

He studied medicine at the Middlesex Hospital medical college, where he graduated with honours in pathology, midwifery and gynaecology. The second world war coincided with his graduation and, after a brief house post at the EMS Central Middlesex Hospital, he joined the RAFVR and served from 1940-46.

On demobilization he became chief assistant in medicine at the Harefield Hospital in 1947, and subsequently consultant physician at the Peppard Chest Hospital in 1950. In 1953 he was appointed physician-superintendent and consultant physician to Milford Chest Hospital and held this post until his retirement.

He married Marian, daughter of Percy Dawson Dutnell, a Lloyds underwriter, in 1947, and they had four children, two sons and two daughters. Horticulture was his chief recreation and he took a great interest in growing orchids.

GW/VL

[Photo]

KEELE, Cyril Arthur *b.*23 Nov 1905 *d.*22 Oct 1987
MRCS LRCP(1927) MB BS Lond(1928) MRCP(1929) MD(1930) FRCP(1948) FFARCS(1958)

Cyril Keele, who was to become professor of pharmacology in the University of London, was the second son of Dr David Keele and his wife Jessie. Three of their sons followed their father into medicine: one became a general practitioner and another, Kenneth Keele (q.v.) who died some five months before his brother Cyril, was a distinguished consultant physician and medical historian who made valuable contributions to historical studies of Leonardo da Vinci and William Harvey.

Cyril went to Epsom College, obtained a scholarship and went on to the Middlesex Hospital where he fell under the spell of Samson Wright [*Munk's Roll*,Vol.V.,p.463]. After qualifying in 1928 he held house appointments at the Middlesex and the Brompton, and was RMO at the Royal Masonic Hospital. He returned to the Middlesex in 1930 to hold the medical registrarship of George Ward [*Munk's Roll*,Vol.VI,p.449] and Evan Bedford [*Munk's Roll*,Vol.VII,p.28].

At the Middlesex Cyril undertook a large scale clinical study of the heart in hypertensive patients, which whetted his appetite for research, and in 1933 he became demonstrator in physiology, succeeding Michael Kremer (q.v.). He developed pure and applied physiology and gave an increasing number of pharmacology lectures backed by clinical demonstrations. The advent of pharmacology was heralded by the University of London instituting an examination in applied pharmacology and therapeutics in the third MB BS. With enthusiasm, Samson Wright advocated the establishment of a new department of pharmacology and Cyril became lecturer in 1938 - but still in the same room!

During the war Cyril accompanied the department to Bristol, then back to the Middlesex, and afterwards to Leeds for two years. He taught pharmacology '...to preclinical and clinical students wherever they happened to be.' As clinical students were at the Middlesex, Central Middlesex, Mount Vernon and Tindall House at Aylesbury - as well as Leeds - this entailed giving his lectures and demonstrations in quadruplicate.

After the war he was given a laboratory and three rooms at the Middlesex and then things 'took off'. He was joined by Arnold Burgen, whose rise to fame is well known. But Arnold went to McGill in 1948 as professor of physiology, returning in 1962 to become professor of pharmacology at Cambridge. In 1971 he became head of the National Institute for Medical Research and later Master of Darwin College. Keele once wrote that Burgen was the brightest student he had ever had.

Keele, together with David Slome, produced a technique for measuring renal blood flow. Later, in 1948, working with Hewer, and his brother Kenneth, he developed an interest in 'pain-producing substances'. He was appointed reader in pharmacology in 1948 and professor in 1962. The department had been strengthened by the appointment of Franz Hobbiger who came from Vienna for a year in 1948 but fortunately stayed on, succeeding Cyril as professor in 1968. Desirèe Armstrong began in 1961 and devoted herself to the 'PPS' with

Theo Chalmers (q.v.) and others.

In 1951 Keele and his team introduced the technique of the blister base for the subjective assessment of pain in man. These results are given in the Physiological Society monograph *Substances producing pain and itch*, C A Keele and Desirèe Armstrong, London, Arnold,1964. This monograph was substantiated by Rocha de Silva of Brazil and John Margolis of Oxford and Sydney, both having come to the UK and worked with the team. They demonstrated that PPS was formed by a two-stage process analogous to that which leads to the clotting of plasma which contacts glass. Keele became FFARCS in 1958.

At this time Cyril was much involved in speaking in symposia abroad. He was also responsible for the organization and editorship of the International Symposium on the assessment of pain, published in 1961. His lectures were appreciated in Europe, the United States, Singapore, Fiji, and in Australia where he was visiting professor of the University of Sydney. His activities abroad did not prevent him from serving in Britain. To list but a few: he was on the committee and editorial board of the *Journal of Physiology*, president of the section of experimental medicine of the Royal Society of Medicine, a member of the Cohen Committee on the categorization of drugs, a member of the Poisons Board of the Home Office, and of the British National Formulary Committee *et seq.* He was also a member of the committee of the Pharmacological Society, 1949-52 and 1955-56, and honorary secretary of the medical library at the Middlesex Hospital Medical School until 1968.

In 1968 he retired from the chair but was immediately appointed to the directorship of the Arthur Stanley Institute of Rheumatology at the Middlesex Hospital - a post he held until 1972. He worked hard to develop this subject.

As well as the monograph already referred to above, his publications included *Recent advances in pharmacology*, with the late J M Robson, London, Churchill, 1950 (2nd edition 1956). With the death of Samson Wright in 1956, Keele took on *Applied Physiology*, London, Oxford University Press, together with Eric Neil, and they published four editions ending with the 13th in 1982.

One wonders whether such a busy life left time for leisure. Cyril felt that his interest in rugger - he was captain of the hospital team - and cricket, of which he knew a phenomenal amount, and athletics, all contributed to keep him happy. Much of his strength was due to his very happy marriage to June in 1942. Their three sons, Gerard who practises medicine in Norfolk, Richard who has a doctorate in geology and John who went into business, provided him with immense satisfaction.

Cyril was a wise and kind man, and deeply religious.

FH/EN

[*Brit.med.J.*,1988,**296**,94; *Middx.Hosp.J.*,Aug 1968,**68** (4)125-27; Feb 1969,69 (1) 8-9; Photo]

KEELE, Kenneth David *b.*23 Mar 1909 *d.*3 May 1987
MRCS LRCP(1932) MB BS Lond(1932) MD(1934) MRCP(1934) FRCP(1951)

Kenneth David Keele was born at Highbury, London, into a family with a strong medical tradition; his father, grandfather and eldest brother George being general practitioners and his other brother, Cyril (q.v.), a professor of pharmacology. He was educated at Epsom College and won a scholarship to St Bartholomew's Hospital medical school where he won the Harvey prize for physiology. After junior appointments he became casualty physician at Bart's and was elected to the consultant staff of St Pancras Dispensary, the Evelina Hospital for Children and King George Hospital, Ilford. From 1933-39 he was clinical assistant to Sir John Parkinson [*Munk's Roll*,Vol.VII,p.443] at the National Heart Hospital.

In 1939, at the beginning of the war, he was appointed consultant physician, Emergency Hospital Medical Service, St Bartholomew's Hospital, St Albans. From 1942-46 he served in the RAMC in India, where he attained the rank of lieutenant colonel. In addition to clinical and administrative responsibilities he was head of the sprue research team in Poona. It was here that he met Mary Thrussell, then in the QAIMNS/R, who became his second wife. She was sister in the sprue research team and worked with Kenneth both before and after they were married.

After the war Kenneth was appointed full time consultant physician to Ashford Hospital, Middlesex, (1946-71), where he developed his great interest in cardiology. It is not perhaps generally known that he was the pioneer in this country of the use of angiocardiography to provide accurate diagnosis of congenital heart disease to the benefit of both cardiac surgeons and their patients. This work was published in 1948. His experience in general medicine led to the publication, with his surgical colleague Norman Mathieson, of a book entitled *Intra-abdominal crises*, Washington, Butterworth, 1961.

Kenneth was an excellent doctor. He had all the qualities needed by the modern physician, a deep knowledge of the science of medicine combined with a very human rapport with his patients. It was through his empathetic understanding of his patient's complaints that he became involved in the study of pain, to which he made notable contributions. He appreciated the difficulties of communication and recording of this common symptom due to its subjective nature. However, insofar as pain has a time course and varies in intensity it can be represented objectively and in 1948 this led Kenneth to introduce the Pain Chart in which intensity of pain, as assessed by the patient, is plotted against time. This has proved to be a most valuable tool in clinical and experimental studies of pain, and a modified form of pain chart, the Visual Analogue Scale, is now used the world over.

Kenneth had long been greatly struck by the variations in response of different persons to noxious stimulation. As a cardiologist he noted that some patients with myocardial infarction suffered little or no pain. To study this phenomenon further he developed, in 1954, the pressure algometer, a simple device to measure the mechanical pressure required to evoke pain when applied to the forehead. Patients hyposensitive to the pressure algometer had suffered the least pain at the time of the infarction.

Among other contributions to the study of pain Kenneth and his colleague, P R S Stern, showed that patients with myocardial infarction or renal colic had marked falls in serum beta-lipoproteins and cholesterol concentrations which were closely related to the severity and duration of the pain. Stress *per se* raised serum lipid levels.

The medical historian in him found expression in *Anatomies of pain*, Oxford, Blackwell, 1957, a scholarly work in which concepts of pain from prehistoric times to the twentieth century were admirably reviewed. Kenneth was indeed a most distinguished medical historian with wide-ranging interests and an international reputation, especially in the USA where he was visiting professor at Yale University in 1958. At home he was research fellow at the Wellcome Institute for the History of Medicine from 1970-75.

His book *William Harvey, the man, the physician and the scientist*, London, Thomas Nelson & Sons, 1965, was a masterly work in which he quoted evidence from Harvey's own writings, showing that the growth in Harvey's ideas was very largely shaped by the creative thought of Aristotle both with regard to the subjects which Harvey studied as well as his method of approach. Harvey may be pictured as '...having one foot in the ancient Greek library of Aristotle and the other in the operating theatre of the modern cardiac surgeon.'

In 1960-61 Kenneth gave the FitzPatrick Lectures at the College, which were published c.1963 as *The Evolution of clinical methods in medicine*, Springfield, Ill.,Thomas. The subject gave him full scope to reveal his wide knowledge and understanding of medical history. Kenneth's lifelong study of the anatomical works of Leonardo da Vinci was a passionate commitment which was favoured by the close proximity of his home in Staines to HM The Queen's collection of Leonardo's anatomical drawings at Windsor. In spite of steadily deteriorating eyesight he produced his magnum opus in three volumes: *Leonardo da Vinci: Corpus of the Anatomical Studies in the collection of HM The Queen at Windsor Castle*, Kenneth Keele and Carlos Pedretti, New York, Harcourt Brace Yovanovich, 1978-80, and *Leonardo's 'Elements of the science of man'*, New York:London, Academic Press, 1983. Kenneth founded the Leonardo da Vinci Society in 1986 and became its first president shortly before his death. A fuller account of his historical activities appeared in the journal *Medical History*.

Kenneth was my brother and my best friend. I give here a few of many thoughts about him. He combined great intellectual and artistic gifts with a most lovable nature. Like his mother he was a good pianist and his love of the arts, especially paintings, showed through all his interests in medical history and gave him the insight to interpret the perplexing anatomical works of Leonardo. Though modest and unassuming he was a very good and enthusiastic communicator, in conversation, in lectures and in his writings. His hallmarks were clarity of thought and of expression allied to a determination to make others see his point of view. He was always delighted to give a helping hand to junior colleagues, with whom he was most popular, and in my view he would have been an excellent teacher of clinical medicine to undergraduate medical students. In later years he was much handicapped by failing sight due to glaucoma, and his inability to read the many medical historical books which he had acquired for his retirement was a galling experience which he endured bravely, with solace from playing his beloved Chopin.

His domestic life brought him great happiness and two tragedies. His first wife Dorée, née Doris Berther, inspired his purpose and achievements before her sudden death when eight months pregnant in 1941. His second wife, Mary, helped him to fulfil his life's aims. They had a daughter Caroline and two sons, Peter and Tony. Peter was killed in a traffic accident. Kenneth and Mary showed great courage in facing this terrible event. Kenneth was survived by Mary, Caroline and Tony and Caroline's son and daughter.

CAK

[*Brit.med.J.*,1987,**295**,338; *Medical History*,No.4,**31**,Oct 1987; Photo]

KEMPTHORNE, Renatus *b.*5 July 1907 *d.*23 Mar 1984
MRCS LRCP(1932) MA BM BCh Oxon(1934) MRCP(1934) FRCP(1964)

Renatus Kempthorne was born into an old-established Cornish family of farmers and professional people. His grandfather was a farmer, from whom he inherited the family farm, and his father William Granville Kempthorne was a solicitor. His mother Jane, née Goodman, was the daughter of the village blacksmith at the Lizard.

Renatus was educated at Blundell's and University College, Oxford. He began by reading mathematics and achieved first class honours in moderations; he then changed to medicine, winning awards in anatomy, pharmacy and pathology, and graduating with second class honours in physiology. His clinical studies were undertaken at University College Hospital, London, where he served as house physician to Charles Bacon in 1933. He was appointed RMO in 1935, and became an assistant on the medical unit in 1936 - a post he held for three years. In 1939 he was appointed physician to the North Middlesex

Hospital where he remained until his retirement in 1967, except for a brief period from 1946-47 when he served as medical specialist to the Duchess of Kent Military Hospital, Catterick, during his two years of military service.

He was a delightful colleague, naturally self-effacing, but a tower of strength when needed and a dedicated defender of the principles of medical etiquette. He had a dry sense of humour: when addressing an audience of some 500 doctors on coronary thrombosis he first showed a slide relating the increase in deaths from heart disease to the consumption of cholesterol containing foods, immediately followed by another slide relating the rise in mortality to the increased possession of television sets throughout the country.

His interests outside medicine included the visual arts and history, both of which he exploited to the full during his holidays. One holiday was spent rambling about Spain, if not exactly on a donkey at least by such haphazard means of transport as were available to him when off the beaten track.

Renatus married Kathleen Margaret, daughter of Hugh Thomas Whitworth, a shipbroker. His wife died in 1978, but he was survived by their son, also Renatus, a Clerk in Holy Orders presently living in New Zealand.

DF

[*Brit.med.J.*,1984,**289**,328,632; Photo]

KERR, John Duke Olav *b.*23 May 1911 *d.*27 June 1987
MB ChB Glasg(1934) FRFPS(1946) MRCP(1946) FRCPG(1962) FRCP(1963) FRCPE(1971)

Olav Kerr was born in Glasgow before the outbreak of the first world war. He imbibed high principles and a caring outlook in his home which no doubt influenced him towards the profession to which he was to devote his life. At school, the Glasgow Academy, he inclined to the classics side rather than the sciences, a pathway not unusual in the Scottish medical tradition in those days. Ill health prevented active participation in sports, something he always regretted. He developed a great love of the countryside and open-air activities. This was fulfilled by annual family holidays in Speyside and, in later years, a cottage at Kinlochard under the shadow of Ben Lomond. His other lifelong interest was in motor cars.

He studied medicine at Glasgow University and had a distinguished record there, gaining a number of prizes and an honours degree in 1934. After a year as house surgeon and house physician in the Western Infirmary, he became an assistant to the professor of physiology. Here he gained experience in research and an opportunity of developing his talents as a teacher. Two physicians had a considerable influence on him at this time, George Allan [*Munk's*

Roll,Vol.V,p.7] in his own hospital and John Parkinson, later Sir John [*Munk's Roll*,Vol.VII, p.443] in the London Hospital, both of whom were founder members of the British Cardiac Society. Though physiology was a possible career, his heart was set towards clinical medicine and the exciting developments in the new specialty of cardiology. In the Scottish situation then, a consultant physician was a general physician, usually with a special interest; in Olav Kerr's case this was to be cardiology.

The war years were a wider experience of medicine and of people. He became a lieutenant colonel in the RAMC in charge of a medical division of a general hospital. This was a time of preparation for what was to be this life's work - a physician in the Western Infirmary, Glasgow.

Olav Kerr was an outstanding teacher. He excelled at the bedside and in the clinic. His enthusiasm and high standards of medical practice inspired even the most apathetic student. His use of anecdotes, not always flattering to himself, is part of the folklore of the hospital. He showed by precept and example that therapeutics were not solely the treatment of diseases but the care of the patient as a person and often the family as well. In those days of euphoria about new drugs this was a much needed corrective and not always popular. Perhaps this was the basis of his undoubted success in private consultant practice. He gave counsel and friendly understanding as well as good medical management. Over the years many students, colleagues and their families came for help, often in regard to matters quite outside the field of medicine. Olav was always available.

He took an active part in the affairs of the British Cardiac Society and was proud to chair its annual meeting held in Glasgow shortly before he retired. He was largely responsible for establishing the cardiac department in the hospital, and enjoyed training younger men in this field. Administration did not appeal to him, but in the early years of the National Health Service he gave wise leadership in the developing period of the new hospital service. He was recognized as a person of complete integrity.

Perhaps his most lasting work was the influence he had on the students and young doctors who were attached to his wards. He demanded upright character, courtesy and kindness, as well as a sound knowledge of medicine. He did not tolerate slackness in these qualities, as some found to their cost. Many in our profession today look back with gratitude and a sense of privilege in having been taught and influenced by one of the great teachers of the Glasgow School.

Olav Kerr was a family man. He married Jenny Gossip in 1937 and they had four sons, and later an adopted daughter. Two boys went into the Royal Navy, one into the Diplomatic Service, and the fourth became a historian and university lecturer. This close-knit family was a source of strength and joy to Olav, especially in the years of ill health following his retirement, and sustained him after the tragic death a few years ago of his third son, Robbie, while serving in the Royal Navy.

He and his wife accepted the Christian faith in childhood and their convictions grew with the passing years, not merely as a set of beliefs but as a way of life. If we seek to know from whence came the motivation for his life this was undoubtedly its source.

CDA

[*Brit.med.J.*,1987,**295**,677; Photo]

KESSEL, Israel *b.*16 June 1917 *d.*10 Nov 1985
MB BCh Wits(1939) DCH(1947) MRCPE(1948) MRCP(1948)
FRCPE(1970) FRCP(1972)

Israel (Rael) Kessel qualified at the University of Witwatersrand and served as a captain in the South African Medical Corps for the duration of the second world war. He first came to England immediately after the war and completed his postgraduate qualification in paediatrics at the Hospital for Sick Children, Great Ormond Street. On his return to South Africa in 1949 he became assistant physician at Baragwanath Hospital where he participated in much of the pioneer work in kwashiorkor and other related diseases. The majority of patients came from the neighbouring area, now known as Soweto, where most of the inhabitants lived in considerable squalor. All varieties of infections, gastrointestinal disturbances and malnutrition, were rife. There was an abundance of perinatal and neonatal problems and a consequent high infant mortality rate. Rael Kessel was very involved with these patients and his efforts contributed to the subsequent reduction in neonatal morbidity and mortality.

In 1950 he was appointed as the first full-time consultant at the Transvaal Memorial Hospital for Children with an attachment to the University of Witwatersrand. He was responsible for the clinical teaching of medical students and nurses. At the same time, he took over the care of neonates at the Queen Victoria Maternity Hospital.

Rael left South Africa in 1963 and was appointed as consultant paediatrician to the Watford and Hemel Hempstead Group of hospitals in 1964. He resumed his association with the Hospital for Sick Children, where he held the post of honorary associate physician. He continued in both these posts until his retirement in 1984; for most of this period he worked single-handed. He worked closely with the large obstetric department of a busy district general hospital and was particularly interested in neonates. He was largely instrumental in setting up a purpose built neonatal unit, and later in his professional career he was able to persuade the Regional Health Authority to appoint a second paediatrician with a special interest in neonatology.

In spite of his heavy clinical commitment, he found time to write a standard textbook on *The Essentials of paediatrics for nurses*, Edinburgh, E & S Livingstone, 1957, which ran into five editions. At the time of his death he was in the process of preparing a sixth edition.

He excelled in committee work, serving as chairman of the medical staff committee and as consultant representative of the district management team at a critical stage in the commissioning of the new Watford General Hospital.

Rael was a very modest and kindly person; qualities which were appreciated by his young patients and their parents. His opinions on matters both clinical and administrative were greatly respected by his colleagues. He also enjoyed family life to the full. He was twice married; first to Elizabeth Farrell, by whom he had a son and a daughter, and then - after her untimely and tragic death in 1973 - to Margaret Wood. Margaret shared with him a passion for antiques and they enjoyed exhibiting at antique fairs together. His special interest was in antique scientific instruments, scales, weights and calculators. His retirement in 1984 did not in the least detract from his enthusiasm both for paediatrics and antiques. He shouldered the burden of unexpected local vacancies for paediatricians at various hospitals in the Home Counties and in Cornwall, which held a special place in his heart, and he was still at work a week before his death, which characteristically, and probably in accordance with his wishes, came suddenly during the course of an antiques fair.

Rael Kessel was especially respected by his junior staff, many of whom subsequently settled in general practice in Watford. In particular, many young doctors from overseas will remember him with great affection as someone who gave them tremendous help when starting their careers in paediatrics. Many of them attended the funeral ceremony, some travelling great distances, and there was hardly a dry eye amongst them.

JHA

[*Brit.med.J.*,1986,**292**,276; *Watford Observer*,15 Nov 1985; Photo]

KILPATRICK, John Alexander *b.*22 Feb 1920 *d.*24 May 1987
MB ChB NZ(1942) MD(1946) MRACP(1947) MRCP(1948)
FRACP(1959) FRCP(1969)

Apart from four years' absence, gaining postgraduate experience in Auckland and the United Kingdom, John Kilpatrick - called 'Jack' by most of his colleagues - lived in the city of Dunedin in the Otago province of New Zealand, where he was born and educated, where he practised as a leading physician of his time, and where he died.

His early upbringing was in modest circumstances. His father, also John Kilpatrick, was a prison superintendent and his mother Iris, née Ramsay, was the daughter of a contractor, Alexander Ramsay. Jack's wife, Effie Forsyth McKnight, came from a farming family in Southland.

They were married in 1943, a year after Jack qualified in medicine. Educated initially at Otago Boy's High School, he completed his medical course at the University of Otago, graduating in 1942.

Jack's long and distinguished association with the Otago Medical School and the Dunedin Hospital, as well as with the people of Otago and Southland, began in 1943 and was to continue throughout his practising life. It was the year in which he was quickly spotted by the professor of medicine, Frederick Horace Smirk, later Sir Horace, who had begun to revitalize the Dunedin scene since his arrival in 1939. Jack Kilpatrick was one of the earliest recruits into Smirk's team of young turks. He was appointed as an assistant junior physician, becoming a research fellow from 1944-46. At the end of his baptism into clinical research he gained his doctorate. His academic career was then interrupted temporarily when he contracted tuberculosis. On recovery he determined to work as a 'tuberculosis officer', which he did for a short period in 1947 at the Green Lane Hospital, Auckland, before coming to the United Kingdom.

In 1948, no doubt on Horace Smirk's strong recommendation, he was appointed house physician at the Hammersmith Hospital, under John McMichael, later Sir John. This was followed by two years as a lecturer in medicine at the University of Sheffield, under Stuart-Harris, later Sir Charles. Despite this early postgraduate experience in cardiopulmonary disorders, he became strongly influenced by a fellow New Zealander, Russell Fraser, who eventually steered Jack towards endocrinology, in particular to the clinical problems of the thyroid and pancreas.

In 1951 he returned to his home town, Dunedin, and rejoined Smirk's then flourishing department of medicine as senior assistant physician and lecturer in medicine. For the next five years he demonstrated his abilities both as a stimulating teacher and a most able physician, one who combined scientific vigour with a caring approach to his patients. Although further academic advancement was one option, he chose to enter part-time consultant practice in 1956, but retained his hospital appointment as a senior physician and his university academic status. His practice became increasingly orientated, though not entirely, to endocrinology, and in particular to the field of diabetes. The development of his diabetic clinic was a major commitment and his approach, in the 1960s, was innovative at that time: using the multidisciplinary model incorporating a podiatrist, dietitian, nurse educator, and a consultant ophthalmologist who was jointly involved in a model diabetic retinopathy screening programme. As a researcher Jack had a degree of insight and a capacity for originality which enabled him to make a significant contribution as a clinical collaborator with endocrine and autoimmunity research groups in the Otago School.

Jack was to become a leader in the field of diabetes in New Zealand, and served as chairman of the medical advisory panel of the New Zealand Diabetic Association, and as a delegate on the executive of the International Diabetes Federation. He was a life member of the Society for the Study of Diabetes and the Society of Endocrinology, being a founder member of the latter in 1963.

Throughout his working life Jack remained first and foremost a good physician, except that he could not adapt to the non-smoking model. He was certainly not a ponderous physician, and his ability to make prompt and clear decisions was especially appreciated by his surgical colleagues who sought his medical advice on a wide range of patients' problems. The discovery of a rare and unusual 'case' excited him, but never to the point of exploiting or diminishing his concern for the patient.

Jack Kilpatrick played his part in several roles for the Royal Australasian College of Physicians: as censor, as chairman for the New Zealand grants advisory committee, and then as vice-president of the College from 1976-78. He served on many other national committees, such as the New Zealand Board of Health committee on drug dependency and abuse.

Despite his involvement in a busy practice, Jack maintained his university attachment throughout his life and when, in 1977, the University of Otago first established 'clinical readers' in recognition of outstanding contributions to teaching, he was among the first to be so recognized. On his retirement in 1985 the medical faculty felt his loss keenly and the dean referred to his humanity, his orginality of mind, and his diligence in pursuing new insights and more effective care for his patients. It was minuted in a faculty valedictory resolution that: '...he was one of those valuable colleagues who believed in and practised the best type of personalised academic medicine in both his private practice and the public hospital system... The large number of patients he saw from throughout Otago and Southland were treated in an interested, professional manner with financial reward relegated to a low level of priority.'

To colleagues and patients alike, Jack's friendliness, optimism and good humour dominated relationships. He was always full of ideas, always willing to speak up at meetings, with plenty of wit, and always fully of energy until his obstructive pulmonary condition sadly caught up with him in later years.

Outside medicine, Jack had several interests which kept him active whenever he could find the time to indulge them; from trout fishing and growing orchids to supporting amateur swimming and surf lifesaving clubs. He was also an active member of the Presbyterian Church.

He and his wife Effie offered generous hospitality to a host of visitors from abroad as well as to their friends, colleagues and students. They were pleased when one of their two sons followed his father's footsteps into medicine.

JDH

[*Otago Daily Times*,26 May 1987; *NZ med.J.*,12 Aug 1987,509; Photo]

KIRK, James Balfour *b.*7 Apr 1893 *d.*26 Mar 1984
CMG MB ChB Edin(1917) DPH Oxon(1921) DTM&H(1922) MRCP(1933) FRCP(1942)

Balfour Kirk was born at Falkirk and lived to the age of 90. He was educated at George Watson's College and Edinburgh University, and like so many of his generation of medical students joined the Army, in the ranks, as a combatant at the outbreak of war in 1914, before being sent back to complete his medical training in 1917 and then rejoin the RAMC. After the war he took a DPH and the DTM&H and joined the Colonial Service as MOH of Port Louis, Mauritius. He rose to be director of medical services in Mauritius and was transferred to the Gold Coast (Ghana) as director in 1941. He served with UNRRA in Greece, and later in Germany, and was awarded the CMG. On retiring from the Colonial Service he joined the Ministry of Health and served there until 1962.

Balfour became a member of the College in 1933 and a Fellow nine years later. He had a broad knowledge of medicine and a clear, succinct way of expressing it. He was a patient and valuable reviewer and collator of drafts in the Ministry of Health; abilities which were used to the full in the editing of its medical publications, including the CMO's annual report *On the state of the Public Health* and the special reports then known as 'Grey Books', which ranged from reports on maternal deaths to the testing of very young handicapped children by Mary Sheridan. He succeeded Alison Glover, also a Fellow [*Munk's Roll*,Vol.V,p.152], in this work.

He was used as a confidant and adviser by colleagues at all levels without ever becoming a partisan of any particular group, or a passer-on of anyone else's opinions, no mean achievement in that kind of cloister. Balfour ran the *Monthly Bulletin* of the Ministry and the Public Health Laboratory Service, which became the forerunner of *Health Trends*.

Balfour was a shrewd, laconic Scot with a lively sense of humour. He was physically active, regularly playing five holes of golf before breakfast - on the course which was at the foot of his garden in Surrey - and then walking the mile and a half from Victoria to the office in Saville Row. He gardened, was something of a gourmet, an expert on wines, and a careful but assiduous operator in the stock market.

He had married on qualification in 1917; a marriage that lasted the rest of his life. He enjoyed 22 years of his second retirement till his death in 1984.

GEG

[*Brit.med.J.*,1984,**288**,1769; Photo]

KNAPTON, Peter James *b.*14 Sept 1934 *d.*29 Jan 1984
MB ChB Leeds(1957) MRCP(1962) FRCP(1979)

Peter James (Jim) Knapton died at the early age of 49. He was born near Bradford where he went to school in his early years, moving on to Clifton College, Bristol. He entered Leeds Medical School in 1952 and qualified in 1957, gaining a distinction and the Mayo-Robson prize in surgery, and the Hardwicke prize in clinical medicine. His main interest, however, was in medicine rather than surgery and after completing house appointments in Leeds he moved to the London Chest Hospital in 1959, where he met his wife Betty who was nursing there. They married in 1960.

After further appointments at Hillingdon and Birmingham, he worked as rotating registrar at Bristol, subsequently going to Canada as an internist at the Brantford Clinic, Ontario. This was a year of great interest and experience for Jim and Betty but, although they loved Canada, England drew them back in 1968. On his return Jim obtained a post as research fellow at the Royal Marsden Hospital in Sutton where he worked for three years in conjunction with the late Gordon Hamilton Fairley [*Munk's Roll*,Vol.VI,p.215]. It was during this time that Jim developed his increasing enthusiasm for oncology and haematology, subjects which were to become his special interest during his consultant career. In 1972 he was appointed consultant physician to the Croydon Group of hospitals. It was at the Croydon General Hospital that most of his time was spent and for which, and the people who made it what it was, he had a great affection. Jim loved his clinical work and was a most conscientious and caring physician, taking endless trouble with the problems of his patients and of the staff who worked with him. As well as general medicine he was able to enjoy his interest in oncology and his opinion was often sought, and highly regarded, by his colleagues both in hospital and general practice. Apart from his clinical load, he was an active committee member; chairman of the Croydon BMA in 1980 and for many years secretary of the medical division. His interests were wide and he and Betty had many friends within the church community. He became an authority on church architecture, but his interest in the church extended far beyond that of architecture alone.

Eighteen months before his death Jim realized that he was suffering from motor neurone disease, which became rapidly progessive, and in coming to terms with his illness he showed a remarkable courage, such that most of us would find hard to equal.

He was survived by his wife Betty and their two children, and it was thanks to their love and his own deep faith that he was able to say - in words that were true in every sense - 'I have had a good life.'

JAP

[*Brit.med.J.*,1984,**288**,944; Photo]

KNOX, John *b.*6 Dec 1926 *d.*20 June 1987
MB ChB Glasg(1949) MRCP(1954) MRCPG(1972) FRCP(1974) FRCPG(1974)

John Knox, consultant physician to Raigmore Hospital Inverness and senior clinical lecturer in medicine in the University of Aberdeen, was born at Whitley Bay, the son of a civil servant. He was educated at Hamilton Academy and the University of Glasgow from which he graduated with commendation. He became successively a research fellow and then assistant lecturer in the department of *materia medica* and therapeutics in the University of Glasgow, and was also a medical registrar at Stobhill Hospital. In 1955 he moved to Aberdeen as senior medical registrar on the professorial unit. He was appointed consultant physician in geriatric medicine at Woodend Hospital, Aberdeen, in 1958 and some four years later moved to the North Regional Hospital Board as consultant physician at Inverness.

Having assumed the responsibilities of a general physician, John Knox was at pains to increase his knowledge of the developing specialties and devoted much time and travelling to do so. He frequently attended clinical meetings at teaching centres throughout the country and arranged for leading specialists to visit Inverness. On these occasions many enjoyed his company and hospitality, and observed his appreciation of good food and good wine.

Knox was a greatly respected member of the medical profession and a generous and loyal colleague. His main interest was in gastroenterology but he was an able and skilled general physician whose opinion inspired confidence. Always his care for patients extended beyond the immediate clinical problems to encompass their social and emotional circumstances - reflecting his great concern for those of his fellows less fortunate than himself. With him, service to his patients came first, ahead of his own interests, and as he practised so he taught.

John Knox retained close links with Aberdeen and the medical school there. He played an important role in bringing medical students to Inverness for instruction during their clinical years. He had a well earned reputation as an excellent teacher at the bedside, emphasizing always the basic approach and the importance of making a diagnosis on clinical grounds. He was a very fair undergraduate examiner, and if occasionally he asked an inappropriately awkward question he was quick to realize it and more often than not gave the answer himself. On the other hand, he regarded his role as examiner in the membership in a different light and his approach could be regarded as distinctly 'hawkish'.

Despite his busy clinical commitment John was constantly involved in administration. He was at one time the deputy representative for the northern region to the Central Committee for Postgraduate Medical Education. He was secretary to the Highland Medical Society for seven years, and a member of the Hospital Medical Services Committee for more than a decade. In 1972 he was scientific secretary of the BMA's first postgraduate meeting held in Inverness. He was a member of the working party which reported on the integration of general practice in the hospital service in Scotland. And, until shortly before his death, was chairman of the medicine sub-committee of the National Medical Consultative Council. He was also a member and past president of the Scottish Society of Physicians, at the meetings of which he was a regular attender.

John Knox's abounding energy was also displayed in many activities outside his professional work. He was a County Scout Commissioner, an elder of the Church, and an active participant in BMA affairs.

Although gentle and unassuming, John Knox was a man of much determination who would never compromise if a principle was at stake, but ordinarily this trait lay hidden behind his charm and courtesy. With a family background in choral and instrumental music, he was an informed musician and in private an accomplished pianist. His first wife, Marion Aitken, a fellow graduate in medicine at Glasgow, died in 1984. In 1987 he married Fiona MacLennan, a consultant anaesthetist in Aberdeen. She survived him, as did two sons of his first marriage. RLH

[*Brit.med.J.*,1987,**295**,278; Photo]

KREMER, Michael *b.*27 Nov 1907 *d.*1 Mar 1988
BSc Lond(1927) MRCS LRCP(1929) MB BS(1930) MD(1932) MRCP(1932) FRCP(1943)

Michael Kremer was one of the most distinguished and best loved physicians of his generation and his active career in the practice of neurology extended over nearly half a century.

He began his medical career at the Middlesex Hospital medical school at an unusually early age. By the age of 20 he had an honours degree in physiology and qualified in medicine three years later, having been elected Broderip Scholar at the Middlesex in 1929. He became a member of the College in 1932 and gained his MD in the same year. He held house appointments at the Middlesex and worked as a demonstrator in the physiology department with Samson Wright [*Munk's Roll*,Vol.V,p.463] and then returned to clinical medicine as medical registrar, and later as registrar at the National Hospital, Queen Square.

His first appointment to the staff of a hospital as a physician came early in his career when he was appointed to the Hampstead General Hospital. At the outbreak of the war he joined the RAMC and served as neurologist to a neurosurgical unit in the Middle East. Later, he returned to the United Kingdom to work in the neurological and neurosurgical unit established for the Army and the RAF at St Hugh's in Oxford by Cairns, Symonds [*Munk's Roll*,Vol.VII,p.563] later Sir Charles, and Riddick. He left the RAMC with the rank of lieutenant colonel and later continued his association with the Army as civil consultant in neurology. While still in the Army he was elected to the Fellowship of the College and delivered the Oliver Sharpey lecture in 1958 on the subject of sitting, standing and walking. Soon after his return to civilian life he was appointed to the consultant staff of the Middlesex Hospital as a neurologist, and to the Maida Vale Hospital for Nervous Diseases as a physician. He resigned the latter appointment when he joined the National Hospital, Queen Square. He remained on the staff of the Middlesex and the National Hospital until his retirement in 1973. After retirement he continued in practice, busy and active, until the day of the tragically sudden onset of his fatal illness.

Both the hospitals he served owed him a great debt. He was devoted to the Middlesex, which had nurtured him, and served his patients and his colleagues there most faithfully. He was responsible for the foundation of the department of neurological studies in the medical school, endowed initially by the family of one of his patients who wished to show their gratitude for his skill and kindness. The department was later supported further by a generous endowment from yet another grateful patient. The clinical science building in which this department is housed, and which also houses the other clinical academic departments, was built in large measure as a result of his efforts.

At the National Hospital, Queen Square, he was dean of the Associated Institute of Neurology from 1954-62 and during this time, largely due to his own work and enthusiasm, great developments in university departments began, which have since borne much fruit in the advancement of knowledge.

He was a first class, most astute clinician but in addition he had remarkable personal qualities which endeared him to his many patients in the wards, clinics, and his private practice, those whom he had cured and also those to whom he could only hope to bring some relief from the possibility of lifelong suffering and disability. To anyone associated with him in his work it was quite clear that his very presence was therapeutic, and his patients knew that behind all his professional skill and sophisticated techniques there was a doctor who would listen with compassion, would understand, and would always do his best. It was, perhaps, his special ability to listen patiently and with understanding, completely unhurried in spite of his immensely busy life, that inspired such devotion in his patients - and devotion is not too strong a word. He was also a great teacher and inspiration to his younger colleagues. Many of his former house physicians and registrars went on to pursue a career in neurology, which perhaps they had never considered until they had seen Michael at work. They were probably attracted to the subject by his own great enthusiasm for it, and the way in which this enthusiasm enabled him to carry on working when they, many years his junior, were wondering how they could keep going for another five minutes. He published papers on a wide variety of neurological topics and even after his retirement he continued working on the effects of immunosuppressive treatment on multiple sclerosis.

Michael bore the trials of his final illness with fortitude and was for many months a patient in his own ward at the Middlesex. He was survived by his wife, Anne, his daughters Susan and Jane, his son David and a grandchild, Tasmin.

CJE

[*The Times*,11 March 1988; *The Middx Hosp.J.*,1974; Photo]

KRISHNANKUTTI, Pathiyil Karthiyayani

*b.*14 Dec 1919 *d.*14 Feb 1985

MB BS Madras(1944) MD(1947) MRCP(G)(1952) MRCPE(1952) MRCP(1952) FRCPE(1954) FRCP(1970)

Krishnankutti was born in the State of Kerala, India, and studied medicine at the Madras Medical College where he obtained the MB BS degree of Madras University in 1944. He was awarded the MD degree in internal medicine in 1947. In 1950 he went to Great Britain where he worked for two years, receiving extensive training in cardiology. He gained his membership of the London, Edinburgh and Glasgow Colleges in 1952, the last year of his stay in the UK.

On return to India he served as an honorary assistant physician at the Government General Hospital, Madras, from 1953-55, the hospital being one of the oldest government hospitals in India to be run along modern lines. Krishnankutti was later promoted to honorary physician in the same hospital, a position he occupied from 1956-65. In 1965 he was appointed to the chair of clinical medicine in the Government Stanley Hospital and Stanley Medical College, which he held until 1979.

Krishnankutti was well known as a physician and internist, and became honorary consultant in medicine to the Southern Railway Hospital in Madras. He was appointed honorary physician to the president of India in 1967, who was at that time the renowned philosopher Dr S Radhakrishnan.

In 1970 Krishnankutti returned to the UK as a Commonwealth senior medical fellow and worked for about three months at University College Hospital, London, with Lord Rosenheim [*Munk's Roll*,Vol.VI,p.394]. He visited the USA and worked at the Strong Memorial Hospital, Rochester, NY.,extending his interest in cardiology by working on echocardiography and other non-invasive techniques.

Krishnankutti was a fine clinician with a deep interest in cardiology and a popular teacher. His special interests were coarctation of the aorta, toxic myocarditis and haemodynamic changes in anaemia. He wrote two textbooks of medicine, and published papers in the fields of internal medicine and cardiology.

He married Gita Nair, daughter of a medical practitioner, and they had four sons.

VR

[Photo]

LACK, Victor John Frederick *b.*29 Sept 1893 *d.*17 Mar 1988
MRCS LRCP(1918) FRCS Edin(1921) MB BS Lond(1925)
MRCP(1925) FRCP(1934) FRCOG(1935)

Victor Lack, a nephew of Lambert Lack, a former honorary surgeon to the ENT department of the London Hospital, entered the medical college at the London in 1910 but his career was interrupted by the outbreak of the 1914-18 war. He joined the ranks of the RAMC while still a medical student and was soon commissioned as a combatant officer, seeing action as a machine gunner, and later as a rear-gunner in the Royal Flying Corps (now the Royal Air Force). Victor soon appreciated that attack is one of the best methods of defence; if he could prevail on his pilot to point the aircraft towards rather than away from the immediate enemy it ensured that he himself, in the tail, was that much further away. After he was severely wounded he was released in 1917 to continue his medical studies and returned to the London.

After qualifying in 1918 he chose a career in obstetrics and gynaecology. In 1921 he obtained the FRCS Edinburgh and in 1925 his MB BS, becoming a member of the College that same year. He was appointed lecturer, tutor and assistant to the chair of midwifery and diseases of women in the University of Birmingham, and assistant obstetrician to the Queen's Hospital. When Russell Andrews [*Munk's Roll*,Vol.IV, p.527] retired in 1926, Lack returned to the London as assistant obstetrical and gynaecological chief - after an absence of little more than a year. He became second in command on the retirement of Gordon Luker in 1930 and succeeded Eardley Holland, later Sir Eardley [*Munk's Roll*,Vol.VI,p.249], as head of department. He held this post until his retirement in 1958.

Victor Lack placed great emphasis on clinical teaching, using a lively and practical approach to his subject mixed with a little humour to maintain interest. In the operating theatre his technique was safe, neat, quick and efficient, and his post-operative care was personal and sympathetic. He was an authority on ectopic gestation.

During the second world war, he cheerfully took on a post in the EMS as medical superintendent of the King George V Hospital, Ilford, in addition to his other duties.

He was a founder member of the Royal College of Obstetricians and Gynaecologists and was elevated to the Fellowship in 1935. He later served as a member of Council and as vice-president. He was president of the obstetric section of the Royal Society of Medicine, represented the RCOG on the Central Midwives Board, and was a member of the General Medical Council. At the London he was chairman of the medical council and of the academic board. He published many papers, chiefly on clinical aspects of midwifery and gynaecology and was a contributor to the popular *Midwifery, by ten teachers...*, London, Edward Arnold, 4th-6th eds.,1931-61,and also to *Diseases of Women, by ten teachers...* London, Edward Arnold.

Victor Lack had a genial personality and was a charming companion. Outside medicine his main interest was in farming. GW/VL

[*The Times*,25 Mar 1988,10 May 1988;
Lond.Hosp.Gaz.,Mar 1959,62,(1); Photo]

LANCASTER, William Macfarlane *b.*30 Oct 1919 *d.*9 Jan 1988
MB BCh Glasg(1943) FRFPS(1947) MRCP(1949) FRCPG(1967) FRCP(1979)

William Lancaster was born at Lennoxtown in Stirlingshire and educated at Kilsyth Academy and the University of Glasgow, where he graduated in medicine. Shortly after qualifying he joined the RNVR, serving as a surgeon lieutenant from 1944-46.

After demobilization he began his clinical training, initially as a medical registrar at the Western Infirmary Glasgow with the legendary D K Adams [*Munk's Roll*,Vol.VI,p.4], and subsequently as a senior registrar at Dumfries Royal Infirmary, previously the Victoria Infirmary, in Glasgow.

In 1962 he was appointed consultant physician to the Northern Regional Hospital Board at Inverness. His responsibilities, which he shared with three colleagues who all predeceased him, were extensive and involved much travelling in all weathers on poor roads. He and his colleagues covered an area of 10,000 square miles in the service of the Board. In addition, his duties as a general physician included the care of the elderly until the appointment of a geriatrician in the early 1970s. With the passage of the years, William Lancaster, who had always been highly regarded for his medical knowledge and integrity, became an elder statesman within the Highland medical establishment. His outstanding qualities as a man and a physician were recognized appropriately in his appointment by the Secretary of State for Scotland to the membership of the Highland Health Board. His contributions to the Board's discussions were always carefully considered and delivered with a delightful sense of humour. He was a highly valued and influential member.

He developed a serious illness in 1983, from which he never fully recovered. He retired from his post as consultant physician in 1984.

In private life he was essentially a family man, his busy professional life leaving little time for interests outside his home, although he enjoyed fishing and also served in the RNVR for over 20 years.

William will be remembered by many Highlanders for his skill as a physician, complemented by his kindly, reassuring manner. But he will be remembered especially by those who were privileged to be his friends for his complete integrity and dependability, supported by his Christian faith.

In 1954 he married Anne Macleod, a nursing colleague and daughter of a Highland headmaster. They had a daughter who died in infancy and two sons. William died in Raigmore Hospital, Inverness, where he had been a consultant physician from 1962 until his retirement. His wife and sons survived him.

ARM

[*Brit.med.J.*,1988,**296**,580]

LASS, Ronald Wilson *b.*9 Feb 1914 *d.*17 Mar 1987
MRCS LRCP(1938) MRCP(1944) FRCP(1973) LMCC(1979)

Ronald Lass was born in London and died in Canada. He first entered dental school, but later studied medicine and qualified at University College Hospital, London, in 1938. On the outbreak of the second world war he joined the Royal Air Force medical service and was stationed in Iraq. During this time he met and married Anne Fisher, who was a nursing sister at the camp. On demobilization he joined a general practice in Maidstone, at the same time being appointed an honorary assistant physician to the West Kent Hospital, Maidstone. With the advent of the National Health Service he resigned from general practice and became consultant physician to the Mid-Kent group.

In 1966 Ronald took up the post of medical specialist to a hospital in Tripoli, Libya. After the revolution he had to leave Libya and returned to England. He was appointed consultant physician to the Royal Albert Edward Infirmary, Wigan, in 1971. Ronald was now in his 50's but was very quickly integrated into the hospital and made a considerable impact. Soon after his arrival he became tutor to the undergraduate students from Manchester University and made a success of this venture. He was very popular with the students and never at a loss for medical anecdotes from his past wide experience. He also became postgraduate tutor, a post he again tackled with considerable enthusiasm, and was both respected and liked by all the doctors in the area.

When he eventually retired from the NHS Ronald was very reluctant to give up medicine and, at this stage, he studied for the Canadian LMCC examination, was successful, and decided to emigrate to Canada where his daughter Rowena was working. He subsequently worked in a district hospital in Ontario.

Though Ronald only spent a short period of his professional life in

Wigan he made a valuable contribution and will be remembered with affection by his colleagues and patients. He had an infectious enthusiasm which characterized his professional and social life and made him an ideal colleague. His clinical acumen was outstanding and his astute diagnoses, and endless repertoire of medical anecdotes, were based on his wide experience and love of the art of medicine.

His main interest outside medicine would sometimes result in his absence, and trying to find Ronald on a Friday afternoon was often difficult - he had gone fishing.

Ronald was a lifelong inveterate cigarette smoker and it came as no surprise to his former colleagues that he eventually succumbed to carcinoma of the bronchus.

BJF

[*Brit.med.J.*,1987,**294**,1698]

LEES, Frederick *b*.5 Dec 1920 *d*.7 May 1988
MPS Edin(1943) MRCS LRCP(1952) MB ChB Sheff(1952)
DCH(1954) MRCP(1954) FRCP(1971) LLB(1976)

Frederick Lees was born in the town of Hemsworth in the South Yorkshire coalfield. His father was an electrician and his mother was the daughter of a farmer. He married Mary Constance Richardson in 1948, whose father was a colliery winder.

Frederick attended Hemsworth Grammar School and at the age of 16 became apprenticed to a pharmacist, becoming a member of the Pharmaceutical Society of Edinburgh in 1943. He served in the infantry in the Army between 1942-47, first in the United Kingdom and later in France, Belgium, Holland and Germany. Subsequently he transferred to the education corps.

After demobilization he went to Sheffield University as a medical student and gained prizes in clinical medicine, paediatrics and psychiatry, with distinction in public health in the examination. After qualifying he became house physician to E J Payne at the Royal Infirmary, Sheffield, and house officer to R S Illingworth in the Sheffield Children's Hospital. He took both the diploma in child health and his membership of the College two years after qualifying. He was then appointed senior house officer and registrar to H P Brody (q.v.) in Sheffield, where he spent the next three years. This period was interrupted by two visits abroad, one on a Ciba Foundation award in 1955 to Paris, and another to Copenhagen on a Winston Churchill scholarship when he visited Professor Meulengracht. He left Sheffield in 1957 to become senior registrar to Fergus Ferguson [*Munk's Roll*, Vol.VI,p.175] at Manchester Royal Infirmary. Three years later he transferred to London, becoming senior registrar to Aldren Turner [*Munk's Roll*,Vol.VII,p.582] until 1963 when he was appointed consultant neurologist of the north east Thames regional hospital board. This appointment covered visits to Colchester, Chelmsford, Billericay, Tilbury and Romford hospitals. After a few years he gave up Tilbury and Billericay and concentrated on the two major district hospitals at Chelmsford and Colchester.

As a registrar, Fred wrote many important neurological papers and also a chapter on the nervous system in *Fundamentals of current medical treatment*, ed. C W H Harvard, London, Staples Press, 1965 (now *Current medical treatment*). In 1970 he wrote a two-volume book *The diagnosis and treatment of disorders affecting the nervous system*, London, Staples Press ; American edition published New York, American Elsevier Publishing Co. This book did not get the full recognition it deserved.

Fred much enjoyed teaching, and a firm of medical students from the London Hospital used to attend his ward rounds at Chelmsford Hospital. He was an indefatigable teacher and his rounds would often go on until 7 p.m. He was clinical tutor at both Colchester and Chelmsford postgraduate centres, and president of the medical societies in both towns. He attended neurological meetings at the regional centre of neurology and neurosurgery at Oldchurch Hospital regularly, and his contributions at these combined meetings were very much appreciated because of his clear thinking, astute criticism and wide knowledge of neurology. With his thorough grounding in general medicine few diagnosticians could fault him. He investigated all his patients very thoroughly, and seemed to have a flair for finding operable lesions inside the cranium; many patients are grateful for Fred's intervention on their behalf. He enjoyed doing locum appointments, which he did even during his working years, and after he reached 65 he continued to work at Chelmsford for two years until his successor was appointed.

Fred was elected a Fellow of the College in 1972. He had amazing energy and this, combined with high intelligence, enabled him to obtain a degree in law with honours in 1976 while still working full-time. He later read for the Bar and obtained a BA from the Open University in 1984; two years later he took an honours BA.

His wife Marie had been a close friend since childhood, when they both lived in Hemsworth. They had a son and a daughter; the daughter became a nurse at St Thomas's Hospital and married an Australian doctor who practises near Sydney. Fred and Marie used to visit them each winter, for a period of six weeks, and it is sad that he did not see his daughter's child who was born a few weeks before he died. His son has two children.

Fred lived a full life; he was very knowledgeable on art and literature and had an astounding number of hobbies. He enjoyed playing snooker in his own home and attended race meetings; with his great sense of humour he made an annual visit to Ascot a real treat. In later years he took up oil painting and his house is full of his pictures, which reveal great talent. In earlier years he had enjoyed tennis and golf. He also collected 18th century literature and antique furniture, and was a keen Freemason.

Fred was a great party man and extremely sociable; he gave up alcohol after hepatitis but this failed to impair his sociability. He was full of droll stories, and having little sense of time his parties would go on well into the night. When attending a friend's house on a Friday evening he would often leave at 2 a.m. to drive up to Yorkshire where he had a house, and where he and his wife had to look after her mother. Looking back on Fred's career and his many hobbies, one can confidently call him 'A man for all seasons'.

KWGH

[*Brit.med.J.*,1988,**297**,130; Photo]

LEGGETT, Humphrey Abraham *b.*16 May 1910 *d.*3 Nov 1988
MRCS LRCP(1936) MB BS Lond(1937) MD(1939) DPH(1939) MRCP(1943) DCH(1946) FRCP(1969)

Humphrey Leggett, consultant emeritus in general medicine to the Redhill group of hospitals, was the son of an engineer, Walter Abraham Leggett and his wife Emmeline Blanch. He was educated at Perse School, Cambridge, and graduated in medicine at Guy's Hospital, London.

After house appointments at Royal South Hampshire and Southampton hospitals, he obtained the diploma in public health after studying at the London School of Hygiene and Tropical Medicine. He was appointed deputy medical superintendent to Redhill County Hospital in September 1940.

In 1942 he joined the Royal Air Force Volunteer Reserve and served in Ceylon, India and Singapore, as a medical specialist. During this time he obtained his membership of the College.

On demobilization in 1946 he rejoined the Redhill County Hospital staff. With the inception of the NHS he became a consultant physician in 1949 and physician superintendent in 1951. During the next two years he obtained his diploma in child health and his doctorate. For many years he carried a heavy clinical burden, caring not only for adult patients but also for children, and also for patients with mental disorders.

In 1945 he married Mary Ellen Geraghty, a former nursing sister at St Luke's Hospital, Guildford, and they had a son and a daughter. Humphrey was a devoted family man and lived in accordance with Christian ethics. At school he had been an athlete and captain of boxing, somewhat surprising in view of his gentle manner. As a young man he played rugby, and in later years he enjoyed watching cricket and hill walking.

On the recommendation of Admiral of the Fleet Sir John Cunningham, for whom he had cared, he was elected in 1951 an honorary member of the Royal Naval Association, which gave him great pleasure for the ensuing 20 years.

Humphrey Leggett was a true general physician, whose opinion was valued by his colleagues and their families. Above all, he was much loved by patients, hospital staff and his medical colleagues. Sadly, in retirement he was disabled by a severe stroke and heart disease. He bore these disabilities with much fortitude, supported to the end by his devoted wife and family.

JPHD

[*Brit.med.J.*,1989,518; Photo]

LEHMANN, Hermann *b.*8 July 1910 *d.*13 July 1985
CBE(1980) MD Basle(1934) PhD Cantab(1938) ScD(1958)
MRCP(1958) FRCP(1964) FRCPath(1964) FRS(1972)

Hermann Lehmann was born in Halle, in what is now East Germany, where his father was a publisher. His mother died when he was only three and the family moved to Dresden, the then beautiful capital of the province of Saxony which he always remembered with affection and where he received his schooling. He studied medicine principally at the Universities of Freiburg and Heidelberg, but his studies were affected first by the hyperinflation and secondly by the coming to power of the Nazis, with its attendant antisemitism. This prevented his taking his MD at the University of Heidelberg; instead, he went to Switzerland and took his MD at the University of Basle in 1934.

Hermann Lehmann began his professional life as research assistant to Otto Meyerhof at the Kaiser Wilhelm Institute in Heidelberg. In 1935 Meyerhof sent him to visit the department of biochemistry at Cambridge, then headed by Sir Frederick Gowland Hopkins. It is recorded that, when Lehmann was about to return to Heidelberg, Gowland Hopkins told him to leave his white coat hanging in the laboratory to await him whenever he wished to return. In 1936 he took up this offer and left Germany permanently. In Cambridge he became a graduate student, joining Christ's College, with which he remained associated for the rest of his life, and obtaining a PhD in biochemistry in 1938. He remained in Cambridge as a Beit memorial fellow for medical research until 1942.

From 1942-47 Lehmann served in the RAMC, rising to the rank of lieutenant colonel and becoming assistant director of pathology to the North East India Command. His lifelong concern with the anaemias began in India, where he was concerned with iron deficiency found in the Indian troops. After the war he worked in Uganda as senior nutrition research officer at Makerere College, where he showed that the commonly prevalent form of anaemia was due to hook worm infection. In Uganda he also became interested in sickle cell anaemia and investigated the incidence of the sickle cell trait, both in Uganda and subsequently in various other parts of the world. For his studies in the Andaman Islands he was awarded the Rivers Memorial medal for field work in anthropology in 1962. His studies on the distribution of abnormal haemoglobins in many different human populations made him one of the founders of the new subject of 'molecular anthropology'.

Lehmann returned to England in 1949, initially as consultant pathologist at the Pembury Hospital in Kent, and from 1951 as senior lecturer (later, reader) in chemical pathology at St Bartholomew's Hospital, where he remained until 1963. While at Bart's he described pseudocholinesterase deficiency as the cause of idiosyncratic hypersensitivity to suxamethonium, and he began on the studies of abnormal haemoglobins which were to occupy him for the rest of his professional life.

He returned to Cambridge in 1963, first as University biochemist to Addenbrooke's Hospital and, from 1967, as the University's first professor of clinical biochemistry. The Medical Research Council created the abnormal haemoglobin research unit for him in 1963. This unit was highly productive in describing abnormal haemoglobins and an account of some of their work was published in book form: *Man's haemoglobins...*,Hermann Lehmann and R C Huntsman, Philadelphia, Lippincott, 1966. Lehmann was able to collaborate with Max Perutz in studying the relationship of structural variation in the haemoglobin molecule to its function as an oxygen carrier - work of high elegance and great importance. He was also the first to demonstrate that the genetic code, derived at that time from work on prokaryotes, applied also to eukaryotes. He combined his research work with the running of the chemical pathology laboratories in Cambridge, a task to which he brought both dedication and managerial skill. His outstanding quality was his concern for people, and his personal interest in his patients and all his staff. He possessed to a high degree the quality of humanity, which is lacking in so many institutions. He would visit a friend who was ill at any hour of the day or night, undeterred by medical convention that no one except the GP should do so. Nothing was too much trouble for him when others were in trouble. When he retired in 1977 he continued to run the WHO haemoglobin reference centre and continued research both on haemoglobins and on myoglobin, and other muscle proteins.

During his productive and fruitful career Hermann Lehmann acquired many honours. He was elected a Fellow of the College in 1964, just six years after having become a member by published work. He became an ScD of the University of Cambridge in 1957, and honorary professor in the University of Freiburg in 1964. In 1965 he was elected a fellow of Christ's College, becoming an honorary fellow in 1982. In 1972 he was elected FRS, and he was appointed CBE in 1980.

Hermann Lehmann was always much devoted to Christ's College

and took an active part in its affairs being, at various times, Fellows' Steward, editor of the College magazine, and president of the College medical society. He was highly esteemed by the undergraduates, and by his colleagues, for his breadth of interest, his warmth of character and his great charm. In 1942 he married Benigna Norman-Butler and they had four children: two sons and two daughters. Sadly, one son died in adolescence but he was survived by the other son, also a biological scientist, and his two daughters, as well as by his wife. PJL

[*Brit.med.J.*,1985,**291**,288-89; *Lancet*,1985,**2**,284,341; The Times,16 July 1985; *Biog.Mems.roy.Soc.*,1988,**34**,405-449; Photo]

LEVINE, Philip *b.*10 Aug 1900 *d.*18 Oct 1987
MD Cornell(1923) Hon DSc Michigan(1967) *FRCP(1973)

Philip Levine, who discovered the cause of rhesus haemolytic disease of the newborn, was born in Kletsk, Russia, but settled in Brooklyn with his parents, Morris Levine and his wife Faya Zirulick, whose father was a rabbi. After obtaining his BS at the City University of New York, Philip proceeded to Cornell University medical school where he graduated MD. From 1923-25 he was an intern at the Beth Moses Hospital, now the Maimonedes.

Philip Levine became an internationally known researcher in the field of immunohaematology and for 20 years was in charge of the Ortho Research Division at Raritan, New Jersey. In later years he was director emeritus of the Philip Levine Laboratories, a showpiece to his memory. His most important contribution to medicine, the discovery of the Rh factor, was a model of simplicity based on clinical observation and logical deduction.

Jaundice of the newborn has been known for hundreds of years, and was reported in twins in 1609. It was noticed later that babies with the familial form of the disease were also anaemic, and when in 1900 Karl Landsteiner discovered the ABO blood groups a mechanism became clear - there could be haemolysis because of ABO incompatibility between mother and baby. In white populations ABO haemolytic disease is usually not serious, and Philip's great discovery came because he treasured his exceptions. He noted that some babies were jaundiced and seriously anaemic even when the parents were compatible on the ABO system. Further, mothers who needed blood near term often had a transfusion reaction when given their husband's blood. Therefore Philip argued that there must be incompatibility of another blood group system.

In the USA, Landsteiner and Alexander Wiener had shown that rabbits injected with the blood of rhesus monkeys formed antibodies which agglutinated the red cells of human bloods carrying M antigen. When this anti-M antibody was absorbed it was found, c.1940, that the rabbit antibody still agglutinated 85% of human red cells. Since it was the monkey's antigens which were responsible for the antibodies, Wiener [*Munk's Roll*,Vol.VII,p.592] called this system 'rhesus', and the terms 'rhesus positive' and 'rhesus negative' entered the language. On the clinical side, Wiener related his discovery chiefly to blood group reactions, but Levine immediately realized that here was the explanation for those unexpectedly jaundiced babies. The ones with ABO compatible parents had been damaged by rhesus incompatibility, the rhesus-negative mother having made antibodies against the baby's paternally derived rhesus-postive cells.

This posed a question: 'Why, with a birthrate in the UK of around 700,000 a year, were there not more affected babies?' Theoretically there should have been around 60,000, but in fact there were only about a twentieth of that number. Levine in New York, and Race (q.v.) and Sanger (Mrs Race) in England produced the answer: some protection against rhesus immunisation was given by ABO incompatibility between mother and baby, a common occurrence. The mother's naturally occurring anti-A (or anti-B in the rarer case of a B baby) destroyed the RH-positive foetal cells that had crossed into the maternal circulation. It was only in the ABO compatible pregnancies that the mother was at risk of being immunized against the rhesus antigen.

A generation later a British team found that rhesus-positive cells in the mother's circulation could be destroyed by giving the anti-rhesus antibody, anti-D. Within 30 years rhesus disease had been both discovered and rendered preventable. In the prevention story, the USA and the UK were in competition, but Philip helped both sides equally, characteristically putting aside any thought of rivalry. Ronald Finn, in the UK, well remembers how, in the early days, Philip was ready to explain all the knotty problems with infinite patience.

The discoveries of Wiener and Levine were complementary, and both deserved the Nobel prize but, sadly, this was not to be. Though Philip was a delightful character, perhaps the Stockholm committee sensed that Wiener would not be happy with a joint award. Philip was a hard worker and in his later years wrestled with the P blood group system and its possible relation to cancer. Autoantibodies were another of his interests.

Philip Levine instituted and funded the Philip Levine lecture, presented annually on the pathogenesis of malignancy by Cornell University medical college, the Rockefeller Institute and the Memorial Sloan-Kettering Institute for Cancer Research.

It was for Levine's work on rhesus haemolytic disease that the College elected him to the Fellowship in 1973. Philip was delighted and came specially to London to receive the award.

Philip Levine was a gentle man yet firm in his approach to science. His fascination for the unknown never waned and he had the admiration and respect of his peers. His dedication was untiring, and his genius

for solving problems of incredibly complexity was unquenchable.
He married Hilda Perlmutter in 1938 and they had four children, three sons and a daughter, and five grandchildren. His wife and one son predeceased him. He was survived by their daughter and two sons. CAC

[*Lancet*,1987,2,1039-40; *The Times*,21 Oct 1987; Johnson & Johnson,2 Nov 1987; *The Med.Technologist*,July 1973,Vol.3,7,1; *Triangle*,Sandoz Journal of Med.Science, Sept 1961,3,168-70; Photo]

LEWIN, Denis Cyril *b.*31 July 1915 *d.*26 June 1984
BSc Lond(1937) MRCS LRCP(1939) MB BS(1939) MD(1946) MRCP(1946) FRCP(1972)

Denis Lewin was born in Croydon, Surrey, the son of George Lewin, a medical practitioner, and his wife Georgina Frances, née Pecker. It was George Lewin who succeeded in indentifying the Croydon epidemic of the 1930s as typhoid fever.

Denis was educated at St Anselm's, Croydon, and Tonbridge School. In 1934 he entered the medical school at Guy's Hospital and by 1936 he had distinguished himself by coming first in his class, and winning prizes for general proficiency and anatomy. He was selected as a student-demonstrator in anatomy and pathology, and passed his preliminary FRCS(Eng) examination at his first attempt. After obtaining a BSc in physiology in 1937 he began his clinical studies. He was a consistently outstanding student and would have been assured of a house appointment at Guy's but for the outbreak of war in 1939. Denis was a keen member of the Senior OTC at Guy's and immediately accepted his call-up papers, joining the RAMC and serving with the 1st and 8th Armies as RMO with the Honourable Artillery Company in Africa and Italy. He recalled these years of comradeship, and the demands on his medical skill, with a sense of fulfilment and satisfaction. In keeping with his character, he toiled on through the long years of the war, always giving excellent service but never allowing himself to be promoted or transferred.

On demobilization in 1945, while at the RAMC College at Millbank, he obtained his MD and his membership of the College. From February-July 1946 he served as house physician at Guy's, working with Marshall [*Munk's Roll*,Vol.VII,p.381] and Hampson [*Munk's Roll*,Vol.VI,p.216], both of whom he had come to respect greatly as a student. He could have stayed on at Guy's and taken his place on the staff in due course, but he preferred to launch out on his own. He was appointed registrar at Churchill Hospital, Oxford, and this enabled him, after only two years, to obtain a consultant post in general medicine for the Rhymney and Sirhowy hospital management committee. He found himself in a highly populated mining area which was relatively deprived of the medical services available in more favoured places. Denis set to work to improve the standard of health care for the country people in the valleys. He devoted the whole 35 years of his professional life to this cause and was never deflected from it, as some others might have been, by offers of work elsewhere or - in particular - by the chance of a move to an academic career for which he was well equipped. His insistence on the need for reform at first provoked the hostility of his local hospital management, yet the same administrators later gave him the respect and support he deserved. Unfortunately, their efforts could not grant him the fulfilment of his hopes and advice. For years he worked single-handed, on call 24 hours a day, seven days a week, with no relief. Even when away with his wife for the weekend they would often have to return following an urgent telephone call. They never went anywhere without leaving a phone number where they could be contacted. He had medical beds both at Tredegar and Caerphilly Miners' Hospitals, and had hoped that a second consultant might be appointed to share the burden of routine and responsibility in the valleys and, also, for the integration of the two hospitals with the teaching hospitals in Cardiff, so that specialist visits, junior staff interchange and student teaching might be developed. Unfortunately, Tredegar Hospital was taken out of his area and so he remained without a consultant colleague until he retired in 1979, but he lived to witness the start of the integration of specialties and staff in the Cardiff hospitals with those of Caerphilly.

Denis was twice married. His son Michael, by his first marriage, is an executive with British Airways. His second wife, Sylvia, whom he married in 1947, is a consultant gynaecologist, now retired, and they had three children: Peter, Gillian and Jacqueline. Peter is a chartered accountant in Jersey, Gillian took a PhD in clinical psychology and is now an Australian citizen, and Jacqueline is a medical practitioner specializing in anaesthetics.

For most of his life, following his first serious illness at the age of 39, Denis was dogged by ill health but he never allowed it to interfere with his work, and never complained. He was a retiring man with few social skills. He was shy of recognition and praise. He made diabetes his special interest for study and research, but had few hobbies or activities outside his work. Until his illness in 1939 he had been a keen sportsman, but following a series of operations he ceased to take an active role, apart from an occasional game of golf. However, he continued to take an interest in all sports, to work in his garden and to enjoy walking. His other interests, outside work and family, were history and listening to good music. He suffered a succession of distressingly painful conditions during the later years of his life and was often depressed and upset by the heavy sedation in his last illnesses, but few people were aware of this.

Denis Lewin retired reluctantly at the age of 63 and moved back to live among the people he knew and loved around Hengoed, near his first home in the valleys. During his last illness he was admitted to Caerphilly Hospital, his own base where he had worked for many years and where he was known, respected and loved by many of the staff. In their devotion to him during his last days it was plain that they were paying their own tribute to his work, and their own debt of gratitude to him personally for all he had done for their folk in the valleys.

He was survived by his wife Sylvia, his four children, and his sister Audrey.

FL

[*Brit.med.J.*,1984,**289**,329; Photo]

LEWTAS, Norman Arthur *b.*10 Nov 1917 *d.*12 Sept 1987
MB ChB Manch(1941) MRCP(1949) DMRD(1951) FFR(1958) FRCP(1970) FRCR(1975)

Norman Arthur Lewtas, a thoroughbred Mancunian, died peacefully after a short illness bravely borne, at his home in the Derbyshire Dales. It is ironic that he died from a pathology - cerebral tumour - to the investigation of which he had devoted much of his professional life.

Norman was born in Manchester, schooled at Manchester Grammar School and graduated from the University of that city. It was at Manchester that he served his residential appointments before entering the RAMC for the latter years of the war, during which he served in Europe, India and the Sudan, and was mentioned in despatches. After the war he returned to Manchester and its environs for further clinical appointments, gained his membership of the College, and in 1950 began his radiological training. He became fascinated by the discipline, technique and interpretative skills required in neuroradiology, which already had a very strong tradition in Manchester under the leadership of E Wing Twining and Sir Geoffrey Jefferson [*Munks'Roll*,Vol.V,p.213].

In 1955 Norman was appointed as the first consultant neuroradiologist to the United Sheffield Hospitals and for the next 28 years he rendered distinguished service to those hospitals. He was a superb neuroradiologist with great manual dexterity and interpretative ability, coupled with excellent clinico-radiological judgement. Norman was a tremendously hard worker, for nearly 30 years he carried the major responsibility for the huge neuroradiological workload generated by over 2,000,000 people in the North Trent region. He was a keen collaborator in many clinical imaging projects developed with his colleagues, and he contributed several original articles and book chapters to the neuroradiological literature. He was always very tidy, disciplined and practical in his philosophy, taking the minimum of films necessary to establish and confirm the diagnosis; providing the community with a fine neuroradiological service of high efficiency at very low cost. He was one of the first to be allocated an EMI headscanner in 1974, and he developed an excellent CT service, with one of the highest daily caseloads in the country.

Norman was a patient and sympathetic teacher and an excellent lecturer, both of neuroradiology and general radiology, and many radiologists in this country and abroad learned their skills from him and his colleagues at Sheffield. Although primarily a neuroradiologist, he also undertook general radiological work, and was a much admired administrative head of the department of radiology at the Sheffield Royal Infirmary from 1959 until its closure, when it was absorbed by the Royal Hallamshire Hospital in 1980. Under his relaxed but dynamic leadership, the Sheffield department of neuroradiology and the department of radiology of the Sheffield Royal Infirmary thrived, with a devoted staff - medical, technical and clerical - totally loyal and devoted to their leader and adviser.

Handsome, charming, tall and slender, Norman was always elegantly dressed. He was greatly respected throughout the hospitals where he worked for his sincerity, sound advice, diligent hard work and leadership. He was a staunch supporter of the British Institute of Radiology, being a member of Council, and of the Royal College of Radiologists. He had the distinction of being the vice-president of the latter from 1978-79, and also of its predecessor the Faculty of Radiologists from 1972-74.

By nature he was a fast mover, whether at angiography, reporting, walking, driving, or racing his favourite souped-up Escort in the competitive hill climbing rallies which he enjoyed so much. He was a natural and enthusiastic athlete; enjoying tennis, squash, golf and, in later years, microlight flying. He gained his university colours in golf, with a low handicap, and after retirement he resumed the game with renewed enthusiasm and much pleasure. 'It's magic', he wrote to the writer. As well as excelling in physical sports, Norman was fond of music and was an avid reader, with a particular penchant for philosophical and biographical essays, and enthused over his favourite authors. Schumacher's *Small is beautiful* became one of his favourite sources of reference.

Norman Lewtas was a gentleman - a truly courteous, gentle man, who was totally dedicated to his family, his patients and his colleagues.

In 1942 he married Dorothy, the daughter of Wilson Harold Hey, a surgeon, and together with their son and daughter they established a delightful home; latterly in Curbar in the Derbyshire Dales, near Sheffield, where they maintained a beautiful garden in which they laboured with skill and devotion. They also enjoyed and cherished their walks through the neighbouring moors and dales. Dorothy, his two children and his centenarian mother all survived him.

RGG

[*Brit.med.J.*,1987,**295**,1147; *Brit.J.Radiol.Bull.*,1987,B77; *Clinical Radiol.*,1988,**39**,236; Photo]

LIGHTWOOD, Reginald Cyril *b*.28 Jan 1898 *d*.26 May 1985
MRCS LRCP(1921) MB BS Lond(1922) MD(1924) DPH(1924) MRCP(1925) FRCP(1936)

Reginald Lightwood was born in Croydon, the son of John Mason Lightwood, barrister, whose somewhat austere bewigged features looked down from a large sepia photograph in the office from which Reggie directed, with much skill, the paediatric unit at St Mary's Hospital Medical School.

Lightwood was educated at Monkton Combe School. After service in the Royal Artillery, during which he was severely gassed in 1917, he studied medicine at King's College Hospital and qualified in 1921. He was a formidable athlete as a student and he retained a spare and vigorous physique throughout his long life. His distinguished career as a paediatrician spanned almost 60 years and in each phase he made particular and distinct contributions to the specialty.

In 1939 Reggie Lightwood was cast in the mould of the successful London consultant. He had gained the MD in 1924, his MRCP in 1925, and was elected Fellow in 1936; he held honorary appointments at the Hospital for Sick Children, Great Ormond Street, and at St Mary's Hospital, Paddington. He was recognized as an able clinician and had published several well regarded research papers. He was an influential member of the eleven year old British Paediatric Association and was established in private practice - an occupation for which he did not greatly care. He was never a charismatic teacher, but did not entirely deserve his placement within the ranks of those given the title of 'shifting dullness'. The 1939-45 war was a time of heavy clinical and teaching commitments for Reggie, and he added to his inner London work a regular and beneficial paediatric presence at several Emergency Medical Service hospitals in the North West Thames region.

The second phase of Reggie's paediatric career began when St Mary's Hospital Medical School made the innovative decision to create an academic unit of paediatrics with Lightwood as part-time director. Under his management the infant unit soon became a model for excellence in teaching and research, and demonstrated that paediatrics could stand worthily with the giants of medicine and surgery. Reggie's particular skill was in identifying important issues, whether to do with the details of the immediate management of a patient, the selection of staff members, the arrangements of a visit overseas or, more broadly, about the development of child health services and the teaching of paediatrics. Once satisfied (often after much consultation) as to the correct course of action (or masterly inaction) he was happy to delegate implementation to a degree regarded as unconventional at that time.

Important investigations into metabolic disorders in infants, and original descriptions of renal acidosis and infantile hypercalcaemia, gained him a well deserved international reputation as a man of paediatric science. His far-sighted interest in preventive and social aspects of paediatrics brought many invitations to advise on child health services and paediatric education in developing countries. In due course he became honorary treasurer and then president of the British Paediatric Association - responsibilities that he carried out with meticulous attention to detail and yet with considerable flair for innovation and necessary change. Curiously, some of his British colleagues regarded him as somewhat indolent, indecisive and devious; characteristics which could not have been at greater variance from his true personality and performance. He was a shy, reserved man who did not readily disclose his feelings or thoughts; he could appear pedantic and obsessional, and his humour was quiet and concealed. Any deficiency in communication skills vanished when he was abroad, and he was a most delightful person with whom to travel, being relaxed, well informed, generous and resourceful.

He retired from his hospital and university appointments in 1963, and for the next 20 years had an extraordinary career as a global peripatetic professor of paediatrics. From 1964-65 he was at the American University of Beirut; from 1966-69 he was at the University of Rhodesia (now Zimbabwe) from whence he became visiting professor at the University of California Los Angeles. A spell as consultant paediatrician to the International Grenfell Association, Newfoundland, took him into the 1970s, and he later became paediatrician to the Missionary Training School in Drogheda, Republic of Ireland. Before finally 'retiring' to England he spent some years in Menorca, where he is said to have still practised. At the age of 85 years he attended Comitia for the election of the president, and was an active participant at the annual general meeting of the BPA in York, the year before his death at a retirement home in Surrey, near to one of his sons who became a consultant surgeon. He married Monica, daughter of Laurance G Ray, in 1937 and they had two sons.

TEO

[*Brit.med.J.*,1985,**290**,1991-92; *Lancet*,1985,**1**,1461; The Times,10 June 1985; *St Mary's Hosp.Gaz.*,1964,**70**,26-27; Photo]

LINDEBOOM, Gerrit Arie *b*.4 Jan 1905 *d*.5 June 1986
DM Amsterdam(1929) FRCP(1974)

Gerrit Lindeboom was born in Ridderkerk, Netherlands, where his father, Cornelius Lindeboom, was a minister in the Dutch Reformed Church. His mother, Anthonia Margareta, was the daughter of Gerrit Arie de Jong, also a minister of the church. All through his life Gerrit Lindeboom stuck firmly to the Calvinistic principles in which he was reared. He was educated at the Gereformeerd Gymnasium, Amsterdam and Kampen, and the University of Amsterdam, pursuing his clinical work at the Wilhelmina Gasthuis, Amsterdam. On qualification he obtained a house post under I Snapper at the Wilhelmina Gasthuis, and specialized in internal medicine and radiology. In 1934 he established a practice in Amsterdam.

During the second world war, when there was a shortage of hospital beds, he founded a private clinic. The Germans had occupied the country and there was a severe shortage of almost everything. But Lindeboom's boundless energy and perseverance ensured that the 80-bed clinic survived, and prospered rapidly in the post-war years. When the Free University of Amsterdam (Calvinist) established a medical school in 1950, Lindeboom's clinic became its first department of internal medicine and he was appointed professor of general pathology, clinical propaedeutics, and the encyclopaedia of medical sciences. Four years later he also became professor of internal medicine. He was an outstanding clinician and a pioneer in the use of radioactive materials for diagnostic purposes.

In 1939 he married Catharina Sophia Kruysveldt, a physician. There were no children of the marriage; their son being adopted.

Lindeboom was an eminent medical historian in the Netherlands. From his student days onwards he was interested in the history, philosophy and ethics of medicine. When he became a professor he founded the Medical Encyclopaedic Institute, encompassing the three disciplines. He also arranged that courses in these disciplines should be obligatory for medical students, a novelty at that time in the Netherlands. When medical philosophy and ethics became independent disciplines this Institute, although continuing to carry its original name, became a department of medical history; the first of its kind at any university in Holland. Lindeboom became its first director, a position he held in conjunction with his other commitments. He was made a Knight in the Order of the Dutch Lion in 1968.

He was elected a Fellow of the College in 1974, and retired from his clinical assignments in 1975, but continued in his post as director of the Institute.

Lindeboom was a born teacher. His courses and conferences on the history of medicine attracted audiences from all over the country and greatly enhanced the prestige of the discipline. His lectures could be brilliant. Even his clinical teachings were enlivened with historical details. His successor in the chair of medicine, C Vermeer, in his biographical introductory chapter to the *Festschrift* presented to Lindeboom upon his becoming a septuagenarian, aptly wrote: 'He lived simultaneously in the past and the present'.

Gerrit Lindeboom was a prolific writer. He wrote some 50 monographs, more than 180 medical publications, and almost 300 publications on medical history and ethics. These were, for the greater part, written in somewhat archaic but beautiful Dutch. He did not always have good translators for his English publications. His main medical-historical publications were on Boerhaave, which earned him international recognition. His masterpiece was undoubtedly *Herman*

Boerhaave: The Man and his work, London, Metheun, 1968, a scholarly work of 452 pages which took him many years to prepare.

Lindeboom should be regarded as a medical historian of distinction, who happily combined his interest in medical history with clinical practice in the firm belief that the one complemented the other. He served on many committees and several editorial boards.

He had a strong-willed, dominant personality which made him somewhat difficult to deal with, but one had to admire his talents and achievements, and his formidable zest for work.

GW/VL

[*Bull.Hist.Med.*,1986,No 4,**60**,559-60; Photo]

LIVINGSTONE, James Livingstone *b.*8 May 1900 *d.*21 Apr 1988
MRCS LRCP(1922) MB BS Lond(1923) MRCP(1925) MD(1925)
FRCP(1933)

James Livingstone Livingstone was born in Paisley, Scotland. His father, William Livingstone, was a medical practitioner who practised in Seaford, Sussex; his mother Martha, née Hamilton, was the daughter of a bank manager and laird.

He went to school at St Wilfred's, Seaford, and Worksop College, and entered King's College, London, in 1917. For 10 months in 1918-19 he was a flight cadet in the RAF. Going on to King's College Hospital Medical School in 1920 he qualified with the conjoint diploma in 1922 and graduated the following year, proceeding MD in 1925. After house surgeon and house physician appointments at King's College Hospital, he was appointed medical registrar in 1924. The following year he obtained his membership of the College.

At that time morbid anatomy was still regarded as an essential study for budding physicians and so, for a year, he held a combined post as medical tutor and demonstrator in pathology. He was appointed assistant physician to King's College Hospital in 1927, and physician in 1934. He developed an interest in respiratory disease and in 1932 was appointed to the staff of the Brompton Hospital. Thereafter, King's and the Brompton remained the centres of his professional life. He was elected a fellow of King's College, London, in 1949.

During the second world war, Livingstone served in the Emergency Medical Service as physician to the chest unit at Horton Hospital, Epsom; this was a major centre for thoracic surgery and dealt with many civilian and armed forces casualties. Another professional interest was in insurance medicine; he was for many years chief medical officer in the UK for an Australian insurance society.

Livingstone was an able, conscientious and kind physician. While essentially a clinician, he was active in the application of advances in basic knowledge to medical practice. He wrote on the use of respiratory efficiency tests in asthma in the middle 1930s, and was an advocate of physiotherapy in broncho-pulmonary disease, notably of breathing exercises in asthma, long before these procedures became routine. He was joint author of important early papers on pulmonary eosinophilia and on diffuse interstitial fibrosis of the lungs. His clinical judgement was respected by consultant colleagues, and was so modestly expressed that they did not hesitate to ask his opinion on clinical problems which puzzled them. He was a valued and willing contributor to the teaching of undergraduates at King's, and of postgraduates at the Brompton.

He took an active part in the affairs of the Asthma Research Council for many years and was a founder member of the Thoracic Society, of which he was president in 1956. In the College he served as councillor from 1945-49, and as censor in 1962-63. He gave the Mitchell Lecture in 1954, on the treatment of pulmonary tuberculosis.

It is as true of Jimmy Livingstone as it can be of any man that he was well-disposed towards everyone and that everyone liked him. To his juniors, his attitude was informal, almost avuncular; it was not unknown for the discussion after a round at the end of the day, at the Brompton, to be completed in the bar of the pub round the corner. This geniality was balanced by a quietly forceful personality; one knew where one stood with him, and he was constant in his loyalties. In appearance he was rather above average height, with an upright bearing, strong features, and dark hair which he retained into old age. After retirement, he withdrew from London and went to live in the other home he and his wife had established in the place where he had spent his childhood, Seaford. Here, for a time, he continued to do a few locums in chest clinics and in his son's general practice, but later concentrated on his leisure interests. The chief of these was golf; he was also interested in bird-watching and in fly-fishing, and in athletics in his youth. He continued to play golf, and to be hopeful of improving his game, until shortly before his sudden death at the age of 88.

Jimmy Livingstone was fortunate in a long and happy marriage, starting in 1935 when he married Janet Muriel Rocke, who was at that time registrar in paediatrics at the Royal Free Hospital. They had two sons, both of whom followed their parents into the medical profession and entered general practice, and a daughter who became a physiotherapist. Of their six grandchildren, one graduated in medicine from King's College Hospital shortly after his grandfather's death, and another has started medical studies.

JGS

[*Lancet*, 1988, 1, 1679; *King's Coll. Hosp. Gaz.*, Summer 1965, 44(2), 112-26; Photo]

LLOYD, Thomas Wigram *b.*19 May 1910 *d.*15 Nov 1984
BA Oxon(1931) BM BCh(1934) DM(1941) MRCP(1937)
FRCP(1968)

Thomas Lloyd was descended on his father's side from the Quaker Lloyds, six generations of whom were landowners at Dolobran, near Welshpool, before the Civil War, after which they moved to Birmingham where they became first ironmasters and then bankers. In 1765 Sampson Lloyd and Sampson Lloyd, father and son, and John Taylor and John Taylor, father and son, established the private banking firm of Taylor & Lloyd in Birmingham; a century later this became a limited liability company, Lloyds Banking Co Ltd., and until a few years ago a member of the family continued to serve on the board of Lloyds Bank. Tom's father, Thomas Zachary Lloyd, studied mechanical engineering at King's College, London, and became a director of the Midlands industrial giant Guest, Keen and Nettlefolds. On his mother's side, his grandfather was a land agent at Collingham, near Newark, and his great-grandfather, Wigram, was Bishop of Rochester.

Tom was educated at West House School, Clifton College, and Lincoln College, Oxford. He returned to Birmingham for his clinical undergraduate training and graduated in medicine from Oxford in 1934. There after he held house officer posts in medicine and surgery at the General Hospital, Birmingham, and was house physician at the Children's Hospital, Birmingham, and the Radcliffe Infirmary at Oxford. Subsequently he held the Caroline Harrold research fellowship in the University of Birmingham, where he studied the anaemias of childhood at the Birmingham Children's Hospital under the guidance of Sir Leonard Parsons [*Munk's Roll*,Vol.IV,p.588] and H S Baar. A number of valuable publications resulted; the work provided the material for his DM thesis and he was elected a fellow of the International Society of Haematology. He became a member of the College in 1937.

It was obvious that a career of great distinction lay before him, for he was a man of high intelligence and great energy, with enormous charm and courtesy, the gift of warm friendship and a delightful sense of humour. Moreover, he had wide interests. His handicap at golf was two, he held a commission in the 5th Battalion of the Royal Warwickshire Regiment (TA), was very widely read, and was greatly interested in music - playing the piano and the violin, but especially enjoying singing in choirs and madrigals, the works of Peter Warlock and Vaughan Williams being particular favourites.

When the second world war arrived an apparently assured and brilliant future was in jeopardy. Under the scheme for the redeployment of medical manpower he was sent to assist in a general practice in Herefordshire, but after only two or three weeks there he was found to have extensive pulmonary tuberculosis. He went to

Midhurst, where he had a three stage thoracoplasty, and during convalescence was an assistant medical officer first at Midhurst itself and then at Winsley Sanatorium. When he was further recovered he became chief medical officer at the National Sanatorium at Benenden, and his work there was so impressive that he was invited to become physician superintendent at St Wulstan's Hospital, Malvern Wells; a new hospital for the treatment of tuberculosis. Under his direction the hospital ran smoothly and the patients were happy and received superb treatment. Tom's interest in research was rekindled and he published important papers on the new antituberculous drugs, streptomycin, PAS and isoniazid, which were just coming into use.

He became a JP for Worcestershire and bought a beautiful house with a fine garden opposite the Abbey School and this gave him a great deal of happiness. It looked as though he was destined to become one of the world authorities on tuberculosis, but in a few years the new drugs on which he had done such excellent work virtually removed tuberculosis from the western world.

At the age of 49, Tom therefore had to start a new career as physician with a special interest in the elderly at the Cheltenham and Gloucester hospitals. He quickly established a prodigious reputation in the south-west, for he was a superb physician with a wide knowledge of medicine and a wholly altruistic approach to it, and his own sufferings gave him the capacity to understand the anxieties and difficulties of sick people; to support their morale as well as to heal their ills. Although he worked incessantly his interest in research never left him and he took a prominent part in devising the ambulift. He was elected a Fellow of the College in 1968.

The Gloucester hospitals have a fine choir and participating in its activities gave him great pleasure. When he reached retiring age he went to live in Alicante. The warmth and the sun suited him and he welcomed the opportunity to spend more time gardening, playing golf, reading and listening to music, and he took up painting at which he became more than competent. He gathered around him a coterie of retired intellectuals living in Spain and their company kept his ever fertile brain active and interested. Sadly, this blissful existence lasted only six years, for then he developed carcinoma of the stomach. A gastrectomy left him with severe dysphagia which he courageously overcame with a mercurial bougie. As the months went by hopes of cure increased but in 1984 there was widespread recurrence and his last months were distressing in the extreme, but he never complained and remained cheerful, optimistic and totally devoid of self-pity to the end. Few men have such courage.

Tom was the youngest of three brothers, one of whom was headmaster of Uppingham for 21 years, and the other a director of Guest, Keen and Nettlefolds and the son-in-law of Neville Chamberlain. Tom met and married Margo Beasley whilst at Midhurst.

Their daughter became a member of the College and their son was at the Chancery Bar.

AGWW

[*Brit.med.J.*,1985,**290**,81; Photo]

LLOYD-HART, Vincent Edwards *b.*23 June 1909 *d.*10 May 1987
MRCS LRCP(1933) MRCP(1937) FRCP(1961)

Vincent Edwards Lloyd-Hart, known to his many colleagues and friends as 'John', died on 10 May 1987. He had asked, and was given, a very quiet, private funeral, befitting a quiet man who described himself as intensely shy.

John was born in Nailsworth, Gloucestershire, the son of a prosperous draper. After preparatory school he was educated at Wycliffe College, Stonehouse, Gloucestershire, 'a Wesleyan establishment with vegetarian inclinations'. He won prizes, played rugby football and shot for the school, and was a champion boxer. As a prefect, he claimed total inability to keep order - perhaps unsurprisingly: among boys he may well have been too gentle. He then went to London to study the basic sciences and gained admission to the London Hospital medical school. Student life was happy. He described entry into the wards as 'opening a new world'. His shyness did not prevent, indeed was perhaps relieved by personal contact with individual patients. Meanwhile, he boxed for the United Hospitals, and played both rugby football and cricket in the second teams. He qualified in 1933 and obtained his membership of the College a little later. He held various junior appointments at the London Hospital.

By way of variety, perhaps a respite from study, and no doubt expressing an altruism which he would not have admitted verbally, he went for a time to drive an ambulance in the Spanish Civil War.

Returning to seek locum work whilst considering the future, he was asked by Sir John Parkinson [*Munk's Roll*,Vol.VII,p.443], to whom he had been house physician, to act on his behalf as personal medical attendant to Pierpoint Morgan on his estate in Scotland. This must have opened a vista on another new world. John reported that this was, fortunately, an uneventful period.

As a locum in a well established practice in Aylesbury he met his future wife, Marcia Read. Also, happily, he was invited to join the practice, so marriage was soon possible and must have fortified him for the war years. The practice was busy, with attendance as physician to the Royal Bucks Hospital, and later to the old 'workhouse' which became Tindal General Hospital and was in the wartime EMS sector with its apex at the Middlesex Hospital. The arrival of students from the latter created a demand for teaching.

The end of World War II and the inauguration of the National Health Service forced a choice between an established general practice and full commitment to the hospital service. He chose the latter.

His contribution to medicine in Aylesbury was indeed great. Particular examples to be recalled are his painstaking participation in the extension and eventual transfer of the Aylesbury medical unit to Stoke Mandeville Hospital, as the latter was gradually transferred from the Ministry of Pensions to the NHS; his clear vision of the need to marry medicine with surgery; despite local difficulties his development of a regular commitment to visiting the county psychiatric hospital, St John's, at Stone, as general physician; and his personal contribution to the organization of care for coronary patients. His own special interests were in cardiology and thyroid disease. He published relatively little of a technical sort, but his reading was avid and wide. His opinion and advice were much sought and respected. His knowledge of general practice made his domiciliary visits of peculiar value to general practitioner and patient, and his hospital letters were humane communications. His awareness of psychological and social factors in illness was highly developed.

His general interests were wide and included European and local history. He wrote two small books, *Health in the Vale of Aylesbury*, a lively review of local health care and social history, and *John Wilkes and the foundling hospital at Aylesbury*, Aylesbury, HM+M Publishers, 1979. Both, the result of many hours spent in libraries and among files, blend serious and comic material and betray a taste for the picaresque among the virtuous. He enjoyed owning fast cars (which he drove with discretion) but also driving a tractor on his farm. With his wife, he furnished their home with pleasant things, including fine china. Before two serious physical illnesses and their complications he was an active village cricketer and an enterprising dinghy sailor; he was also experienced with a shotgun. He enjoyed good food and wine, but frugally.

As a man, despite his shyness and a diffidence sometimes resulting in an over elaborate courtesy, and occasional difficulty in choosing words, for which he had a proper respect, he was good company. The most common word used of him by articulate patients, then and now, has been 'gentleman'. The less articulate called him, in the modern vernacular, 'a lovely man'. His own commendation of others was often to say 'an honourable man'. He had no lack of humour, as is illustrated in the selection of anecdotes in the two books mentioned above. When forced by political circumstances into a slightly devious course, he would say apologetically that he had been 'a little Machiavellian'. His charm earned forgiveness for occasional unpunctuality. As a patient he was extraordinarily considerate of his medical attendants and nurses.

Throughout his married life he was supported by his wife, with strength and affection. The two were excellent hosts in their own home, and he treated his patients as guests in the hospital.

John's last years were clouded by ill health. He left much unfinished, but much was achieved. Friends and colleagues will

remember him with affection, colleagues with respect, and very many patients with gratitude.
WBA

[*Brit.med.J.*,1987,**295**,613; Photo]

LOPIS, Sol *b*.4 Nov 1916 *d*.1 Dec 1985
MB BCh Wits(1940) MD(1944) MRCPE(1948) MRCP(1948) FRCPE(1963) FRCP(1970)

Sol Lopis was a man whom everybody respected and many loved. A dedicated physician who treated his patients with old-world courtesy, he showed that happy frame of mind and good humour which made him a loving husband and father in his home.

Sol was born in Johannesburg, of a good middle-class family, his father Isaac being an ornithologist whose love of birds determined his starting a pet shop business which progressed rapidly, while his mother Anne supplied the moral support and home atmosphere which added to his happy upbringing. He was educated at Spes Bona Primary School, Parktown Boys High School and the University of Witwatersrand, from where he graduated in 1940. This was followed by postgraduate work at the Johannesburg Hospital,and he later presented a thesis on 'A survey of gall bladder disease in European and Bantu races' which gained him an MD in 1944. A spell in the Army as a captain in the South African Medical Corps was recognized by a 'Certificate for good services' in 1945. In 1947 he received further recognition of his postgraduate work when he was awarded the Nuffield Scholarship, which enabled him to study in Britain until 1949.

Before leaving for Britain, he married Hodda Todes who later presented him with a daughter, and two sons who are following in their father's footsteps. In London he studied under several prominent clinicians including Sir Max Rosenheim [*Munk's Roll*, Vol.VI,p.394], later Lord Rosenheim, Sir Harold Himsworth, R D Lawrence [*Munk's Roll*,Vol.VI, p.275] and Wilfred Oakley. He also obtained his membership of the Edinburgh and London Colleges.

On his return to South Africa he rejoined the visiting staff of Johannesburg Hospital, ultimately becoming senior physician and later honorary visiting physician, and lecturer at the University of Witwatersrand. He was also examiner in medicine for the University, and for the College of Medicine of South Africa. His special interest in endocrinology and diabetes was marked by his attachment to the diabetic and thyroid clinics, and he was the founder and physician in charge of the endocrine metabolic unit. He published a number of scientific papers both in his own name and in collaboration with others. His work was recognized by election to the fellowship of the Edinburgh and London Colleges. He was a founder member of the South African College of Physicians and Surgeons (now the College of Medicine) and also of the South African Diabetes Association.

His dedication to his work ultimately led to his health becoming affected. In spite of this he continued hospital and consultant practice, much to the benefit of both patients and students. His patients were his friends, for he lavished both care and kindness on them, while his students and all members of staff associated with him appreciated greatly his approach to patients and his knowledge, both of general medicine and of endocrinology.

There was never very much time for relaxation or sporting activities, but he never neglected his home and family, and his wife and children fully reciprocated the affection he showered on them. It was in fact his love for his family which determined the long trip to Australia, along with his wife, to see his sons and grandchildren before he died. Although he survived the air trip, he collapsed and died suddenly just one week after his arrival in Australia. But he had accomplished what he set out to do; which was typical of Sol Lopis.

TS

[Photo]

LURIE, David *b.*1 Dec 1916 *d.*4 Mar 1984
MB ChB Cape Town(1938) MRCP Lond(1948) FRCP(1979)

David Lurie was born in Stellenbosch, South Africa, and died suddenly in Johannesburg. He was educated at Stellenbosch Boy's High School and matriculated with distinction in 1932. He obtained his medical degree at the University of Cape Town in 1938, being awarded gold medals in physiology and public health. In 1939 he became an intern at Addington Hospital, Durban.

During the war years 1940-43 he saw active service in North Africa and Egypt, particularly in Sidi Rezek and Tobruk. He was one of the youngest majors in the South African medical services. He had a particularly good talent for telling stories, and he told of his war experiences with a simplicity, yet with a clarity, that made the whole era seem alive and terrible.

On returning from North Africa, he was in charge of the chest unit at Baragwanath British Military Hospital from 1943-46. He then spent two years in London, obtaining his membership of the College in 1948. He returned to South Africa and settled in Johannesburg where he had an illustrious and extremely distinguished career as a specialist physician for the next 36 years. He was consultant physician to the South African Airways and Railways and will always be remembered with great awe by numerous friends and patients in this sector. He was also consultant physician to the Pneumoconiosis Medical Bureau, and the Department of Mines and Mines Benefit Society. He was a part-time physician at the University of Witwatersrand from 1949 onwards. Among other things, this involved teaching most of the medical students and registrars and they enjoyed his quaint sense of humour. He was a founder member of the South African College of Physicians in 1953 (now the College of Medicine) and was elected a Fellow of the RCP in 1979. He was extremely active in medical practice until the day he died.

David belonged to that rare breed of doctors, the 'old-time' physician; unfortunately this type of doctor is becoming extinct. His sense of humour was energizing and he had an infectious laugh that was wonderful company, but his understanding of the human psychology and the complexities of human relationships must have been his greatest gift. His observation of people was magnificent, and because of this he never ministered to the symptom only but in every sense to the person as a unique being.

He married Ethne Cohen in 1946 and they had a son and two daughters. Virtually all his children, and their spouses, have followed his proud example in the medical profession.

He had a great love for sports and, although he was no great sportsman, he was always the centre of attraction on the golf course every weekend. He also played tennis, and jogged for fun for at least 40 years; that was how he died, quite suddenly, while running on the track, aged 67.

APG

[Photo]

LUXTON, Reginald William *b.*12 Aug 1905 *d.*17 July 1984
BSc Manch(1926) MB ChB(1929) MD(1932) MRCP(1932) FRCP(1952)

Reg, as he was known to everybody, was born in Shipley, near Bradford in Yorkshire; one of eleven children. He was deeply attached to his family, never married, but acted as 'father' to his many nieces and nephews. Although all his working life was in Manchester he remained a Yorkshireman.

After early education at Bradford Grammar School, he came to Manchester University on a scholarship provided by the Methodist Missionary Society. He had a distinguished undergraduate career, obtaining distinctions in anatomy and pathology together with a number of prizes. He was considered medically unfit to go abroad and so, on graduation, he had to repay the loan from the Missionary Society. Thus he had to refuse the offer of a house physician post at the Manchester Royal Infirmary and went instead to Hope Hospital. There he was assistant then resident medical officer, and after a period as demonstrator in anatomy at Manchester University he even went back to Hope Hospital as resident obstetrical officer. In 1934 he started his long association with Crumpsall Hospital, at first as resident medical officer and later, in 1936, as consultant physician; combining this with a chief assistant appointment at the Manchester Royal Infirmary and, in 1938, with the post of visiting physician at Christie Hospital. During the war years he even acted also as visiting physician at Oldham Royal Infirmary.

With the coming of the National Health Service Reg confined his work to Crumpsall and Christie Hospitals, retiring from Crumpsall Hospital at the age of 65. A little later he also retired from Christie Hospital.

He was an excellent general physician, well respected as a clinician and teacher. His unofficial undergraduate teaching sessions were soon regularized by his appointment as honorary lecturer in medicine, with senior students being attached to his unit, and he also lectured to the dental students.

Reg Luxton became a Fellow of the College, a member of the Thyroid Club, an associate member of the Cardiac Society, and was proud to be made a member of the Association of Physicians of Great Britain and Ireland in 1955.

His contributions to medical research were threefold. Firstly, he did much work on the use of sulphonamides when they originally became available. Secondly, he was one of the first to recognize that Hashimoto's thyroiditis was part of a general disorder, and he was one of the founders of the concept of auto-immune disorders. Lastly, as a result of his observations at Christie Hospital, he recognized the damage that could be done to kidneys by radiation. His article 'Radiation nephritis' appeared in the *Quarterly Journal of Medicine* 1953,**22** p.215-42, and he also published a chapter on the subject in *Diseases of the Kidney*, 2nd ed.,M B Strauss and L G Welt, Boston, Little, Brown, c.1971, and in *Structural basis of renal disease*, ed. E L Becker, New York, Hoeber Division, Harper, 1968.

He played a full part in medical activities in the city, being secretary, then treasurer, to the Pathological Society, and president 1959-60 of the section of medicine of the Manchester Medical Society. Before the NHS he was honorary secretary of the committee coordinating the work of the Manchester and District Hospitals. He was a member of the Regional Health Authority for many years. His work for the hospice movement in Manchester, Yorkshire, and even Australia, was very considerable; it was inspired by his religious convictions and association with the Moral Rearmament Movement. He was a keen fisherman, and he loved to walk in the Yorkshire dales, although progressive breathlessness eventually prevented this. In his final years he retired to a lovely home overlooking Airedale, where he died peacefully.

SO

[*Brit.med.J.*,1984,**289**,708,1012; *Lancet*,1984,**2**,475; *Proc.roy.Soc.Med.*,1957,**50**,943; *Lancet*,1956,**2**,105; *Quart.J.Med.*,1953,**22**,215; Photo]

LYLE, Thomas Keith Selfe *b.*26 Dec 1903 *d.*9 May 1987
CBE(1949) K St John(1960) BA Cantab(1926) LMSSA(1928) MB BChir(1930) MRCP(1930) FRCS(1932) MA MChir(1933) MD(1935) FRCP(1963)

Keith Lyle was born at Bromley, Kent, the son of an ophthalmic surgeon and physiologist, Herbert Willoughby Lyle, who was dean of King's College Hospital medical school for 20 years and also a JP. His mother Elizabeth was the daughter of James Dawking who was in the shipping industry. He was educated by a private tutor until he entered Dulwich College, where he gained an exhibition scholarship to Sidney Sussex College, Cambridge. After taking the natural science tripos, parts I and II, he proceeded to King's College Hospital in 1926 with a Burney Yeo scholarship and was later awarded the Todd prize for clinical medicine. He took an active part in the life of the medical school and played rugby football in the first fifteen.

After appointments in the neurological department at King's he went to the Royal Westminster Ophthalmic Hospital in 1934 as house surgeon, which formed his interest in this subject. He later became surgeon to Moorfields Eye Hospital, and ophthalmic surgeon to King's College Hospital and the National Hospital for Nervous Diseases.

Keith Lyle's clear thinking, enthusiasm and organizing ability, made him a fine teacher and lecturer. As a teacher, his registrars and house surgeons much appreciated his instruction in surgical technique and the opportunities he gave them for personal experience, and valued his advice. As a lecturer, he would put his thesis audibly and clearly, talking rather than reading to his audience. His penetrating analysis of the problems of strabismus and binocular vision led to him being regarded as an international authority on the subject.

During the second world war Keith Lyle served with the Royal Air Force Volunteer Reserve and reached the rank of air commodore, being mentioned in despatches. After the war he continued as civilian consultant to the RAF and the Department of Civil Aviation. His distinguished service was recognized by his being appointed CBE in 1949.

Keith gave much devoted service to the Order of St John of Jerusalem and became a knight grand cross of the Order in 1981, and a knight of the Holy Sepulchre. He spent much effort in maintaining the excellence of the ophthalmic hospital in Jerusalem.

Over very many years he also showed a keen interest in the Society of Apothecaries of London. He was elected a member of the Court in 1954 and was Master from 1962-63. A fellow of the Royal Society of Medicine, he was a former president of the united services section and a vice-president of the ophthalmic section. He was also a former president of the British Orthoptic Board, and the high regard in which orthoptists are held in this country owes much to his concern with its work and organization. He examined for the English and Edinburgh fellowships, and many other boards, and was known to be a fair, if exacting, examiner.

Keith Lyle was honoured by many learned societies, both at home and abroad. He was awarded the Doyne medal in 1953, the Nettleship medal in 1959 and the Richardson cross medal in 1972.

He was co-author, with Sylvia Jackson, of *Practical Orthoptics in the treatment of squint*, London, H K Lewis & Co, 1937, which ran into several editions; co-editor of *May and Worth's diseases of the eye*, London, Baillière,Tindall & Cox, llth ed.1954 and 12th ed.1959, and also co-editor of his father's book *Applied Physiology of the eye*, London, Baillière,Tindall & Cox, 1958. He published numerous articles in journals of ophthalmology.

Keith Lyle married Jane Maxwell, daughter of an Army officer, in 1949 and they had three daughters and a son. His friends much appreciated the warm hospitality of his charming wife and family at his delightful home at Henley-on-Thames.

All that Keith Lyle did was done with energy and enthusiasm. He did everything 'at the double', whether it was catching a train or walking through London. He was an excellent tennis and squash player, a keen horseman, and a member of a ski club.

His wife and family survived him.

GW/VL

[*Brit.med.J.*,1987,**294**,1698; *King's Gazette*,Spring 1969,**48**(1)8-9; Photo]

MACAULAY, Douglas Blair *b.*17 June 1904 *d.*5 Apr 1985
MB ChB Liverp(1926) MRCP(1948) FRCP(1966)

Douglas Blair Macaulay was a well known consultant allergist to the Liverpool teaching hospitals, having previously spent some time as a general practitioner in Crosby. He was born in Perth, Scotland. When Blair was only thirteen weeks old his father died and, because his mother had a job, young Blair was brought up by his grandparents and aunts who all held deeply Calvinistic principles so that, as Bryan Walker wrote in the *Transactions and report 1979-80* of the Liverpool Medical Institution: 'Everything that was fun was sinful: no cards, no smoking, no drinking, no dancing and church three times on a Sunday. It was therefore no wonder that when he arrived at the Medical School he fell in with a particularly jolly band of students, one of whom he later married.' She was Mary Hope Simpson (her father an MP, her brother second master at Rugby) and they established an excellent practitioner partnership, with other associates, which lasted from 1927-47. Blair gave time and thought to allergy, while 'Maisie' was largely interested in women's problems, writing two good books, *The Art of Marriage*, London, Delisle,1956 (a revised and enlarged edition, and later published as a Penguin Handbook No.PH35 in 1957), and

Marriage for the Married, London, Delisle, 1964. Both of them volunteered for Army service during the war but were rejected on the ground of high blood pressure. Nevertheless, when not engaged in his arduous practice, Blair was apparently well enough to act as a major in the Home Guard, in charge of the first aid post at Seaforth Docks - no sinecure in the bombings of 1941.

Blair's career changed when, in 1946, he was invited to run an allergy clinic at the ENT hospital. In 1948, very sensibly, he obtained his MRCP; he was a regular attender at the writer's Sunday morning teaching sessions. He therefore qualified as consultant allergist to the United Liverpool Hospitals with the coming of the NHS. He soon developed a thriving and widespread practice, and further appointments followed at the Chest, and Bootle Hospitals. In 1946 allergy had not yet gained official approval as a specialty and many of his senior colleagues equated it with quackery, but that word very soon disappeared from their vocabulary. Blair taught well, and he was very knowledgeable about the importance of psychological factors on allergic disorders.

In presenting Blair for life membership of the Liverpool Medical Institution in 1979, Bryan Walker recalled an interesting episode in his life: 'While on a visit to the USA as British representative on the International Committee of Allergology, Blair received a note at his Washington hotel saying that President Kennedy wished to see him urgently. Blair, suitably flattered, hastened to The White House and was ushered along innumerable corridors by armed gaurds, each with a 'marksman' badge on his uniform. At each stage of what seemed an interminable journey, Blair could not help but wonder what particular allergy he was going to have to deal with - asthma? hay fever? or merely a wasp sting? At last he arrived at a small ante-room. The door of the next room was open and there he could see the most powerful man in the world talking to his physician, an eminent New York woman practitioner. Presently she came out and very apologetically told Blair that they had been wrongly informed that he was a rheumatologist and that the President, after all, did not wish to consult him. And so a somewhat crestfallen Blair Macaulay had to retrace his steps, still with his armed guards, all the way back again.'

In his spare time Blair wrote good poetry, knew a great deal about flowers and plants, and cultivated a perfect garden. A great sadness overtook him in 1983 when his son, Michael, died suddenly of cardiac arrhythmia. He was survived by his wife and elder son, a pathologist in Canada.

CAC

[*Brit.med.J.*,1985,**290**,1593; Liverpool med.Inst.,*Trans.and rep.,1979-1980*; Photo]

MACAULAY, Jack Cameron *b.*30 Jan 1918 *d.*22 July 1987
MB BS Lond(1942) MRCP(1949) MD(1950) DCH(1950) FRCP(1972)

Jack Macaulay was born in Hong Kong where his father was serving in the RAMC. Hugh Montague Cameron Macaulay CBE became one of the great figures in community medicine and was the first senior administrative medical officer of the North-West Metropolitan Regional Hospital Board. His mother, Julia Garner, came from a farming family.

Macaulay was educated at Merchant Taylor's School, Northwood, and St Bartholomew's Hospital. After house appointments at the Central Middlesex Hospital he joined the Royal Navy, serving as a surgeon lieutenant RNVR in destroyers in the western approaches and later in the far east. He was particularly delighted to have been in the first RN squadron to re-enter Hong Kong after the Japanese surrender. After demobilization he became registrar to Sir Francis Avery Jones at the Central Middlesex. His paediatric training started at Hillingdon Hospital, and he later became house physician and then registrar at the Hospital for Sick Children, Great Ormond Street. He completed his training as chief assistant to the children's department at St Thomas's Hospital.

Jack Macaulay took up his appointment as the first consultant paediatrician in Shrewsbury early in 1952. The two children's wards were in an old isolation hospital some 2½ miles from the Royal Salop Infirmary. He quickly appreciated that to achieve a better service it was essential to become involved in administration. He later became the chairman of the hospital medical committee, and was an active member of the hospital management committee for many years. In less than 10 years the children's wards had moved to their present site in the refurbished military hospital at Copthorne. The battles which, at that time, had to be fought for children in a general hospital were won with tact and skill. He realized the need for services to the handicapped and was the district paediatrician on the Carnegie Committee for Services to Handicapped Children in the early 1960s. He was also visiting paediatrician to the Condover Hall School for the Blind.

In 1943 Macaulay married Sylvia Grace, daughter of the Rev Henry Edward Dodd, and they had two sons and a daughter; Anthony, Angus, and Gail.

Jack Macaulay was well liked and respected by his patients, their parents and his staff. He was not an easy colleague, being at times prickly and jealous of his department. He had few close friends, none of them on the staff of the hospital. His great loves were salmon and trout fishing, and rough shooting. He was an amusing raconteur, and a shrewd judge of men.

During the first half of Jack's time in Shrewsbury he lived in the centre of the town, inside the duodenal loop of the River Severn. He later moved to Longnor, a village near the Stretton Hills, where he and Sylvia became interested in arboriculture.

Cutting his links with paediatrics, he became president of his local branch of the British Legion, and was one of the founders of the Shropshire Nightingale Nursing Fund. Unusually, at the time of his retirement both his parents were alive. He died at Longnor. FPAC

[Photo]

McCARTAN, William *b.*22 May 1899 *d.*3 July 1984
MB BCh BAO(1922) MD Belf(1925) DPH(1927) MRCP(1935) FRCP(1963) FRCPsych(1971)

William McCartan was born in Castlewellan, Co Down, where his father John Bleakley McCartan was in the linen business. His mother Margaret was the daughter of William Moore, a farmer. William was educated in Castlewellan, going on to Queen's University, Belfast, and undertaking his clinical studies at the Royal Victoria Hospital.

From 1918-19, during the first world war, he served as a surgeon sub-lieutenant in the RNVR, and on demobilization he took house posts first at Belfast City Hospital and then at the Seamen's Hospital, Greenwich. Having decided to specialize in nervous diseases, he spent a year at the Hospital for Nervous Diseases, Maida Vale, London. From 1925-38 he worked in LCC mental hospitals with a break of a year, from 1927-28, at the Maudsley. He was appointed physician superintendent to St Francis and Hurstwood Park Hospitals, Haywards Heath, in 1938 and continued in this post until 1957; during this period he was also adviser to the mental health department of Brighton Corporation and consultant psychiatrist at Brighton General Hospital. In 1957 he was appointed honorary consultant in psychiatric medicine to St Francis and Lady Chichester Hospitals.

McCartan was a member of the British Medical Association and president of the Royal Medico-Psychological Association (now the Royal College of Psychiatrists) from 1961-62. In 1961 he published an article 'Monism and dualism: new lamps for old' in *The Journal of Mental Science*,107,no.450.

He married Edith Lord, daughter of Richard Lord, a dentist, in 1929 and they had one daughter. His interests outside medicine and his family included bee-keeping, gardening and fishing. GW/VL

[Photo]

MacCARTHY, Dermod de la Chevallerie
*b.*15 Mar 1911 *d.*12 July 1986
MRCS LRCP(1934) MB BS Lond(1937) MD(1940) DCH(1940)
MRCP(1943) FRCP(1958)

Dermod MacCarthy was a paediatrician, best known for his efforts to encourage mothers to be with their children in hospital. His father, Sir Desmond MacCarthy, was the foremost literary and dramatic critic of his day: familiar with all the notables of that era, such as Shaw, Wells, Chesterton, Bennett and Belloc, and at a later date with those of 'the Bloomsbury set'. His mother Mollie Warre Cornish was the daughter of the vice-provost of Eton, and the author of a highly entertaining book, *A Nineteenth-century childhood*, 1924. Both parents feature prominently in memoirs of their times; Desmond because of the legendary charm of his conversation, and Mollie for what one contemporary decribed as 'her varnish of conformity which concealed an original and unexpected response that was close to genius'.

Dermod was educated at Gresham's School, Holt, where W H Auden and Benjamin Britten were among his schoolfellows. From Gresham's he went straight to medical school at Bart's, qualifying in 1934. Between house jobs he signed on as ship's doctor for a voyage to the Far East. He had early decided to devote himself to children's medicine, and the start of the war in 1939 found him working at Great Ormond Street Children's Hospital, where he became resident assistant physician; and when the bombing of London began he was responsible for evacuating the children from Great Ormond Street to outlying hospitals. In 1942 he joined the RNVR and served as surgeon lieutenant until 1946. An interest in meteorology was a by-product of that service and was later a useful asset in his sailing.

When the NHS was set up in 1948 paediatrics in this country (the word was then hardly in use) was still poorly developed. Fortunate was that generation of paediatricians who found themselves with the responsibility for the setting up of children's units across the country. McCarthy was appointed consultant paediatrician to a group of hospitals in Buckinghamshire, centred on Aylesbury, and here he worked until his retirement in 1976.

People tend to divide themselves into two sorts - those whose natures are predominately practical, and those with the gifts of imagination and intuition that we usually associate with the artist. Medicine seems to attract mainly practical people and it is comparatively rare to find a doctor with the temperament of the artist. McCarthy was in this sense an artist. His understanding of his child patients was exceptional, as was his insight into the impulses and emotions of their parents. In the view of some of his peers, he was the 'compleat' paediatrician. When, in the 1950s, a movement arose to liberalize the arrangements then customary for children in hospital wards (visiting by parents was restricted, or even forbidden) it was inevitable that he should come to the fore. He helped James Robertson to make the second of two influential films: *Going to hospital with mother*, which was filmed in the children's ward at Amersham Hospital, and when the National Association for the Welfare of Children in Hospital was formed he became its friend and adviser.

McCarthy's influence was also strong in bringing together clinical paediatrics and child psychiatry, not necessarily the easiest of bedfellows, and he was successful in moving effectively in both these fields. In 1974 he was president of the paediatric section of the Royal Society of Medicine; while he was also consultant paediatrician to the Institute of Child Psychology in London and convened the paediatric discussion group which met regularly at Anna Freud's home in Hampstead. His influence extended abroad and in 1975 he was elected president of the Confederation of European Societies of Paediatricians.

A clinical problem particularly apposite to his skills was that of growth retardation, sometimes seen in children who are the victims of a disturbed family environment. He made some of the first observations on this condition (psycho-social deprivation or PSD, as it came to be known) and wrote an extended and authoritative account of it in *Scientific Foundations of paediatrics*, ed. John A Davis and John Dobbing, Philadelphia, Saunders, c.1974.

His influence on his contemporaries and juniors was great, surprisingly so perhaps as he wrote comparatively little. It was this influence that came to be recognized by his colleagues when, in 1982, the British Paediatric Association awarded him its highest honour - the James Spence medal.

In 1947 he had married Marie-France Geoffroy-Dechaume, who had gained the Croix-de-Guerre for her work in the French Resistance. They had known one another from childhood and Dermod was never able to explain to himself why he had taken so long (he was then 36) to realize that here was his ideal partner. Hilaire Belloc, himself half French, had been a sailing companion and a formative influence in his youth, so the marriage enriched his already strong affection for all things French. From then on, family holidays in Brittany were often to be happily dovetailed with his participation in French medical meetings. Lear House, their home in Weedon overlooking the Vale of Aylesbury, was one which their countless friends were always hoping to find some good excuse to visit. After retirement he continued to be in demand as a lecturer, and he was also able to pursue a lifelong love of sailing and to paint.

What will his friends remember him for? His charm, an inheritance from his father, his wit: 'Take care, I think they're going to gybe!' as we were walking behind a pair of Dutch nuns with large, starched, sail-like coifs; his enviable ability to recall his early childhood and the literary circles in which he had been brought up? Fortunately something, or all, of these qualities are preserved in his book *Sailing with Mr Belloc*, written and prepared for publication just before his death. Dermod McCarthy must be counted fortunate among mortals for he was richly endowed with notable talents, and was given the means to put them to good effect.

DMTG

[*Brit.med.J.*,1986,**293**,396,511; *Lancet*,1986,**2**,234; *The Times*,15 July 1986; *Arch.of Diseases in Childhood*,1982,**57**,565-66; Photo]

MACDONALD, Ranald Ian *b*.23 May 1904 *d*.20 Aug 1987
BA Dalhousie(1926) MD CM(1930) MRCP(1933) FRCPC(1947) FRCP(1955) FACP(1947) Hon LLD Dalhousie(1968)

Ian Macdonald was born in Sydney, Nova Scotia. His father, Donald John Macdonald, was a physician and surgeon and his mother Marion, née Longworth, was the daughter of a lawyer. He was educated at King's College School, Halifax Academy and the University of Dalhousie, pursuing his clinical studies at the Victoria General Hospital. Following house posts at the Toronto General Hospital, he took further postgraduate training in London, obtaining his membership of the College in 1933.

On returning to Canada, Ian Macdonald began his academic career as an instructor in medicine, and demonstrator in biochemistry and physiology, at Dalhousie, and as physician at the Halifax Visiting Dispensary. In 1936 he returned to Toronto and embarked on a teaching career which was to span 38 years. He was appointed a professor in the department of medicine, University of Toronto, and was also director of the division of postgraduate medical education from 1956-68. For four years prior to his retirement from the University he carried the additional responsibility of associate dean.

Macdonald was an attending physician at the Toronto General Hospital from 1936-68, and from then until his retirement in 1974 he was a consulting physician. During a 40-year period, 1934-74, Ian Macdonald's work was primarily in the clinical field, including some private consulting practice which enabled him to keep fully abreast of the problems of practitioners and the changing attitudes of the public.

During the second world war, Ian served for five years with the Royal Canadian Army Medical Corps; in Canada for one year and four years overseas, 1941-45. Initially, he arrived in England as a medical specialist to No 2 casualty clearing station and No 7 Canadian General Hospital. He was later promoted to lieutenant colonel in charge of medicine at No 9 Canadian General Hospital in England and to No 21 Canadian General Hospital in England and France. Following demobilization and return to Canada, he served as chief of medicine at Christie Street DVA Hospital, and continued in this capacity at Sunnybrook Hospital from 1947-66, where he was instrumental in creating an outstanding department of postgraduate medical education and training. He also served the DVA as adviser in medicine to the director general of treatment services, retiring from this position in 1974.

Ian Macdonald was elected president of the Academy of Medicine, Toronto in 1954, served eight years on the council of the College of Physicians and Surgeons of Ontario, and was president from 1963-64. He was a senior life member of both the Ontario and Canadian Medical Associations, and was for many years a member of the OMA committee on education. When the Ontario Council of Health was formed in 1966, the government appointed Macdonald as a charter member and he was subsequently reappointed, serving with distinction as the medical profession's representative until his retirement.

Ian married Janet Marjorie Mackinnon in 1932. Her father, Clarence Dunlop Mackinnon was a clergyman and principal of a theological college. There were three children of the marriage, Margaret, Isabel and Donald. Ian spent his retirement at his family home in Wolfville, Nova Scotia, where he continued his special interest in colonial and maritime history, regularly attending courses at Acadia University. He was an ardent salmon fishing enthusiast and, in later years, became recognized as an extremely knowledgeable and competent apiculturist.

LMB/GW

[Photo]

MACFARLANE, Robert Gwyn *b.*26 June 1907 *d.*26 Mar 1987
CBE(1964) MRCS LRCP(1933) MB BS Lond(1933) MD(1938)MA Oxon(1948) FRS(1956) *FRCP(1960) FRCPS Glas(1986)

Gwyn Macfarlane, the only child of Robert Gray Macfarlane and his wife Eileen, née Sanderson, was born at Worthing in Sussex. Sadly his father, Bangkok manager of the Eastern Trades Corporation, had succumbed to a bite from a rabid dog earlier in the year. After private school and Cheltenham College, Macfarlane entered St Bartholomew's Hospital in 1926 but did not qualify until 1933. Not that he was lazy or stupid but there were so many other interests - sailing, ice skating at which he was an expert, music and ballet, and the pleasure of fast motorcycles and later motor cars - all of which he continued to enjoy throughout his life.

As a medical student he had two emotional experiences that affected his career and achievements. As a pathology clerk to the Horder firm he had a 'brush' with Sir Thomas, later Lord Horder, [*Munk's Roll*, Vol.V,p.198] over the accuracy of haemoglobinometry, which resulted in a decision to become a pathologist. Then, as a dresser he looked after a small boy with a cut on his chin which would not stop bleeding. He was known to be from a family of 'bleeders', two of his brothers having died from biting their tongues, and in spite of every effort the cut continued to bleed. Finally, as the boy had lost so much blood he was given a blood transfusion and almost immediately the bleeding stopped. Macfarlane was both fascinated and horrified and in 1931 this 24-year old student wrote 'Haemophilia - a short survey', a remarkable, critical and comprehensive study of this condition, of which little was then known. He concluded that the only effective treatment for stopping the haemorrhage was a fresh transfusion of human blood or plasma, something not generally appreciated at that time.

Shortly after qualification Gwyn received a legacy of 25 pounds and approached Edgar Kettle [*Munk's Roll*,Vol.V,p.229], in the hope that he might work on a cure for haemophilia in the department of pathology at Bart's. Kettle, a shrewd judge of character, felt that this young man deserved encouragement and agreed to his joining the department.

After some initial experiments which came to nothing, Macfarlane learned from Hartridge that it was believed that some snake venoms affected blood clotting. Macfarlane immediately consulted the Zoo's curator of reptiles, Burgess Barnett. They tested the coagulating powers of a large number of snake venoms and found that they did affect blood clotting, and that the Russell viper was far the most effective. The venom was purified and tested for the first time on a haemophilic patient on 19 July 1934 with entire success; it was later introduced commercially as 'Stypen'.

By now Macfarlane was a junior demonstrator of pathology at Bart's, and also a Sir Halley Stuart research fellow, trying to find a cure for the disease. In 1936 he became clinical assistant pathologist at the new British postgraduate medical school, Hammersmith, under Janet Vaughan. Here he learned the full scope of haematology and the patterning of scientific experiments. He was also preparing a London MD thesis on 'The normal haemostatic mechanism and its failure in the haemorrhagic state', which was awarded the university gold medal in 1938; it was offered as a review in *The Quarterly Journal of Medicine* and revealed a completely new approach to the problem of haemostasis and its defects.

Macfarlane then spent a little over a year as assistant bacteriologist at the Wellcome Physiological Research Laboratories learning about the biological significance of enzymes, which was going to be important to him throughout his scientific career, but the oppressive commercialism was not to his taste and in December 1940 he was appointed clinical pathologist to the Radcliffe Infirmary, Oxford, a post in which he remained until his retirement a little over 25 years later.

At the Radcliffe he had to organize a department of haematology to meet the needs of the new Nuffield postgraduate scheme for research, as well as the training of pathologists and undergraduate teaching both in the clinical school and in the university department of pathology in particular.

In 1941 the Medical Research Council asked Macfarlane to organize a team to look into the discrepancies that appeared to exist between the normal haemoglobin levels in Britain and the United States. This was found to be due to an irregularity in the national Haldane standard but it also became apparent that there were very considerable observer errors in haemoglobinometry and a new, and more satisfactory, instrument - the MRC Grey-Wedge photometer - was introduced in 1948.

Macfarlane was appointed Radcliffe lecturer in haematology in the University of Oxford in 1943 and the following year spent six months, at the request of the Medical Research Council, in the RAMC working in France with Major Maclennan on gas gangrene, in which they were able not only to elucidate the pathogenesis of this condition but also recognized a new enzyme toxin 'collagenase' which had played a significant part in the local lesion.

When Macfarlane returned to Oxford he began an extensive study of fibrinolysis which was proving significant in relation to a wide range of biological processes and to tissue injury in particular. This research work attracted Rosemary Biggs to haematology, who had been studying the crush syndrome in morbid anatomy, but she so much enjoyed the department that she stayed as Gwyn Macfarlane's assistant and collaborator and, when he retired, she succeeded him as the director of the Blood Coagulation Research Unit.

By now the Radcliffe haematological department was becoming recognized as a power house of research and exciting new ideas. Young assistants from all over the world were anxious to work there and, although there were many who stayed just for a week or more, Macfarlane was never prepared to have more than three attached research workers at one time; otherwise the department would have become impersonal and no longer stimulating.

The recognition of errors in haemoglobinometry was the start of a general study of the reliability of all haematological methods; now called quality control. The methods of controlling anticoagulant therapy were next investigated as it was found that the Quick test was not always consistent, and this in turn revealed that the classical theory of coagulation was no longer tenable and that there were other factors previously unrecorded involved in the process. However, it was probably the revelations of the errors of red cell counting and similar tests which caused real alarm on the part of clinical pathologists throughout the world.

To overcome this, Macfarlance had the idea of producing an automatic electronic red cell counter and with a young technician, Heinz Wolff, he built the first equipment of this type, which proved to be entirely reliable and efficient. Unhappily, the Ministry of Health decided that there was no need for such a machine and that therefore it should not be developed commercially. A few years later the American Coulter electronic red cell counter, working on a different principle, was introduced and became essential equipment for all diagnostic laboratories.

The fields of interest in the Radcliffe department of haematology were broadening; bone marrow tissue culture initiated the study of cytogenetics of the marrow cells, and there was also a unit of immunohaematology concerned with haemolytic anaemias and, particularly, haemolytic disease in the new born.

In 1950 circumstances within the department caused Macfarlane to decide that he could no longer carry on with research work on fibrinolysis, although during the 15 years in which he had studied it he and his colleagues had made important contributions which are well recognized. On the other hand, there were indications that the blood factor lacking in haemophilics could be identified, so Macfarlane and his colleagues decided to study the clotting of whole blood, something which had not been undertaken since the days of Lord Lister. It was discovered that in addition to the well known extrinsic coagulation system there was an intrinsic system, equally powerful, which was initiated when blood came in contact with a foreign surface. It also became apparent that it was possible to identify and quantify the various clotting factors, including that which was lacking in haemophilia. Further, it was discovered that there was a haemorrhagic disease clinically and genetically identical with haemophilia but due to the lack of a different clotting factor, and this became known as Christmas Disease - named after the surname of the first patient recognized to be suffering from this condition.

A skilled enzyme chemist, Dr Edith Bidwell, now joined the team and began work on the purification and concentration of the various blood clotting factors but in particular that which was lacking in haemophilia, variously known as antihaemophilic globulin or Factor VIII. It became possible to offer haemophilics replacement therapy and by 1953 it was not only emergency operations which could be performed with a reasonable degree of safety but also formal orthopaedic and other major operations. In that same year *Human blood coagulation and its disorders*', Oxford, Blackwell Scientific Publications, 1953, appeared and was immediately highly popular and rapidly went through a number of editions.

Macfarlane was now receiving the recognition that he richly deserved: he became a fellow of the Royal Society, professor of clinical pathology at Oxford, a fellow of All Souls College, and was awarded the CBE. He was elected a Fellow of the College in 1960. Despite these honours he remained the same delightful, witty, friendly person who took no particular pleasure in the banquets, speeches and prize giving at international congresses.

In 1959 the Medical Research Council established a blood coagulation research unit at Oxford and this facilitated the inter-relationship of investigation and treatment, but inevitably there were administrative difficulties and Macfarlane was happy to occupy himself at the Sir William Dunn School, studying the early stages of thrombosis as seen through the electron microscope.

By now it was beginning to be recognized that there were a large number of blood clotting factors, with a confusing nomenclature until a committee was set up - in which Macfarlane played a considerable part - which suggested that they should be clearly defined and allotted roman numerals for identification, thus making communication much easier. It was also thought that many of these factors were either enzymes or substrates and, in 1965, Macfarlane put forward 'The basis of the cascade hypothesis of blood coagulation', *Thromb. Diath. Haemorrh.*,15:591-602,15 May 1966, in which he suggested that there was a sequence of enzyme reactions from the initial contact with a foreign surface, step by step accelerating and amplifying until the great energy release required in the transformation of fibrinogen into fibrin. His idea of a trigger mechanism was quite novel in biology and has been shown to extend to quite a number of other physiological systems.

Macfarlane had always intended to retire at the age of 60, and the administrative difficulties which the Ministry of Health were introducing in relation to the haemophilia unit gave him no reason to change his mind, for he already had a delightful house in Wester Ross. On the occasion of his retirement in 1967, a special number of the *British Journal of Haematology* appeared, with papers by many of his pupils, and a bibliography of his writings, as well as an acknowledgement of a delightful talk which he had given: 'Russell Viper's venom'.

Hilary and Gwyn Macfarlane found great contentment in Malla's Cottage, a converted croft in the small West Highlands village of Opinan which at that time had very few incomers. Macfarlane found the opportunity to write, something that he had always wished to do for he had a natural ability and elegance of style. He first wrote a philosophical fantasy in the form of a novel, which has not yet been published, but then he wrote a life of Howard Florey, (Lord Florey), [*Munk's Roll*,Vol.VI,p.178], *Howard Florey: the making of a great scientist,* Oxford, OUP, 1979, for whom he had a great admiration. It is a marvellous account of the develoment of a great scientist, with an ability to coordinate others, and the problems he had to face and the success he was to have. But Macfarlane was disappointed that the public did not see it like this; from their point of view Fleming and penicillin were synonymous and Florey did not exist. Macfarlane always felt that there was some sort of newpaper plot against Florey; so he wrote a life of Fleming, *Alexander Fleming: the man and the myth*, London, Chatto and Windus, 1984, and found that there was no plot, it was just that Florey was not very gracious to the media and Fleming thoroughly enjoyed publicity. Somewhat to Macfarlane's chagrin, his Fleming biography - which is the only accurate account of his life - was a great success, in contrast to the one on Florey which was much the better book. Macfarlane gave a brilliant 'Horizon' television programme,'The Mould, the Man and the Myth', and he was writing his own autobiography, but this was not completed.

In 1981, Sir Weldon Dalrymple-Champneys [*Munk's* Roll,Vol.VII,p.133] a former deputy chief medical officer of health, who had taken a great interest in the care of haemophilics and had been president of the Haemophilia Society for many years, died and Macfarlane - who had been vice-president for some time - was asked to succeed Sir Weldon. He agreed to do so provided he was not required to make any speeches, but he was most assiduous in his work for the Society and needless to say was bitterly disappointed that the Ministry had ignored the accurate estimates which he had made as to the amount of Factor VIII which would be required for the treatment of haemophilics, and instead had purchased from America virus contaminated material; at first with hepatitis virus - which was bad enough - but then with AIDS which was really serious. Macfarlane was much saddened to know that the work that he and his colleagues had done to relieve haemophilics had, through no fault of theirs, become a vehicle of death and disease. Somewhat bitterly, he would quote Andrew Mellon's 'No good deed goes unpunished.'

Although he had one or two illnesses, Macfarlane enjoyed his retirement and celebrated his golden wedding anniversary quietly with his family. On the evening of 25 March he developed severe myocardial ischaemia and died peacefully, in his sleep, in the early hours of the following day.

Gwyn Macfarlane had married Hilary Carson, then a medical student, on 25 January 1936, and it was a perfect marriage. They had five children who, in their different ways, reflected the interests and skills of their parents. One son, a doctor working in the field of blood disorders; another in the world of motors, two sons who combine engineering and boat building, and a daughter who is a skilled photographer as well as being a wonderful mother with the charm of both her parents.

AHTR-S

[*Lancet*,1987,1,932; *The Times*,30 Mar 1987; Wellcome Inst.Hist.Med.,*Contemp.med.archives*,RGM'S Working Paper; Photo]

MacKENNA, Robert Merttins Bird *b.*16 Nov 1903 *d.*12 Nov 1984
MRCS LRCP(1926) MA Cantab(1928) MB BChir(1928)
MRCP(1928) MD(1931) FRCP(1941)

Robert MacKenna was the only child of Robert William MacKenna and his wife Amelia, née Bird. Both his parents were physicians who graduated from Edinburgh University in 1898 and settled in Liverpool, where 'Mac' was born. His father started in general practice but in 1907, after working in several skin departments in Europe, went into partnership with Stopford Taylor, an established dermatologist.

As a boy, Robert became fascinated by ships in the Mersey and he early decided to make his career at sea. In 1917 he entered the Royal Naval College, Osborne, as a Naval cadet but, although Robert enjoyed life in the Navy, his father foresaw that the number of junior officers would be greatly reduced after the war and advised him to consider changing his career. His father was a powerful persuader and Robert selected medicine, entering Clare College, Cambridge, after attending classes at Liverpool University for one year; moving on to St Thomas's Hospital for clinical training. After qualifying he worked as clinical assistant to S E Dore [*Munk's Roll*,Vol.IV,p.568] in the dermatology departments of St Thomas's and of St John's Hospital for Diseases of the Skin.

In 1927 he married Helen Dora Todrick and in the same year he returned to Liverpool, with his wife, because both his parents were ill. He worked in his father's dermatological practice, and in the male venereal department of the Liverpool Stanley Hospital - where he later became honorary dermatologist. He always regretted that he had no opportunity to widen his experience at this stage of his career by working in various skin departments in this country and abroad. He resigned from his previous appointments in 1936 when he was appointed honorary dermatologist to the Royal Southern Hospital, Liverpool, where he also had teaching responsibilities. He was an enthusiastic member of the small but active North of England Dermatological Society which met in rotation, six times a year, in Leeds, Liverpool, Manchester and Sheffield.

At the outbreak of war, having previously volunteered for the RAMC, he was mobilized as a lieutenant. Liverpool was considered to be a likely bomb target, so his wife and her two daughters returned to her family home in Scotland. Robert served under rather makeshift conditions at Catterick and in Lincolnshire, and it was at this time that his organizing ability became manifest. In 1941 he was appointed adviser in dermatology to the War Office. He was to take care of dermatology, now separated from venereology, since up to that time the two disciplines had been linked under the control of one senior officer. Although the total number of trained dermatologists in the United Kingdom was small, he was able to select a few of them to form a nucleus, each in charge of a UK Army command. They were reinforced by young medical officers interested in dermatology who, after carefully planned 'crash courses', were appointed graded dermatologists. Many of them became successful consultants in the newly formed National Health Service after the war. The system was expanded to cover all the overseas Army groups and when the invasion of Europe began it was stated that MacKenna had organized the best dermatological set-up of any army Britain had ever had. In 1943 he became War Office consultant in dermatology with the rank of acting colonel, local brigadier, an appointment he retained until 1964. With the end of the war in Europe he travelled extensively in India and the Far East, supervising the care of skin disease problems. In December

1945 he was released from military duty, being made an honorary colonel RAMC in 1954.

'Mac' had been actively concerned in the many discussions which took place during the war on the future place of dermatology in the anticipated National Health Service, and he became the first chairman of the newly formed dermatological group of the British Medical Association. Together with a small group of like-minded colleagues he ensured the proper recognition of dermatology in the emerging NHS.

Soon after his release from the Army he was invited by A C Roxburgh, [*Munk's Roll*,Vol.V,p.361] then past retiring age, to apply for his post as honorary dermatologist to St Bartholomew's Hospital. This he did, and he was appointed physician in charge of the skin department in 1946. MacKenna was one of a few dermatologists at that time who felt very strongly that when dealing with problems affecting the skin it was essential to consider the patient as a whole. He accordingly built up a department which not only had close links with general medicine but also particularly with psychiatry and biochemistry. His collaboration and friendship with senior dermatologists in other allied medical services, especially with the Americans and Canadians, enabled him to arrange for many of his post-war trainees to spend time in appropriate departments in North America.

His administrative abilities, which had blossomed in the Army, were well utilized when he was appointed to the boards of both St John's and St Bartholomew's Hospitals. These were important decision making bodies and it was work which gave him great pleasure. He was involved in re-establishing in London the meetings of the International Congress of dermatology which had been disrupted by the war. He was a successful president of the British Association of Dermatologists in 1967, and of the dermatological section of the Royal Society of Medicine in the same year. He retired in 1968 but he kept an active interest in dermatology - always being ready to help and advise the young, especially those who had been 'his boys' in the Army or at Bart's.

He wrote well and concisely. He edited and revised the last three editions of his father's book *Diseases of the skin...*, R W MacKenna, 5th ed.,London, Baillière,Tindall & Cox,1952, which ran into five editions. His own book *Aids to dermatology...* , 2nd ed.,London, Baillière & Co.,1939 also ran into five editions, the last two being in collaboration with his friend and colleague E L Cohen. He edited *Modern trends in dermatology,* Series 1 appearing in 1948, Series 2 in 1954 and Series 3 in 1966. From 1944-63 he contributed annually a synopsis of important current articles on dermatology to *The Medical Annual.* He encouraged his juniors to study problems in depth and to take an active part in dermatological meetings. He was a Councillor of the College 1956-58, Watson-Smith lecturer in 1957 and Examiner 1964-70.

'Mac' had a very busy professional life, and he was also a great bibliophile. He had a deep knowledge of historical London, as many of his overseas friends discovered. The value of his contribution to dermatology was recognized by his honorary membership of fourteen overseas dermatological societies, as well as being a special honorary fellow of the American Medical Association, and honorary member of the Alpha Omega Honor Medical Association. It is a matter of regret to British dermatologists that he never received any national recognition in his own country in spite of his outstanding contribution to dermatology both in war and peace. This may have been due to his natural modesty and self-effacement.

His first wife divorced him in 1943, and he then married Margaret Hopkins who became closely associated with all his activities. He found great relaxation at home and tried to live according to his deeply held Christian faith. He was very happy when the two sons of his second marriage were ordained clergymen of the Church of England. There were two daughters of the first marriage. HRV

[*Brit.med.J.*,1984,**289**,1547,1703; *Lancet*,1985,**1**,61; *The Times*,16 Nov 1984; *J.Amer.Acad.Derm.*,1982,**6**,1,135; Photo]

McKEOWN, Thomas *b.*2 Nov 1912 *d.*13 June 1988
BA Br Col(1932) PhD McGill(1935) DPhil Oxon(1938) MB BS Lond(1942) MD Birm(1949) MRCP(1952) FRCP(1958) Hon FFCM RCPI(1980) Hon DSc McGill(1981) Hon FACP(1981)

Tom McKeown's approach to health care brought public medicine into the modern era. He was a leading exponent of the basis of the new approach - that public medicine must exhibit a comprehensive concern for the medical needs of society. As foundation professor of social medicine at Birmingham University from 1945-1977 he, and a few like-minded contemporaries, could claim to have created the subject. The 32 years that he held the chair included three when he was also pro-vice-chancellor.

In 1940 he married Esmé Joan Bryan Widdowson, daughter of a dental surgeon, and they had two children, a son and a daughter.

Tom was born in Portadown, Northern Ireland, the son of William McKeown, a building contractor, and his wife Matilda Duff, daughter of a journalist. He graduated BA in British Columbia and then came to Oxford as a Rhodes scholar. By the age of 25 he had been awarded two doctorates (McGill and Oxford), and he graduated in medicine from Guy's Hospital medical school in 1942. He was a demonstrator in biochemistry at McGill, and a demonstrator in physiology at Guy's, before being appointed to the chair at Birmingham. Between 1960 and 1980 he lectured in Canada, the USA and Australia, and in 1976 he was Rock Carling fellow of the Nuffield Provincial Hospitals Trust.

Before the NHS, and indeed up to 1974, 'public health' was the professional concern of Medical Officers of Health employed by local government. In practice, their main duties were defined by the environmental protection regulations set out in the Public Health Act of 1936. Tom and his colleagues saw this as a crippling constraint upon any fully integrated medical service for the community as a whole. The approach he advocated and practised was based upon the scientific study of health problems in the community combined with social, administrative, economic, demographic and historical studies. These ideas were reflected in the creation of the specialty of Community Medicine in 1973, which eventually unified the activities of scattered professional groups pursuing different aspects of their subject. Social medicine from then on would have a global rather than a narrow specialized role.

Tom's early scientific career was based upon the laboratory, in physiology, and his later scientific work upon epidemiology, but his most influential commentaries upon medicine and health owed as much or more to his demographic and historical skills. McKeown argued that the main improvements in health during the last century owed less to consultative and therapeutic medicine than to improvements in the economy, in diet, in public-health engineering, and in the evolution of social and working conditions. Before 1900 doctors probably did more harm than good; between 1900 and 1930, they broke even; but only since 1930, by which time the major health improvements of the present era were established, was it clear that doctors were beginning to win. Although Tom McKeown's conclusions were not always welcome to doctors, his international reputation in public health circles was among the highest. The medical service had now to be judged almost entirely upon the humanity and quality of care offered to the sick, including the chronic sick, the elderly, the mentally ill and handicapped, and the physically handicapped. These criteria have all been accepted, at least in principle, and it is perhaps difficult now to recall how awful some of these services were, or the low level of priority accorded to them. Here was a philosophy which ultimately required the evaluation of all aspects of medical practice by independent scientists and specialists, and a judgement of worth centred on the most neglected areas. It was not entirely painless.

Tom McKeown wrote two influential books, *The modern rise of population*, London, Edward Arnold, 1976, and *The Role of medicine: dream, mirage or nemesis*, Oxford, Blackwell, 1979. In the latter he argued that set against its own curative impotence, medicine must not try to create health but aim to '...assist us to come safely into the world and comfortably out of it, and during life to protect the well and care for the sick and disabled.' The full implications of this simple statement have yet to be realized.

McKeown had an elegant and polished, but formally academic style of writing and oral presentation, which often elicited resistence. But though his audience might resist his ideas, they could find no logical escape and little room for manouevre. It led to a good deal of wishful misrepresentation. Although he received universal academic recognition, there would be no civil honours for a man as forward looking and disturbing as Tom McKeown. Tom's teachings will, however, be remembered, certainly until such time as they cease to be other than a part of the common understanding of the true nature of medicine.

EGK

[*The Times*,20 June 1988; *The Independent*,21 June 1988; *The Guardian*,18 June 1988; *Brit.med.J.*,1988,**297**,129; *Lancet*,1988,**2**,58; *Guy's Hosp.Gaz.*July 1988; Photo]

MACLAGAN, Noel Francis *b.*25 Dec 1904 *d.*16 July 1987
BSc(1925) MSc(1930) MB BS Lond(1932) MD(1935) MRCP(1933) DSc(1946) FRCP(1952) FRCPath(1964) FRSC(1948)

Noel Francis Maclagan was born in north London, the son of Oscar Frederick Maclagan, a company director, and his wife Ada Sarah Nielson. He attended Maidenhead College in his early childhood and later University College School, London, where he was an above average pupil but failed to distinguish himself in games - except at chess, at which he excelled for much of his life. Not until late in his boyhood was it realized that severe myopia accounted for his poor performance in sport; when this was corrected, he greatly enjoyed tennis. At London University he read chemistry, with physics as a subsidiary subject, and graduated with first class honours. At this time jobs in science were not numerous and it was not uncommon for new graduates to accept jobs for which their university careers were inappropriate. It was not until 1926 that Noel was offered, and accepted, a job as a laboratory technician, later as research assistant, in Charles Dodds' - later Sir Charles [*Munk's Roll*,Vol.VI,p.151] - department of biochemistry in the Middlesex Hospital medical school. There were few posts available either in clinical biochemistry or chemical pathology and the department of biochemistry was unique, not in only being concerned with the teaching of medical students during their preclinical studies but also the provision of a proper chemical pathology service to the clinical department of the Middlesex Hospital.

Noel was encouraged to attend the medical school classes while at the same time carrying out his research activities in the biochemistry department - by no means an easy task, as the writer well knows - but in 1930 he gained his MSc and in 1932 his MB BS.

After a post as house physician at the Middlesex, he became research assistant at the Courtauld Institute of the Middlesex Hospital, until 1934. By this time he must have felt rather more secure in his search for a permanent post for he married Annemarie Wilhelmina May Herzog, a union which proved to be a happy and stable one for the rest of his life. During the following year he was a full-time worker for the Medical Research Council, and obtained his MD. He was then appointed biochemist to the Westminster Hospital where he remained until his retirement in 1970.

At the time of Noel's appointment to the Westminster, complicated biochemical analyses were rarely undertaken and, apart from Charles Dodds' department at the Middlesex and G A Harrison's department of chemical pathology at Bart's, only simple and uncomplicated analyses were carried out by the departments of clinical pathology. Noel immediately began to change this situation. Initially his laboratory in the old medical school in Caxton Street was small and separate from the hospital but with the building of a new medical school attached to the hospital, which was opened by their Majesties the King and Queen in 1939, he gained a new laboratory which he found more acceptable. It was during this period that Noel greatly increased the available analyses and became especially interested in liver function.

When the war began Noel was transferred to Ashford County Hospital, which was part of the Emergency Medical Service. His laboratory work and teaching continued and it was here that the liver function tests, including the well known thymol turbidity test, were developed. In 1945 his work on liver function and other contributions to clinical chemistry were submitted successfully to the University of London for his DSc. In 1948 he was elected a Fellow of the Royal Institute of Chemistry (now the Royal Society of Chemistry).

In 1946 the clinical biochemistry staff returned to London, to the no longer new medical school building, and Dr Pulvertaft, who prior to the war had been clinical pathologist, became professor of clinical pathology. Until that time, heads of pathology departments had always been professors of morbid anatomy and the staff of pathology departments had always been subordinate to the morbid anatomists. Some haematologists had been independent heads of their departments, but most other branches of pathology were under the administrative control of the morbid anatomist. In 1947, Noel Maclagan was appointed the first medically qualified professor of chemical pathology in Britain, but was officially under the administrative control of the professor of clinical pathology. Relationships became somewhat strained and a small pathology committee, with some administrators, was formed which led to the chemical pathology department becoming autonomous and resulted in more harmonious relations.

Noel's earlier work had been concerned with the relationship between appetite and control of body weight and blood lipids - especially in relation to blood coagulation and atherosclerosis and coronary heart disease. He greatly extended his work on hepatic function, especially in relation to enzymes in blood. In these researches he skilfully guided his junior colleagues and 27 successfully presented theses for their doctorates. His department was a very happy one; medical and scientific staff working harmoniously. Of the 27 successful PhDs, nine ultimately became professors.

The first five years following the end of the war saw biochemistry, both clinical and non-clinical, just beginning to recover from the rigid limitations imposed on theoretical biochemical research, as opposed to the heavy practical demands made on the relatively small number of trained biochemists. Initially, scientific meetings were small and infrequent. The first international meeting of biochemical societies was held in Cambridge in 1949 and was well attended and very fruitful. At that time the fragmentation of biochemistry, due to the enormous extension of knowledge in this subject, had not yet occurred. It was then possible to profit from attending international meetings in Europe and America without experiencing the difficulty of deciding which field of knowledge it was essential to attend. Noel and I, together with our wives, frequently met at these meetings and I rather think our wives were responsible for our subsequent close friendship.

Noel was highly regarded in Westminster Hospital and medical school, by both students and staff. His popularity was apparent in the warmth with which he was greeted on the many social occasions, both formal and informal, to which he was invited both before and after his retirement. For many years he was examiner in pathology for the Conjoint Board and for the University, and was always regarded as an excellent examiner. He was frequently invited to examine postgraduate theses not only for London University but also for other British universities. For several years he was examiner for the medical faculty of the University of Ibadan, Nigeria, and when I followed in his footsteps a year later I was informed that he was much appreciated by the students as a kind, considerate and competent examiner - nevertheless, the examiner was put on his plane before the examination results were published. Noel was always a welcome participant in many clinical chemical pathology meetings in European countries, and he was especially pleased when he was invited to tour most of the major clinical chemical pathology centres in South America. He was asked to postpone a lecture at Caracas University, Venezuela, because a revolution was planned to begin that day - October 24, 1960 - but as he was due in Bogatá, Colombia, the following day it was the revolution that had to be postponed: an unusual honour for a British professor.

Noel played an important role in the creation of the Association of Clinical Pathologists and was its first president from 1953-56. He was chairman of the board of studies in biochemistry at London University from 1960-63 and was appointed honorary secretary of an advisory committee on hospital biochemistry set up by the Ministry of Health in 1961. From 1976-79 he was chairman of the Nuffield Project on Clinical Chemistry Laboratories.

After his retirement Noel was for many years a welcome guest at many of the hospital and medical school functions. His marriage was a very happy one and he fully enjoyed the childhood of his children, Andrew and Jennifer. In later years, Andrew lived abroad but frequently visited his parents, together with his daughter, although at irregular intervals. Jennifer obtained an honours degree in physiology from University College London, followed by a doctorate, and was appointed lecturer in pharmacology at the Royal Free Hospital. She married and had four children, three daughters and a son, but her husband died while still a young man. During this sad period, Noel and Anna were able to provide their daughter with wonderful support. They purchased a small country house in Barrowden, Rutland, which they used with their daughter and grandchildren, and which provided an important sheet anchor at that time.

Noel adapted to retirement well, strongly supported by Anna. He enjoyed music and reading, and numerous activities around his home. He also greatly enjoyed driving - especially to Barrowden. About this time Anna had to have a hip replacement, which was highly successful and accepted with fortitude.

This peaceful period came to an end when Noel was found to have a small tumour in the scalp. In the preliminary preparations for its removal he was found to have lymphatic leukaemia which necessitated a long course of various treatments. Noel accepted this philosophically, but after some months became progressively weaker and unable to leave his bed. He was transferred to the Royal Free Hospital where he was found to have a severe osteomalacia - presumably due to his long course of treatment for leukaemia. This was efficiently cured within a few weeks but, just as he was responding well to treatment, he developed herpes zoster of an ophthalmic nerve which persisted for over a year and gave him appalling bursts of pain. During this time he was virtually sightless because of a scotoma in the other eye. After several months the herpes became quiescent, and it was decided that the damaged eye might respond to treatment by laser. This was very successful and within a few more months Noel was able to walk to the local shops, and to read if the print was sufficiently large.

Anna was then obliged to have a second hip replacement, which was also highly successful, but before she had fully recovered from her operation and was still walking around the house with two sticks, Noel suffered a severe cerebral haemorrhage and was admitted to University College Hospital where he died about two weeks later.

Shortly before his death he was made an honorary fellow of the Royal Society of Medicine. As his colleague, J R Hobbs wrote: 'The world of clinical biochemistry owes a great deal to this modest man, a gentleman not only of science but also in the fullest meaning of the word.'

CHG

[*Brit.med.J.*,1987,**295**,337,385; *Lancet*,1987,**2**,288; The *Times* 20 July 1987; *Festschrift*,Feb 1985,Charing X & Westminster Medical School; Photo]

McNEE, *Sir* John William *b.*17 Dec 1887 *d.*26 Jan 1984
Kt(1951) DSO(1918) MB ChB Glas(1909) MD(1914) DSc(1920) MRCP(1920) FRCP(1925) FRFPS Glas(1938) LLD Glas(1954) FRSE(1939) Hon MD NUI(1954) Hon FRFPS Glas(1954) LLD Toronto(1955)

Born at Mount Vernon, Lanarkshire, John William was the only son of John McNee of Glasgow and Newcastle-upon-Tyne. Little is known of his family background; the death of a sister in 1939 from streptococcal septicaemia is noted in his correspondence. He was educated at the Royal Grammar School, Newcastle-upon-Tyne, and at Glasgow University, where he qualified in medicine with honours. After graduation he held appointments as lecturer in pathology under Sir Robert Muir [*Munk's Roll*,Vol.V,p.298], and as assistant to the regius professor of the practice of medicine in Glasgow. Awarded a McCunn scholarship in 1911 and a Carnegie research fellowship in 1912, he went to Freiburg where, under the direction of Aschoff he worked on the aetiological classification of jaundice and the role of the reticuloendothelial system in the metabolism of bile. In 1914 he obtained his doctorate with honours, and the John Hunter and Bellahouston gold medals.

During the first world war he served in the RAMC, attaining the rank of major and acting as asistant adviser in pathology to the First Army in France. He undertook important clinical and pathological studies on trench fever and gas gangrene with J Shaw Dunn; on war nephritis; and on chlorine poisoning - in which he carried out one of the first, if not *the* first, autopsy on poison gassed soldiers in France. His work pointed the way to more effective preventive and therapeutic measures for ameliorating conditions among the troops. For his war services he was awarded the DSO, mentioned in despatches, and in 1920 was made a commander of the Portuguese Military Order of Avis.

Having obtained his DSc, he moved to London where the new medical unit had been established at University College Hospital under the direction of T R Elliott [*Munk's Roll*,Vol.V,p.119]. It caused some surprise that the two first assistants appointed were McNee, regarded essentially as a pathologist, and F M R Walshe, a neurologist, later Sir Francis [*Munk's Roll*,Vol.VI,p.448]. The latter soon moved away into consulting practice, but McNee remained, to continue his research on diseases of the liver and the gall bladder for which his training in pathology had provided a sound basic scientific background. His considerable skills as a physician and a teacher soon became apparent. In 1924 he obtained a Rockefeller fellowship and went to Johns Hopkins as an associate professor, at a time when the syndrome of coronary artery thrombosis was being defined. On his return he published in the *Quarterly Journal of Medicine*, 1925, the first description in Britain of this condition, with the encouragement of Elliott and Sir Thomas Lewis [*Munk's Roll*,Vol.IV,p.531]. At this time he elected to abandon a purely academic life and moved into full-time consulting practice, while remaining a physician to UCH. His reputation as a sound and sympathetic clinician brought rapid success and his advice was widely sought, especially on difficult cases of liver and biliary tract disorders. He was called upon to give prestigious lectures by Royal Colleges and learned societies, and further consolidated his reputation by the publication, jointly with Sir Humphrey Rolleston [*Munk's Roll*,Vol.IV,p.373], of the classical textbook *Diseases of the liver, gall bladder and bile ducts*, London, Macmillan & Co., 3rd ed. 1929. It appeared that his future in London was secure and permanent.

But, in the middle 1930s, the regius chair of the practice of medicine in Glasgow fell vacant and McNee was invited to accept it. Locally there was some resentment that the post should not be filled by a distinguished Glaswegian physician. Moreover, in Glasgow McNee was remembered, even respected, as a pathologist rather than a clinician. McNee himself had two reservations: firstly, he would wish, if he moved north, to keep an option open to continue private consulting practice and, secondly, he insisted that he should be given an institute in which he could establish a research team. The University agreed to his first condition, and the second was made possible by a handsome bequest from Gardiner, a local shipowner, after whom the institute would be named. Thus it was that McNee, satisfied on both counts, accepted the regius chair and moved back to Scotland.

Back in his *alma mater*, despite a lukewarm welcome, McNee aimed at becoming a first class professor, and at setting up a first class department. His efforts to raise the standard of medical education and examinations in Glasgow met with some antagonism, and his appointment of a paediatrician as his senior lecturer was not well received. His insistence that all his juniors should obtain the MRCP(London) rather than the Glasgow FRFPS, although forward looking, did not go down well.

McNee threw himself enthusiastically into the creation of a new department of medicine attached to his wards at the Western Infirmary. Here he had magnificent offices, consulting rooms, six research beds with their own diet kitchen, and very ample laboratory space. He was probably the first professor of medicine to establish an animal laboratory within his department, and to encourage his staff to work on experimental pathology. Although, as had been promised to him, the

Institute provided the necessary facilities, in the event he saw few private patients there. The scene was therefore set for the establishment of a first class research unit, and McNee's expertise as a lecturer, clinical teacher and 'beloved physician' were at last beginning to be appreciated when war broke out in Europe.

In 1938, McNee had accepted with alacrity a request to become a naval consultant, with the rank of surgeon rear-admiral (given to few, if any, as he would recount in later life, who had previously been ranked as major, RAMC). He entered the senior service as consulting physician to the Navy in Scotland and the Western Approaches. He was proud of his new status and was not inhibited from sweeping up to the hospital in his WRNS-driven Rolls-Royce - gifted, incidentally, by a colleague, - to deliver a lecture or conduct a ward round in all his gold-braided glory. During the first year of the war, especially, McNee refused to sit idle, or only partially occupied at the naval base hospital near Aberdeen; it made more sense to him to remain in Glasgow and continue as much teaching and research as possible. This was barely acceptable to Navy top brass, who refused to sanction travel expenses to and from Glasgow since officially - officiously - he was permanently stationed in Aberdeen. But he worked hard at his naval duties, publishing useful data on wartime afflictions, especially immersion foot. He took an active part in the organization of medical staff and equipment for support ships accompanying North Atlantic convoys, and his contributions to this are recorded in *The Rescue Ships* by Vice-Admiral B B Schofield and Lieutenant Commander L F Martyn, Edinburgh, William Blackwood & Sons,1968.

When the turmoil eventually ceased, McNee was faced with the task of starting up his research unit virtually from scratch. But the war had taken its toll and he had a limited time left in his chair. So, although he continued unabated with his teaching and clinical commitments, few research projects of significance were accomplished before his retirement in 1953.

He had always loved the countryside and sporting activities, especially field sports - though he was an enthusiastic golfer. He moved south to Hampshire where he and his wife nurtured a lovely garden. The death of his wife in 1975 left him lonely, and his gradually failing health caused him frustration. But he remained mentally alert to the end, and he could astonish his visitors to Winchester with an amazing flow of vivid reminiscences.

McNee was always a controversial character. His aristocratic appearance, patrician manner, and lavish lifestyle distanced him from a number of his colleagues and associates, both in London and Glasgow, but many of his students discovered him to be warm and sympathetic, and to his assistants he remained helpful, loyal and supportive. His lectures to undergraduates were memorable, delivered with clarity of expression and without notes, and laced with anecdotes which made the subject come alive. His ward teaching rounds are still remembered for the lucid exposition of the difficult case. He had an excellent rapport with his patients, towards whom he showed every kindness and consideration. As an examiner - and in this role he was very much in demand - he earned the nickname of 'the smiling ploughman' on account of his kindly and very fair approach, coupled nevertheless with a determination to maintain standards.

From 1929-48 McNee served the *Quarterly Journal of Medicine* in an editorial capacity. With D M Dunlop, later Sir Derrick [*Munk's Roll*, Vol.VII,p.170] and L S P Davidson, later Sir Stanley [*Munk's Roll*, Vol.VII,p.136], he produced the popular *Textbook of medical treatment*, Edinburgh, E&S Livingstone, 1939, which ran to several editions, the 6th being published c.1953. He also continued to update his textbook with Sir Humphrey Rolleston on liver disease, to its 6th edition in 1955.

His many prestigious appointments included physician in Scotland to two sovereigns, president of the Association of Physicians, 1951, and the British Medical Association, 1954, and master of the Barber-Surgeons' Company of London, 1957-58.

In 1923 he married Geraldine Z Le Bas, MSc London, who was herself active in the field of medical science. There were no children of the marriage.

John McNee will be remembered as a benefactor of the College. Shortly before his death he made a gift of his Portuguese decoration, and in his will he bequeathed to the College one half of his residuary estate, together with his portrait by James Gunn, a smaller version of that which hangs in Glasgow.

SH

[*Brit.med.J.*,1984,**288**,494,725; *Lancet*,1984,**1**,408; *The Times*,4 Feb 1984; 12 July 1954,31 Mar 1975; Portrait; *J.roy.nav.med.serv.*,1984,**70**,124; *Glas.Med.*,1984,**2**,4; Photo]

MAITLAND, Charles Titterton *b.*6 July 1894 *d.*26 Sept 1984
CBE BSc Lond(1915) MRCS LRCP(1917) MB BS(1920) MD(1921) DPH(1921) DTM&H(1921) MRCP(1922) FRCP(1936) MFCM(1974)

Charles Maitland died, aged 90, at his home at Gerrards Cross where he had lived for more than half a century. He was the son of missionaries and had spent part of his youth with them in China, where he later did some medical work of which he seldom spoke. He was closely involved with the work of his church, to the point of participation as a chorister into his late sixties. His religion strongly influenced his work throughout his life.

He received his medical education at Bart's, whence he qualified in 1917. His bent was strongly towards preventive medicine and he became MOH of the Metropolitan Borough of Stoke Newington. He was recruited by George Newman, later Sir George, [*Munk's Roll*,Vol.IV,p.559] to the Ministry of Health and achieved his first promotion there, exceptionally for that time, within ten years. He was Isabella Cameron's deputy in the maternity and child welfare section and seemed destined for the highest level. In 1939 he was one of the five principal medical officers designate for a regionalization scheme, which did not reach fruition because of the outbreak of war.

Maitland spent the war years at the Birmingham wartime regional office dealing with the West Midlands, and then returned to become head of the Ministry's section for hospital planning, where he remained until his retirement. He was awarded the CBE.

Maitland had much detailed knowledge of the fields in which he worked, coupled with a great urge to impart it at length to any interlocutor, and complete certainty in his views. By contrast, he wrote with great difficulty and incessant revision, so that his literary output was late and small. He did draft the report of the Birmingham Hospital Survey for the surveyors, to whom he had given great help, but his inability to satisfy himself with his product made it the last of all to be completed. The same inhibitions were to lead to his failure to complete the treatise on hospital planning which was intended to be the outcome of his last two year's work.

Maitland was not an easy colleague, especially for members of other professions, and despite the real kindness which was hidden behind his abrupt manner and rather forbidding appearance, he had few close ties at the end of his long service with the Ministry. He had been utterly loyal to his department, and his uncompromising assertion of official doctrine ruffled feathers outside it too. So one must record that a good, kind and able man did not find the place or opportunity that his knowledge and intellect should have made available to him. Perhaps the system should be blamed for that, rather than the man. He did have real friends, including the writer, but by the end of his 30 years of service he was an isolated man. Perhaps he would have been happier and gone farther as an academic, for he enjoyed teaching 'Public Health' at St Thomas's and his exposition was always lucid and well informed.

In retrospect one must record that a highly talented man was given by circumstance too small an opportunity to make his mark - he would have been a good collaborator for Chadwick. Yet he did not evince disappointment or embitterment as other younger colleagues went past him, and that shows a quality of character that is all too uncommon. He was devoted to his family, who never failed to support him, and that reward may have been what he would have wanted most.

GEG

MARKUS, Hettiya Kontehelage Nihal Ian
*b.*18 May 1935 *d.*28 Apr 1986
MB BS Ceylon(1961) MD(1965) MRCPE(1968) MRCP(1969) FRCP(1983)

Nihal Markus was born in Colombo, Sri Lanka, where his father, Charles Haddow Markus was an accountant. He was educated at St Joseph's College, Colombo, and what was then the University of Ceylon (Faculty of Medicine). His clinical studies were pursued at the General Hospital, Colombo. After house appointments at the Lady Ridgeway Hospital and the General Hospital, Colombo, he became senior house officer at the latter and also demonstrator in pathology at the University. In 1965 he was appointed resident physician at the General Hospital. He was awarded the Commonwealth Scholarship in medicine in 1967 and for the next two years he worked as an honorary registrar in the professorial medical unit of the Royal Infirmary, Edinburgh, and in the gastroenterology department of the Western General Hospital in the same city.

For the next 12 years, on his return to Sri Lanka, he served as a consultant physician at various Government Hospitals, with a short break between November 1981 and March 1982 when he visited Denmark, Germany and the UK as WHO Fellow in cardiovascular diseases. In 1983 he was appointed to the General Hospital Galle, a teaching hospital.

It was as a teacher that Nihal Markus was most widely acclaimed. He was able to kindle in others his own joy in medicine and fire them with enthusiasm. Lucidity of expression and clarity of thought marked his clinical lectures. He was a fine diagnostician, with a rigorously logical appraisal of clinical evidence.

In his student days he evinced much interest in religious and social activities. Though a devout Christian, he found himself taking part in Buddhist brotherhood activities with equal zest. His actions were reciprocated by some of them when he became the president of the Students' Christian Movement and invited Buddhists to participate in its activities. His wife, Dr Vineetha Manel Kusumsri, whom he married in 1963, was a Buddhist - their union being witness to the fact that religion is intended to bring people together and not keep them apart. They had three children, two sons and a daughter. He was a devoted husband and a loving father, and his untimely death was a severe blow to the family.

Outside medicine, Nihal Markus' interests lay in cricket, music and social welfare work. It was in the sphere of personal relationships that he excelled. Everyone with whom he was associated felt him to be a trusted friend.

NA/GW

[Photo]

MARTIN, James Purdon *b.*11 June 1893 *d.*7 May 1984
BA Belf(1915) MA(1918) MB BCh BAO(1920) MD(1922)
MRCP(1922) FRCP(1930) Hon DSc Belf(1982)

James Purdon Martin was born at a farm in Jordanstown, County Antrim, and died at the National Hospital, Queen Square, London. He entered Queen's University, Belfast, in 1912 with a scholarship in modern languages and graduated three years later with first class honours in mathematics. Because of widespread psoriasis he was graded medically unfit for Army service when he attempted to join the Army during the first world war, so he chose to study medicine and qualified in 1920. His first house appointment was in Liverpool, but within a year he had been elected to the staff of the National Hospital for Nervous Diseases and never really left it. He gained his membership of the College in 1922, his fellowship in 1930, and was an examiner from 1939-43. He was appointed to the consultant staff of the National Hospital in 1925, and to the consultant staff of the Royal Postgraduate Hospital, Hammersmith, in 1935. He also had consultant appointments to the Seamen's Hospital in Greenwich, the Bolingbroke Hospital and Whipps Cross.

From 1940-45 he was neurologist to Eastern Command, for he was still not regarded as fit for military service, and he was also neurologist to the St Mary's Hospital sector. He was elected dean of the National Hospital, Queen Square, in 1946. In 1948 he became the first dean of the newly constituted Institute of Neurology. There had of course been a postgraduate school of neurology at Queen Square since the days of Hughlings Jackson [*Munk's Roll*,Vol.IV,p.161] but it became officially part of the British Postgraduate Medical Federation in 1948.

Purdon Martin delivered the Lumleian lecture in 1947 on 'Consciousness and its disturbances...', and the Arris and Gale lecture of the Royal College of Surgeons in 1963 on 'Basal Ganglia and locomotion'.

For some years he was joint editor of *Neurology*, and after he retired he was appointed visiting professor to the University of Colorado. In 1982, to his great delight, he was made an honorary doctor of science at Queen's University, Belfast. He was a member of the Association of British Neurologists, and of the French and Canadian Societies.

In 1922, while still a resident at Queen Square, he married the resident medical officer, Marjorie Ada Blandy, and there were two sons of the marriage. Sadly, his wife died in 1937. In 1949 he married Janet Smiles Ferguson, née Nichols, who had been a co-student of mathematics at Queen's University. Janet died in 1978.

At the time of his resident appointment in Liverpool, Purdon Martin came into contact with the first cases of encephalitis lethargica that occurred in the British Isles. The epidemic spread southwards and when Purdon arrived in London, one year later, he had exceptional clinical experience of the sequelae of this disease, which the London neurologists were only just beginning to see. His special interest in basal ganglia was aroused and remained his main concern for the rest of his life.

His first major contribution, in 1927, established the association between hemiballismus and a partial lesion of the corpus luyisii. In 1948 he was the first to report the therapeutic effects of penicillin in neurosyphilis, and he was the first neurologist who had the courage to treat neurosyphilis simply with penicillin. In spite of the spectacular results of penicillin, well into the 1950s many of Purdon Martin's seniors continued to use methods which had always been inefficient and had never cured the disease; which would no doubt have raised eyebrows among many young neurologists. One of Purdon's major original contributions was the collation of evidence showing that the basal ganglia had a principal function relating to postural mechanisms.

After retirement, Purdon continued a study that he had commenced in earlier years; of patients with post-encephalitic Parkinsonism, who had been inpatients at the Highlands Hospital, Winchmore Hill, for over 40 years. He had a hut built for him in the grounds for this special study, where he could record their reactions to displacements of posture. He collated all these findings, together with other material, in a book entitled *The Basal Ganglia and posture*, Philadelphia, Lippincott, 1967, which is a model of English prose. The central idea was novel; such examination and recording methods had not previously been part of the armamentarium of British neurology. His work was increasingly noticed worldwide and a society was set up in Europe for the study of posture. He became the first president at the age of 85. This work brought him recognition among his peers which had not been granted before. He was particularly gratified by the notice his work received in the recent edition of the *Handbook of Physiology*, Bethesda, American Physiological Society, 1981, and by a request to prepare a second edition of his book. He contributed the section on neurology in *Price's Textbook of Medicine*, 12th ed.[London] English Language Book Society, 1978, regarded by the majority of postgraduate students as the best textbook in clinical neurology.

These are the bare bones of the remarkable career of a remarkable man. He was a man of strong principles, totally honest, and who expected those who worked with him to have the same high principles and the same devotion to work. Consequently, he was regarded by many as a daunting figure but in fact was a man with a total lack of affectation, great tolerance and an open and enquiring mind which he applied to all his activities - not only to his work, but also to the study of the financial pages of the Stock Exchange and to ballroom dancing. To his patients and those who worked with him he was unvaryingly courteous and kindly. Unthinking colleagues and juniors were inclined to regard him as being slow and ponderous. Admittedly he lacked the charisma of many of his contemporaries on the staff of the National Hospital, but his apparent slowness was merely a manifestation of his determination to think things out before he opened his mouth, so that whenever he did express an opinion it was invariably well formed and to the point. An afternoon spent on a teaching round with Purdon was always a lesson in new ideas, new observations and new challenges. At any hospital, the pathologist always has a much clearer idea of the accuracy of the observations and diagnostic acumen of his clinical colleagues. Godwin Greenfield [*Munk's Roll*, Vol.V,p.164], when discussing his observations on those patients who had come to post-mortem, always claimed that 'Purdon's' diagnoses were confirmed some 15% more frequently than any of his other colleagues, who included all the giants of the middle quarter of the century.

Purdon Martin retained his active and enquiring mind up to his 90th birthday, in spite of a series of almost catastrophic illnesses. He remained up to date in his knowledge of his favourite subject, the disorders of the basal ganglia, and was able to debate with younger men and women who might be regarded as experts in the field of movement disorders. Future historians will come to regard Purdon Martin as one of the greatest clinical contributors of his day.

Louis Sachs, an American professor of neurology, wrote: 'Purdon Martin was endlessly thoughtful and ingenious in designing a variety of mechanisms and methods that made it possible for even the most incapacitated patient with Parkinsonism to achieve an artifical normality in gait and posture; lines painted on the floor, counterweights in the belt, loudly ticking pacemakers to set the cadence for walking, and these he always learned from his patients to whom his great book is dedicated. He was indeed a great human pioneer, and understanding and collaboration were central to his medicine. Patient and physician were co-equals, on the same level, each learning from and helping the other and between them arriving at new insights.'

REK

[*Brit.med.J.*,1984,**288**,1698; *Lancet*,1984,1,1135-6; Photo]

MAYCOCK, *Sir* **William d'Auvergne** *b*.4 Feb 1911 *d*.19 Feb 1987
Kt(1978) CBE(1970) MBE(1945) LVO formerly MVO(1961) MD McGill(1935) MRCS LRCP(1935) FRCPath(1963) MRCP(1967) FRCP(1972)

When the second world war began William (Bill) Maycock had the prospect of a distinguished career in physiological and pathological research. He joined the RAMC on the first day of the war, 3 Septemebr 1939, and completed his military service, with the rank of colonel, only at the end of November 1945. Throughout more than six years he contributed greatly, both as a scientist and administrator, to the efficiency and safety of the Army blood transfusion service on all fronts. For the rest of his professional life Bill was recognized as an international authority on blood transfusion and blood products.

Born in London, Bill was the second son of William Perren Maycock MIEE and his wife Florence Marion, née Hart. His father, born in 1866 and the author of several textbooks, was a consulting electrical engineer who had entered this field in the last decade of the nineteenth century, an exciting time when the application of electrical energy to domestic and manufacturing processes, and to transport and communication, was growing dramatically: he died in 1918, shortly before the introduction of wireless broadcasting for entertainment, a development he had foreseen in his writings during the first decade of this century. Bill's father had been one of ten children, of whom four were boys, only one of whom lived almost to the psalmist's allotted span; he himself died of cancer of the oesophagus at the age of 52; a second at age 29 from cholera in Calcutta; and another died of multiple sclerosis at 42. By contrast, the sisters reached ages between 72 and 94.

Bill's paternal grandfather was 'country manager' of the London Joint Stock Bank, and one of his great-grandfathers was George Perren, a well known singer in the third quarter of the nineteenth century who had trained at the Scala, Milan, and took some part in the formation of the English Opera Company. No musical talent descended to Bill or to his brother, an architect. Bill's maternal grandfather was a pharmacist, but his mother was an only child and Bill did not know any members of her family, who came from Sussex. His father's family came from the village of Wellesbourne in Warwickshire, where they were recorded in the early 1600s, and where one member had been vicar of a neighbouring village.

Bill showed early promise. He was a senior King's Scholar at The King's School, Canterbury, and became a member of the rugby XV, captain of boats, and captain of the school, before winning a school-leaving Lord Milner memorial exhibition to any university of his choice. Since his mother and brother were living in Montreal, he chose McGill University, where he studied medicine between 1929-35, gaining a Birks scholarship, the Francis J Williams fellowship in clinical medicine, the Lieutenant Governor's bronze medal in pharmacology, the Sutherland gold medal in biochemistry and, on graduation, the Holmes gold medal. He turned at once to pathology, as demonstrator at McGill and house pathologist at the Royal Victoria Hospital, 1935-36. From 1936-39 Bill was a Leverhulme Scholar at the Royal College of Surgeons in London, also attending the National Institute for Medical Research as a visiting worker. He published two papers with Lindor Brown on vascular reactions in the sympathectomized cat, and had just been appointed to the physiology department at St Thomas's when war was declared. Bill soon found himself in France in charge, first of a mobile blood transfusion unit and then as second in command of a short-lived blood transfusion and surgical research laboratory. On return to England in 1940 he joined L E H Whitby, later Sir Lionel [*Munk's Roll*,Vol.V,p.444], at the Army

Blood Supply Depot, Bristol, and was second in command from 1942-45, when he took over command. He remained a specialist in pathology.

During the war years there were many crises in supplies and occasional fierce periods of tension and controversy over the appropriate use of blood and plasma, and in regard to problems of compatibility - including the newly discovered rhesus factor - and transmission of infections, especially hepatitis. Throughout, to all those concerned with transfusion and resuscitation on world-scattered British Army fronts, Bill never ceased to be a trusted, respected and much loved colleague.

In 1946 Bill joined the Lister Institute, and from 1949-78 was either superintendent of the Institute's laboratories in Elstree, or head and director of the Blood Products laboratory. Throughout the same period he was a consultant in blood transfusion to the Ministry (later Department) of Health, and honorary consultant in blood transfusion and resuscitation to the War Office, later the Ministry of Defence. He was also a recognized teacher of pathology in the University of London and was president of the British Society for Haematology.

The many problems to which Bill turned an expert, acute, disciplined and practical mind included the used of plasma substitutes, the preparation of anti-D immunoglobulin for clinical use, the prevention of post-transfusion hepatitis, detection of hepatitis B surface antigen and its antibody, the danger of varicella to immunosuppressed patients, the purity of antihaemophilic factor, criteria of suitability for blood donation, and higher training in haematology. Bill never played the role of an 'authority'; modesty and a considerable sense of fun tended to conceal the real depth of his knowledge.

On leave from France at the beginning of 1940, Bill married Muriel Macdonald of Toronto, who had joined up with the First Aid Nursing Yeomanry (FANY) at the outbreak of war and who was a driver at the Army Blood Supply Depot. From a particularly close and happy marriage came two sons, one of whom became a road engineer and the other a solicitor.

In 1976 Bill was diagnosed to be suffering from muscular dystrophy. Physically it became a frustrating handicap. but mentally he met the challenge with stylish courage and humour. He refused to regard himself as less than fully active in mind and body, and maintained all his normal interests in medical science, local history, and the cultivation of plants - except that actual work in the garden was for others.

Bill was a slight, always handsome man, invariably courteous, well-mannered almost to the point of old fashion, and quietly spoken except for delightful eruptions of uproarious laughter.

GW

[*Brit.med.J.*,1987,**294**,979; *Lancet*,1987,**1**,818; Photo]

MAYON-WHITE, Richard Mayon *b.*15 Dec 1915 *d.*2 Nov 1984
MRCS LRCP(1940) MB BS Lond(1940) MD(1949) PhD(1949)
MRCP(1949) DCH(1949) FRCP(1964)

Richard Mayon Mayon White was born in Boreham, Essex, the son of R H White, an electrical engineer, and Jane, née Henshaw, a mining engineer's daughter. He was educated at University College School and Eastbourne College, and entered St Thomas's Hospital medical school. He did his first house jobs there and also became house physician to the children's department.

During the war he served on the staff at the War Emergency Medical Research Council. Five years of his time there was spent in the physiological laboratory at Lulworth Camp in Dorset, doing research on the effect of explosion on tanks and tank crews which led to his MD and PhD thesis entitled 'Tank crew casualties'.

He was determined to pursue a course in paediatrics, and was awarded a Rockefeller travelling fellowship in paediatrics, spending a fruitful and happy year as a visiting fellow at the University of Minnesota. The completion of his medical training was guided by Douglas Gairdner at Addenbrooke's Hospital, Cambridge, who had recently arrived there from the Newcastle paediatric department, at that time so ably chaired by James Spence [*Munk's Roll*,Vol.V,p.386], later Sir James.

With the coming of the National Health Service the stage was set for the expansion of paediatrics in this country. The NHS needed to recruit physicians of stature and proselytising enthusiasm, to establish the facilitating environment for the young specialty. Dick was admirably suited for this role, and was appointed consulting paediatrician to the East and West Suffolk Hospitals in 1949. He was not only an advocate for children and their families but also the planner and designer of the purpose built accommodation for both inpatients and outpatients, and neonatal services. It is hard now to imagine Dick doing ward rounds in a stetson hat and western boots, but he knew it was important to establish the fact that both children and paediatricians were different. He was also an active member of the hospital Christmas concerts group. It is easier to imagine that the establishment of the department took up most of his energies and thoughts, but in addition he took on local and national commitments for the British Paediatric Association, being a member of Council from 1961-64, and he also had an outpatient commitment to Great Ormond Street from 1961-63.

He was a lifetime member of the British Medical Association, and was active locally. He served on the Joint Consultants' Committee for 10 years or so and was president of the Wholetime Consultants' Association. He claimed to have worn out two bowler hats on behalf of the profession, using the technique of courteous argument and subtle negotiation with various health ministers. Although he was saddened to leave the JCC, he later admitted that he would not have done well with the confrontation tactics which later had to be used to safeguard the profession's position. He was delighted, towards the end of his career, to be elevated to the fellowship of the BMA in 1980.

Dick remained a single-handed consultant until 1967, when he was joined by a colleague who also had a major commitment to Colchester Hospital - in another district and another region. He was an astute guide and counsellor to the newcomer, and would always have a refreshingly lateral approach to any problems, both medical and political. He found it more difficult to share clinical control and remained steadfastly and consistently on call for his patients except when he was away on leave, or officer of the day to the Royal Ipswich Yacht Club. But his prime consideration was for his small patients and their families, and he was always available for consultation.

Dick was a private man, whose common sense and integrity were supported by his love for his family and his strong Christian beliefs. He was a founder member of the Institute of Religion and Medicine, and ran the Monday group with the hospital chaplain for many years - to which doctors, nurses and other hospital workers came to discuss problems of ethical and moral difficulty in their work. These strengths allowed him not only to counsel the distressed, and the bereaved child and his family, but also to inspire a sense of purpose and high standards in those with whom he worked, and in those whom he helped to train. His prose was meticulously and economically chosen. He had early fallen under the influence of Richard Asher [*Munk's Roll*, Vol.VI,p.16] and was dismayed by his colleagues' profligacy and carelessness with words. He read widely, particularly the poets of the 1930s and '40s. He preferred the solitary sports of fly fishing and bird watching, but shared a love of the sea with his family and friends.

He was twice married, first to Ellen Mary Townroe, by whom he had five children - one of whom is a consultant epidemiologist and another a sister tutor. His first wife died in 1961 after a long illness. A year later he married the Hon. Mrs Mary Musker, daughter of Robert Fitzroy, 2nd Viscount Daventry, and she brought him three stepsons and 2 stepdaughters. The solution was to buy an old preparatory school near Ipswich in order to accommodate the large and growing family. He and Mary entertained, with almost Whig-like enthusiasm, the nurses, students and staff of the paediatric department and invited them to share their grounds and their swimming pool. Shortly after Dick retired, Mary died after a short illness, from the desolation of which Dick never recovered.

Dick was a colossus in his time, bestriding two district health authorities. Where once he consulted single-handed, he is now succeeded by five consultant paediatricians and one consultant community paediatrician. He had established a new specialty successfully in the fertile soil of East Anglia. He retired at a time when he was becoming increasingly dismayed by the dilution of consultant responsibility. He was suspicious of shared-care, and critical of the insincerity of interdisciplinary cooperation and integration. At its birth paediatrics needed single-minded, dedicated godparents to guide its development. Where Dick Mayon-White led, others gladly followed. CN

[*Brit.med.J.*,1984,**289**,1464; Photo]

MEDAWAR, *Sir* Peter Brian *b*.28 Feb 1915 *d*.2 Oct 1987
Kt(1965) OM(1981) CH(1972) CBE(1958) FRS(1949) BA Oxon(1935) MA(1939) Hon FRCP(1974) Nobel Prize 1960

There were many extraordinary facets to Peter Medawar's personality and life. Not least among them was the fact that, following his first and very serious cerebral haemorrhage in 1969 and a series of subsequently progressively debilitating strokes, he not only continued to lead an experimental research programme at the Clinical Research Centre in Northwick Park but published six books that received much acclaim and had a wide readership. Although his most innovative and creative period as a scientist ended tragically and prematurely at the age of 54, he thus added immensely to his reputation as a writer who invariably discussed matters concerning science, the scientific method and the relationship between science and society with wisdom, insight and wit, and who set down his ideas with a lucidity and elegance of style rare among scientists.

Peter Medawar was born in Rio de Janeiro to an English mother, Edith Muriel Dowling and a Lebanese father, Nicholas. Anxious for their son to receive an English education they sent him to Marlborough College, where he was deeply unhappy but where he nevertheless developed his interest in biological science. He went on to Magdalen College, Oxford, to obtain a first class degree in zoology in 1935. The next three years were spent in the Sir William Dunn School of Pathology, and there he was exposed to the influence of Howard Florey, later Lord Florey [*Munk's Roll*,Vol.VI,p.178], whom he came to admire greatly. Peter became a lecturer and fellow of Magdalen College at the age of 23, having married a fellow student, Jean Shinglewood Taylor, the year before. Thus began a partnership that proved to be a vital factor in his life, for Jean not only gave her husband the unstinting support he needed to be able to function in top gear but sustained him in his last 18 years of increasing disability - years that put his patience, courage and determination to keep going severely to the test. She was also co-author of two later books, *The Life Science...*, London, Wildwood House, 1977, and *Aristotle to Zoos....*,London, Weidenfeld and Nicolson, 1984, which was originally published in Cambridge,Mass.,USA, by Harvard, in 1983. She also saw his last book, the autobiographical *Memoirs of a thinking radish*, Oxford, Oxford University Press, 1986, through to publication.

Soon after graduating Medawar became interested in mathematical logic and a mathematical approach to biological problems, having been strongly influenced by the writings of Bertrand Russell. The first laboratory project that came his way involved the use of tissue culture in an attempt to identify a substance in malt extract that inhibited cell growth. He discontinued this in favour of a mathematical study of scalar growth of cells, again in culture, followed by an analysis of growth data already available on the human foetus, in an attempt to test the ideas of D'Arcy Wentworth Thompson on growth and form. He then turned to the question of why tissue grafts transplanted from one individual to another of the same species (allografts) are rejected, and it was in this field that he was to establish his towering reputation as a pioneer and a brilliant experimental scientist. Graft rejection had become an urgent concern early in the war because of the large number of airmen who received severe and extensive burns demanding treatment with skin grafts, and the Medical Research Council therefore invited Medawar to study it. Much work had been done since the turn of the century to elucidate the problem of graft rejection, and both skin and kidney allografts had been used experimentally in several species. Although there was tacit, but not universal, agreement that all allografts failed as a result of some sort of genetically determined incompatibility, it was left to Medawar to demonstrate beyond doubt that rejection is brought about by a specific immunological response against the histocompatibility antigens of the donor. Complementary evidence was provided by Peter Gorer, working at Guy's Hospital, and the American George Snell in Bar Harbor.

Medawar's conclusion derived from two separate studies, both bearing the hallmark of his subsequent work: simplicity of design, meticulous planning and perfect execution, followed by careful analysis and lucid exposition. The first study, published in 1943, was done with human skin allografts in collaboration with a Scottish surgeon, Thomas Gibson. It showed that skin grafts are rejected very promptly after a latent period, that the body develops 'memory' for a second graft from the same donor, and that the accelerated rejection of a second graft occurs only when it comes from the donor who had also provided the first graft, i.e. the rejection phenomenon has the latency and specificity typical of immunological responses. The second study (1944-45) was carried out in rabbits and, apart from confirming the human data in every detail and on a larger scale, it added a careful and systematic description of how skin grafts are rejected macro- and microscopically. Furthermore, it drew attention to the presence of large numbers of mononuclear cells (mainly small lymphocytes) in the graft at and just before the time of rejection, raising the possibility that rejection was cell-mediated rather than caused by the production of antibodies. Any doubts there may have been about the immunological nature of the rejection process were now wholly dispelled, and the field of cellular immunology was opened up. It was mainly for this work, as well as for some other interesting observations (for example, that the anterior chamber of the eye and the brain are to some degree 'privileged sites' in which allografts are not rejected as rapidly as might be expected, and that cortisone has a marked effect in prolonging skin graft survival - work done with P L Krohn) that Medawar was elected to the fellowship of the Royal Society at the unusually early age of 34.

He had meanwhile been appointed to the Mason chair of zoology at the University of Birmingham and had taken with him a postdoctoral fellow, Rupert E Billingham, who had completed his doctoral thesis in Oxford after war service in the Royal Navy. It was in Birmingham that the scene was set for the later studies on immunological tolerance. Sponsored by the Agricultural Research Council, Medawar and Billingham (aided by G H Lampkin) set out to devise a simple skin graft test that would distinguish clearly between dizygotic and monozygotic twins, on the assumption that the former but not the latter would reject skin grafts exchanged between them. Agriculturally this was of some importance because a female calf twinned with a male was known to lose both its secondary sexual characteristics and its fertility. Identification of such 'freemartins' was therefore both necessary and difficult. The results were unexpected in that *all* twins, regardless of sex, accepted each others' grafts, and Medawar and Billingham soon realized that the reason for this must have been the fact (known to embryologists since the beginning of the century) that cattle twins usually share their blood circulation *in utero*. This, as the American immunogeneticist Ray D Owen had demonstrated as early as 1945, leads to an exchange of blood from an early enough foetal stage for the establishment, from blood cell precursors, of red cell chimerism. By showing that the unresponsiveness to red cell antigens extended to skin they established the concept of immunological tolerance, and much of Medawar's energy in the '50s was directed at proving experimentally its existence and characteristics. They were encouraged in this enterprise by the far-sighted speculations of two Australians, Macfarlane Burnet (q.v.) and Frank Fenner who, in their 1949 monograph on *The Production of Antibodies*, 2nd ed., Melbourne, Macmillan, developed a complex theory of self and non-self recognition, (Burnet, who was to share the Nobel Prize with Medawar in 1960 and who, too, was later knighted, had attempted to demonstrate the induction of tolerance to viral antigens but had failed.)

The beginning of the tolerance studies in 1951 coincided with Medawar's appointment to the Jodrell chair of zoology at University College London, and he took with him Rupert Billingham and Leslie Brent - the latter an ex-serviceman who had just taken his first degree in Medawar's department and who was to commence his postgraduate studies. Medawar led this small team with great verve and geniality, and set out to reproduce the naturally occurring cattle chimerism and tolerance in mice by the inoculation of first foetal, then neonatal, mice with viable allogeneic cells. Their task was considerably simpified by the use of inbred strains of mice obtained from G D Snell at Bar

Harbor. After many failures they succeeded in showing that an immunologically immature mouse thus inoculated became a cellular chimera and failed to reject skin grafts transplanted later in life, that this unresponsiveness was strictly specific for the histocompatibility antigens of the donor strain, and that it was brought about not so much by the development of a positive controlling/suppressing mechanism as by some form of clonal deletion or permanent inactivation of lymphocytes. Later, when suppressor cells became demonstrable, this hypothesis fell into disfavour but it has recently, with the benefit of modern knowledge of the T cell receptor and of histocompatibility antigens, received significant support. The experimental demonstration of tolerance, which was first published in *Nature* in 1953, and in greater detail in the *Philosophical Transactions* of the Royal Society in 1956, caused a great stir among immunologists and transplant surgeons because it showed dramatically and for the first time that the barrier standing in the way of allotransplantation could be overcome after all, and because it provided a cogent explanation of the distinction made by the body between self and non-self molecules, thus ensuring that auto-immune responses and disorders are the exception rather than the rule. Further, it led directly to the discovery by Billingham and Brent of graft-versus-host disease (GVHD), a condition frequently suffered by experimental animals rendered tolerant by cell populations comprising significant numbers of T lymphocytes. In their tolerance studies GVHD was avoided by the use of F^1 hybrid donor cells, but the recognition of GVHD pointed to the difficulties - still not entirely overcome - of transplanting allogeneic bone marrow to patients.

Another area opened up by Medawar and his colleagues at UCL was the study of the antigens responsible for allograft rejection, for they were the first to show that non-viable cells, and tissue extracts, can be immunogenic. Here, for once, their hypothesis (that the histo-compatibility antigens were DNA-proteins) proved to be erroneous. During this highly creative period of his life, in which Medawar headed a thriving and multifarious department of zoology with an exceptionally talented staff, he found the time to publish some brilliant speculations on the possible reasons for the survival of the mammalian foetus in its immunologically alien environment - speculations that continue to be of interest today.

Medawar's contribution to transplantation immunology was recognized in 1960 by the award of the Nobel Prize for physiology and medicine, and honours both scientific, civil and royal, were heaped upon him. The former included a Royal medal in 1959, the Copley medal in 1969, and, in his last year, the Royal Society's Michael Faraday medal, not to mention 17 honorary doctorates from universities in Britain and abroad. He was elected an Honorary Fellow of the College in 1974. It says much for Medawar's strength of character that all this did not affect him adversely: whilst he derived great pleasure from the recognition he received, life went on normally and he did not allow himself to be deflected from his real and abiding interest, laboratory based research, which continued even when, in 1962, he became director of the National Institute for Medical Research in Mill Hill.

He had seven years as director before he became partially incapacitated by his first haemorrhage. In that period his intellectual brilliance and integrity, his continuing research with colleagues (for example, on the effects of irradiation on graft rejection, the use of cell-free tissue extracts in the induction of unresponsiveness, the properties of antilymphocyte serum, and an *in vivo* tissue typing test) and his outstanding personality as a leader who had the knack of drawing the best out of people without imposing himself, coupled with an extraordinary capacity for relentlessly hard work, made him an outstanding director. Indeed, in his time at the NIMR, the Institute became renowned internationally as a centre of excellence, and its reputation in immunology soared. This led to a constant stream of visiting workers who added further to the Institute's intellectual stature and creativity. He nevertheless found - or made - time to sit on all kinds of public bodies (the Medical and Agricultural Research Councils, the University Grants Committee, the Governing Body of the BBC), he was widely in demand as a lecturer (his lectures, always carefully prepared, were invariably brilliant, and his BBC Reith Lectures on *The Future of Man* in 1959 aroused a great deal of interest), and he published several books in which he established his reputation as a writer and thinker with a wider public. These included: *Uniqueness of the individual,* London, Methuen, 1957, *The Art of the soluble...*,London, Methuen & Co.,1967, and *Induction and intuition in scientific thought...* Philadelphia, American Philosophical Society, 1969. His analytical ability as well as a certain streak of intellectual ruthlessness were well demonstrated in his devastating critique of the work of Teilhard de Chardin.

Although, following the disastrous brain haemorrhage in 1969, which happened when he was reading the lesson in Exeter Cathedral in his capacity as president of the British Association for the Advancement of Science, his research never again reached the heights recorded in the previous quarter century, he was now able to devote more time to writing. Apart from the two books written with his wife, and his final autobiographical volume, he published *Advice to a young scientist,* New York:London, Harper & Row, 1979, *Pluto's republic,* Oxford, Oxford University Press, 1985 (the title with a typical Medawarian touch), and *The Limits of science,* Oxford, OUP,1985, originally published in New York by Harper & Row in 1984. The elegance of his prose and his perceptive analysis, linked to a self-deprecating humour, won him a wide readership and gave much pleasure to many, scientists and non-scientists alike.

All this time he was battling, together with his wife, children and friends, against the cruel fate of becoming increasingly handicapped, and his resilience and sense of humour, even in his darkest days, were both touching and exhilarating. He continued to preside over the transplantation biology section of the Clinical Research Centre, where he was strongly supported by the director and his staff, and where he was surrounded by loyal colleagues who regarded him with great affection and respect. His celebrated forays in his battery-powered wheelchair frequently caused hilarity both to him and to others.

What kind of man was Peter Medawar, apart from being a great scientist, thinker and writer? He had a tall, imposing presence and his handsome, dark features, his *joie de vivre,* boyish charm and self-evident intelligence, won him instant admirers. He enjoyed life to the full: good companionship, music, and sporting activities such as cricket and squash, provided him with the emotional and physical outlets that were essential to him in a life packed with intellectual activity. He had a passion for opera and knew the works of Verdi and Wagner intimately. He enjoyed chess, but did not mind losing. He was loyal to his colleagues and friends - a loyalty that was mutual - and he was generous to other scientists, especially those who were still making their way. His mind was wide open to new ideas and he made his own thoughts freely available to others. Medawar in his prime did not suffer fools; small talk was not for him, though he was not averse to hearing about minor scandals. He felt protective to those working in subordinate positions, such as his secretaries and technicians. Characteristically, it was under his directorship that the NIMR had its dilapidated sports pavilion restored and a common room, with a bar, established.

With such a busy and creative life it was perhaps not altogether surprising that he found it difficult to devote enough time and attention to his four children, Caroline, Charles, Louise and Alexander, a fact to which he referred with regret in his memoirs. He was nevertheless much loved and admired by them.

Peter Medawar was buried in Alfriston, Sussex, and a memorable Service of Thanksgiving was later held at Westminster Abbey. With his death there closed an important chapter in the history of immunology. LB

[*Brit.med.J.*,1987,**295**,1072-73; *Lancet*,1987,**2**,923; *The Times*,5 Oct & 8 Dec 1987; *The Daily Telegraph*,5 Oct 1987; *Roy.Soc.News*,June 1985 7; *New Scientist*,1975,27 Feb; *Med.News-Tribune*,20 Aug 1971; *Mayo Clinic Proc.*,**39**,Sept 1964]

MELICK, Roger Aziz *b.*4 Aug 1925 *d.*15 Nov 1986
MB BS Sydney(1947) MRACP(1951) FRACP(1962) MD(1972) FRCP(1975)

Roger Melick was born in Sydney and died in Melbourne. His father was Aziz Nicholas Melick, a merchant. Roger was educated at King's School, Sydney, one of Australia's oldest schools. He qualified in medicine from the University of Sydney and did his early post-graduate training in that city, at the Royal Prince Alfred Hospital. He obtained his membership of the Royal Australasian College of Physicians in 1951 and was elected a fellow in 1962.

Roger went to Boston, USA, for his early training in endocrinology, and he worked at the Massachusetts General Hospital from 1954-56 with the legendary Fuller Albright, and one of his earliest papers was as a co-author - describing the use of the then 'miracle' drug cortisone in the treatment of Addison's disease.

Roger Melick first arrived in Melbourne in the middle 50s to join the newly created department of medicine at the Royal Melbourne Hospital under the chairmanship of Richard Lovell. Because of his developed interest in endocrinology he was warmly welcomed into this small but active community of endocrinologists. He was among the first of a number of medical graduates from the University of Sydney who migrated to Melbourne and added great lustre to academic medicine there.

On his arrival, Roger immediately established a vigorous programme of scientific research in endocrinology. One of his early papers established the variations in the normal patterns of the growth of body hair, which was an important contribution to clinical diagnosis in patients with hirsutism. At this time he also started quite basic research on what was to remain a lifelong interest - bone structure and function - using fairly sophisticated cell culture methods in these studies.

Roger's research on bone and calcium regulating hormones extended over more than 30 years and his contributions to this field made him a highly respected international figure in bone and calcified tissue research. Some of his early work on the mechanical stressing of bone is only now exciting a newly awakened interest, since methods for making such studies have improved. It was typical of Roger that he undertook this work at a time, in the early '60s, when it was technically difficult but was nevertheless a problem which needed to be solved. This attitude characterized his approach to research over the years.

A sabbatical period in the UK, in Oxford, at the MRC Bone Research Laboratory in the early 1970s with Maureen Owen and her colleagues was spent, during the early days, working on the proteins of bone. It was a difficult area to which he made substantial contributions, and in which he continued to work until his death. He was a pioneer in the development of assays for parathyroid hormone, following his work with Gerald Aurbach and John Potts. Roger made a number of original contributions in connexion with his studies of the metabolism of parathyroid hormone. Roger was a wonderful colleague to work with closely; he loved to work in the laboratory and continued to do so in a way which few of his seniority were able to maintain.

Perhaps the least known, but probably one of the most important aspects of his work was his partnership with Lester Van Middlesworth from Memphis, USA, on the detection of atmospheric pollution with

Cecil David Alergant

Roger Edward Collingwood Altounyan

Sir Christopher Andrewes

Thomas George Armstrong

Cyril William Curtis Bain

Eric Tom Baker-Bates

Harold Wykeham Balme

Robin Mowat Bannerman

Enrique Teofilo Bassadone

Ali Bedri

Simon Behrman

Robert Charles Stirling Benson

Trevor Baylor Benson

Leslie George Blair

Patrick Joseph Aloysius Blaney

John Henderson Bolton

Harben Jay Boutourline-Young

Percival Thomas Bray

Denis Hubert Brinton

Wilfred Danvers Brinton

William Alexander Briscoe

William Brockbank

Jan Brod

Sir John Brotherston

Stanley George Browne

Sir Graham Bull

James William Douglas Bull

Sir Frank Burnet

James Kenneth Cameron

Leslie Henry Capel

Wilfrid Ingram Card

Cedric Oswald Carter

Theodore Moir Chalmers

Harish Chandra

Melville Ernest Chinner

Ronald Victor Christie

Archibald Edmund Clark-Kennedy

Sir Stanley Clayton

Vicky Veronica Clement-Jones

Archibald Leman Cochrane

Robert Greenhill Cochrane

Nigel Dean Compston

Julius Hiram Comroe

John Bowen Cook

Samuel Harold Cookson

Christine Elizabeth Cooper

James Donaldson Craig

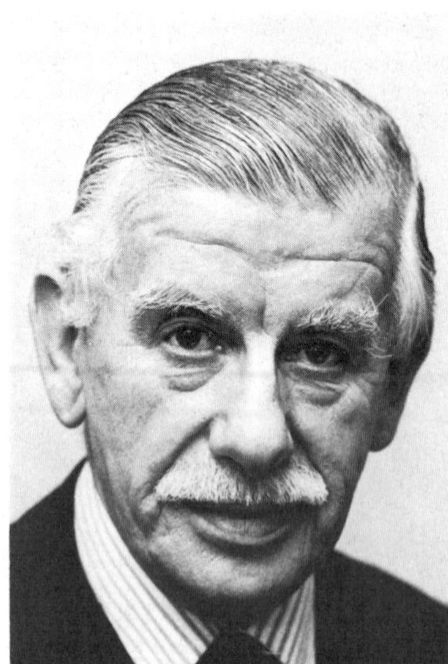

Sir John Croom

John Norman Cruickshank

Richard Desmond Curran

Jacques George Léon Dagnelie

Keshavrao Rao Krishna Rao Datey

Patience Elizabeth Davies

Harold Sydney Davis

John Dean

Percival Cholmondeley Chalmers de Silva

Robert Clark Dickson

John Patrick J Donaghy

Ian Donald

Allan Watt Downie

Ross Graham Dreadon

Otto Gustaf Edholm

Thomas Archibald Watkin Edwards

Philip Raymond Charles Evans

William Evans

Roland Henry John Fanthorpe

Derrick David Felix-Davies

Sir Clive Fitts

Oliver Patrick Noel Alexis Fitzgerald

Sir Edward Ford

Louis Forman

Alfred White Franklin

Raymond Basil Franks

Peter Kennedy Fraser

Sir John Frew

Wilfrid Fletcher Gaisford

Walter Ganado

Edgar Norman Gauld

Raymond Tom Gaunt

Michael Gelfand

Paul Ghalioungui

Alan Morton Gill

Annis Gillie

Frederick Campbell Golding

Frank John Goodwin

Hugh Walker Gordon

John Lee Greaves

Harry Greenburgh

Robert Amos Griffiths

Allan Seaverns Grimble

Hyman Grusin

David Robert Wigston Haddock

George Dickinson Hadley

Pentti Ilmari Halonen

Max Hamilton

Patrick John Sinclair Hamilton

Ronald Harris

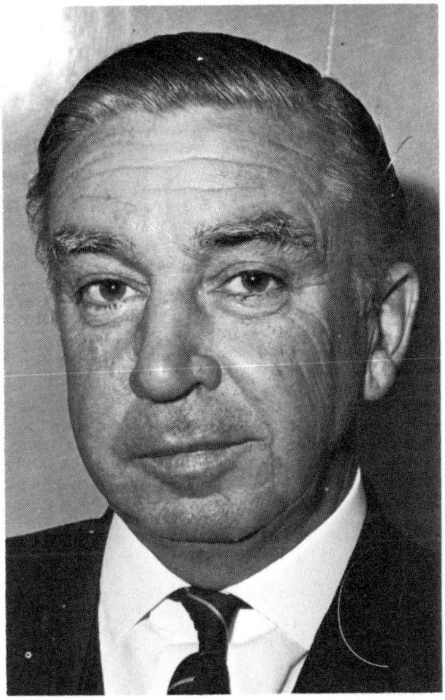

John Norman Harris-Jones

Robert John Harrison

Edward Watson Hart

Mahmoud Ahmed Hassanein

Frank Hawking

Leslie Alan Hawkins

Francis Findlay Hellier

Richard Anderson Hickling

James Maurice Hill

Cecil John Charles Guy Hodson

John Herbert Humphrey

Jack Naylor Hunt

Lord Hunt of Fawley

John Vernon Hurford

James Holmes Hutchison

Solomon Jacobson

Saroop Chand Jain

Rupert Desmond Charles Johnstone

Hugh Reginald Jolly

Francis Edwin Joules

Cyril Arthur Keele

Kenneth David Keele

Renatus Kempthorne

John Duke Olav Kerr

Israel Kessel

John Alexander Kilpatrick

James Balfour Kirk

Peter James Knapton

John Knox

Michael Kremer

Pathiyil Karthiyayani Krishnankutti

Victor John Frederick Lack

Frederick Lees

Humphrey Abraham Leggett

Hermann Lehmann

Philip Levine

Denis Cyril Lewin

Norman Arthur Lewtas

Reginald Cyril Lightwood

Gerrit Arie Lindeboom

James Livingstone Livingstone

Thomas Wigram Lloyd

Vincent Edwards Lloyd-Hart

Solly Lopis

David Lurie

Reginald William Luxton

Thomas Keith Selfe Lyle

Douglas Blair Macaulay

Jack Cameron Macaulay

William McCartan

Dermod de la Chevallerie MacCarthy

Ranald Ian MacDonald

Robert Gwyn MacFarlane

Robert Merttins Bird MacKenna

Thomas McKeown

Noel Francis Maclagan

Sir John MacNee

Hettiya Kontehelage Nihal Ian Markus

James Purdon Martin

Sir William d'Auvergne Maycock

Richard Mayon Mayon-White

Sir Peter Medawar

Roger Aziz Melick

Sir Ashley Miles

Stanley Miles

Peter John William Mills

Louis Minski

Joseph Stanley Mitchell

Minocher Bamanji Mody

Raymond Andrew Moir

Robert Montgomery

John Kempton Morgan

Robert Mowbray

Isaac Muende

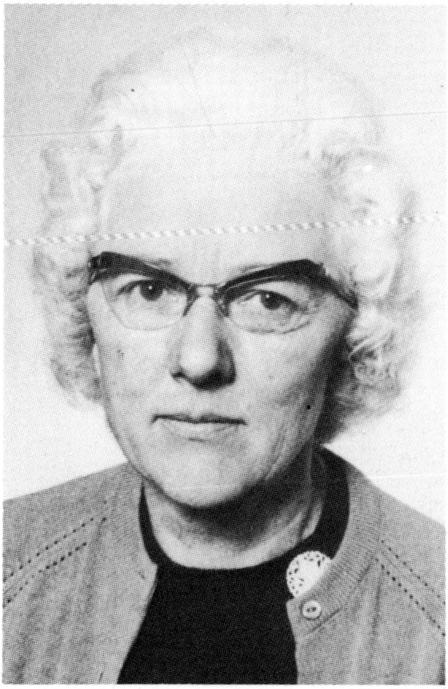

Sheilagh Mary Cecilia Murray

Robert Walsh Mussen

Gordon Norman Myers

George Henry Newns

Claude Scott Nicol

John Washington Orr

Hugh James Orr-Ewing

Richard William Parnell

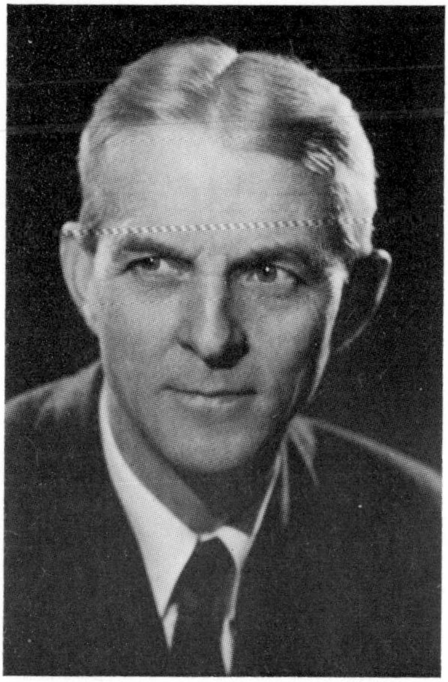

Robert Hughes Parry

Hubert Edward Spencer Pearson

Kenneth Murray Allan Perry

Charles Plowright Petch

Douglas Pickering

Arthur Bairstow Pollard

Cyril John Polson

Sir Desmond Pond

John Stobo Prichard

Sydney Augustine Propert

Brian John Prout

Cedric Barrington Prowse

Jack Sydney Pryor

David Rabin

Robert Russell Race

Peter Knowles Renshaw

Sir John Revans

Edward Gregson Grant Rhind

George Richardson

Michael John Riley

John Alexander Fraser Roberts

John Clifford Roberts

Sandy Edwin John Robertson

Robert Ewing Rodgers

Howard John Rogers

Mark Andrew Rugg-Gunn

Norman Lloyd Rusby

Harold William Salmon

Eric Sanders

Gerard Sanderson

William Walters Sargant

Ramchandra Vishwanath Sathe

Harold Scarborough

Bernard Edward Schlesinger

George Albert Scott

Douglas Ronald Seaton

Sir Thomas Holmes Sellors

Edmund Day Sever

Robert Semple

Robert Andrew Shanks

Harold Leeming Sheehan

Joanna Sheldon

Samuel Shone

Peter Josef Gabriel Simons

Cedric Keith Simpson

John Simpson

Eric Walter Skipper

James Stuart Ian Skottowe

Robert Arthur Smart

Bernard Hooper Smith

Kenneth Shirley Smith

Louis William Smith

Ian Bruce Sneddon

John David Spillane

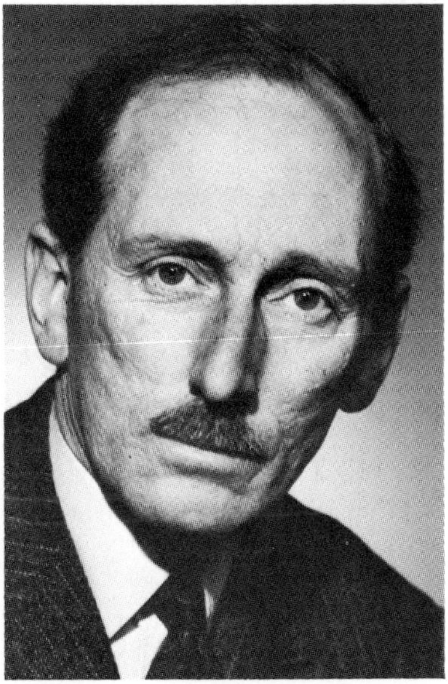

Edmund Anthony Spriggs

Jules Stahl

William Percivale Stamm

Ian McDougall Guthrie Stewart

Roy Summerly

Nanna Charlotta Svartz

Cecil Symons

Paul Szekely

Bashi Lal Taneja

Selwyn Edward Tanner

Peter John Taylor

Lord Taylor of Harlow

Harold Ernest Thomas

Ian Oriel Thorburn

Bezley Thorne-Thorne

Sir Geoffrey Todd

Hugh John Trenchard

Sir Ronald Tunbridge

Ronald Houstoun Vasey

Robert Harold Vines

William Walker

Hugh John Wallace

Reginald Lawson Waterfield

Robert Ford Whelan

Allan George Williams Whitfield

Claude Howard Whittle

Khushwant Lal Wig

Eric Gordon Wilkins

Richard Robert Willcox

Cyril James Williams

David Aelwyn Williams

Norman Bruce Williamson

Brian Douglas Raymond Wilson

Sir Graham Wilson

Dame Albertine Winner

Sir Ian Wood

Eluned Woodford-Williams

David Henry Morgan Woollam

Carmichael Aretas Young

radioactive iodine by the analysis of sheep thyroid glands. He dispatched his collections every two or three weeks for over 29 years - a total of more than 11,000 glands. This work helped determine relationships between the size of atomic explosions, their latitude, their dilution and the time lag in the passage of radioactive fallout between the northern and southern hemispheres. The data he collected were instrumental in ensuring that tests in the southern hemisphere were at least conducted underground. Roger continued to collect thyroid glands during his last illness, and some of these glands showed a small but significant increment of radioactivity in the southern hemisphere following the Chernobyl disaster.

When Roger took on the responsibility for students, as the Associate Dean (clinical) at the Royal Melbourne Hospital, he did so energetically and with great devotion to the welfare of the students. He became well known and well loved by a generation of students from that hospital. He had a very strong sense of fair play and his position as dean enabled him to exercise this to defend the cause of a student when he believed that some injustice was being done. He was indeed a defender of causes, and could be extremely cantankerous and difficult when confronted with what he believed to be pomposity or bureaucracy, and he took no heed of either rank or popular opinion in making his particular form of protest. He served for some years on the council of the Royal Australasian College of Physicians and took on many College sacred cows.

Roger Melick's unobtrusive, but unreserved service to the hospital, to the University of Melbourne, and to his profession will probably never be catalogued adequately. His immense dignity and cheerfulness during the long months facing his certain death were an object lesson to all. He was more concerned for his family, and for those who came to visit him, lest they be upset by his illness. He was survived by his wife Mardi, whom he had married in 1960, and their daughter and two sons.

BH

[Photo]

MILES, ***Sir*** **Arnold Ashley** *b.*20 Mar 1904 *d.*11 Feb 1988
Kt(1966) CBE(1953) MRCS LRCP(1928) MRCP(1929)
FRCP(1937) MA MD Cantab(1952) FRCPath(1969) FRS(1961)

From the outset of his career, Ashley Miles' rapid progression to the summit of medical microbiology was marked by a series of milestones indicative of his outstanding academic abilities. A Yorkshireman, his education began at Bootham School, whence he won an exhibition in natural sciences to King's College, Cambridge. During his premedical studies there his bent towards microbiology and pathology was reinforced by the influence of E G D Murray and H R

Dean [*Munk's Roll*,Vol.V,p.98]. He qualified in medicine at St Bartholomew's Hospital, and in the following year achieved the remarkable feat of gaining membership of the College while still a house physician. He once remarked, airily but somewhat disingenuously, that he did so because he must get the right trade union stamp on his bottom before he could talk to the clinicians on equal terms.

In 1929, Miles became a demonstrator at the London School of Hygiene and Tropical Medicine, and it seems clear that the influence of those two great bacteriologists, W W C Topley [*Munk's Roll* ,Vol.IV,p.561] and G S Wilson (q.v.),later Sir Graham, finally decided the course of his life's work. At the School of Hygiene he undertook his first researches, on the antigens of *Brucella,* and continued when, in 1931, he returned to Cambridge as university demonstrator in pathology. There followed a readership at Hammersmith postgraduate medical school and then, at the early age of 33, the chair of bacteriology at University College Hospital medical school, London. Soon afterwards, however, the outbreak of war brought many additional responsibilities and provided the first example of Miles' prodigious capacity for work and of his ability to tackle several jobs at once, not just adequately but supremely well. In addition to his teaching and research, in temporary accommodation in Hertfordshire, he was in charge of the pathological services in a sector of the Emergency Medical Service, acting director of the Graham Medical Research Laboratories in London and, in retrospect most important of all, director of the MRC wound infection unit at Birmingham. There he became concerned with hospital infections of war and industrial wounds by streptococci and staphylococci: his studies on the ways in which they are transmitted resulted in practical and effective recommendations for their control. These researches had a further outcome, less apparent at the time but of great future significance, for they sowed the seed of Miles' interest in the mechanisms of inflammation that was to dominate his laboratory work for the rest of his life.

The war over, Ashley Miles moved to the National Institute for Medical Research, where he was both deputy director and head of the department of biological standards. Although much of his time was taken up with this rapidly developing field, he continued his studies of infection, in some of which his wife Ellen collaborated. These investigations related both to the toxins of anaerobic pathogens and to the use of experimental skin lesions for studying inflammatory responses, a technique that was to be cleverly exploited in later years.

In 1952 Miles became director of the Lister Institute of Preventive Medicine, an independent, privately funded organization that housed a variety of somewhat disparate activities, ranging from esoteric researches in microbiology, biochemistry and biophysics, to the manufacture and sale of vaccines, antisera and blood products. The Lister already enjoyed a high reputation for biomedical research, and the unusual mixture of people and laboratories that Miles inherited seemed perfectly to suit his own talents and personality. Here he was to remain for the next 20 years, and during this period his ability for intensive work at the highest level was again tested to the full. In his own department of experimental pathology he both directed and personally undertook investigations into the early tissue responses to injury, particularly those that increase capillary permeability, and established that the outcome of many bacterial infections is largely determined by the efficacy of the host's defences during the first few hours after the initial invasion - what he termed 'the decisive period'. These researches alone would have kept the average research worker fully occupied, but Miles was far from average. His genuine interest in the scientific activities of others took him from time to time into every other department, not so much in a directorial capacity but as a colleague whose clear thinking and quick insight into problems often far removed from his own field earned him the respect of his staff. Inevitably, these qualities made him much in demand on many boards and committees, including those of the World Health Organization, the Medical Research Council, the British Pharmacopoeia Commission, and the Public Health Laboratory Service. Elected a fellow of the Royal Society in 1961, he served for five years both as a vice-president and in the demanding position of biological secretary. Despite all these preoccupations, his literary output was, to say the least, impressive. He was a superb writer of clear, concise English that always bore his individual stamp, however impersonal the subject. As well as publishing some 140 papers on his own work, he was a general editor of no fewer than five editions of Topley and Wilson's *Principles of Bacteriology and Immunity*, London, E Arnold & Co, from 1929 onwards, and to ensure that his own scientific and literary standards were met he added to his seemingly impossible workload by personally vetting every paper published from the Institute.

Ashley Miles' contributions to biomedical sciences, both in the United Kingdom and overseas, brought him many distinctions, including the CBE and a knighthood. He was elected to honorary fellowship of many distinguished academic bodies; the one that perhaps gave him most pleasure was that conferred by his former college at Cambridge, King's.

After his official 'retirement' in 1972, he spent the next four years at the bench as a MRC grant-holder at the Clinical Research Centre at Harrow, and then, *mirabile dictu*, continued to work as deputy director of the department of medical microbiology at the London Hospital medical college where, despite a severely disabling stroke, he carried on with research and teaching until a few months before his death.

The best scientific experiments are often marked by an elegant simplicity, and this was certainly a feature of all Miles' laboratory work; it was particularly well illustrated by one of his earliest papers, in which he and S S Misra [*Munk's Roll*,Vol.VI,p.340] described the method of counting viable bacteria that still bears their names. Like many of his generation, he was rather wary of high technology, and indeed once confessed to feeling that he had 'lost his amateur status' when obliged to use a radioisotope in an experiment.

Especially in his later years, he was held in some awe by junior staff and his postgraduate students; he was intolerant of stupidity and implacably opposed to anyone who contravened his own standards of personal and scientific integrity. To those unacquainted with him, these characteristics might imply a stern and unbending personality, but nothing could be further from the truth, for underlying them was a deep kindliness that, in fact, left little scope for the ruthless pruning of dead wood that is often necessary in any organization.

His engaging sense of humour and wide knowledge of music, literature, and the countryside, made him a delightful companion and conversationalist. It is sad that his last years were marred by illness, during which however he was greatly supported by the devotion of his wife, who died just a month before him. There were no children of the marriage.

LHC

[*Lancet*,1988,**1**,425; *The Times*,12 Feb 1988; *Independent*,19 Feb 1988; *MRC News*,No.**39**,June 1988; *Bull.roy.Coll.Pathologists*, No.**63**,June 1988; No **50**,April 1985; Audio Tape; Photo]

MILES, ***Surg.Rear-Admiral*** **Stanley** *b.*14 Aug 1911 *d.*9 July 1987
CB(1968) MSc Sheff(1934) MB ChB(1936) DTM&H(1949) MD(1955) MRCP(1970) FRCP(1971) FRCS(1971)

Stanley Miles was born in Sheffield, the son of T C Miles, a company director. He was educated at King Edward VII School and Sheffield University, graduating in medicine in 1936. In the same year he joined the Royal Naval Medical Service and served in China, West Africa, and with the Pacific and Mediterranean fleets.

In 1952, as a surgeon lieutenant commander, he was appointed to the chemical defence establishment, Porton Down, as naval medical liaison officer and was attached to Bernard Lucas' section, working on protection against nerve gas. Their particular interest was artifical ventilation and, following a series of experiments on monkeys, they turned their attention to mouth-to-mouth resuscitation, which soon superseded the traditional methods of ventilation (Schafer, Silvester) recommended in manuals of first aid, and awoke in Miles an abiding interest in physiology.

In 1956 he was appointed to the Royal Naval Medical School for duty at the RN physiological laboratory as specialist in physiology, where he investigated a number of problems connected with submarine operations and naval diving, in particular shallow water blackout - sudden loss of consciousness in assault swimmers - hitherto attributed to carbon dioxide poisoning. Miles was convinced that the root cause of the problem was cerebral vasoconstriction resulting from breathing pure oxygen and persuaded the Royal Air Force Institute of Aviation Medicine to allow him to expose naval divers to G-forces on the human centrifuge to the point of unconsciousness, breathing either oxygen or air. He succeeded in demonstrating that oxygen-breathing reduced G-tolerance and concluded that his theory was confirmed.

In 1957 he was awarded the Gilbert Blane medal of the Royal College of Surgeons for his contribution to naval medical research and was made a consultant in physiology. In 1961, as surgeon captain, he was appointed medical officer in charge RN Medical School and director of medical research. His book, *Underwater medicine*, Philadelphia, Lippincott, 1962, became the standard textbook for doctors associated with sport diving.

In 1966 he was promoted to surgeon rear-admiral and appointed medical officer in charge of the Royal Naval Hospital, Plymouth. Rather to his surprise he found that he enjoyed this appointment more than any he had held previously; the hospital seemed to run itself and he found it very relaxing. Out of working hours he liked to slip into the pub just outside the gate, where he could chat informally with naval ratings over a pint of beer. He regarded this as a pastoral duty and it gave him a useful insight into current thinking on the lower deck, but it came to the notice of the commander in chief, who considered such informality quite inappropriate for a flag officer. In retrospect, Miles felt that this had mitigated against his further promotion.

After retiring from the Navy in 1969, he became postgraduate dean at Manchester University, at a time when such appointments were beginning to be made nationwide. He showed himself adept at building bridges between the central teaching hospitals and the district hospitals throughout the large northwest region, a task which involved a great deal of travel, when Barrow-in-Furness was still within the region and the journey took longer than one to London. The close liaison which he helped to forge was roughly coincident with the disappearance of the Board of Governors, and it was to prove of great value when the district hospitals were called upon to play a major part in undergraduate as well as postgraduate teaching.

In 1976 he decided to exploit his experience in diving medicine by setting up Gaelic Health Guard at Aberdeen. Sponsored by Gulf Oil, it was the first medical centre in Scotland for divers working in the North Sea oilfields. Two years later, however, the premises were taken over by Offshore Medical Services, sponsored by Shell and British Petroleum.

In 1969 Miles had joined the Medical Commission on Accident Prevention and later he became chairman. In 1982 he became vice-president of the International Trauma Foundation. He retained both these interests until his death.

Miles was married in 1969 to Mary Rose, and they had a son and a daughter. He was a man of great charm, with an acute and fertile intellect and a ready wit. As a research worker he was guided more by inspiration and intuition than by scientific discipline. When he believed he had the solution to a problem he would set about devising an experimental programme to demonstrate its validity. He was not much concerned with statistics, and those subjects or results which failed to conform to the general pattern he dismissed as anomalies. This irritated his more conventional colleagues, but Miles was concerned with finding practical answers to the problems which beset divers, submariners and others who operated in hostile environments, and they generally proved to be right.

Miles will be recalled with affection by those who were privileged to hear him speak at conferences, where his enthusiasm was infectious and his communication effective. He possessed a robust sense of humour which delighted his audience at the annual Divers' Dinner; no one who heard his hilarious speech consisting entirely of quotations from the Royal Naval Diving Manual is ever likely to forget it.

He was essentially a caring doctor, who loved life and his fellow men. The diving fraternity were his chosen parishioners - and they will remember him.

JR/DAKB

[*The Times*,17 July 1987; *Lancet*,1987,**2**,347; Photo]

MILLS, Peter John William *b.*15 June 1913 *d.*18 Aug 1984
MRCS LRCP(1936) DA Eng(1940) MRCP(1943) FRCP(1967)

Peter Mills was the son of William Henry Mills, who was a surveyor and a partner in the firm of George Trollope & Sons, London. His mother, Alice Jane Clarke, was the daughter of a civil servant in the India Office. Peter was born in London and educated at Dulwich College Preparatory School, and subsequently at Emmanuel School and by a private tutor. His medical training was undertaken at St Bartholomew's Hospital whence he qualified in 1936. After qualifying he worked at Bart's, the Royal Northern Hospital, Redhill County Hospital, and then the Borough General Hospital (now Southampton General Hospital) from 1938 onwards. It was there that he met staff nurse Bronwen Margaret Harris, whom he married in 1945.

During the war Peter Mills was a flight lieutenant in the RAFVR, later becoming chief assistant to R V Christie (q.v.) at Bart's, and then resident assistant physician at the Hammersmith Hospital. In 1945 he was appointed consultant physician to the Lister Hospital at Hitchin, and became visiting consultant physician at St Alban's and honorary physician at the Mildmay Mission Hospital in London in 1947. In 1964 he gave up his post at St Albans to become consultant physician to the Queen Elizabeth II Hospital, Welwyn Garden City. For some years he served on the merit award committee.

Peter Mill's particular interests in medicine were thyroid and heart disorders, which were well represented in his publications. 'Uncle Peter' was always enthusiastic about medical education and took a lot of time and trouble teaching generations of medical students and house physicians. His teaching interest and clinical acumen were reflected in his book *The Significance of physical signs in medicine,* London, Lewis, 1971. In 1967 he was appointed associate teacher in medicine at St Mary's Hospital Medical School in recognition of his teaching of students at the Lister Hospital. He was also interested in the teaching of nurses and helped in the formation of the Lister Hospital School of Nursing, where he continued to teach until shortly before he died. For several years he was medical superintendent at the Lister Hospital in Hitchin and was deeply involved in the planning of the new Lister Hospital at Stevenage, where he was senior physician from the time of its opening in 1972 until his retirement in 1979.

On his ward rounds nothing was too much trouble. He took detailed histories, followed by extremely careful examinations, and in the days before modern investigations were available he had an uncanny accuracy in diagnosis. He was always a raconteur and his various anecdotes frequently enlivened more serious matters. He showed unfailing courtesy to staff and patients, and was always considerate to all around him.

His main outside interests were golf and sailing, both of which he enjoyed until his death, but his primary interest was medicine and in retirement he did frequent hospital locums and also worked in general practice for many months. He would have hated to give up medicine completely. By a happy accident, in the last week of his life he had the pleasure of doing a consultant locum, and he was actually away on a weekend of sailing at the time of his death.

SMW

[*Brit,med,J,*1984,**289**,1011-12,1235; Photo]

MINSKI, Louis *b.*6 June 1902 *d.*11 Sept 1986
MB BS Durh(1923) DPM(1926) MD(1927) MRCP(1929)
FRCP(1943) FRCPsych(1972)

Louis Minski, one of the outstanding psychiatrists of his generation, died at Cheam, Surrey, aged 84. He was born in Sunderland, the son of Henry Minski, an estate agent, and his wife Anne Jacobs.

Minski was educated at St Kitts School and Argyle House School, Sunderland. He studied medicine at Durham University Medical School, where he had a distinguished undergraduate career, being awarded the Pears scholarship in arts, the Gibb scholarship in pathology and the Dickinson scholarship and gold medal in medicine, surgery and midwifery. He continued his high level of academic achievement as a postgraduate by winning the prestigious Gaskell prize and gold medal of the Royal Medico-Psychological Association in 1933 (which became the Royal College of Psychiatrists in 1971, and of which Louis Minski was a foundation fellow).

He had decided shortly after qualifying to pursue a career in psychiatry and to this end he joined the staff of Bootham Park Hospital, York, where he served from 1923-29.In 1929 Minski was appointed second assistant medical officer at the Maudsley Hospital, London, where he came under the tutelage of Edward Mapother [*Munk's Roll*,Vol.V,p.266]. It was about this time that he became interested in mental symptoms associated with organic brain disease and,together with Erich Guttman, he undertook one of the first surveys of families suffering from Huntington's chorea.

In 1938 he was appointed deputy medical superintendent at St Ebba's Hospital, Epsom, a position he held only briefly. The second world war seemed inevitable and, in anticipation of the host of psychiatric casualties which were bound to arise, particularly in the London area, it was imperative that preparation for their reception and treatment should be made. Minski was called upon to set the wheels in motion. He took over the Sutton Training Centre and converted it to the Sutton Emergency Hospital (later to become Belmont Hospital) which functioned under the Emergency Medical Service (EMS). In 1939 Minski was appointed physician superintendent, a position he held until 1948.

The challenge that the running of this hospital presented was awesome but Minski rose to it in heroic fashion and, in the exercise of his duties, his capacity for organization, administration and improvisation was given full rein. Included among the casualties admitted to his hospital, both military and civilian, were evacuees from Dunkirk and survivors of the London 'blitz'. Minski was assisted from a clinical standpoint by a team of high-flyers such as, for example, William Sargant (q.v.), Maxwell Jones, and Eliot Slater [*Munk's Roll*, Vol.VII, p.541]. After the war Minski's talents were widely acknowledged, as witness his appointment in 1946 to the consultant staff of St George's. the Royal National Throat, Nose and Ear Hospital, and the Sutton and Cheam hospitals.

It was at the Royal National Throat, Nose and Ear Hospital that Minski was to produce some of his most important work. He pin-pointed the need for small, but highly staffed and home-like, residential units in order to diagnose the cause of non-communication in children; with the aid of the Regional Board and the Nuffield Provincial Hospital such a unit came into being. Subsequently it was discovered that the cause of non-communication was far more complex than had been imagined hitherto. Apart from subnormality of intelligence, other conditions such as autism, psychosis, brain damage and emotional disturbance could be implicated. In 1970, with Margaret J Shepperd as co-author, Minski published *Non-communicating*

Children, New York, Appleton-Century-Croft, c.1970, which received loud acclaim and was translated into Spanish. He had seen the need for an authoritative textbook of psychiatry when he published his *Practical Handbook of psychiatry for students and nurses*, London, Heinemann, 1946, which was an immediate success and ran into six editions. In 1978 a seventh edition entitled *Minski's handbook of psychiatry for students and nurses*, by R G Priest and G Wolfson, London, Heinemann Medical, 1978, was published.

Minski was a keen and active sportsman until arthritis began to take its toll. In his day he played rugby and represented Yorkshire Gentlemen at cricket. He was a devoted family man and never really recovered from the death in 1973 of his wife Violet, whom he had married in 1932. He was survived by their only daughter, Margaret. HRR

[*Brit.med.J.*,1986,**293**,1446-47; *The Times*,6 Nov 1986; Photo]

MITCHELL, Joseph Stanley *b.*22 July 1909 *d.*22 Feb 1987
CBE(1951) BA Cantab(1931) MB BChir(1934) MA(1935)
PhD(1937) DMR(RCS 1943) FRS(1952) FFR(1954) MRCP(1956)
MD(1957) FRCP(1958) FRCR(1975) Hon DSc Birm(1958)

Joseph Mitchell was born in Birmingham, where he spent his early years. The son of a schoolmaster, he was educated at Yardley Road Council School and King Edward's High School, Birmingham. Although he started his medical education at the University of Birmingham medical school in 1926, he won a scholarship to St John's College, Cambridge, in 1928 and pursued his studies there. He was awarded first class honours in Part 2 of the natural sciences tripos in physics. Lord Rutherford [*Munk's Roll*,Vol.V,p.362] advised him to return to Birmingham to complete his clinical training, which he did.

He became house physician to K D Wilkinson [*Munk's Roll*,Vol.V,p.548] at the hospital where he had trained. His success in physics, and his great interest in the physics of radiation, led to his appointment to a Beit Memorial fellowship which he held at the Colloid Science Laboratory, Cambridge, under Sir Eric Rideal FRS. Joseph Mitchell was elected to a fellowship of St John's College, and under Sir Eric he investigated the effects of irradiation on protein mono-layers, obtaining his PhD.

Mitchell's flair for physics was already determining his subsequent clinical career. He was appointed radiological officer at the Christie Hospital and the Holt Radium Institute in Manchester in 1937. A year later, he pursued his interest in radiotherapy as assistant in research to the regius professor of physic, J A Ryle, [*Munk's Roll*, Vol.IV,p.595] in the new department of medicine at Cambridge.

With the outbreak of war, he was installed as radiotherapist to the Emergency Medical Service in Cambridge. In 1943 he was sent to

Montreal in charge of medical investigations at the National Research Council Laboratories where the British and Canadian Atomic Energy project was sited. He later described his horror when he found that no one was taking care of the medical protection of the workers. He decided that the quickest way to make the laboratory safe was to order that a foot of concrete be laid over the entire laboratory floor. He stayed there until the end of the war, not only ensuring that those who worked with uranium took proper precautions but also studying the biological effects of radiation, including the action of neutrons.

It was at that time that he realized the value of the gamma-emitting isotope cobalt-60 in the possible treatment of cancer. His recommendation led to its worldwide use. After the war he returned to Cambridge where he was elected to the new chair of radiotherapeutics, which he held with distinction from 1946 until he was invited to be regius professor of physic in 1957. It was during that time that he became internationally known for his work on radiotherapeutics and the treatment of cancer. In 1952 the Royal Society elected him to their fellowship.

Mitchell was director of the radiotherapeutics centre of Addenbrooke's Hospital, Cambridge, and concentrated on clinically orientated research. His determination to find more effective ways of treating patients with cancer was the stimulus to his exploring new lines of approach including the high energy x-rays generated by the synchrotron which, as a result of his pioneering spirit, he had had installed in his department. For the last 20 years of his life he concentrated on the idea of finding molecules which would specifically seek out cancer cells. He then endowed the molecules with radioactivity to destroy the cells which took them up. He stuck rigidly to the concept and developed the radioactive analogue of vitamin K, Synkavit, but it must be said that the number of successes he had - though sometimes dramatic - fell far short of his hopes.

At the time when Joseph Mitchell was appointed regius professor of physic in 1957 he immediately took up the advice of the Goodenough Report that Cambridge should build up a postgraduate school of medicine so as to establish the academic basis for a subsequent clinical school. It was unfortunate that the precise number of six chairs was quoted as this was subsequently used to limit the growth of the postgraduate school. Although, in 1963, he was a signatory to the university report that said they did not envisage Cambridge having a clinical school to add to its renowned preclinical school, he did give energetic support once the new professor of medicine (the establishment of whose chair he had organized) launched an attempt in 1965 to get the idea of such a school accepted.

Joseph Mitchell was essentially a kind and gentle man, but he was aggressive in his dedication to treat his patients. His compassion sometimes led him to tears when he found that his treatment failed to cure their cancers. He showed the same concern and kindness to all his staff, and it was this quality which made it difficult for him to chair with an authoritative approach the board of the faculty of medicine when he was regius professor of physic. As a result, his colleagues voted for an outside chairman, much to Mitchell's dismay. He was treated similarly when the university set up the clinical school planning committee without including him on it, at a time when the Todd commission recommended that Cambridge should have a clinical school. It took the combined efforts of the three clinical professors to force the university to install all three of them on it. Nevertheless, in 1974 Joe made the enormously generous gesture of offering to resign from the regius chair so that a new regius professor of physic could be in the post in 1975, a year before the clinical school was to have its first intake of students. He then reverted to the professorship of radiotherapeutics until he retired in 1976. After retirement, true to his real role as a scientist, he continued his research and trained PhD students until a short time before his death.

In 1934 Joseph Mitchell had married Lilian Mary Buxton MB ChB, who was some eight years his senior. They had two children, a boy and a girl. Mary Mitchell devoted her life to supporting her husband at work as well as at home. In addition, she did great work on her own account as medical adviser to the Cambridge Association for Social Welfare. She inherited her father's outfitting business, and Joseph helped her direct it.

In 1958 Joseph Mitchell was both Dunham lecturer at Harvard, USA, and Withering lecturer at Birmingham. He was awarded the Pirogoff medal of the USSR Academy of Medical Sciences in 1967, and was Sylvanus Thompson lecturer of the British Institute of Radiology in 1968. In 1970 he was appointed honorary consultant to the Atomic Energy Authority, and he was appointed Linacre lecturer at St John's College, Cambridge, later that same year. The latter led to the publication of his book *Cancer: if curable, why not cured?*, Cambridge (Eng), Heffer, 1971. His previous book was *Studies in radiotherapeutics*, Oxford, Blackwell Scientific Publications, 1960. In 1965 he edited *Treatment of cancer with special reference to radiotherapy and chemotherapy*, Cambridge (Eng), Cambridge University Press.

Mitchell was an honorary member of the German Roentgen Society and a foreign fellow of the Indian National Academy of Sciences. He was a founder member of the English section of the Anglo-German Medical Society and president from 1959-68. He spoke and wrote fluently in German, which was no doubt related to the fact that he had a German grandfather. Each year he spent some time in laboratories in Freiburg, and while there he and his wife often took the opportunity to pursue his main hobby of walking.

It may be surprising to some that Joseph Mitchell took an active part in local politics,and for a time was president of the South-east Cambridgeshire Liberal Association.

Mary Mitchell died in 1983, but his two children survived him. He left behind an enormous army of appreciative patients, not only because of his dedicated clinical, social and psychological care but also for the improvements in the treatment of cancer to which he had devoted his life.

IHM

[*The Times*,24 Feb 1987; *Lancet*,1987,1,639; *Biog.Mems.roy.Soc.*,**34**,583-607; Contemporary Scientific Archives Centre,*Progress Report* No.**25**,1985-86; Photo]

MODY, Minocher Bamanji *b.*16 Nov 1900 *d.*7 Sept 1987
MB BS Bombay DTM MD MRCP(1929) FRCP(1983)

Minocher Mody was born into a middle class Parsee family in Bilimora, Gujarat, India. His father, Bamanji Mody, was a ship's doctor. When Minocher was 10 years old the family moved to Bombay, where he went first to the Esplanade High School and then to a Jesuit College, St Xavier's, for premedical studies. He went on to the Grant Medical College, pursuing his clinical studies at the J J Group of Hospitals, Bombay.

After graduation he came to London, on borrowed money, and dedicated himself to further study, obtaining his London MB BS within a year of arrival. He then spent some time at the Liverpool Tropical School, where he obtained his DTM, and his membership of the College while studying for his doctorate.

On his return to Bombay he was appointed honorary assistant physician at the G T Hospital and some seven years later transferred to the parent institution, the Sir J J Group of Hospitals, where he served until his retirement. He continued to serve the Grant Medical College for many years as honorary physician, teacher and examiner, and was also attached to three private hospitals in an honorary capacity. He had a rapidly growing private practice in Bombay and became known as one of Bombay's leading physicians; a compassionate and caring physician, for no patient was ever turned away for any reason and many were treated free of charge. Every month he took weekend trips out of Bombay to visit a tuberculosis sanatorium and an old people's home, where he attended the poor. In an age of declining moral values and increasing avarice, he was a doctor who truly represented the Hippocratic tradition. He was capable of a prodigious amount of work which regularly covered a 12-14 hour day.

Mody was a voracious reader, with an acutely receptive mind and retentive memory, which enabled him to keep pace with the changing modes of medicine. A great deal of his reading was done while he was being driven to hospital, or to the homes of patients, and he was equally at home with all branches of medicine, and able to discourse knowledgeably on most subjects.

As a teacher, he attracted students from all over the College. He was an astute clinician in an environment where special investigations were seldom available or, at best, unreliable. Generations of students trained by him have distinguished themselves, not only in India but throughout the world.

In 1932 he married a fellow medical student, Piroja, daughter of Nusservanji N Mody, a merchant, and they had three children; a son who followed in his footsteps to become a consultant physician, and two daughters - one of whom is a psychiatrist. His wife both supported and maintained his home, since he had little regard for money and financial records. As well as running the house, she also ran a GP clinic.

Minocher Mody always had a great love of cricket, a game he played with skill, being in his college team and, later, always ready to participate in staff/student matches at J J Hospital.

He had a long and productive life, well fulfilled.

GW/VL

[Photo]

MOIR, Raymond Andrew *b.*19 Apr 1919 *d.*19 Jan 1984
MRCS LRCP(1941) MB BS Lond(1941) MRCP(1942) MD(1951) FRCP(1970)

Raymond Moir was born at Chatham, Kent, the son of Andrew William Moir, civil servant, and his wife Louisa Irene, daughter of Roger Muncey, a farmer. He received his early education at Sir Joseph Williamson's Mathematical School, Rochester, where he distinguished himself by his academic brilliance and his considerable ability as a wicket keeper of the school cricket eleven. From there he proceeded to King's College Hospital, London, obtaining the MRCS LRCP and MB BS in 1941. He became a member of the College in 1942 and a Fellow in 1970.

At King's he held various house appointments, was appointed RMO, and then there followed a brief period as assistant medical registrar before he joined the Army in August 1942, where he served as a medical specialist. As consultant physician to Western Command he rose to the rank of lieutenant colonel. On demobilization in 1947 he was appointed assistant physician to St Bartholomew's Hospital, Rochester, and on the inception of the NHS he became senior hospital medical officer to the Medway and Gravesend Group of hospitals. His appointment was regraded to consultant physician in 1952. He served as a member of the old hospital management committee, and later as chairman of the medical executive committee. His principal interest was rheumatology.

He first married when very young and the marriage was subsequently dissolved. In 1948 he married Dora Winifred, daughter of Frederick Henry Kirkton, a master printer, and they had two sons and a daughter. Raymond was a man of wide interests but very few intimate friends. He was very widely read and keenly interested in classical music. He was also a keen Freemason, being first master and then a master of ceremony of the Pentangle Lodge, Rochester, where he was highly esteemed. His memory was prodigious, and according to local practitioners he could remember details of individual patients long after he had last seen them.

Towards the end of his life his health failed and he entered Guy's Hospital cardiac unit early in 1983. Shortly after his return home he had to be admitted to the intensive care unit at the Medway Hospital following a haematemesis which proved to be due to oesophageal varices. He died in King's College Hospital the following January. He was survived by his wife and family.

JDC/GW/VI

[Photo]

MONTGOMERY, ***Brigadier*** **Robert** *b.*21 Sept 1916 *d.*28 Feb 1987
MB ChB Glasg(1939) MRCP(1947) FRCP(1964) FRCPE(1967)

Robert Montgomery was born in Neilston, Renfrewshire, where his father was a school teacher. His mother Edith Winifred was the daughter of John Thomas Wood, a foreman wood turner. 'Monty', as he was widely known, was educated at Glasgow High School and Glasgow University. His clinical studies were undertaken at the Royal Infirmary Glasgow, where he was house physician for some six months before joining the RAMC in 1940.

During the war Monty served in Iceland and later in northwest Europe - in field ambulances, field dressing stations, and general hospitals. In 1946 he attended a senior officers course at the Royal Army Medical College, Milbank, and was graded as a specialist physician in 1947. He then saw service as officer in charge of the medical division in military hospitals in Gibraltar, Kenya, Malaysia, the UK and Germany. He was mentioned in despatches in 1955 while serving in Nairobi. His last post before retirement was as consultant physician, in the rank of brigadier, at HQ British Army of the Rhine.

Monty married Mary Meiklem (always known as Mollie) in 194 and they had a son and three daughters. He was a devoted husband father and grandfather. Together they formed a team whose friendship and hospitality were shared by many in the Army.

His main interests, outside the Army and medicine, were gardening and golf. His enthusiasm for golf was shared by his wife, and Monty was never happier than when taking part in a four ball at the golf course at RAF Bruggen. His call 'I'll see it in' put many an opponent off his stroke - all in good fun. While in Malaysia he made full use of the golf course at Kuala Lumpur, where he played with his friend Tom Fowler.

Monty was a wise and caring physician, with a wide knowledge of

medicine in both tropical and temperate regions.
His wife, children and grandchildren survived him.

WSM/TF

[*Brit.med.J.*,1987,**294**,1238; Photo]

MORGAN, John Kempton *b.*13 July 1920 *d.*12 Nov 1988
MRCS LRCP(1944) MB BS Lond(1944) MRCP(1949)
FRCP(1971)

John Morgan was born in Sydenham, London. His father, Charles Henry Morgan, was an accountant whose wife Lucie Raeburn Ellis was the daughter of an actor, George Ellis.

John was educated at Raby Lodge and Whitgift School, and went on to Guy's Hospital medical school, University of London. After graduation, he joined the Royal Army Medical Corps and served until December 1947. On demobilization, he held house posts as outpatient officer at Guy's and then as house physician. He developed an interest in dermatology and in 1950 he was appointed registrar in the skin department at Leeds General Infirmary, and subsequently tutor and senior registrar. From 1952-56 he was senior registrar at Manchester and Salford Hospital for Skin Diseases, and at the end of his final year he was appointed consultant dermatologist to South Teeside Hospitals, taking administrative charge of the team in 1969. He was also clinical tutor in postgraduate medicine, South Teeside, from 1968.

He married Elizaberth Verity, daughter of Ralph Franklin Eminson, a doctor, in 1947, and they had three children. Outside medicine and his family, his interests lay in bird watching, boating, sea fishing and painting. From 1964 he was a member of the Runswick and Staithes branch of the Royal National Lifeboat Institution, and chairman 1966-69.

GW/VL

[Photo]

MOWBRAY, Robert *b.*23 May 1914 *d.*20 May 1986
MB BS Durh(1936) MD(1938) DCH(1938) MRCP(1939)
FRCP(1960)

Robert Mowbray was born in Newcastle upon Tyne. His father, also Robert Mowbray, was of independent means and had his son educated at the Royal Grammar School. Robert had a distinguished career as a student in the medical school and Royal Victoria Infirmary, Newcastle, graduating with first class honours from the University of Durham, at that time the parent university. After resident appointments at his teaching hospital he went, again as a resident, to work at Paddington Green Children's Hospital, London, under R H Miller [*Munk's Roll*,

Vol.IV,553] who had a special interest in rheumatic heart disease. The cardiological interest thus inspired remained with Robert all his life and later prompted Sir William Hume [*Munk's Roll*,Vol.V,p.206] to invite him to become the first secretary of the newly formed Newcastle Cardiac Club.

Robert returned to Newcastle as medical registrar, but the second world war soon supervened and as a reservist he was called up on the first of September 1939. Soon promoted to the rank of major, RAMC, he served as a medical specialist in France, the UK and the Middle East. He was in charge of a casualty clearing station in Italy and Austria, and mentioned in despatches. His interest in things military continued into peace time and he was, successively, lieutenant colonel in charge of the medical division and colonel in command of No 1 Northern General Hospital, TA. On demobilization he had returned to his post as medical registrar and from 1947-50 he worked under F J Nattress [*Munk's Roll*,Vol.VII,p.421] as first assistant and lecturer in medicine. While in this post he published some papers on varied medical topics and at the same time developed his interest in books, and in the history of medicine.

In 1950 Robert was appointed senior physician to the Durham City group of hospitals, based principally on Dryburn Hospital, where he developed an excellent medical unit with a prominent cardiological component. The medical students seconded to him from Newcastle found him a gifted teacher, and the various maladies in his wards a refreshing contrast to the over-specialization which was creeping into some of the wards of the teaching hospitals in Newcastle. In his early years he also did private practice but, like several of his contemporaries in the region, paid the penalty of providing a prompt, efficient and kindly hospital service by finding private practice not worth while in terms either of time or money. However, his reputation as a physician spread outside Durham. In 1956 he was given the freedom of the Worshipful Society of Apothecaries, and for several years he served on the panel of examiners for the membership examination of the Royal College of Physicians.

Within the Durham group of hospitals his influence extended beyond his own department into the development of Dryburn as a self-sufficient district general hospital, with its own department of medical physics and a fine postgraduate centre and library. It was not surprising therefore that in 1964 he was invited to become a member of the Regional Health Authority, of which he soon became vice-chairman and, most importantly, chairman of its medical advisory committee. His service with the RHA continued beyond his retirement from clinical work in 1979, until 1982. He was also a member of a DHSS working party on hospital development.

After his retirement, Robert left his lovely house in Durham, with its incomparable view across the city to the cathedral, and settled in the village of Little Salkeld, near Penrith in Cumbria, where he and his wife tended a very beautiful garden and where, when not fishing in the River Eden, he could browse among his books in a spacious library. These books, by now a large collection of early editions, both medical and non-medical, reflected the breadth of his interests and after his death they sold for a large sum. He died suddenly and unexpectedly.

Mowbray was a man of medium height and solid build, for whom in later years the wise abandonment of a pipe seemed a physical loss. As a clinician he was well read, competent and kind, and inspired much confidence in his patients. Possessed of much geniality and charm, he was also firm and occasionally even obstinate in his opinions and purposes. It was these qualities combined with his considerable intelligence which made him so successful a chairman and medical adviser. His marriage, in 1938, to Ivrene Violet Stenhouse, a fellow medical graduate and daughter of an ENT surgeon, was a very happy one and they had two sons. The elder inherited his parents' medical genes and entered practice as a primary care physician in Newfoundland. The younger inherited his father's gift for medical administration and became director of the management advisory service to the NHS.

Mowbray's two disappointments in his career were a narrow failure to be appointed to the consultant staff of his teaching hospital, and the foundering of his plans to establish a medical faculty in the University of Durham after it had separated from Newcastle. Having no nagging itch for research he may well have used his considerable gifts to better advantage as a regional physician than he would have done in a teaching hospital, but the failure wounded him deeply. Plans for a medical faculty in Durham probably never had much chance of success and, since Newcastle opened in 1984 a new medical school too big for its needs, a competitor 15 miles away could only have been a planner's disaster. His memorial is likely to be Dryburn Hospital but, apart from his family, his most rewarding happiness probably came from teaching medical students and young graduates; they liked him very much indeed.

HAD

[*Brit.med.J.*,1986,**293**,144; *The Times*,29 May 1986; Photo]

MUENDE, Isaac *b.*9 Oct 1900 *d.*29 Sept 1987
BSc Lond(1920) MRCS LRCP(1926) MB BS(1927) MRCP(1940) FRCP(1952)

Isaac was born in London and lived there all his life. He died peacefully at his home where he had lived for more than 60 years. From school he went to King's College, London, to study engineering and obtained a BSc with honours, but he then decided that he preferred medicine and entered the medical school of Charing Cross Hospital. He qualified with the Conjoint and graduated MB BS in 1927.

At that time dermatology was grouped with ophthalmology and otorhinolaryngology as a 'specials' post in the student curriculum. Many regarded it as almost voluntary, or paid minimal attention to it. One day the student Muende crept silently into a 'skin outpatient clinic' - to find himself alone. As he turned, about to leave, he was spotted by the dermatologist J M H Macleod [*Munk's Roll*,Vol.V,p.260], who called him over to 'look at' some feature on a patient's skin. There was no escape, and Muende remained there as the only student for the entire morning. But Macleod had sown a seed of interest which grew steadily.

After his qualification, they met again and Macleod encouraged him to specialize in dermatology. He took him to St John's Hospital for Diseases of the Skin in Leicester Square, where Macleod was a physician and director of pathology. Muende naturally became interested in skin pathology, and soon afterwards was awarded the Radcliffe Crocker travelling fellowship and gold medal. This fellowship and medal was endowed by Radcliffe Crocker of University College Hospital [*Munk's Roll*,Vol.IV,p.321] to enable a 'dermatologist in training' to travel abroad in order to further his education. Muende went to work in Bruno Bloch's department in Zurich, where he found himself among several future world leaders in dermatology, including Marion B Sulzberger and Sam Peck. He also spent some time at the Allgemeinerkrankenhaus in Vienna and the Hôpital Saint Louis in Paris.

On his return to London in 1931 Muende was appointed physician and pathologist to St John's Hospital for Diseases of the Skin, where he spent every afternoon in the laboratory. About the same time, a full-time technician was appointed, Reginald Syms, with whom he had a lifelong friendship. Isaace Muende and Reg Syms were the only laboratory staff until July 1948 when, with the introduction of the NHS, Muende was replaced by a full-time consultant pathologist. But during the previous years they had updated Macleod's textbooks on *The Pathology of the skin*. Muende obtained his membership of the College on his published work in 1940, and was elected a Fellow in 1952.

Muende also worked as a clinical assistant to Sir Ernest Graham-Little [*Munk's Roll*,Vol.IV,p.475] at St Mary's Hospital, where he was soon recognized by the students as an outstanding teacher. He became consultant dermatologist at Edgware General and Willesden Hospitals, as well as St John's Hospital in Leicester Square. At all of them he had a large entourage of trainees who always found it worthwhile to attend his outpatient clinic. He enjoyed teaching and shared his knowledge willingly. He would often take the trainees in his car, in which the talk was always on dermatology. Many will remember his custom, at the end of an outpatient clinic, of taking a gold case out of his waistcoat pocket and offering cigarettes to everyone in the group.

He was unfortunate in not obtaining a consultant post at a teaching hospital, but he was an enthusiastic teacher to numerous post-war registrars. He read the dermatological literature widely and was happy to discuss any subject with them. He was the author, with J M H Macleod, of *A practical Handbook of pathology of the skin*, London, H K Lewis & Co, 1940, which went into several editions.

Isaac Muende had one brother and three sisters. He remained an elegant bachelor, always neat and well dressed, a meticulous person. He played tennis as a student, was a keen chess and bridge player, and a linguist. Although he had a large private practice, he gave up all dermatology with his retirement from hospital work at the age of 65. He spent a lot of time at his second home in Israel, developing a keen interest in history and archaeology. He was a member of the British Archaeological Association.

Muende was a generous man in numerous ways. In 1963 he set up the Isaac Muende Charity Trust with a donation of 10,000 pounds of investments from his own portfolio. The Trustees were his accountant, Thomas E Fay FCA, J Mazure (since deceased), and his nephew J H Muende, also a doctor. Annual donations were made to some 30 charities, including Boys Town Jerusalem, Yeshivat Kerem B'Yavneh Foundation, the British Technion Society, the Society of Friends of Jewish Refugees, Friends of the Hebrew University, the British Council of the Shaare Zadek Hospital, Jerusalem, and the Jewish Blind Society. He showed a special interest in charities benefiting children and Boys Town Jerusalem was probably the recipient of the largest donations. By April 1987 the Charity's funds had increased to 222,000 pounds. It was his wish that the Charity should continue to make donations for a further ten years after his death, at which time the Trust would be wound up and the capital distributed to various charities.

CDC

[*Brit.med.J.*,1987,**295**,1149; Photo]

MURRAY, Sheilagh Mary Cecilia *b*.23 May 1914 *d*.27 Apr 1984
MB BS Durh(1937) MD(1956) FRCPath(1963) MRCP(1966) FRCP(1976)

Sheilagh Murray came from a medical family in north west Durham. Her father was the last of the archetypal family doctors; immaculate blue velour coat with a freshly picked buttonhole, bowler hat, leather gloves, spats, a little brown bag, and of course chauffeur driven. Four out of five children became doctors; ranging from 'a Man from the Ministry' to a GP in the 'Outback'. The fifth married a doctor.

Sheilagh qualified in 1937 (the youngest that year) and served in the RAF throughout the war. She married an RAF pilot, but when she returned to civilian life was a widow with two young children. With such responsibilities, Sheilagh sought a job in the newly formed

National Transfusion Service and on her first day had to have explained the difference between the ABO Groups and the Oxford Group. But within months she became the local authority on all blood group serology and, in particular, developed an interest in haemolytic disease of the newborn. Meanwhile, she had discovered a flair for administration, and the service she built up for antenatal prediction, instant diagnosis and provision of blood at a minute's notice, did much to establish Newcastle as a leading centre in the care of this condition. It was no surprise when, in 1950, she became the first full-time director of the Newcastle Regional Transfusion Centre.

As a co-author she was devasting, everything - even the English - had to be correct, and to write a paper with Sheilagh was a salutary and maturing experience. She wrote her MD thesis in 1956 on the Rh genotypes and their different antigenic properties, which proved the forerunner of renewed interest in this phenomenon.

In 1959 she received the Oliver Award for blood transfusion, and in 1960 a Council of Europe fellowship to study tranfusion facilities in Europe. This was a fruitful experience for all involved and she formed many friendships and links, particularly with the Scandinavian countries, which survive to this day.

When tissue typing became possible she was quick to realize its importance and, with the late Pat Dewar, established the Newcastle Transfusion Centre as one of excellence and reference in this field. Her collaboration with David Kerr and the Newcastle renal unit was one of the success stories of her time.

Sheilagh never enjoyed robust health and in her latter years was handicapped with respiratory insufficiency, and was therefore rarely to be seen at international congresses, but she hardly ever missed the Transfusion Director's meetings. Her precise views and dogged determination could be very effective, and it was this attention to detail which determined the efficiency of the Newcastle Transfusion Centre. Almost every member of the staff, whether a driver, donor attendant or deputy director, was chosen with extreme care and the final decision was usually Sheilagh's.

She had many other interests, particularly literature, and in later years developed an expertise in antique glass and frequently contributed articles to journals. She published an authoritative account of Newcastle glass, *The Peacock and the lions: a history and manual for collectors of pressed glass of the north east of England,* Stocksfield, Oriel, 1982. After her retirement, having become a national figure in glass, she took an interest in paintings, particularly those by local artists. Whenever one visited her at home there was never any shortage of conversation: transfusion topics and personalities, glass and painting. One could easily find oneself out of one's depth. You would forget that she had serious respiratory insufficiency but for the fact that from time to time she would break off, perhaps in the middle of a sentence, to take a whiff of oxygen in order to enable her to continue the conversation with renewed vigour. You were very likely to be joined by her son and daughter, with their respective children, and find yourself involved in a family gathering which revolved around Sheilagh. WW

[*Brit.med.J.*,1984,**289**,120; *Lancet*,1984,**2**,474-75; Photo]

MUSSEN, ***Surg.Rear-Admiral*** **Robert Walsh**
b.13 May 1900 *d*.12 Nov 1985
CBE(1949) CB(1945) MRCS LRCP(1922) MB ChB BAO(1922)
MRCP(1935) FRCP(1949)

'Paul' Mussen, who retired from the Royal Navy as a surgeon rear admiral, was the son of a distinguished Ulster solicitor and was brought up at Bangor, County Down. After attending Campbell College, Belfast, he received his medical education at Queen's University, Belfast, and Charing Cross Hospital, London.

Paul entered the Royal Navy in 1922. His early years in the Navy were spent at sea and he had a number of foreign commissions: the East Indies, the Persian Gulf, and twice in the Mediterranean. His ships included HMS *Birmingham* 1923, Dido 1926, *Danae* and *Calliope* 1928, *Shropshire* 1937-37, *Maylaya, Barham,* and *Hood* in 1938-39, as squadron medical officer. Fortunately his medical ambitions were catered for by extensive periods in RN Hospitals and by attachment to Charing Cross Hospital, with the opportunity to obtain his MD and MRCP. In 1939 he was appointed to the Royal Naval Hospital Haslar as the medical specialist, and remained there during the early years of the war. In 1942 he was selected for a newly created wartime job with attachment to the surgeon general of the United States Navy in Washington DC. In this prestigious appointment he forged most valuable links between the RN medical service and the USN medical service; not only did his work prove invaluable to both navies during the war but it also established a special relationship with the United States Medical Service which continued after the end of hostilities. He was made a surgeon captain (acting) in 1943 and was confirmed in this rank in 1944.

In 1945 he took over as medical officer in charge, RN Medical School, Clevedon, Somerset. Later he moved this establishment to Alverstoke; it is now the Institute of Naval Medicine and it always gave Paul great satisfaction throughout his later years to know that he had laid the foundation for this Institute, which does so much to support the modern Navy in teaching, training and research.

Paul Mussen was promoted to surgeon rear admiral in 1952, for duty as medical officer in charge RNH Chatham and as command medical officer on the staff of commander-in-chief Nore. This was a very happy period, not only for Paul Mussen and his family but for all who served in this hospital. He maintained high standards himself and expected the same of others, in whom he inspired admiration and loyalty.

In 1932, at RNH Chatham, Paul Mussen married Mary, daughter of Surg.Rear Admiral and Mrs H E R Stephens. They had three children who married and lived in England.

On retirement from the Navy in 1955 he was pressed to stand as the MP for County Down, but finally settled for a second career with the Ministry of Health where he had a special interest in polio cases and the provision of respirators. He retired from the Ministry in 1965. In retirement he wrote his autobiography *The Story of a Naval doctor*, [Ditchling], 1983, and was involved in local government, being chairman of Chailey Rural District Council 1972-74.

GM-T

[*Brit.med.J.*,1985,**292**,1209; *The Times*,25 Nov 1985; Photo]

MYERS, George Norman *b.*18 Feb 1898 *d.*6 Sept 1981
BSc Durh(1922) MSc(1923) MB BS(1927) MD(1930) PhD(1933) MRCP(1935) FRCP(1941)

Norman Myers was born and educated in Newcastle-upon-Tyne. His mother was an emancipated woman; running a large family, two shops, and yet still managing to travel abroad to remote parts. He graduated from Durham University, then served as house physician and later house surgeon at the Royal Victoria Infirmary, Newcastle. It was at this time he met his future wife, Karen Danielson, who worked there as a physiotherapist. He gained the gold medal for his MD thesis in 1930, and was Beit Memorial fellow 1930-33.

During the first world war he served as an infantry officer and was wounded in the right hand. This resulted in tetanus and ended his service in the Army. He quickly learned to write with his left hand, and remained ambidextrous even when his right hand had regained its use.

In the 1930s he came south to Cambridge to start research with W E Dixon [*Munk's Roll*,Vol.V.,p.104] on toxaemia, its effects, and on substitutes for morphine and heroin. He had some interesting tales to tell about repeated theft of the genuine drugs by a young male addict, and of how he and his laboratory assistant managed to foil him, and catch him. During this time he became a Freeman of the City of London, a Liveryman of the Worshipful Society of Apothecaries, and a member of the Heberden Society, the Phsyiological Society, and the British Pharmacological Society.

The second world war interrupted his career and he served as a local practitioner, stepping in to fill the gap left by the younger men who were called up for service in the Forces. He combined this with the supervision of medical studies at Cambridge University, and active participation in the Home Guard as a medical officer. His family soon became accustomed to being surrounded by the tools of war and mock manoeuvres: explosives were stored beneath the Wendy House, barbed wire and empty oil drums encircled the garden, meetings were held deep into the night in a hut in the lane, complete strangers were carried off, protesting, on stretchers, under cover of smoke bombs on Grantchester Meadows, when Dad's Army pounced. Plots for foiling an invasion by the use of canisters of poison concealed down rabbit holes were mentioned, but no clear details were ever made known to his bemused wife and children.

He was a man of great charm and a vivid raconteur. When he left Cambridge and general practice in 1947 it was a loss to many of his patients, who remained his good friends. He was appointed director of research in rheumatoid arthritis at the Regional Medical Research Centre, Harrogate, and was consulting physician to the Royal Bath Hospital, Harrogate; York County Hospital; Bradford Royal Infirmary; Scarborough General Hospital, and Wakefield and Pontefract General Infirmary. Visiting each of these hospitals on a weekly, or fortnightly, rota he became a familiar figure driving on the roads of the West Riding, often accompanied by his family who made use of his regular 'taxi' service.

He was a pioneer in the use of cortisone, using it sparingly as it was both a new drug and in short supply, and finding better results with lower doses. To his child patients in the rheumatology wards he was known as 'Uncle Norman' and he always spent an hour or two on Christmas morning acting as Santa Claus. He was much loved by patients and staff. Reluctantly, he retired in 1968. He remained very active however, enjoying his grandchildren and his hobbies of carpentry and polo - the latter strictly as an observer. He was killed in a road accident outside Harrogate in September 1981. He was survived by his three children: his elder son carrying on the tradition of medicine, his second son as an accountant, and his daughter as a professional artist.

DEM

[Photo]

NEWNS, George Henry *b.*27 July 1908 *d.*20 Jan 1985
MB BS Lond(1931) MRCP(1932) MD(1933) FRCP(1951)

George Newns was born in Dartford, Kent. His father died when he was eleven years old and his early life was spent in circumstances that were financially difficult, although his mother later married again. He was educated at Whitgift School, Croydon, and then went as a medical student to King's College, though when a boy he had ideas of becoming a missionary.

He graduated in 1931 and a year later obtained membership of the College, and within two years his MD at the age of 25, a truly remarkably feat that indicated his high intelligence and his capacity for sustained hard work; a capacity he retained throughout his life. He became registrar at the Royal Northern Hospital and then in the children's department at King's College Hospital. In 1935 he was appointed to the post which nowadays seems such a curious one: medical registrar and pathologist at the Hospital for Sick Children, Great Ormond Street. It was about this time that the walking tour in Norway took place, with R A McCance, later professor of experimental medicine at Cambridge, about which many probably mythical stories used to be told. Walking, usually in the Yorkshire dales, remained an abiding pleasure for him, and the Lake District, the Italian Alps and, of course, his wife Deirdre's home country of Tipperary and Connemara were the scene of many holidays. For all that, he liked to claim that his idea of a perfect holiday was an hotel with a first class cuisine, where he could lie on his bed on his own with the windows of his room wide open.

In 1946 he was appointed consultant paediatrician to Barnet Hospital and in the same year to the consultant staff of the Hospital for Sick Children, where he remained until his retirement in 1973. He became dean of the Institute of Child Health in 1949 and played a major part in the development of the Institute buildings in Guilford street. It was he who emphasized the importance of a first rate lecture theatre with really comfortable seats, seeing no reason why it should be less comfortable than a cinema; he succeeded in obtaining a grant from the Marks-Kennedy Trust to finance this. His relationship with the professor of child health, Alan Moncrieff, later Sir Alan, [*Munk's Roll*,Vol.VI,p.343] was excellent, both going their own way and rarely meeting but both having similar radical attitudes and rarely disagreeing. In earlier days, George Newns' radical feelings made him one of a small group at Ormond Street who were unhappy with the very conservative attitudes of most of the then board of governors and senior consultant staff of the hospital, and the group met regularly for a while to plan changes, most of which never came about. Later he supported Alan Moncrieff enthusiastically in the introduction of daily, and then free, visiting by parents in place of the old Sunday afternoon ritual; he was among the first to encourage mothers to stay in hospital day and night with their sick child.

His training and experience were essentially in general paediatrics but he early saw the need for specialization and he recognized the importance of endocrinology in paediatrics. With his capacity for hard work he soon learned what was then a new discipline, and rapidly developed one of the first major clinics for paediatric endocrinology in Britain, himself becoming a member of the European Society of Paediatric Endocrinology.

It was after the retirement of Moncrieff in 1964 that Newns was able to emerge as a dean with administrative skill and a tough approach in committee when he wished to get a decision through, or an appointment made which he wanted. Yet on other occasons he would quietly back away from a matter which seemed to require his decision. He was a good and hard-working dean; his postgraduate students admired and respected him, and indeed many formed a long lasting friendship with him. He listened patiently to their problems, and helped many of them to achieve progress in their own countries.

His listed publications number 77, starting in 1935 with 'Portal cirrhosis in a child' and ranging through all the usual preoccupations of a children's physician of the '40s and '50s until 1962. He then reported on 'A human intersex with xx/xxy/xxyy chromosomes' with Martin Bodian and two others, and from then on most of his publications concerned endocrine disorders. He also got on very well with the redoubtable Donald Paterson [*Munk's Roll*,Vol.VI,p.365] and edited the 10th edition of Paterson's *Modern Methods of feeding in infancy and childhood,* London, Constable & Co.,1955.

His secretaries and nursing staff were very fond of him in spite of, or perhaps because of his impatience and violent tempers that were over as quickly as they erupted. One of his secretaries writes that working for him was quite unlike working for anyone else, and never dull, with sudden explosions of laughter and outrageous asides. He had a favourite 'O&S' file into which went all correspondence which bored him. Needless to say, he was an equally impatient and unpredictable driver of a car.

In most respects he disliked change and over many years had his hair cut at the same time in the same place; when in the Institute he had an unvarying lunch of ham sandwiches, two jam doughnuts and Lincoln biscuits, with a pot of strong tea.

George was a kind and generous man who, with his friend and colleague, the surgeon George McNab, and Gordon Piller - then secretary to the Institute of Child Health - formed the Trust which, with the aid of gifts from colleagues and friends, provided for the education of the children of Martin Bodian after the death of that remarkable man, morbid anatomist to the Hospital for Sick Children.

George was a great reader, mostly of non-fiction, and was passionately fond of music and painting, and with a special liking for the Romanesque churches of southern Europe.

After his retirement, and until 1984, he remained chairman of the Leukaemia Research Fund to which he had beeen elected in 1963, having been chairman of the medical scientific and research panel from 1962-73. He was president of the paediatric section of the Royal Society of Medicine 1966-67; paediatric consultant to the Army from 1966-67, and from 1962-74 to the Navy.

He was a very private person; few of his colleagues knew him well and he kept his hospital life quite apart from his home life. His family consisted of a son and a daughter, and his wife Deirdre who respected and loved him dearly. She was a Roman Catholic as, eventually, was George himself.

The impression he gave to those who knew him for the first time, after his retirement, was radically different from the terse, turbulent and impatient man of earlier years. Perhaps, when the strain of work and responsibility was lifted, another character was able to emerge and they saw a quiet, gentle, scholarly and thoughtful elderly physician. APN

[*Brit.med.J.*,1985,**290**,561; *Lancet*,1985,1,293; Photo]

NICOL, Claude Scott *b.*23 Feb 1914 *d.*17 Feb 1984
CBE(1977) TD MRCS LRCP(1936) MB BS Lond(1938)
MRCP(1946) MD(1946) FRCP(1962)

Claude Nicol was born in Dublin, the son of Claude Gordon Nicol, barrister-at-law and physician. He was educated at Harrow and St Mary's Hospital, London. He was a postgraduate student at St John's College, Oxford, working under Howard Florey, later Lord Florey [*Munk's Roll*,Vol.VI,p.178]. After qualifying at St Mary's, and before moving to Oxford, 'Nic' held house appointments in obstetrics and gynaecology, medicine, surgery, neurology, and dermatology and venereology. While in Oxford he met his future wife, Janet Bosworth Smith, an undergraduate at St Anne's. There is no doubt that this was a whirlwind romance, as when they married in 1939 they had known each other for only three months, and indeed had become engaged three days after their first meeting. Their marriage was outstandingly happy.

When the second world war began Nicol, already a keen territorial, was soon called up. He had a distinguished war record in the RAMC which included service at St Nazaire, where he played a leading part in the *Lancastria* rescue operation in 1940, as a result of which many lives were saved. He specialized in venereology, eventually becoming adviser in the subject to the Eighth Army. On returning to civilian life he continued in the specialty, as well as resuming his pre-war interest in the TA. In 1947 he was appointed to Bart's, where he ran the department until 1974, and then became consultant to the London Hospital in 1950. He left the London to take over the clinic at St Thomas's in 1956, taking early retirement in 1976 due to increasing disability from Parkinson's disease. He fought this with characteristic undemonstrative courage right to the end.

In the TA, he commanded 217 (London) General Hospital and rose to the rank of colonel. He continued to maintain his interest in the TA after his retirement, and was appointed guest of honour at an RAMC dinner only a year before his death. In 1967 he was appointed an honorary physician to the Queen.

At Bart's and St Thomas's he built up two superbly organized departments and trained many future consultants. He was uncannily good at spotting promising juniors, and was a master of the chess moves involved in hospital careers. Outwardly he was a modest man, almost shy, but combined this with an unusual degree of determination, and he could be formidable at appointment committees. Nic's approach to his work was undoubtedly formal but he was a very good man at a party as well, whether this was in the Lydia department at St Thomas's or at his home in Hampstead. He would often be seen talking to his registrar's wives and the conversation would always include his plans for their husband's future career; all this in his marvellous, inimitable drawl.

He was an outstanding clinician due to an encyclopaedic knowledge of his subject, but perhaps even more to his mastery of differential diagnosis. When seeing a patient he was imperturable, and his characteristic thoroughness was unaffected either by the time of day or by how busy the clinic happened to be. He expected his junior staff to be as painstaking as he was himself, and he had an uncanny knack, simply from seeing a patient's notes, of spotting when things looked as if they might be going wrong. He had absolutely no time for hurried or skimped work and his own scholarly clinical notes, some of them dating back many years, remained a pleasure to read. By contrast, he did not waste unnecessary words with general practitioners and hospital colleagues, and his splendidly terse short reports were a welcome relief from the voluminous accounts which became fashionable.

His influence in genito-urinary medicine spread far beyond Bart's and St Thomas's, and he was secretary - and later president - of the Medical Society for the Study of Venereal Diseases. He was appointed consultant venereologist to the Army in 1967, and in the following year became adviser in the specialty to the Ministry of Health. His reputation was worldwide, perhaps especially in the USA where he had spent a year as a Fellow in medicine at Johns Hopkins Hospital, from 1949-50, working with the late Earle Moore. He was also secretary- general and later a vice-president of the International Union against Venereal Diseases and Treponematoses. During the '60s and '70s he visited many European centres, and went to India for the British Council.

Nicol published a considerable amount of original research and his book *Venereal Diseases*, London, Cassell, written with Ambrose King, and later also with Philip Rodin, was first published in 1964, becoming a standard work. In addition, he contributed sections to Price's *Practice of Medicine,* Roxburgh's *Common Skin Diseases,* and several other textbooks. He examined for the diploma in venereology for the Society of Apothecaries.

Undoubtedly his greatest contribution to the development of genito-urinary medicine was his special interest in the training of junior staff, even more important than his outstanding clinical, teaching and administrative abilities. His open office door became quite famous and there was often a little queue of people seeking his help on one subject or another. Here again he demonstrated his enviable quality of refusing to be hurried and one could always be certain of sage advice given with avuncular charm. Inevitably, with so many people seeking his opinion, he built up a considerable private practice, but I often felt that this was to some extent defensive and that he derived his real pleasure from hospital medicine. Indeed, he would often advise patients wishing to see him privately to attend his department instead.

As a young man he had been a good games player and played squash for United Hospitals, as well as occasionally representing Oxford University. In his later years it was his wife and family who gave him his greatest pleasure and interest but he continued to be fascinated by travel, and particularly archaeology. He loved India, and it was during a visit there with his wife that he died suddenly, just a few days before his seventieth birthday.

MAES

[*Brit.med.J.*,1984,**288**,1387; *Lancet*,1984,**1**,692; *The Times*,24 Feb 1984; *St Bart's Hosp.J.*,1974,138; Photo]

ORR, John Washington *b.*5 Aug 1901 *d.*17 June 1984
MB BCh BAO Belf(1923) BSc(1924) DPH(1924) MD(1926)
MRCP(1940) FRCP(1950)

John Orr was born in Belfast, the son of Frederick William and Elizabeth (née Strathdee) Orr. He received his scholastic education at the Royal Belfast Academical Institution, proceeding to Queen's University, Belfast, where he won many scholarships. As a medical student he became firmly committed to a career in pathology, and after graduation was successively Riddell demonstrator and Musgrave student in pathology at Belfast. In 1926 he was appointed first assistant pathologist and assistant curator of the museum at St Mary's Hospital, Paddington, where he remained for six years. By this time he had become deeply interested in the pathology of malignant disease, and had published many valuable papers relating to his research in this area of medicine. He then moved to Leeds as assistant director of cancer research and lecturer in experimental pathology, and five years later he was promoted reader.

At the outbreak of the second world war he was designated an EMS pathologist, and he continued as such until the termination of hostilities, also serving as a battalion medical officer in the Home Guard. In 1946-47 he was president of the Leeds Pathological Club, and in 1948 he was appointed director of cancer research and professor of pathology in the University of Birmingham and honorary pathologist to the United Birmingham Hospitals. At that time his fellow professors in Birmingham included Solly Zuckerman, later Lord Zuckerman, Lancelot Hogben, Tom McKeown (q.v.), John Squire [*Munk's Roll*,Vol.VI,p.412], Alastair Frazer [*Munk's Roll*,Vol.VI,p.186] and Melville Arnott, later Sir Melville. Orr complemented this brilliant team most effectively for, in addition to high intelligence and great industry, he had by this time acquired a wide experience of pathology, and in the field of malignancy he was the acknowledged national expert.

He was a member of the British National Committee on Cancer and of the committee of the British Association of Cancer Research, and was visiting scientist at the National Cancer Research Institute USA. He was also Commonwealth Visitor at the Universities of Malaya and Hong Kong. For many years he was an editor of the *Journal of Pathology and Bacteriology*, and he held office in the Pathological Society of Great Britain and Ireland, the Association of Clinical Pathologists, the Royal Society of Medicine and the British Medical Association.

When Orr reached retiring age in 1966 he went for a year to the Detroit Institute of Cancer Research as a senior research pathologist, and thereafter he spent three years as a research pathologist at the Royal Victoria Hospital, Bournemouth, before finally retiring to Ealing where he lived for the remainder of his life. He was opposed to the foundation of the Royal College of Pathologists and never took any part in its activities.

Orr was a friendly, modest man, a most helpful colleague, and was intensely loyal and supportive to everyone in his department. A strong sense of humour and wide interests made his company always a pleasure. Throughout his life he was a devout member of the Presbyterian Church and, in Birmingham, he played the organ and was an elder of the church in Moseley. Music was one of his main leisure interests, and he played the viola in one of the quartets at the Univerity of Birmingham. He also enjoyed golf at Ganton and Edgbaston, and was a keen bridge player.

He married Nora Margaret, the second daughter of David and Margaret Carmichael. His wife died following cardiac surgery in 1965. There were two children of the marriage, one of whom followed a career in biochemistry.

AGWW

[*Brit.med.J.*,1984,**289**,57; *Bournemouth Evening Echo*,17 Nov 1967; Photo]

ORR-EWING, Hugh James *b.*1 May 1891 *d.*28 Mar 1987
MC MRCS LRCP(1913) MB BS Lond(1913) MD(1917) MRCP(1918) FRCP(1931)

Hugh James Orr-Ewing was born in Northamptonshire but his family moved to Weston-super-Mare when he was 12 years old. He entered the Bristol medical school in 1908 and qualified with the Conjoint in 1913. He graduated in the same year with distinction in the clinical subjects and was awarded the University gold medal. After holding house appointments at the Bristol Royal Infirmary, he joined the RAMC at the outbreak of the 1914-18 war and served in France until 1916, when he was invalided home. He was awarded the Military Cross after the battle of Loos.

In 1917 Orr-Ewing passed the London MD, again with the gold medal, and a year later was admitted to the membership of the College. A devout Christian, he became a medical missionary and in 1920 was posted to Jerusalem. He served for about 12 years as medical superintendent of the English Mission Hospital, and was also appointed honorary consulting physician to the government of Palestine. During this time he achieved considerable distinction and was elected to the Fellowship of the College.

Orr-Ewing was long remembered in medical circles in Jerusalem, with admiration and affection. Returning to England in 1934 he was appointed assistant physician to the Bristol General Hospital and also physician to Clifton College. He became a full physician in 1945, and retired in 1956.

Hugh Orr-Ewing was president of the Bath, Bristol and Somerset branch of the British Medical Association for the session 1946-47 and gave his presidential address on 'The History of Arabian medicine'. Three years later he was elected president of the Bristol Medico-Chirurgical Society; his address was on 'Medicine in 18th century Bristol'.

After his retirement he continued to lead an active life and was a member of many committees and chairman of not a few. Many of these were connected with the Church of England of which he was an active member. He was a lay reader for 58 years.

Orr-Ewing was a great raconteur with a never-ending fund of stories. It has been said that all the best of these started 'When I was in Palestine...'. A keen and enthusiastic golfer, it was sad that a few years before his death he lost one eye in a golfing accident. This loss was borne with much fortitude, as was - for over 20 years - his gradually increasing deafness. Before his death he made a tape recording, which he wished played at his funeral, in which he affirmed his Christian faith and his confident belief in a future life.

BP

[Photo]

PARNELL, Richard William *b.*2 Oct 1911 *d.*26 Dec 1985
MRCS LRCP(1936) MA BM BCh Oxon(1939) MRCP(1941)
DM(1951) FRCP(1969)

Richard (Dick) Parnell was born in Rugby and educated as a day boy at Rugby School. He was a cricketer and played in the Rugby School XI at Lord's. His father was a farmer at West Haddon, near Rugby, as had been his father's forbears for several centuries; his mother was the daughter of the professional cricket coach at Rugby School, who had himself played for England. At the age of 29, Dick Parnell married his first cousin, Dorothy Parnell, daughter of a plant geneticist who spent his working life in India and Africa.

From Rugby, Dick went to St John's College, Oxford, and hence to the London Hospital where between 1936-39 he held a succession of junior posts including those of house physician to Donald Hunter [*Munk's Roll*,Vol.VII,p.288], and to the cardiac department under Sir John Parkinson [*Munk's Roll*,Vol.VII,p.443]. In 1940 he was admissions officer at the North Middlesex County Hospital, arranging the admission of no less than 9,000 patients. There followed a year as registrar at the Westminster Hospital before he joined the RAF. In the last two years of his war service he was medical specialist in a mobile field hospital in India and Burma. He successfully applied new methods he had devised to promote healing of indolent leg ulcers, a very common problem in servicemen in the Far East at that time. Men were nursed in hammocks, the ulcers exposed and air funnelled towards the lesions to dry them, and to counteract the effects of humidity.

On return to civilian life Dick continued to work among young people as physician in charge of a recently formed student health service at the Institute of Social Medicine in Oxford. He became a pioneer in the field and was particularly interested in W H Sheldon's work on somatometry, [*Munk's Roll*,Vol.VII,p.531], later Sir Wilfrid. He applied this initially to Oxford undergraduates, and later worked at the Warneford Hospital on a Nuffield research grant followed by support from the Medical Research Council. He made original observations on the constitutional aspects of psychology and of psychiatric medicine. The title of his DM thesis was 'The health of undergraduates', and the book which embodied his later observations was entitled *Behaviour and physique: an introduction to practical and applied somatometry*, London, Edward Arnold, 1958. He was intrigued by the possibility of using precise physical measurements to help in the assessment of the mental and emotional capabilities and susceptibilities of individuals.

At the age of 52, after many years of research in essentially unrelated fields, he became a senior registrar in geriatric medicine at Cowley Road Hospital, Oxford, and he was soon appointed as physician in geriatrics in North Birmingham, working mainly at Highcroft Hospital and at Good Hope General Hospital - a new district hospital, the building of which had recently started in Sutton Coldfield. For 12 years, until his retirement, he worked solo as the only geriatrician for a large population with charge of about 550 beds. He set himself a formidable task in providing the services to the district of 'A compleat geriatrician', the title of his contribution to the inaugural meeting of the West Midland Institute of Geriatric Medicine and Gerontology in 1972. In combining psychogeriatric and geriatric work, at a time when further consultant support was not forthcoming, he provided an excellent service, but it was a pattern of work which could not be sustained or emulated in the long term by a single individual working alone.

Dick Parnell retired with a sense of relief, tinged with some disappointment that he had remained a voice in the wilderness, with only limited recognition by others of the demographic trends and portents concerning the number and needs of old people. Adopting the traditional attitude of a farmer to the weather, he was both philosophical and mildly pessimistic, but never daunted by the tasks which confronted him. His manner was one of gentleness and charm. In addition, he possessed a streak of doggedness and determination which was so necessary for him not only to sustain a very high order of local commitment but also to enable him to communicate nationally on organizational issues in geriatric and psychogeriatric medicine.

In his retirement he found enjoyment in gardening, fly fishing, and travels with a caravan. He loved watching cricket and coached his grandson in bowling offbreaks. He also continued his exposition begun in *Behaviour and physique...* with a second book, *Family physique and fortune: a study in multifactoral inheritance,* Sutton Coldfield, 1984, in which he applied the principles of somatotyping to family structure and family life.

DDG

[*Brit.med.J.*,1986,**292**,904; Photo]

PARRY, Robert Hughes *b.*3 Jan 1895 *d.*1 May 1986
MRCS LRCP(1918) MB BS Lond(1921) DPH(1924) MD(1933) MRCP(1934) FRCP(1943)

Few in this country's health services can equal Robert Hughes Parry's string of successes. He was Bristol's medical officer of health for 26 years, and for much of this time also held the University chair in preventive medicine; a not uncommon combination in pre-war days, which could advance collaboration between academia and public health but at the risk of conflicting demands on the holder's time. After the war he was appointed KHP and QHP, and in 1946 he was elected president of the Society of Medical Officers of Health. After his retirement he was High Sheriff of Caernarvonshire from 1958-59, the county to which he retired, and in 1959 he undertook a WHO assignment to help establish a higher institute for public health in Iran - one of many tours abroad which take up almost half of his autobiography *Under the Cherry Tree,* Llandysul, Gomerian Press, 1969.

Parry's family had long been farming in Caernarvonshire, where he was born. At the age of 17 he entered University College, Aberystwyth, and two years later he went to Middlesex Hospital medical school, where he acquired two clinical scholarships and the gold medal in surgery. He qualified at the age of 21 and held several house posts at the Middlesex, including those of Sampson Handley and Webb Johnson. He was RMO at the time of the influenza pandemic of 1918:'...never a more hopeless task than mine...'. He then spent a year in the medical branch of the RAF, and three in cancer research at the Middlesex until he was warned by his former chief, Sampson Handley, '...that he must not remain a "lab" man.'

A curious sequence of commitments followed: less than a year as school medical officer in Bristol, a job which he found '...monotonous and unexciting...', and three years in general practice with his wife in Pwllheli - near the place of his birth. He describes this period as the most exciting in their lives, and does not explain his return in 1928 to public health work in Bristol as chief assistant MOH, an appointment which was much resented by contemporaries in public health posts. Two years later he was promoted to the principal MOH post, the Ministry of Health waiving the requirement to have the post advertised.

Of the many developments for which he was responsible two merit special mention: the William Budd Health Centre in the centre of Bristol, the first in the country, and a new public health laboratory under the retiring professor of pathology in the joint University and City Health departments headed by Parry. For the rest of his life he retained his dedication to the laboratory and his opposition to the national public health laboratory service inspired by W W Topley FRS [*Munk's Roll*,Vol.IV,p.561] and G S Wilson FRS,later Sir Graham [q.v.], and expressed it vigorously at the age of 90 in a letter to the *BMJ*. Parry retired at 56, '...at the earliest possible moment.'

In his local setting in Bristol he was a controversial figure, fighting primarily in and with the University, for example over the role of the City's hospital vis-á-vis that of the University, and with Bristol general practitioners. His contemporaries describe him as colourful and refer to his cunning and his Machiavellian tactics, and to a caustic and devastating sense of humour, but also to his fierce loyalty to his staff. In this context, the steadfast and inspired support which he received in the post-war period from his deputy, Robert Wofinden [*Munk's Roll*,Vol.VI,p.473] who succeeded and predeceased him, must be recorded.

In 1924 he married Elsie Joan Williams, daughter of Percy Edwin Williams, a chemical engineer, who was herself a medical practitioner. They had two sons and two daughters; one son and one daughter followed their parents into medicine.

WGH

[*Brit.med.J.*,1986,**293**,397; Photo]

PARTINGTON, Thomas *b.*10 Oct 1913 *d.*25 Dec 1988
MRCS LRCP(1937) MB BS Lond(1938) MD(1947) MRCP(1947) FRCP(1962)

Tom Partington was born in Chadderton, near Oldham in Lancashire, the son of a builder, Thomas Albert Partington, whose father before him had founded the firm.

Tom was educated at Hulme Grammar School, Oldham, and went to University College, London, where he did his preclinical training. From 1934-37 he was at St George's Hospital medical school, where he was successively casualty officer, house surgeon, house physician and registrar. These four jobs in a teaching hospital gave him a good grounding in medicine and, with this under his belt, he moved as medical registrar to the wartime EMS hospital at Old Windsor and then to the West London Hospital as clinical assistant to the outpatient department.

He joined the Royal Navy in 1942 and saw much active service, about which he was extremely reticent. He suffered from severe seasickness throughout his four and a half years in the Navy, and had little time for sailors who came to him complaining of the same trouble. He carried out his sick-bay 'surgeries' with a vomit bowl besides his examination couch, for his own use.

Soon after joining the Navy he served in destroyers on the Malta and Russian convoys, the invasion of Sicily and the Salerno landings. He was in a tank landing ship on the D-day Normandy landings and the subsequent evacuation of casualties. He went to the Far East with a Fleet aircraft carrier, *HMS Pioneer*, visiting Australia, New Guinea and the Philippines, and took part in the re-occupation of Hong Kong and the rehabilitation of the medical services. He later became a medical specialist in the colony's RN Hospital, with the rank of surgeon lieutenant commander RNVR. He was demobilized in 1946.

This bare outline of his service career cannot convey his bravery and dedication to his profession in the face of extreme danger. He let slip, occasionally, accounts of some of his adventures, from which he was extremely lucky to emerge unscathed, and said that he was very frightened a lot of the time.

After the war Tom had to cope with the universal problem of too many demobilized doctors chasing too few jobs. No doubt because of his excellent record in the past, St George's took him on and he spent the next three years there, becoming a senior registrar in July 1948 and working under Anthony Feiling [*Munk's Roll*,Vol.VI,p.173] and Hugh Gainsborough [*Munk's Roll*, Vol.VII,p.197], and it was then that he became especially interested in neurology and cardiology, the former being perhaps his first love.

In February 1950 he was appointed consultant physician to Kettering and Northampton General Hospitals, with the majority of his sessions at Kettering. At the time the hospital was emerging from its Cottage Hospital status and was still largely run by local general practitioners, with visits from London and Northampton consultants.

Tom was determined to build it into a first class hospital and, with a few colleagues appointed around the same time, soon built up a district general hospital which was able 'to stand on its own feet' without outside help. It involved a tremendous amount of work, not least of which was due to the fact that he was the sole consultant physician in the district and was called upon to see patients from a wide area - after he had settled his differences with the local practitioners, who were naturally jealous of their appointments but whose opposition ceased when he set up a medical department with junior staff from London teaching hospitals. For nine years he ran the department alone and was uncompromising in his view, as Kettering Hospital grew into its district general hospital status, that no consultant should be appointed without full postgraduate training and a higher degree.

During this time he served on the Kettering and District Hospital management committee, the medical advisory committee - of which he was chairman, lectured nurses, and even found time to read papers at the Oxford regional physicians' meetings, usually with a neurological flavour.

In the late 1950s, Tom was struck down with weakness in the legs, with marked fibrillation of the muscles, which was confidently diagnosed by two neurologists as motor neurone disease. Despite this, and his knowledge of the prognosis, he carried on working and lived another thirty years without obvious deterioration.

Tom retired in August 1976, but by then the fruits of his labours had matured into a medical department with four consultants in a newly built district general hospital.

Tom Partington was never one to sit back and do nothing; he had almost no interests outside medicine and had never married. He once said that he followed the precepts of Sir Thomas Browne in *Religio Medici:* to put off marriage until well established in the profession. Although he had certainly become established, he felt he had put it off too long and left it too late.

On retirement he turned his energies to establishing a hospital in Teheran - the Notre-Dame de Fatima - in the north of the city at the instigation of the Knights of Malta, with help from the Ministry of Health of the Shah's Government. While he was there he studied Farsi so that he could converse with the peasants at the clinics in the countryside covered by the hospital. The hospital, run by British consultants and staff nurses, did great work among the local community before it had to close rather precipitously when the Shah was forced to leave.

Tom learned French to A-level standard; he had gone to Paris with others to learn how to set up an intensive care unit, and was lectured for over an hour by the administrator at the hospital in Suresnes, where there had been such a unit for some time. He was incensed at not understanding a word and was determined to correct his lack of knowledge.

Tom Partington had a retiring personality, with a fiery temper if roused which went with his red hair, but once admitted to his friendship there was revealed a most kindly, sincere and humorous man who engendered great affection from all his professional colleagues and others who knew him well.

His last years were spent sadly alone in Cheltenham where, until he was disabled by his final illness, he would receive visitors with lavish and painstakingly detailed hospitality. He will be remembered for his incisive character and personal charm, and also as a great and dedicated physician.

GSC

[*Brit.med.J.*,1989,**298**,317,520]

PARTRIDGE, Maurice Aubrey *b.*23 OCT 1910 *d.*29 June 1984
LMSSA Lond(1936) MA BM BCh Oxon(1937) DPM(1946)
MRCP(1960) FRCP(1965)

Maurice Partridge was educated at Shrewsbury School and Balliol College, Oxford. He did his clinical studies at Guy's, for which he retained much affection and where he lived in bachelor rooms in the Warden's House for many years, although a consultant at St George's. After Guy's he trained at the Phipps Clinic (Johns Hopkins) where, like many British psychiatrists of that generation, he was deeply influenced by the psychobiological approach of Adolf Meyer. Following service in the RNVR during the war, where he rose to the rank of surgeon lieutenant commander, he spent a short time at St Andrew's Hospital, Northampton, before his appointment to St George's Hospital, where he was one of a triumvirate with Desmond Curran (q.v.) and Sir Paul Mallinson. These men made St George's pre-eminent in psychiatry in the post war years; with a well developed department of psychiatry within a teaching hospital at a time when similar hospitals were in a string-and-sealing-wax stage. Maurice was only involved in one major research project, a follow-up of 300 patients after prefrontal leucotomy, but that was splendidly carried out and led to an influential publication, *Pre-frontal leucotomy*, Springfield,Ill.,Thomas,1950, the only one of its kind at that time. He was co-author with Curran of their textbook *Psychological Medicine*, Edinburgh:New York, Churchill Livingstone, 1980, in which he managed to incorporate some very amusing footnotes.

At St George's, Partridge was a popular teacher, and he and his two consultant colleagues provided lectures on psychiatry to all comers, at Hyde Park Corner, alternating with neurology given by Denis Williams and Hamilton Paterson.

Maurice Partridge was an extremely amusing man: dry, ironic, and with a superb gift of timing when telling stories. He became a well known after-dinner speaker in rather limited circles, because he was basically a shy man and turned down invitations to larger gatherings. Everyone called him 'Bird', and his signature when writing to friends was a little wavy line which looked like the outline of a seagull. His holidays were always spent in extraordinarily remote places. Long before there were package tours to the Himalayas, 'Bird' was already sending cards and letters from places like Nepal.

His appearance was a little eccentric. He was short and rotund, untidy, usually wore thick tweed suits even in summer, and always had a large, coloured handkerchief in his breast pocket which he used to flourish and then mop his brow. He looked rather like Pickwick, but plump rather than fat. He was a good squash player and took part in matches with the students, and others. On the hottest days he would wear a white sweater on the court, and on occasions would leave the court between sets and return wearing an extra sweater; which was usually worth two or three points against a student opponent.

Maurice was a kindly and cultured man, helpful to his juniors, much respected as a clinician and colleague, and his death has saddened many.

PBS

[*Brit.med.J.*,1984,**289**,192; *The Times*,6 July 1984]

PEARSON, Hubert Edward Spencer *b.*21 Oct 1909 *d.*15 Nov 1985
MRCS LRCP(1933) MB BS Lond(1933) MRCP(1935) MD(1936) FRCP(1963)

Hubert Herbert Spencer, known from childhood as Bob, was born at Leighton Buzzard, the son of Reginald Spencer Pearson, a general practitioner. He was educated at Berkhamsted School, where the subsequently famous Greene brothers were also pupils [C R Greene, *Munk's Roll*,Vol.VII,p.228] and their father was headmaster. Bob graduated from St Thomas's Hospital, becoming clinical assistant in the ECG department where he developed a lifelong interest in cardiology. He was a member of the Cardiac Society of Great Britain. The following year he joined the LCC hospital service and, in 1939, became senior assistant medical officer at Lambeth Hospital. From 1945-47 he served in the Royal Army Medical Corps, with the rank of lieutenant colonel, in the Canal Zone, where he developed a deep interest in horses. In 1947 he was appointed by the LCC as senior physician to the Archway Hospital Group. The following year, with the inception of the NHS, he became consultant physician to the Group, now renamed the Whittington Hospital.

Bob Pearson was recognized by his colleagues as being a most able physician. He was particularly respected for his generous and conciliatory temperament, which helped smooth differences that arose between senior clinical staff and administration during the development of the unification of the Whittington Hospital from the three Archway Group hospitals.

He was a very reserved and private individual, and a man of few words. He gave dedicated care to his patients but acknowledged his own limitations in verbal communication with them; confessing that his Great Dane hound, which he brought into the ward on his Christmas round, made a better job of it. However, his quiet, courteous manner concealed deeply held opinions reflecting his strong religious convictions; he was a devout Roman Catholic.

Bob was a lover of animals, especially horses and dogs, and for many years he bred native ponies. Local GPs recall him doing domiciliary visits with bags of feed in the back of his car. He was also fascinated by language; with a particular interest in Spanish literature, Spanish poetry of the 17th century and earlier being his favourite bedtime reading. He also took delight in the English language and was determined always to use it with the greatest clarity.

The several strands and themes of Bob Pearson's personality and interests came together in his work as medical adviser to the British Limbless Ex-Servicemen's Association appeals tribunal. He had always shown a particular concern for the rehabilitation of patients after myocardial infarction, and gained a detailed knowledge of the physical demands placed on them by their particular jobs. For over 25 years, until his death, he provided expert medical opinion opposing DHSS decisions. Thanks to his dedication to the disadvantaged individual, and his lucid advocacy, many members and widows of BLESMA gained a pension or had it increased. For many years the DHSS rejected Pearson's view that there was a link between amputations and cardiovascular disorders but they accepted it in 1979 with the publication of the Hrubec Ryder report from America. This work took up much of his time after his retirement in 1969, and just before he died he completed a booklet for the amputee and his helper entitled *Better Health for the Amputee*, sponsored by BLESMA. It is characteristic of Bob's modesty and reserve that, for a quarter of a century, such an important contribution appears to have been unknown to his colleagues.

Bob Pearson married Ruth Monica Keating, daughter of a factory manager, in 1935, and they had a son and two daughters.

BIH

[Photo]

PERRY, Kenneth Murray Allan *b.*1 Feb 1909 *d.*4 Sept 1984
BA Cantab(1930) MRCS LRCP(1933) MB BChir(1933)
MRCP(1936) MD(1938) FRCP(1948)

Kenneth Perry was born in London, the younger brother of Allan Perry, a well known London Hospital surgeon. He was educated at Christ's Hospital and Queen's College, Cambridge, where he was a Kitchener Scholar. He carried out his clinical studies at the London Hospital, qualifying with both the Conjoint and MB in 1933. Subsequently he held a series of house appointments at the London, and became a medical registrar. He followed this with a period of research as a Dorothy Temple Cross fellow at Massachusetts General Hospital,

Boston, and research fellow at Harvard University. In 1939 he returned to England and during the war worked for the Medical Research Council, where his research included investigation into occupational disease affecting the lungs and other organs. He was appointed a consultant physician to the London Hospital in 1946 and remained in this capacity until his retirement in 1972. In 1949 he also became a consultant to the Royal Masonic Hospital. As an expert in the treatment of tuberculosis Perry was consultant to Papworth, as well as being physician to the Brentwood and Waverley Hospitals.

He was a keen traveller and participated actively in the international medical scene, whether attending congresses of the International Society of Internal Medicine, or as an examiner. He was a corresponding member of the National Academy of Medicine in Buenos Aires, and an examiner in Hong Kong. Wherever he went he made friends, and always took a keen interest in the local and political scene in the countries he visited; it was characteristic of him to have wide interests outside medicine. He enjoyed examining, and was an examiner for the Universities of Cambridge, Liverpool and London, as well as being examiner for the Royal College of Physicians and the Society of Apothecaries.

Perry's specialty was chest medicine, and he published many papers and edited *Diseases of the Chest*, London, Butterworths, 1952, with Sir Geoffrey Marshall [*Munk's Roll*,Vol.VII,p.381] and *Chest Diseases*, London, Butterworths, with Sir Thomas Holmes Sellors (q.v.) in 1963. He was an enthusiastic supporter of the Thoracic Society. He retained his interest in occupational lung disease, yet was always ready to point out the importance not only of 'general medicine' as such but also the need to consider the patient as a whole. Perry was a clinician with a real understanding of his patient's problems, and also an excellent teacher who showed great interest in his junior staff, for whom he would go to infinite trouble. He had a critical eye for both the medical and political scenes, on which he held strong views. He had a particular charm of manner and a ready wit which made him many friends. He was a keen freemason and, besides his consultancy at the Royal Masonic Hospital, served on the board of governors of the Royal Masonic Institute for Boys. He was interested in young people, understanding their problems well. He always supported his juniors, taking great interest in their progress, and his advice on planning their careers was always sound.

His wife Winnie had been a ward sister at the London Hospital, and they were a devoted couple. Although they had no children of their own they took many young people under their wing. While still on the staff of the London Hospital they bought a most attractive flat near the water's edge at Portsmouth, where they eventually retired, and many will remember their generous hospitality.

DTDH

[*Brit.med.J.*,1984,**289**,929; *London Hosp.Gaz.*,March 1973,**76**,(1); Photo]

PETCH, Charles Plowright *b.*2 Sept 1909 *d.*8 Dec 1987
MRCS LRCP(1939) MB BChir Cantab(1939) MRCP(1940) MD(1948) FRCP(1966)

Charles Plowright Petch was born and died in Norfolk. His father was a director of a research institute and his mother, Edith Mary Plowright, was the daughter of a surgeon. He was educated at Gresham's school and St John's College, Cambridge, gaining a scholarship at both. He gained first class honours in the natural sciences tripos. After teaching for a few years at Stowe school, he decided to take up medicine. So he returned to his old college, qualifying from St Thomas's Hospital in 1939. He gained his membership of the College in 1940 and was elected a Fellow in 1966.

During the second world war he was a medical specialist in the RAF, and in 1948 he was appointed physician to the St Helier Hospital, Carshalton, Surrey, retiring in 1975.

Petch was a reticent and punctilious man, activated by a strong sense of duty and common sense, little affected by the vagaries of fashion. These qualities made him a valuable, stable influence in his newly emerging hospital, especially in the early days of the NHS.

Outside medicine his great interest was the countryside of Norfolk, and especially its flora. Together with E L Swann, he wrote the *Flora of Norfolk*, Norwich, Jarrold & Son, 1968, and served for three years on the council of the Botanical Society of the British Isles.

While at St Thomas's he met and married a nurse, Margaret Stirling, the daughter of a solicitor, and they had two sons, the elder becoming a cardiologist at Cambridge.

JBH

[*Brit.med.J.*,1988,**296**,141; Photo]

PICKERING, Douglas *b.*6 July 1937 *d.*24 Jan 1987
MRCS LRCP(1960) MB BS Lond(1960) DCH(1963) MRCP(1966) FRCPE(1979) FRCP(1981)

Douglas Pickering was born at Beckenham, the son of Arthur Pickering, a chartered accountant, and the nephew of four doctors. He was educated at Dulwich College, where he carried off several sports and scholastic prizes, and at Guy's Hospital, graduating in 1960.

His house appointments were at Pembury Hospital and included a post with Norman Jacoby, whose example of the practice and philosophy of paediatrics had a lifelong influence on him. A short period as a ship's surgeon had a happy result; he met Flora, his future wife, who was such a tremendous support to him for the rest of his life, particularly in his long and ultimately mortal illness.

Douglas gained further experience in general paediatrics in Ronald Illingworth's unit at Sheffield, and in cardiology at the Edinburgh Royal Infirmary. In 1967 he came as senior registrar in paediatrics to Oxford, where he spent the rest of his life. His paediatric skills were soon evident. He set about developing expertise in paediatric cardiology with a year's fellowship in John Keith's unit at the Toronto Royal Hospital for Sick Children. In 1971 he was the obvious choice for the newly created post of consultant paediatrician and paediatric cardiologist at the Radcliffe Infirmary. He soon built up a flourishing local and regional service in cardiology - indeed, was one of the few pioneers of modern paediatric cardiology with its new interventionist techniques - but never became divorced from general paediatrics, taking his full share in its practice and teaching. His contributions to the literature of both subjects were considerable, mostly written with colleagues, for he was an easy collaborator, and it was only ill health which all too soon reduced the volume of vigorous but careful clinical research.

After a cruelly brief time in the full flow of the busy consultant practice of a teaching hospital, he suffered a stroke from which he made a good recovery only to be afflicted later by a chronic and finally fatal disease. He died at the age of 49.

There was a nobility about Douglas, and all who knew him in the last few years of his life could only marvel at his cheerful, uncomplaining persistence in adversity. He had a strong Christian faith which was not flaunted but was all the more impressive for that. He had a clear set of values which included loyalty to colleagues, devotion to the best interests of his patients, and a conscientious application to all he undertook. But he was not boring. He had a lively sense of humour, which was often expressed as outrage at the more extreme foibles of the undisciplined in modern society. His strong sense of the need for order and self-discipline was often affronted by what he met in his clinical practice during the last 20 years of his life, but he brought sympathy and, where possible, unsentimental help to such families. Indeed, he well knew the social obligations of the hospital clinician; for instance, he helped found the Children's Heart Circle to enable parents to share experiences and support each other.

He was a slim, upright figure who, bearded as he was for the latter part of his life, must have been a convincing father figure to his young patients. Until prevented by illness he was a keen athlete, playing cricket, hockey and golf with enjoyment. He and his wife had a son and two daughters, in whom he had a proper but unobtrusive pride. BDB

[*Brit.med.J.*,1987,**294**,713; Photo]

PICKERING, Ian George Walker *b.*24 Nov 1915 *d.*18 Jan 1984
VRD(1952) MB ChB Leeds(1939) MD(1947) MRCP(1966)
FRCPsych(1971) FRCP(1972) FFCM(1974)

Ian Pickering was the son of George William Pickering, a professional musician turned manufacturer. He was born at Bradford, Yorkshire, and educated at Bradford Grammar school and Leeds University. After qualification he was house surgeon at St James's Hospital, Leeds, before volunteering for the Royal Navy. He had a distinguished war record, achieving the rank of surgeon lieutenant commander. He ended his war service at a naval hospital in New South Wales, having served with the 2nd Submarine Flotilla in Scotland, the North Sea, North Atlantic and North America from 1939-41, as well as in the Mediterranean and the Indian Ocean on destroyers. He took part in the landings at Anzio and Normandy, and was awarded the Volunteer Reserve Decoration with clasp in 1952. On demobilization in 1945 he returned to his teaching hospital in Leeds, where he joined the medical professorial unit and obtained his MD in 1947. It was at this time that he met Jean Lowthian, a nurse, and they were married in 1948.

In 1947 he joined the Prison Medical Service, in which he was to make his real career and reputation. He served at HM Prisons in Leeds, Wakefield and Durham and, following a Nuffield travelling fellowship in 1961-62 which took him to Europe to study medical aspects of the treatment of offenders, he succeeded Harvie Kennard Snell in 1963 as director of the prison medical services.

For many years the heads of the medical branch of the prison department in the Home Office, whether commissioners or directors, had kept a low profile - it was a rigid, small and detached organization having little contact with medicine outside its own walls. All this was to change under the new director. Ian Pickering's directorship opened up the prison medical service: the number of medical officers, senior medical officers and principal medical officers increased, hospital staffing became more generous, more hospital officers were promoted to principal officer and chief. Psychotherapists, radiographers and occupational therapists were appointed. Equipment, including expensive and sophisticated instruments, was made available on a more generous scale and shelves of standard, up to date medical textbooks appeared in each prison. The director himself adopted a high profile; he attended boards, committees, councils, conferences and symposia. His visits to the various prison service establishments were occasions to be anticipated with interest and, indeed, enthusiasm. His inspections were always thorough and scrupulous - many a pithy comment in the Commissoner's Book (held in each prison) still remind the governor and medical officer in charge of their responsibilities regarding the quality of food, cleanliness of the kitchen, and sanitation for the prisoners in their care. His energy seemed boundless and his enthusiasm immense. In recognition of the contribution he made to the overall running of this difficult and demanding service he was made a full member of the Prisons Board in 1967 - a tribute indeed to the esteem in which he was held by the prison authorities. One of his favourite sayings was 'It never does to go head-on at the Establishment. They always win.'

The prison medical service was Ian Pickering's pride and joy; an attack on it was an attack on him personally and woe betide any member of the service who let the side down for they would get no quarter. Ian held various extramural offices and was especially proud of his role as inspector of retreats for inebriates from 1963 onwards. He was vice-president of the 2nd International Congress of Social Psychiatry in London in 1969, and vice-president of the British Association of Social Psychiatry in 1972. He was a founder member, and past member of council, of the North of England Medical Legal Society, and a founder member of the Faculty of Community Medicine. He was also president of the British Academy of Forensic Sciences 1969-70. All this persistent determination and great enthusiasm were acknowledged by his election to the membership of the College in 1966 and his election to the fellowship in 1972. He was also elected to the fellowship of the Royal College of Psychiatrist and of the Faculty of Community Medicine.

Ian Pickering was a good medical administrator and was always concerned about the welfare of his staff - what touched them touched him. He knew about their individual aspirations, their triumphs and sorrows. In earlier years it seemed that he knew the names of everyone, down to the most newly joined hospital officer. He ensured that his staff got a fair share of recognition for the part they had to play in the service: in 1965 Christine F T Saville was the first woman to be promoted senior medical officer (HM Prison Wakefield) and the first woman medical officer to be awarded the CBE on her retirement in 1968.

In the days when the spheres of work of governor and medical officer were ill-defined, Ian Pickering was influential in establishing the importance of the medical officer involving himself in the prison management structure, as well as taking part in the care and custody of offenders. He anticipated the full-time senior medical staff's administrative role in organizing the NHS specialties to provide the various therapeutic aspects - medical, psychiatric and surgical - on a sessional basis in the prison.

What were the secrets of Ian Pickering's success? Apart from the factors referred to above, I would suggest his prodigious memory, including his endless fund of anecdotes; his wit; his tremendous sense of humour and, above all, his famous infectious laugh. Beyond all this was his power of leadership.

In 1976, at the age of 60, when the rules dictated his retirement from the prison service, he became a consultant forensic psychiatrist at Rampton Special Hospital where, in the absence of a superintendent, he became chairman of the medical advisory committee. Retiring once again in 1983, he did not give up but became a consultant psychiatrist to Queen Mary's Hospital for Children, Carshalton, a post he still held at his death.

Ian was extremely proud of his son, a doctor, and his daughter, a nurse, but the distress of his wife's protracted illness and death in 1975 was something from which he never really recovered. She had been his great love and support, without which the sparkle went out of him. His daughter wrote that he was an enthusiastic numismatist from an early age and earned some repute in this field, particularly in the area of Indian coinage. He took every opportunity to visit local curio shops and antique markets. This resulted in his building up a varied and rather eccentric collection of glassware, china, furniture, pictures and antique medical equipment, in addition to his coin collection. He was also very interested in and knowledgeable about food and was delighted when, in 1975, he received the title of Officier de Jurade et Vigneron, St Emilion, Aquitaine.

In the *Bulletin of the Royal College of Psychiatrists,* Vol.**8**,No.6,, Henry Rollin wrote: '...Ian Pickering was a handsome man with a schoolboy's shining morning face. Before his first coronary some few years ago he had an almost Dickensian figure: he was rotund and across his ample middle he sported a gold watch-chain. He was an excellent raconteur: he told his stories with more than a vestige of his native Yorkshire accent. But what was so idiosyncratic was his infectious belly-laugh, echoes of which will ring round the Halls of Fame for all time.'

DEMS

[*Brit.med.J.*,1984,**288**,798-9; *Bull.roy.Coll.Psychs*,Vol.**8**,no.6,118; *Prison Medical Journal*,Vol.**1**,no.1,1965]

PIERCE, John Wynne *b.8* Aug 1913 *d.24* Jan 1988
MRCS LRCP(1936) MB BS Lond(1936) MRCP(1938) MD(1939) DMRD(1948) FRCP(1968) FFR(1970) FRCR(1975)

John Pierce was always known as 'Benjie'; he was a greatly loved figure and a radiologist of the chest with an international stature. Born the son of a doctor, he had a typical doctor's upbringing: Epsom College and St Thomas's Hospital. Following house jobs at St Thomas's, he became house physician at the Brompton Hospital, where he found his life's interest, and later house physician at Great Ormond

Street where he met his wife, Olga David. He served in the RAF during the war and, perforce, became interested in radiology as he was put in charge of a mass radiography unit. On demobilization he trained at UCH as a radiologist, and was appointed consultant radiologist at St Thomas's in 1950, and at the Brompton Hospital in 1952. With George Simon [*Munk's Roll*,Vol.VII,p.536] he became *the* authority on the chest x-ray and contributed much new knowledge, particularly in immunological disease of the lungs. Benjie was a great teacher and formed one of the clique around E P Sharpey-Schafer [*Munk's Roll*,Vol.V,p.372] who revolutionised the medical side of St Thomas's in the 1950s and '60s. Benjie also took a great interest in gastrointestinal radiology, and gave an expert opinion on oesophageal disease. Not surprisingly he became a leader of the new wave of innovative investigators who were as much a home at the bedside as in his department.

Benjie was a founder member of the Fleischner Society, an Anglo-American club dedicated to radiology of the chest and supported by teaching sessions in Europe and the USA. He became a well known personality and a recognized expert on both sides of the Atlantic.

He was very well read, enjoying all forms of literature - from Tolstoy to Wodehouse. He also had a deep love of music and ballet and was a regular attender at Covent Garden; he had known Anthony Dowell, the dancer, since childhood. As he grew older, he matured into a delightful character, with his white beard and cherubic face, and was usually surrounded by good company; dispensing good conversation and fine wine.

Benjie was at his best when teaching informally, his humanity and humour were much in evidence as he gently mocked the pompous and stiff necked. Passionately anti-bureaucratic, he was a man of liberal thought and attitudes.

He lived all his life in his father's house in Guildford, where he weas a devoted husband and father. Unfortunately, heart disease limited him in his retirement, but the superb views from the house gave him great pleasure and he enjoyed the unstinting support of his wife Olga. BC

POLLARD, Arthur Bairstow *b.*18 Apr 1916 *d.*5 Sept 1988
MRCS LRCP(1939) MB BS Lond(1939) MRCP(1946) MD(1947) FRCP(1969)

Arthur Pollard was a Yorkshireman to the backbone - born at Bradford, the son of Bairstow Pollard, a wool merchant, and his wife Florence, née Cusworth, the daughter of a farmer. He was educated at Bradford Grammar School, and went on to study medicine at St Mary's Hospital medical school, London.

The writer first knew him in 1934, at St Mary's, when Arthur was starting the second year of his preclinical studies. Arthur already showed promise of being a good doctor: he was naturally highly intelligent, an acute observer, and disinclined to talk very much.

After graduation, Arthur held junior posts at St Mary's as house surgeon and house physician, and subsequently became house surgeon at the King Edward VII Hospital, Ealing, and later house physician at the Postgraduate Medical School, Hammersmith. From 1947-49 he was medical registrar at the Central Middlesex Hospital, and then registrar and senior registrar to the medical unit at St Mary's. In 1950 he was appointed consultant physician to the St Helena group of hospitals in Colchester, Essex, where the writer was a consultant anaesthetist.

From 1940-46, during the second world war, Arthur Pollard served in the Royal Navy with considerable distinction as a medical officer, reaching the rank of surgeon lieutenant commander.

On his arrival at Colchester he was already a very well qualified clinician, with a deep knowledge of the problems and application of blood transfusion, as well as all aspects of general medicine. He worked tirelessly, not only at Essex County Hospital but also at St Mary's, Colchester, where he started to build up a geriatric service, and at Black Notley, where he provided the much needed medical support for the rapidly expanding orthopaedic unit; and he still found time for smaller GP hospitals at Clacton, Harwich and Halstead, and for the geriatric unit at Tendring. No one ever heard him complain about overwork.

As a man, he was sometimes brusque when confronted with humbug, but he had unlimited kindness and patience with the seriously ill. He had a dry wit, and the capacity for breaking down problems in committee with a dozen words of basic common sense. None who worked with him failed to appreciate the value of his advice and help, or to realize that they were all the better for having known him.

After his retirement from the NHS, although he was beginning to have his own health problems, he continued to work where he was needed to the limits of his strength, and there are many in and around Colchester who will always be conscious of their debt to him.

Arthur Pollard loved Colchester and the country around, and in his younger days he delighted in spending many of his precious leisure hours sailing at Mersea. He never lost his pride in having served in the Royal Navy, and was an avid reader of anything written about Nelson.

Colchester was the centre of Arthur's professional life and he was a valued contributor to the Colchester Medical Society, one of the three oldest medical societies in the country, which he served as secretary for five years, and as president, and supported enthusiastically.

In 1948 he had married Helena Marie, née Eaton, daughter of a civil engineer, and they had three children, a son and two daughters. His wife and family survived him.

JNF

[Photo]

POLSON, Cyril John *b*.9 Nov 1901 *d*.31 Dec 1986
MRCS LRCP(1924) MB ChB Birm(1924) MRCP(1926) MD(1929) FRCP(1941) Barrister-at-Law

Cyril Polson was born in Edinburgh, into a medical family, and was educated at Wrekin College, Shropshire, and Birmingham University. He was Queen's Scholar in pathology and graduated with first class honours. His experience in clinical pathology commenced at the Crichton Royal Institution, Dumfries, where he was responsible for the laboratory service of the hospital. In 1925 he went as lecturer in chemical pathology to the University of Manchester and worked with Shaw Dunn. He obtained his membership of the College the following year.

In 1928 he moved to Leeds as lecturer in pathology under Matthew Stewart [*Munk's Roll*,Vol.V,p.396] and was awarded his doctorate the following year. While at Leeds he took some responsibility for pathology services at St James's Hospital and, in due course, was appointed pathologist. By this time he had also had some experience in neuropathology at Queen Square, and had visited some continental centres, notably Vienna. He published a number of papers on various aspects of pathology, such as iron poisoning, and developed a particular skill in the pathology of the eye - building up a splendid section on this subject in the departmental museum.

At this stage, his interests were turning increasingly to medico-legal matters. He was admitted to the Inner Temple in 1937, gained honours in the Bar Final in 1939, and was called to the Bar in 1940. He was elected a Fellow of the College in 1941.

In 1947 he was appointed to the chair of forensic medicine at Leeds University. Until that time the chair, established in 1906, had only been occupied on a part-time basis but Polson, as a full-time professor, began to build up a department from very modest beginnings. Over the next 25 years, under his leadership, it grew into one of the largest departments of forensic medicine in the country, with a well established international reputation. Initially the practice of the department was confined to the city of Leeds, but with the passage of time it was extended to cover the whole of Yorkshire apart from the area around Sheffield. Polson was renowned as a skilful dissecter and meticulous observer, and dealt with many notable cases the most famous of which was probably the Moors Murders in 1966. He trained a number of young forensic pathologists who, over the years, have had occasion to be grateful for the careful habits he taught them.

In addition to postgraduate teaching, he was well regarded for his teaching of undergraduates. He not only lectured to medical students but developed courses for law and dental students. His course for medical students was to become the most extensive in England. Long after he had retired from teaching many of his students remembered his humour and skill as a lecturer.

Polson was also strong in scholarship. He wrote several books which came to be regarded as classics of forensic literature. The first, with R P Brittain and TC Marshall, was *The Disposal of the dead,* ed. C J Polson, London, English Universities Press, 1953, followed by *Essentials of forensic medicine,* London, English U.P.,1955, and *Clinical Toxicology,* with R N Tattersall, London, English U.P.,1959. Of these, *Essentials of forensic medicine* went into four editions in his lifetime and has been described as the best textbook on forensic medicine in existence.

Cyril Polson was highly regarded abroad as well as in this country. He became a corresponding member of La Société de Médecine Légale de France in 1950, vice-president of the second international meeting on forensic medicine in New York in 1960, and fellow of the Indian Academy of Forensic Sciences, and had many close friends abroad, particularly in the Scandinavian countries.

For many years he tried to bring the special needs of forensic pathology to the attention of the government, and he campaigned for a national forensic pathology service as he foresaw the inevitable difficulties which would arise from increasing financial pressures on the universities, which were the main sites for forensic studies. He played a large part in providing evidence to the Broderick Committee. His activities on behalf of his subject were recognized when he became president of the British Association in Forensic Medicine for three, instead of the customary two, years. He was also president of the British Academy of Forensic Sciences, a founder fellow of the Royal College of Pathologists, an active member of the Pathological Society, and of the Forensic Science Society.

In his spare time he collected silver, gardened, photographed church architecture, and was president of the local Gilbert and Sullivan society. He was twice married; first to Mary Tordoff, who died in 1961, and two years later to Mary Pullan, who was herself a doctor. There was one daughter of the first marriage.

Polson was short in stature, with a fine head of black hair throughout his life; a kindly man who was much loved by his staff, though sometimes intolerant of others; a first-class pathologist though never really at home in the witness box; with a pronounced sense of fun which he could quickly direct at himself, and a great loyalty to his friends. He was not one to seek the limelight, but he was one of the great scholars of forensic medicine in this century. DJG

[*Brit.med.J.*,1987,**294**,318-9; *Lancet*,1987,**1**,173; *The Times* 10 Jan 1987; Photo]

POND, Desmond Arthur *b.*2 Sept 1919 *d.*29 June 1986
Kt(1981) MA MB Chir Cantab(1945) MRCP(1945) DPM(1947) MD(1951) FRCP(1960) FRCPsych(1972) Hon FRCGP(1982)

It was to psychiatry, and to the alleviation of the form of mental suffering with which it deals, that Desmond Pond devoted the greater part of his many sided gifts and his energies. He had made a lasting imprint upon the subject by the time his life ended. The breadth of his approach, and the imagination and generosity of spirit with which he put it into effect in clinical practice and research was only one expression of the man he was, and of the manner in which he lived.

The view that emerges from his writings encompasses the contributions made by social disadvantage, familial deprivation and the vicissitudes of the formative years of human developments, along with the biological and cerebral factors, as the underlying determinants of mental disorder and the stunting of personality. In some of his writings, and in his efforts to build bridges between psychiatry and religion, most notably in his Riddell lectures, one discerns an attempt to form some conception of, or gain a measure of insight into, the totality of things.

Desmond was born in London in 1919 and educated at St Olave's School. His parents were Thomas and Ada Della Pond, née Clutten. A strong streak of individuality was manifest from an early age. Of his school years he has written: 'My chief distinction was to have survived the English educational system without ever playing a proper team game, though physically fit.' Survival was perhaps an understatement, for he gained a foundation exhibition to Clare College, Cambridge, where he had a distinguished career. There he met John 'Honest to God' Robinson who became a lifelong friend and spiritual influence. Another contemporary at Clare has quite recently described a sense of liberation after initiation into life at the University which was probably similar to that experienced by Desmond: 'To arrive in Cambridge was to discover that the ideal of what had been an intellectual minority was here so widely shared as to carry no stigma of eccentricity.' Desmond obtained a double first in natural and moral sciences, having taken psychology in Part II, an early sign of his burgeoning interest in mental life.

A Rockefeller studentship took him to Duke University medical school in the USA, where he spent two years, 1942-44. After some house jobs, and a preliminary canter in a psychiatric hospital in

Bristol, he arrived at the Maudsley in 1947. His interest in electroencephalography had already been aroused, and with the encouragment of Denis Hill [*Munk's Roll*,Vol.VII,p.264], later Sir Denis, another lifelong friend, he embarked upon his enquiries into epilepsy and brain damage in early life, an area in which some of his most important contributions to science were made. His subsequent research evolved along lines that clearly expressed the wide sweep of his intellectual and scientific interests. His contributions included the EEG of children with brain damage, temporal lobe epilepsy, an enquiry into narcolepsy (for which he obtained an MD from the University of Cambridge), a study of subjects with criminal psychopathy, and his important pioneering survey of childhood epilepsy in general practice. These studies culminated in his influential Goulstonian lectures in which he collated the evidence that parental attitudes, family disruption and other social factors, could contribute to undermine emotional stability, impair curiosity and disrupt schooling, and so deny brain damaged and epileptic children the chance of achieving their full intellectual potential.

His experiences in the course of these surveys paved the way for studies of the role of psychiatry in general practice, and the place that should be accorded to it in the education of family doctors. He had also become interested in the dynamics of the general practitioner's interaction with patients, as delineated in the contributions of Balint. His activities at this interface equipped him some years later for the leading role he was to assume on the joint committee of the Royal College of Psychiatrists and General Practitioners, which made some widely influential recommendations regarding the education of family doctors. Desmond's contributions were recognized by the award of honorary fellowship of the Royal College of General Practitioners.

His interest in the social psychodynamic and biological aspects of personality development, which he considered to be inextricably entwined, found further expression in the mother and child unit he created in the London Hospital. This was more than a multidisciplinary ivory tower. It developed research for investigating the formative years under conditions as natural as could be contrived wihin a hospital. And it became a community service station for the treatment, social support and guidance of problem children and their mothers in Tower Hamlets in the East End of London.

In 1971 he was invited to deliver the Riddell lectures at the University of Newcastle-upon-Tyne. His theme was 'Counselling in religion and psychiatry'. His treatment of the subject matter evolved into an original exploration of the common ground and the divergences between psychiatry, religion, sociology and the Law, in their attitudes to and management of individuals whose conduct brought them into conflict with society and its mores. Desmond saw religion both as an expression and a fulfilment of psychological needs deeply engrained in human personality. These were needs that he believed secular societies to be increasingly unable to satisfy through the weakening of social bonds and the blurring of common ideals, activities and purposes they were liable to engender. Some of these ideas had been previously expressed. But others were fresh, and all were explored here along original lines and with an erudition which drew upon social anthropology, theology, linguistics, psychodynamics and psychiatry. In 1964 he became a founder member of the Institute of Religion and Medicine, and he served as a member of committees and ecclesiastical commissions which explored problems that arose within marriage and from escalating rates of divorce.

His considerable achievements as a scientific member of the Medical Research Council, 1968-72, stemmed from the wide range of his scientific interests, his openness to new ideas and dispassionate guidance. He was to serve again in virtue of his appointment as chief scientist at the DHSS. He found himself somewhat unexpectedly stimulated by policy making and the cut and thrust of medical politics, and he played his part with unobtrusive skill. He succeeded in strengthening some bonds between the DHSS, the Medical Research Council and the Royal Colleges, and in creating some new ones.

His presidency of the Royal College of Psychiatrists between 1978-81 was a memorable success. Desmond's sensibility and tolerance enabled him to encompass the character of the members of various committees, and he was good at anticipating their patterns of interaction. The Council of the College is a larger, more talkative, argumentative, and turbulent body of men than the corresponding committees of comparable organizations. Desmond was a quiet chairman, and he had rare gifts of tact and discretion. He advanced with the aid of a distinctive form of positive non-intervention through a mountainous agenda and towards the resolution of knotty problems. His qualities of leadership were acknowledged when he was elected to chair the Committee of Presidents of Royal Colleges.

Montesquieu has said that knowledge makes men gentle, and Desmond was a gentle man. In retirement he settled in his home and the garden in Devon which he had created over a number of years in close partnership with Helen, his wife, a consultant physician. He loved the countryside around which for him had a luminous quality. It was a place of green and gold enchantment that emerged from their joint labours. There were some months of tranquil happiness, surrounded by his family, in the period immediately after his retirement. His marriage with Helen had been close, and a mutual source of growing emotional fulfilment over 40 years in which they had shared successes and some sorrows. During the year before the end, when an illness of obscure origin was manifest, Helen's exemplary fortitude and forbearance sustained his hope and spirit.

Desmond was a fine musician, deriving deep satisfaction from his piano and from singing in choirs, an activity which he and Helen engaged in together. He was able to see more of his three daughters, who visited as often as they could. Two have proved to have outstanding musical talent, and a third is a gifted zoologist. Plainly the genes of both parents were at work. But Desmond, sceptical of rigid biological determinism, would have been reluctant to invoke heredity as an explanation, and modesty would have inclined him to minimise the contribution made by the rich, intellectual and musical environment he and Helen had created.

There was a brief surge of improvement in the Spring, when his natural vitality, humour, and infectious laughter returned for a while, but relapse soon followed. He endured the painful weeks at the end with legendary courage, and unfailing and unremitting concern for Helen and the family.

He was, perhaps, sustained by an integrity and a faith eloquently expressed by another kind of man from Cambridge. Looking back on his long life, Bertram Russell has written: 'I have lived in pursuit of a mission both personal and social. Personal: to care for that which is noble, for what is beautiful, for what is gentle, to allow moments of insight, to give wisdom at more mundane times. Social: to see in imagination the society that is to be created where individuals grow freely, and where hate, greed and envy die because there is nothing to nourish them. These things I believe, and the world for all its horrors, has left me unshaken.'

MR

[*Brit.med.J.*,1986,**293**,272; *Lancet*,1986,**2**,171; *Bethlem Maudsley Hosp.Gaz.*,Sept 1966,**8** (3); *Brit.med.J.*,Apr 1966; *Lancet*,2 Apr 1966; *Univ. Coll.Hosp.Magazine*,1952/3,**37** (4),153; Photo]

PORTER, Rodney Robert *b.*8 Oct 1917 *d.*6 Sept 1985
CH(1985) BSC Liverp(1939) PhD Cantab(1949) FRS(1964) Hon FRCP(1974) Nobel Laureate Medicine(1972)

Rodney Porter was born at Newton-le-Willows, Lancashire, the son of Joseph and Isabel Porter. He died in a road accident while driving from Oxford to Plymouth just three weeks before he was due to retire as Whitley professor of biochemistry at Oxford. He made outstanding contributions to immunochemistry and was largely responsible for elucidating the chemical structure of antibodies; for this work he was awarded, jointly with G M Edelman, the Nobel Prize in medicine in 1972. He received many other honours, including the Royal Medal (1972) and Copley Medal (1983) of the Royal Society, foreign membership of the American Academy of Sciences (1972) and Companion of Honour (1985).

Rodney was educated at Ashton-in-Makerfield Grammar School and Liverpool University, where he took his BSc honours in 1939. For the next six years he served in the Army, mainly in the Mediterranean theatre with the Royal Engineers, and was discharged in December

1945 with the rank of major, Royal Army Service Corps. He then went to Cambridge where he worked with Fred Sanger, then engaged in elucidating the structure of insulin for which he was awarded the Nobel Prize some 10 years later. It was at Cambridge that Porter developed his lifelong fascination with protein chemistry, and he always thought of himself as a protein chemist devoted to immunology. His particular interest in immunochemistry was stimulated by reading Landsteiner's classic monograph *The Specificity of serological reactions*, revised edition, New York, Dover Publications, 1962, a small book that inspired a generation of immunologists. Since most antibodies have identical overall molecular properties and antigenicity, their endless diversity of combining specificity provided a striking paradox. As Porter remarked in the Nobel lecture [*Science* 180:713-6,1973] 'This combination of an apparently infnite range of antibody combining specificity associated with what appeared to be a nearly homogeneous group of proteins astonished me and indeed still does.'

Porter left Cambridge in 1949 and joined the staff of the National Institute for Medical Research, working first with A J P Martin, soon to be awarded the 1952 Nobel Prize in chemistry for the development of chromatography. On the day of that award the young Dr Porter received an early morning phone call in the laboratory congratulating him on winning the Nobel Prize. The caller was a Swedish journalist somewhat confused by the fact that A J P Martin's middle name happened to be Porter. Rodney later remarked that on receiving this encouraging news his mind had scanned with remarkable speed over the contents of his publication for that year and, excellent as it was, even he had to admit that a Nobel Prize appeared a little excessive! In later life Porter felt that he owed most to Sanger and Martin, and to Sir Charles Harington [*Munk's Roll*,Vol.VI,p.222], then director of the NIMR, who had given him invaluable encouragement.

It was at Mill Hill that Porter initiated the work that, within a decade, was to elucidate the basic structure of antibody molecules and provide the foundation for our current understanding of their molecular diversity and individual specificity. At that time the protein nature of antibodies had been confirmed, but their structure was totally mysterious and presented no obvious means of solution. Porter himself believed , on the basis of N-terminal amino acid analyses, that antibodies comprised a single polypeptide chain and that different specificities might be generated by differential folding of that chain. His inspirational approach to this problem was to fragment the molecule in the hope of obtaining smaller pieces which retained antibody activity and were more amenable to analysis. In 1958 he found that the enzyme papain cleaved rabbit IgG into three pieces of similar size; two were identical and retained the original combining specificity (Fab) while the remaining one crystallized (Fc). That protein quite incapable of crystallizing itself should yield a crystalline fragment was unthinkable, and Porter would later relate with wry pleasure how for several months he had disposed of Fc crystals down the sink believing that they were crystals of cystine derived from the cysteine used to activate papain. However, the finding of common and variable structures within a single protein molecule was the vital key to the solution of the antibody paradox, and the papain technique enabled the properties of antibodies, some common and some individually distinct, to be assigned to discrete portions of the molecule.

In 1960 Porter became the Pfizer professor of immunology at St Mary's Hospital medical school, the first chair of immunology to be created in the United Kingdom. By this time, Gerald Edelman at the Rockefeller Institute had demonstrated that antibodies are multichain proteins. The problem of relating the papain pieces to the chains required isolation of the chains in soluble and undernatured form. This was accomplished by Porter soon after he moved to St Mary's. Subsequent antigenic analysis of the heavy and light chains with antisera to the Fab and Fc fragments provided the data which clinched the four chain structure, first proposed by Porter in 1962. This stucture proved applicable to the antibodies of all vertebrate species. It led to the precise understanding of the chemical relationships which exist between different classes of antibodies as well as between those produced by normal and cancerous lymphoid tissues. The work also heralded a progressive understanding of immunological processes in molecular terms, which has given immunology its currently pervasive role in general biology and clinical medicine.

Porter succeeded Sir Hans Krebs [*Munk's Roll*,Vol.VII,p.325] as Whitley professor of biochemistry at Oxford in 1967; inevitably he became heavily involved in teaching and administration and yet still found time actively to direct the MRC immunochemistry unit with outstanding success. The work of this unit focussed on the interaction of antibody-antigen complexes with the complement system. The molecular structure of C1q and of C4 were elucidated,the unexpected covalent binding of C4 to IgG was demonstrated, and as culmination his Oxford laboratory described the genes for C2, factor B and C4 and established their alignment in the MHC region of human chromosome 6. Porter had latterly become fascinated by the extensive polymorphism of C4 and the possible role of this in determining individual susceptibility to autoimmune disease. As he neared retirement from the Oxford chair, his research maintained momentum and the MRC, with customary wisdom, arranged that he should continue to direct the immunochemistry unit after he had vacated the chair of biochemistry.

Rodney Porter was an outstandingly successful scientist whose creative acitivity was unusually sustained. His early years were academically unremarkable and his first distinction, the fellowship of the Royal Society, did not come until the age of 47. His singular good fortune was, of course, to be a graduate student in Fred Sanger's laboratory at the time when techniques for the detailed study of protein structure were being evolved for the first time. His own capacity to exploit that nascent experience derived from his innate tenacity and enviable capacity for direct and uncluttered thought. He was able to identify and focus upon the central issue of a problem and ignore irrelevant distractions. His laboratory might appear fairly chaotic to the uninitiated, but any promising experimental lead from an always pragmatic approach was consolidated and rapidly expanded by logical progression, based upon great technical expertise and painstakingly precise interpretation.

These same qualities of honesty, simplicity, firmness and understanding, illuminated Rod Porter's personal life. He abhorred any hint of deviousness, pretence or pomposity. His many honours were received with robust enjoyment, but always tempered by a spicy irreverance which proclaimed his unshakeable identity. Over the years his endearing qualities served to transmute colleagues from near and far into lifelong friends. All were taken to Downhill Farm to enjoy the warmth of his wife Julia's open-hearted hospitality, and very likely to be unceremoniously enrolled into apple harvesting, lumbering, bee-keeping, dry-wall construction or tramping many miles across the countryside. Rodney Porter was universally admired and respected as a scientist, but he will be most vividly remembered by countless friends and colleagues for his forthright and heart-warming character, which enriched innumerable lives.

SC

[*The Times*,9 Sept 1985; *Roy.Soc News*,July 1983,**2**,No 4,1; *St Mary's Hosp.Gaz.*,1960,**66**,212; Contemp.Scien.Arch.Centre, *Prog.Report*,**25**,1985-1986]

PRICHARD, John Stobo *b.*16 May 1914 *d.*3 Dec 1986
MC(1941) MRCS LRCP(1938) MB BChir Cantab(1938)
MA(1939) MRCP(1948) FRCP(1966) FRCPC(1972)

John Stobo Prichard was born in Barry, Wales, and trained in England and the United States, but spent most of his medical career in Canada. His father, Thomas Preece Prichard, was a solicitor. His mother was Ethel Jane Frazer Andrew.

John Prichard was educated at Clifton Preparatory School, Clifton College and Cambridge University, where he took his MA and MB BChir, completing his medical undergraduate training at the London Hospital which he left to join the RAMC in 1939. In the same year he married Joan Suzanne Webber, daughter of Sir Robert John Webber. Prichard served in England, and later with the Eighth Army in Africa and Europe. He was awarded the Military Cross for bravery under fire.

After demobilization, he trained at the National Hospital, Queen Square, and in 1949 left England to spend a year in Boston at the Massachusetts General Hospital on a British travelling fellowship. He then went to Toronto, Canada, in 1950, as the first full-time neurologist on the staff of the Hospital for Sick Children. There he began a career which spanned 35 years of dedication, of excellent teaching and, for his students, years of excitement and fun.

Students and Fellows came from across the world to study at the Hospital for Sick Children under Stobo Prichard. They included neurologists who now practise in Japan, Indochina, Micronesia, Australia, India, Africa, Europe, and both North and South America. Wherever he travelled with his wife Joan, they visited old friends and were warmly welcomed.

At the Hospital for Sick Children he established the electroencephalography service, he initiated the combined neurological-neurosurgical rounds, he persuaded the administration that neurological patients would be better served on a single, dedicated ward, and he encouraged the establishment of an active research programme in neurosciences.

The provincial ministries of health, of community and social services and of education, in the province of Ontario, frequently sought his opinion and his help in the development of programmes covering problems as diverse as special education and mercury poisoning of native peoples in the north. He entered these tasks with enthusiasm. He cajoled his colleagues into helping, and he accomplished much good which can never be found in a simple reading of his curriculum vitae.

One of Stobo's great concerns as a physician was the plight of the intellectually handicapped and their families. He played a major role in the development of Surrey Place Centre, the major institution, and in the establishment of their medical services. When he retired from the position of divisional chief of neurology at the Hospital for Sick Children in 1977, he did not actually retire but moved laterally and established the child development clinic, and also the very successful learning disabilities research programme.

Nationally and internationally he was an active member of many scientific societies of neurology and electroencephalography in the role of founding member, president or journal editor. His efforts culminated in 1975 with the founding of the International Child Neurology Association. He became the first president of that organization and directed its first international congress in Toronto. He was also the founding editor of the International Review of Child Neurology.

While spending his active career in his profession, Stobo also managed to raise three exceptional children: Jane, with a doctorate in education from Harvard, is on the staff at the University of British Columbia, Sarah has a fellowship in medicine and is on the faculty at McGill University, and Robert, who has a host of degrees in business and law, is now the dean of law at the University of Toronto. The children were brought up on the same kind of challenges that Stobo normally handed to his students and colleagues. They were expected to think independently, to challenge dogma and to defend their positions. Dinner time at the Prichard household was a special time, well remembered by all who attended. Stobo would drop an apparently casual comment about any subject, political or ethical but always topical, and he then expected everyone to participate in a debate which he would often stimulate by taking a seemingly indefensible position. Joan's excellent cuisine, and Stobo's equally welcome martinis, provided a base for long and remarkable discussions. No one was allowed to sit and observe.

On a personal note, I owe thanks to Stobo for encouraging me to enter a career in science when I was a half-trained paediatrician. When my family and I returned to Toronto, after an absence to complete postgraduate studies, Stobo was there to ask us to dinner, to support my efforts in setting up a laboratory and to encourage me in those first nervous steps of a new investigator on his own. His kindness and help made my early career much easier, and his continued advice was always welcome. Like all his colleagues, I miss him.

JAL

[Photo]

PROPERT, Sydney Augustine *b.*6 Apr 1909 *d.*21 Aug 1985
BA Cantab(1930) MRCS LRCP(1933) MA MB BChir(1934)
MRCP(1935) FRCP(1951)

Sydney Propert was born in London, the son of the Revd Peregrine Sydney Goldwin Propert, vicar of St Augustine, Fulham, and his wife Rose Ethel, née White. Peregrine Propert was also a prebendary of St Paul's, and one time president of the Poor Law Association of England and Wales. The Propert family originally came from the St David's area of Wales. Sydney was quite proud of his Welsh origins and loved to spend his holidays on the west coast of Wales.

He was educated at St Paul's School and later at Trinity Hall, Cambridge, going on to St Thomas's Hospital for his clinical training. Soon after qualification he obtained his membership of the College, at the youthful age of 26; he was elected a Fellow in 1951.

His first post was in 1933, as clinical assistant in the children's department at St Thomas's Hospital, followed by a house physician appointment at the Tite Street Children's Hospital. He was subsequently casualty officer and resident anaesthetist, house physician, medical registrar and tutor at St Thomas's. During this time he was working under R J Tewkesbury, Douglas Firth [*Munk's Roll*, Vol.IV,p.569],J S C Elkington [*Munk's Roll*,Vol.V,p.116], and in particular Sir Maurice Cassidy [*Munk's Roll*,Vol.IV,p.528].

In September 1936, following the death of Sydney Curl [*Munk's Roll*, Vol.IV,p.558], the honorary physician at the Essex County Hospital, Colchester, and at the instigation of his friend of St Thomas's days, Ronnie Reid, FRCS, - recently arrived there as honorary surgeon - Sydney applied for and obtained the post of honorary physician at

Colchester. He was 27 years old at the time, a remarkably young age to hold such a post. He retained this position until the coming of the NHS when, although not enthusiastic about the change from the voluntary system, he became a consultant physician, a post he held until he retired in 1974.

The original appointment was most welcome; an able, active, intelligent young man was a great asset to the hospital and the town, particularly following Curl who, while being a good clinical doctor, was a most eccentric man, fearful of infection, aloof and apparently humourless. Sydney was immediately and subsequently a very busy man, with honorary appointments at the Royal Eastern Counties Hospital, Severalls Mental Hospital, Halstead Hospital, Courtauld Hospital Braintree, St Leonard's Hospital Sudbury, Clacton & District Hospital and Harwich & Dovercourt Hospital. Until the second world war there was no pathologist for these hospitals, and only one laboratory was staffed. This was at Severalls Hospital, where there was a technician and a boy. Sydney acted as pathologist at the Essex County Hospital; here a room was set aside as a laboratory, where relatively simple tests were done by him, anything more difficult being sent to the Clinical Research Laboratory in London or else picked up by the 'boy' from the Severalls laboratory - who came on a bicycle and took the specimen back to Severalls for the technician to handle. His heavy load was somewhat lightened by the arrival of A B Pollard (q.v.) in 1952.

Sydney Propert, with his natural intelligence and his great experience, became a thoroughly sound physician - with that something extra that some call 'sixth sense'. He had very shrewd clinical judgement and made very few mistakes, but he was 'of the old school' and did not change too readily with modern advances. He kept in close touch with his teacher and mentor, Sir Maurice Cassidy, whose niece he married. It is therefore not surprising that Sir Maurice paid regular visits to Colchester, the high spots of which were joint ward rounds. With or without Sir Maurice, it was always a pleasure to be in such a stimulating clinical atmosphere where the diagnoses were thoroughly and coolly considered, and where the patients were properly treated both medically and as human beings. Sydney was a gentleman, in the proper sense of the word, always courteous and tolerant, and his patients and staff loved him dearly. He was among the best physicians the writer has known.

He was not by nature a research worker, lacking the bent for it and that tenacity of purpose which carries one on after the initial enthusiasm has burned low, but he did make one very important medical observation: in 1938 he was the first to describe the relationship between injections of blood or its products and the development of hepatitis in the recipients (*British Medical Journal* 1938,ii,677). He studied an outbreak of severe jaundice with some deaths, affecting about 10 people in a mental subnormality institution, and found by dint of careful questioning that they had all received an injection of convalescent measles serum two or three months previously. The demonstration of this relationship in the causation of disease was of prime medical importance.

As a man and a friend he was always loyal, tolerant and trustworthy. He was a very private person, reticent about matters that concerned him deeply and personally. He did not, as the saying goes, 'wear his heart on his sleeve' and one could not have guessed at the deep personal sorrows he had. However, if one had the honour to be admitted to the inner circle of his mind, one found a man of tremendous courage and high intellect and creativity, but with a sensitive nature which could be easily hurt. Fortunately there was present too a wonderful sense of humour and wit. He was a fine raconteur and public speaker, with a marvellous twisted grin and manner of speech that was quite endearing. He was not really interested in sports, although in his youth he did some rowing. He was in the St Thomas's four, which one year won the inter-hospital races. His leisure was spent in pursuits of a more intellectual kind; writing poems or short stories of delicacy and wit, although some were of a satirical nature. It would be a fitting memorial to him to have his better writings published as a book.

In 1938 he married Ethel Margaret (Peggy) Gow who, as mentioned previously, was Sir Maurice Cassidy's niece. Her father was Andrew Gow RA, one time Keeper of the Royal Academy, and an aunt was Mary Gow RI, another distinguished artist. It was a happy marriage, apart from the premature deaths of two of their four children. Their first home was in Lexden Road, Colchester, and when he retired they converted Crockleford Mill, Colchester, into a beautiful home. Sydney died there, leaving a widow, a son, a daughter, and seven grandchildren.

On April 6, 1987, his old friends, many of whom had been taught by him at various stages of their careers, met in the library of the postgraduate medical centre at his old hospital, the Essex County, Colchester, to unveil a picture and plaque to his memory. It was a gesture of remembrance to a well loved physician.

JBP

[Photo]

PROUT, Brian John *b.*3 Sept 1933 *d.*16 Sept 1985
MRCS LRCP(1958) MB BS Lond(1958) PhD(1963) MRCP(1965) MD(1967) FRCP(1977)

Brian Prout was born in Surrey of Cornish parents. His father, William Prout, was born in Carn Brea and his mother in East Pentire. They moved to London when William was admitted to Hendon Police College, subsequently joining the Metropolitan Police and spending most of his working life in Kingston. Brian was educated at Tiffin Boys

School, Kingston-upon-Thames, and displayed an early enthusiasm for drawing but was attracted to medicine, to a large extent as a challenge because his schoolmasters said he could not do it. He went on to University College Hospital, London, graduating in 1958 after an outstanding student career, during which he was awarded the Lister gold medal, the McGrath prize and scholarship, and the Percival Alleyn award.

At UCH Brian met and married in 1960 Elizabeth Jane Chilcott, daughter of John Chilcott, a general practitioner who was born in Truro, trained at King's, and who eventually practised in Kingswood, Bristol.

After junior house officer posts at UCH and Whittington Hospital Brian went to the Royal Free as research registrar, where his work on the autonomic nervous system was rewarded by a PhD (physiology) in 1963. For two years he was medical registrar at Epsom District Hospital, following which he became a senior registrar at the Radcliffe Infirmary. He gained the MRCP in 1965, his MD following in 1967,and was elected FRCP in 1977. During his senior registrar post he wrote his second doctoral thesis, again on the autonomic nervous system, exemplifying his remarkable energy and tenacity. For most, writing a doctoral thesis is a major endurance test but to Brian Prout it was just another milestone in his overall plan.

In 1968 he was appointed consultant physician to the new Royal Cornwall Hospital (Treliske), fulfilling his lifetime ambition to return to his native Cornwall. There followed many years of dynamic activity. He introduced upper gastrointestinal endoscopy to Cornwall and established his own unit, giving a highly competent and efficient service to the county. His ability and high standards were quickly recognized and he became extremely popular with his general practitioners and consultant colleagues alike, developing a very busy NHS and private practice. Early on he was appointed a clinical tutor and tackled his teaching commitments with tremendous gusto. He initiated and organized regular lunchtime tutorials, and continued to chair these meetings even after his retirement through ill health.

Throughout his career he remained an enthusiastic clinical investigator and published widely, mainly on gastroenterological topics. He was also a keen traveller and participator in clinical meetings, and was in considerable demand as a visiting lecturer. In 1983 he was joint author, with J G Cooper, of a useful text, *An Outline of clinical diagnosis*, Bristol, Wright, based on his tutorials.

In 1976, at the height of his career, he became acutely ill with a highly malignant lymphoma and over the next nine years he endured surgery, radiotherapy and recurrent courses of cytotoxic drugs. Throughout his illness he displayed outstanding courage and strength. Although near death on many occasions he exhibited extreme resilience and stoicism, and was never heard to complain. To the last he maintained his optimism, and always managed to talk about and plan for the future.

Outside medicine, Brian had a plethora of interests; indeed one could say he was hyperactive. Family life was tremendously important and with his wife and two daughters he sailed, surfed, jogged, cycled and swam at every opportunity. He radiated energy, there was an urgency about him, an eagerness to get things done, a delight in activity and wholehearted enthusiasm for whatever he was doing.

His talents were many; he was both creative and practical, and always resourceful. A first class swimmer, he was a founder member of Truro Swimming Club in 1970, at the age of 37. Adept at the horn, clarinet and euphonium, he had considerable musical ability. An appreciation of beauty and an acute eye combined to produce in him a painter of considerable achievement. Good draughtsmanship and a keen sense of humour likewise coupled to make him a skilled cartoonist. His drawings illustrated many hospital publications and a local paper carried his cartoons for many years.

Brian had many endearing qualities, the most striking being his unfailing good humour and easy manner. He never lost the common touch, and his rather laconic hybrid rural accent heightened his approachability and contributed to his popularity.

Brian loved Cornwall and thought it a great privilege to live and work there. The sea, and nature in all its aspects, delighted him and he had a particular affection for Devoran village, where he was a regular churchgoer and took an active part in village life. He died in the hospital where he had worked for sixteen years.

ADP

[*Brit.med.J.*,1985,**291**,1284; Photo]

PROWSE, Cedric Barrington *b.*14 Dec 1903 *d.*2 Mar 1988
MRCS LRCP(1929) MA MB BChir Cantab(1931) MRCP(1933)
MD(1957) FRCP(1959)

Cedric Prowse was the son of William Barrington Prowse, a consultant radiologist, and his wife Mabel Jermyn Ford. He was born in Brighton, Sussex, into a family which had an unbroken tradition in medicine since the late 18th century. He was educated at Tonbridge School, where he excelled in sport, representing his school in cricket, rugby, athletics and hockey, and before he left became head of school. He then went to Caius College, Cambridge, where he was a member of Hawks Club and of the Oxford and Cambridge Achilles Club. He obtained the natural sciences tripos MA and went on to St Bartholomew's Hospital.

After qualification he was appointed casualty house physician, and house physician to Sir Thomas Horder [*Munk's Roll*,Vol.V,p.198] before settling down to practice in Brighton, where he joined the staff of the Royal Alexander Hospital for Sick Children as medical clinical assistant. After obtaining his membership of the College in 1933, he was appointed honorary assistant physician to the Royal Sussex County Hospital and became honorary physician in 1939. From 1954 he was senior physician and physician to the Brighton General Hospital and the Victoria Hospital, Lewes.

He had previously volunteered for military service in 1938 and became a lieutenant RAMC in 1940, being promoted to medical specialist with the rank of major later that year, and rising to the rank of lieutenant colonel in charge of a medical division. During the war he saw overseas service in North Africa with the British First Army, in the Italian campaign with the Eighth and Fifth Armies, and with the British Land Forces in Europe - from France through to north Germany - being mentioned in despatches.

Cedric Prowse was a physician of the old school, always immaculate, with a fresh buttonhole each day from his garden, his youthful figure belying his years, a strict observer of medical etiquette, a rather shy and distant manner, and a ready smile always just in the background.

The large and successful diabetic clinic in Brighton was mostly due to his efforts and he ran it until his retirement with skill and enthusiasm, as well as his general medical clinics. Though not a great writer for the journals, he published work on acute pulmonary oedema, scleroderma and amyloidosis, choosing the latter subject for his MD thesis.

Cedric married Jean Ogilvie Grant, daughter of Selwyn Seafield Grant, a consulting engineer, and they had four children, three sons and a daughter. There was a strong naval tradition, as well as a medical one, in the family and two of his ancestors - both Royal Navy Captains - sailed respectively with Drake against the Armada and with Nelson at Trafalgar.

Cedric Prowse was a man of high professional standards who expected those who worked for him to have the same, and who continued to work long after retirement age. His wife, four children and seven grand-children survived him.

ECBK

[*Lancet*,1988,2,85; Photo]

PRYOR, Jack Sydney *b.*4 July 1933 *d.*21 Dec 1988

Jack Pryor was consultant physician, with a special interest in renal diseases, at the Norfolk & Norwich Hospital, Norwich. He was born in London, the son of Robert William Pryor, who served in the Merchant Navy. He received his medical training at Charing Cross Hospital, where he was later to return as a senior medical registrar for a period of three years from 1966. His interest in renal diseases began when he was a medical registrar at the Institute of Urology from 1960-62, when he was directly concerned with treating patients on the artifical kidney and undertaking renal biopsy studies. His basic interest in fundamental renal physiology was developed in the research he did while there.

For a period of two years he worked as a research fellow and senior medical registrar at the Institute of Urology, where he was directly involved with clinical problems of a renal and metabolic nature, before he moved to Charing Cross Hospital as senior medical registrar. At Charing Cross he was involved with the cardiac surgical team and concerned with the management of the matabolic problems arising after bypass surgery, with particular orientation to the handling of patients with acid-base problems, together with water and electrolyte disturbances, as well as those of a respiratory nature. In 1969 he was appointed senior lecturer in medicine, with honorary consultant status, at Charing Cross Hospital and took over the responsibility of the home dialysis programme, as well as participating in the hospital maintenance dialysis programme.

In 1970 he was appointed head of the department of experimental therapeutics at Sandoz in Basle, Switzerland, and was responsible for designing studies in the early stages of the development of new drugs. While working in this department he was further involved in the development of a number of drugs, including bromocriptine, salmon calcitonin and beta blockers.

Jack Pryor was appointed consultant physician in renal diseases at the Norfolk & Norwich Hospital in May 1971, a post he held until his sudden death in December 1988.

Jack had always been an enthusiastic teacher of clinical medicine prior to taking up the post in Norwich, and he continued in the role of medical tutor for many years. He maintained an active research relationship with the department of biological sciences at the University of East Anglia, where he developed animal organ culture as a model for studying drug reaction and toxicology. His particular interest was in the use of kidney culture for studying renin disease and the action of various drugs on this. In recognition of the important role he played in lecturing to the university students and directing many research workers, he was appointed honorary professor to the School of Biological Sciences at the University of East Anglia in 1980.

On the clinical side, his opinion was often sought in the handling of complicated and seriously ill patients, and his weekly major ward rounds proved to be extremely popular with junior doctors, medical students, and many representatives of the departments of dietetics and pharmacy. Overseas students were particularly welcome and not uncommonly some of the ward rounds would be conducted in French, German or Spanish. He always had time to explain points of clinical interest, as well as those of physiology, biochemistry and pharmacology.

When Jack Pryor first arrived in Norwich there was no renal service whatsoever and, with his enthusiasm and dynamism, he organized the building of two renal units. He was delighted when the opening of the chronic dialysis unit, in May 1978, was undertaken by Sir Douglas Black, then president of the College. An additional extension to the unit was opened in August 1986 by Dr (later Dame) Sheila Quinn, president of the Royal College of Nursing. Further developments were planned, but regrettably he did not live to see these come to fruition.

In 1958, Jack Pryor married Bridget Egan, a nursing sister at the Brompton Hospital, whom he had met at Charing Cross Hospital. They had one son, Robert, who teaches in Madrid.

Jack was a francophile and delighted in spending his holidays in his house in the Dordogne. His visits there could be predicted to coincide with Bastille Day, when he joined the local townspeople in their celebrations. He was full of *joie de vivre,* and enjoyed good wine - on which he was an authority.

DVH

[*Brit.med.J.*,**298**,181-2; Photo]

RABIN, David *b.*25 Jan 1934 *d.*26 Oct 1984
MB BCh Wits(1956) MRCP(1959) MRCPE(1959) MD(1963)
FRCP(1979) FACP(1979)

David Rabin was born in Zastron, Republic of South Africa. He displayed scholastic and scientific brilliance early in his career; when he graduated from Aliwal North High School he won an award for the best chemistry student in South Africa. He held an unparalleled record as a medical student at the University of Witwatersrand Medical School in Johannesburg, and graduated with the highest cumulative grade average ever attained by a student there. He married Pauline Aron, a medical school classmate, in 1956. Their children, Michael, Roni, Dana and Keora, witnessed a marriage and family life unsurpassed in warmth, humour, liveliness, and close friendships made wherever they went.

After two house staff years in Johannesburg, and subsequent postgraduate training with Russell Fraser at the Postgraduate Medical School in London, Rabin moved to Johns Hopkins Medical School where he became one of the most productive investigators in the field of endocrinology in the USA. His work with Kenneth Zierler, using catheters to study the effects of insulin on the metabolism of adipose tissue and muscle in the forearm, was a classical contribution to the understanding of hormone-fuel relationships in humans. He next turned his attention to growth hormone secretion. Collaborating with Tom Merimee, David Rimoin and Victor McCusick, Rabin studied insulin and growth hormone secretion and action in patients with isolated growth hormone deficiency, and the resistence to growth hormone action in the African pygmy. While at Hopkins he contributed seven chapters in the textbook *The Principles and practice of medicine*, 17th ed.,edited by A McGehee Harvey, Appleton-Century Crofts, New York, c.1968. After nine years at Johns Hopkins University, during which he served successively as a fellow, instructor, and assistant professor of medicine, Rabin and his family moved to Jerusalem. From 1968-73 he served as associate professor of medicine at Hebrew University, Hadassah Medical School, and head of the department of chemical endocrinology at Hadassah University Hospital, where he began studies in reproductive endocrinology, focussed on gonadotropins. He was credited with documenting the first cases of hereditary FSH deficiency in women.

Rabin returned to the United States in 1973 as a visiting scientist at the National Institutes of Health, where he worked with Jesse Roth. He became chief of the division of endocrinology at Vanderbilt University Medical School in Nashville, Tennessee, in 1975, succeeding Grant W Liddle who had been appointed to the chair of the department of medicine. He continued his contributions to the field of reproductive physiology and concentrated on the use of LHRH analogues as a male contraceptive. While at Vanderbilt he also played a crucial role in collaborative investigation into intermediary metabolism, which resulted in a series of landmark papers on the hepatic roles of insulin and glucagon in controlling glucose homeostasis. Altogether, he contributed some 206 original articles to medical journals and as chapters in textbooks.

At the height of his investigative career, in 1979, he noted the onset of symptoms of amyotrophic lateral sclerosis (motor neurone disease). This condition progressively deprived him of his power to walk and the use of his upper extremities, and finally reduced him to a state in which he could only move his eyelids. Over a difficult five-year period his slight frame gradually withered away, but his intellect, courage, humour and production never diminished. It was not long before he was unable to go to work at Vanderbilt, where he was the Alexander Heard distinguished service professor in the departments of medicine, obstetrics and gynaecology, as well as director of the division of endocrinology. He carried on at home, where his devoted family cared for his physical problems. When he could no longer speak he learned to operate a computer word processor, using a switch controlled by his eyebrow. He continued to direct his research activities, and to write scientific articles and letters to his friends and colleagues. His encyclopaedic knowledge of endocrinology and the broad scope of his interests and expertise were reflected in the textbook which he co-authored with Joseph McKenna, *Clinical Endocrinolgy and metabolism: principles and practice*, New York, London, Grune and Stratton, c.1982.

David's love of music, and his enthusiasm for people, issues and politics, continued. Musicians from the Blair Academy of Music frequently performed in the Rabin home for the family and their friends. Using the eyebrow switch to operate his word processor, he wrote an article for the *Nashville Tennessean* encouraging support for a public radio station that played classical music and, with his daughter Dana, wrote an article calling attention to the physical dangers inherent in the sport of boxing.

His own experiences as a patient increased his interest in the care of the patient with incurable disease. Sensing that his own experiences were likely to be shared by others in similar circumstances, he published his concern about the attitudes of many physicians and their inability to deal with chronic or terminal illness in an article entitled 'Compounding the ordeal of ALS: isolation from my fellow physicians', *New England Journal of Medicine*, 1984,**304**,699-701. Written with his wife Pauline and his daughter Roni, the paper concluded: '...bear in mind that the absence of a magic potion against the disease does not render the physician impotent. There are many avenues that can be helpful for the victim and his family. I am often surprised and moved by the acts of kindness and affection that people perform. Fundamentally, what the family needs is the sense that people care. No one else can assume the burden, but knowing that you are not forgotten does ease the pain.'

The growing concern of David and Pauline with the attitude displayed by society and physicians towards persons with progressive or terminal illness, dementia and mental illness, culminated in the publication of a book which they co-edited: *To provide safe passage: the humanistic aspects of medicine*, New York, Philosophical Library Inc.,1985, a collection of essays devoted to the role of the physician in the face of incurable illness, and containing guidelines to help patients deal with grief, chronic physical disability, intractable pain and terminal illness. Their daughter Roni wrote an eloquent and inspiring book: *Six parts love - one family's battle with Lou Gehrig's disease*, New York, Charles Scribner's Sons, 1985, chronicling the struggle of the Rabin family in coping with ALS.

David Rabin was widely respected and admired for his broad knowledge, his exciting lectures, and his courage in the face of a ravaging illness. His colleagues at Vanderbilt established an annual David Rabin Lectureship in endocrinology in his honour. But above all, he will be remembered for the warmth of his personality and the twinkle in his eye.

ALG

[Photo]

RACE, Robert Russell *b.*28 Nov 1907 *d.*15 Apr 1984
CBE(1970) MRCS LRCP(1933) PhD Cantab(1948) FRS(1952) *FRCP(1959) Hon MD Paris(1959) Hon MD Turku(1970) Hon FRCPath(1972)

Robert Race described his entry into the world of blood groups - which he was later to dominate - in the following way: having drifted into clinical pathology after an undistinguished career as a medical student he was leafing through the pages of the *BMJ* when an advertisement for a job in the Galton Laboratory Serum Unit caught his eye. Not knowing quite what serology was, he went for an interview, conducted by the formidable R A Fisher. He was bowled over by Fisher's imaginative account of what might be done with blood groups in mapping human chromosomes and joined the unit, which soon moved to Cambridge on the outbreak of the second world war. Work was confined almost entirely to ABO groups until the discovery of the Rh system, in the USA, opened up exciting new possibilites. Soon sera revealing a bewildering pattern of reactions were arriving in the laboratory. Meanwhile, G L Taylor [*Munk's Roll*,Vol.V,p.405] had died and Race was running the laboratory; he discussed his new findings with R A Fisher over beer in the 'Bun Shop'. R A Fisher immediately, on the back of an envelope, wrote out his suggested CDE scheme, showing how antithetical reactions between anti-c, anti-C etc., would explain all that had been found and, furthermore, predicting what else might soon be found. This scheme was promptly accepted with gratitude and relief by a wide circle of people who felt that otherwise they would have no chance of understanding developments in the Rh field. Unhappily, at the same time, A S Wiener in the USA who had played a major role in discovering the Rh system, saw the CDE scheme as both incorrect in principle - since it seemed to postulate three sets of closely linked genes rather than a single complex locus - and a threat to his own position, since it appeared to be taking over a blood group system in which he had a certain proprietary right. The disagreement between the supporters of Wiener, on the one hand, and Race and Fisher, on the other, was to last a very long time. Indeed, up to the time of Race's death no really satisfactory scheme for the genetics of the Rh system had been proposed, although the CDE terminology had almost completely replaced Wiener's.

At the end of the war, at the instigation of A N Drury, later Sir Alan [*Munk's Roll*, Vol.VII,p.169], who had become director of the Lister Institute, the Medical Research Council took on the responsibility for the Galton Laboratory Serum Unit, which became the Blood Group Research Unit and moved into a position on the second floor of the Lister Institute, looking out over the Thames. In this atmosphere Race was extremely happy and the unit now gathered momentum in an impressive way. One of the first new recruits to the laboratory was Ruth Sanger, who had come from Australia on a fellowship from the Red

Cross, and she and Race soon announced an important finding; namely, that the newly discovered antigen S was genetically linked to M and N. This discovery was one of a very large series to be made in the next 20 years, in which a whole host of antigens was fitted into a dozen or more major blood group systems, each system being inherited independently of the others. Race's reputation grew rapidly and soon blood samples were being sent to his laboratory from all over the world, notably from the USA which regarded him as the final Court of Appeal on any problem in his field.

Amongst the many important discoveries made by Race's unit were: the finding of -D-, ce and ceó(V); the discovery of the Oè ('Bombay') phenotype and of the 'null' phenotypes S-s- and Fy(a-b-); the recognition of an acquired form of the B antigen and of the first human chimera and, finally, the discovery of Xgá, determined by an X-borne gene. This very impressive catalogue, which is far from complete, established Race's central role in the development of his subject.

Race did not have formidable intellectual powers but he did have endless tenacity and tremendous enthusiasm, linked to a notable flair for imagining unusual solutions to problems. This last talent was reflected in his conversation which often took surprising turns and was invariably entertaining, and also in his writing, particularly as reflected in his textbook with Ruth Sanger: *Blood groups in man,* Oxford, Blackwell Scientific Publications, 1975, in which a wholly serious approach to the subject was leavened by many droll comments. This textbook, of which the first edition was published in 1950 and the sixth in 1975, was a really outstanding work which made Race and Ruth Sanger (later his wife) known and highly respected wherever in the world there was any interest in blood groups.

Rob Race was a rather retiring man and was happiest with very small groups of people. Throughout his career he carried out much of his serious business in pubs; there he met his scientific friends, speculated about the interpretation of recent findings, and even wrote lectures. At larger meetings he was nervous and did not like to take part in discussions unless to communicate something which he had prepared with great care. When he did give a paper he was masterly, because he had the power to make complex subjects simple, and he had a command of the English language which reflected his strong interest in literature.

Race was married twice: first in 1938 to Monica Rotton, with whom he had three daughters, and after Monicas's death he married Ruth Sanger in 1956, with whom he did so much of his work and who looked after him in his final, protracted illness.

PLM

[*Brit.med.J.*,1984,**288**,1544; *Lancet*,1984,**1**,1085; *The Times*,1 May 1984; Photo]

RAPER, Alan Boynton *b.*25 Oct 1908 *d.*22 Sept 1984
BSc Leeds(1931) MB ChB(1934) MD(1937) MRCP(1937)
DTM&H(1950) FRCPath(1964) FRCP(1965)

Alan Raper was born in Wakefield, the son of John Henry Raper, an ironmonger, and his wife Mary, née Livingstone. His maternal grandfather was a sailor, inventor and entrepreneur, and it is tempting to suppose that Alan inherited some of his gifts, which he displayed in his fertile and original research.

Alan was educated at Wakefield Grammar School and the University of Leeds. After house appointments at the General Infirmary, Leeds, and at Brompton Hospital, London, he began his training as a pathologist in the Yorkshire Clinical Laboratories in 1935. Throughout the war he served in the RAMC in France, Kenya, Ceylon and Bengal, and this experience gave him a lifelong interest in tropical medicine. After demobilization he spent 10 years as a senior pathologist in Uganda. In 1958, after spending two years in London as a staff member of the Medical Research Council, he was appointed consultant pathologist (haematology) to the United Bristol Hospitals, where he remained until he retired in 1973. After his retirement, he and his wife went to live in the Lake District which they dearly loved.

Alan Raper belonged to a time when it was still possible to be an all-round general pathologist, and he was an excellent one. He was a first class microscopist and histopathologist. His wide knowledge embraced parasitology as well as the cytology and chemistry of the blood. He had the eye of an artist and this probably helped him to see things that sometimes escaped others. These talents were evident in his laboratory work. He believed that good microscopy and clinical insight should complement the automated chemical and physical techniques that have developed in haematology since the second world war.

Raper will be particularly remembered for his research on the haemoglobinopathies, especially for his contributions to our understanding of the relationship between sickle-cell anaemia and malaria. There was controversy in the 1950s over the claim that sickling protected against malaria. The work of Raper and his colleagues appeared in three classic papers that are still widely quoted, published in the *Quarterly Journal of Medicine* and the *British Medical Journal*. According to his friend Hermann Lehmann, one of Alan's most important contributions was to show that the protective effect of sickling occurred in children when they lost their maternal immunity to falciparum malaria, and before acquired immunity was established. Sicklers and non-sicklers were infected equally often, but the severity of the infection and the mortality were less in the sicklers. This explained how falciparum malaria selected for the sickle-cell trait in the population.

During his early years at Bristol, Raper published some 20 papers dealing with thalassaemia, the hereditary persistence of foetal haemoglobin, the change over from foetal to adult haemoglobin synthesis, the incidence of transplacental haemorrhage at the time of delivery, and other haematological topics. As a diagnostician the breadth of his experience was invaluable. This was particularly notable when patients with tropical infections were admitted to the hospitals in Bristol. Soon after his appointment Raper instituted regular teaching programmes in haematology to clinical students and these were very popular.

He made another contribution to medicine, as an editor; first of the *East African Medical Journal* (Uganda), and later of the *Bristol Medico-Chirurgical Journal*.

Alan Raper was a friendly and modest man, and an excellent colleague. He had a deep interest in the arts: he read widely and loved music, and was a talented amateur painter. He married Doris Grainger, a farmer's daughter, in 1939, and their family life was happy and rewarding. They had five children, a daughter and four sons: Margaret died in infancy, Martin is an architect, James a solicitor, John works for the Atomic Energy Authority, and Christopher is a doctor who, at the time of writing, is a consultant haematologist at the Kingston General Hospital, Hull.

WAG

[*Brit.med.J.*,1985,**290**,475]

RENSHAW, Peter Knowles *b.*21 Oct 1917 *d.*Dec 1985
MA Cantab(1939) MB BChir(1942) MRCP(1948) FRACP(1970) FRCP(1973)

Peter Renshaw died at Dunedin, New Zealand. He had been educated at Repton and Emmanuel College, Cambridge. Since his father was a doctor, as were many members of his family going back over four generations in the Manchester area, it is not surprising that he elected to read medicine. He proceeded to the London Hospital for his clinical training. There, through his discerning interest in his mentor's methods of teaching, he developed a style of his own which not only held the interest of his students but also encouraged them to think for themselves.

During the war he served with the RAMC in West Africa and Europe. After demobilization he was appointed medical registrar to the Luton and Dunstable Hospital, becoming senior registrar there in 1949. Applications for senior medical posts in the United Kingdom were heavily over-subscribed at this time and, in 1951, with characteristic initiative, he applied for and was appointed senior registrar to Sir Horace Smirk in Dunedin. There he enjoyed such friendship and hospitality from Sir Horace's family and his other colleagues that he never seriously considered leaving New Zealand.

In 1956 he was appointed physician and lecturer in medicine to the

University of Otago, and became the first medical member on the staff of the new Wakari Hospital in Dunedin on his appointment as medical superintendent, a position he held until his retirement. His ability to judge character enabled him to collect a team which established the high reputation of the new hospital. He recruited numerous medical graduates from overseas, many of whom were surprised to get a long distance telephone call long before they expected a reply to their application. By asking a few shrewd questions, which the candidates had not anticipated, he often obtained more candid information than would have been forthcoming from a formal interview. In addition to his administrative duties, he made time to continue the clinical work and teaching which he so much enjoyed. He was made a senior lecturer to Otago University in 1970 and elected FRACP in 1975. He became a foundation member of the New Zealand College of Community Medicine in 1980, and latterly was appointed deputy medical superintendent in chief to the Otago Hospital board. He visited England regularly, returning often to his old teaching hospital, where he used to hold ward rounds for the undergraduates.

Peter was a very much repected and loved member of the medical community in Dunedin. He was always deeply concerned with the health and welfare of his junior staff. He had a keen financial sense and regularly listened to the stock market reports on the BBC World Service. Many charities benefited from his well chosen investments, including the Otago Medical Research Foundation, the New Zealand Medical Education Trust, the Dunedin Diocesan Trust Board and the Crippled Children Society. He generously loaned his house in Wanaka and his beautiful yacht, *Lenticula,* to his friends, including many of his juniors. Only a few weeks before his death he endowed the Peter Renshaw Charitable Trust.

At University Peter's friends gained some insight into his character when he joined the Gliding Club, for this sport enabled him to develop his many skills. He pioneered gliding in the South Island and when the Otago Gliding Club was formed he became its first president and was appointed chief flying officer. Those of us who were privileged to see the superb photographs which he took from his glider readily understood his fascination with the sport.

Since some of his forbears had had heart attacks at a relatively early age, Peter was prepared for the massive cardiac infarction which he sustained when only 46. He philosophically gave up gliding for sailing and soon became an able helmsman.

Peter always remained young at heart. He had an impish sense of humour and loved entertaining children with his conjuring tricks. His disarming charm opened many doors that were closed to the less enterprising.

He was a staunch supporter of St Paul's Cathedral in Dunedin. The funeral service there was a final tribute to one of whom it can be said 'There went a beloved physician and a gallant one withal'. JKW[1]/JH/JL

[*N.Z.med.Journal*,22 Jan 1986; Photo]

REVANS, *Sir* John *b.*7 June 1911 *d.*4 Nov 1988
KT(1977) CBE(1967) MBE(1943) MRCS LRCP(1935)
DCH(1946) MRCP(1964) FRCP(1969) LLD Southampton(1970)
FRCGP(1974) O St J(1947)

John Revans was born in Birkenhead, on the southern side of the Mersey estuary, at the height of Liverpool's maritime influence. His father, Thomas William Revans, was a prominent naval architect and a council member of the Royal Institute of Naval Architects. Indeed it could be argued that John himself was a consequence of a maritime engagement as his mother, Ethel Amelia Charlotte, née White, was the daughter of a master mariner.

John was educated at Battersea Grammar School and the University of London, and received his undergraduate medical training at the Middlesex Hospital. At the age of 24, he married Eileen Parkhurst Mitchell, and almost at the same time took up his first medical appointments as house physician to the neurological and children's departments, and house surgeon to the orthopaedic department at the Middlesex.

He was a young man of much ability and with much to prove. Instead of seeking advancement and experience within the safe confines of his own country, John took up a commission with the Indian Medical Service and in 1936 was appointed medical officer to the 1st Battalion 2nd King Edward's Own Gurkhas. He became civil surgeon to the Roorkee and Saharnpur areas of the United Provinces in 1937.

With the onset of war John found himself in the thick of things. Between 1940-43 he was staff captain HQ and graded physician at the 11th Indian General Hospital. He was mentioned in despatches in 1941, during western desert operations in Egypt, and was later promoted to major. He became a member of the Order of the British Empire (Military Division) in 1943, and was further promoted to the rank of lieutenant colonel on the war staff of the India Office in London. In 1946 he was appointed an officer of the Order of St John of Jerusalem, being demobilized from the Army with the rank of colonel.

Almost as soon as he had retired from the Indian Medical Service, as staff surgeon at Poona, John began a second - and perhaps even more distinguished - medical career. In 1947, with the exciting prospect of a national health service about to become a reality, John took on the challenging appointment of assistant secretary to the British Medical Association. His talents were quickly spotted by the then fledgling service and John was chosen to become the South West

Metropolitan regional hospital board's assistant senior administrative medical officer. The region stretched from central London down to the borders of Devon, and it became clear that a sub-office was necessary in Wessex to coordinate the management of the southern part of the region. John Revans took up this challenge enthusiastically and, in due course, piloted the separation of the Wessex region from the South West Metropolitan region, setting up the headquarters in Winchester. It was a small region and had no medical school but this was all to change, largely as a result of the efforts of John Revans with the University of Southampton. A new medical school was founded and quickly developed a reputation for its progressive approach towards the training of doctors, using all the hospitals in the region. Revans was the region's first senior administrative medical officer and, following the reorganization in 1974, he led the team as regional medical officer for its first two years as a regional health authority. The region profited well under him and had a reputation for progressive educational policies and economic services.

John Revans undertook a wide variety of other duties for the NHS. He was a member of the central midwives board and of the Salmon committee, and did much work to further the development of nursing and rehabilitation services. He also found time to carry out work for the Nuffield Provincial Hospitals Trust on sterilising practices, vocational training for GPs and postgraduate medical education.

During the 1960s he was adviser to the Royal Commission on Health, Newfoundland, and carried out work for the World Health Organization.

He received his CBE in 1967 and, although he retired as regional medical officer to the Wessex regional health authority in 1976, he was knighted for services to medicine and the NHS in 1977.

He married Eileen Parkhurst Mitchell in 1935 and they had two daughters, and later two grandchildren. Although professional commitments meant a heavier than usual workload, he still managed to devote time to his personal interests in sailing and steam trains. It was sad that he was not able to enjoy fully his retirement on account of the stroke he had suffered. He will be remembered throughout the region as the 'Father of Wessex', and for his kindliness and humour, expressed through a fund of stories '...out of India.' WJEMcK

[*Brit.med.J.*,1988,**297**,1399,1538; Photo]

RHIND, Edward Gregson Grant *b.*3 May 1912 *d.*1 Feb 1986
MB ChB Aberd(1935) MRCP(1939) FRCP(1964)

Edward Rhind was born in Huntly, Aberdeenshire, and died in Sheffield where he had spent the greater part of his professional life. His father John Rhind, had a business concerned with the distribution and marketing of beef, one of the products for which the north east of Scotland is rightly famous. On his father's side, his grandfather had been employed by the now amalgamated Union Bank of Scotland, and his maternal grandfather, James Sellar Leith, was also an official in one of the numerous small Scottish banks which were represented in every substantial town throughout the country.

Edward Rhind was educated in the Gordon Schools, Huntly, and following the well established local pattern went to Aberdeen University for his medical education, where he graduated with honours in 1935. As an undergraduate his clinical instruction was therefore within the Aberdeen Royal Infirmary, and other hospitals in the city. With his distinguished undergraduate career, he was a natural choice to become Sir Stanley Davidson's [*Munk's Roll*,Vol.VII,p.136] house physician in the professorial department of medicine. He then came south to Sheffield, where his cousin already worked in general practice, and completed his surgical house officer training in the Royal Infirmary. Thereafter he returned to clinical medicine, becoming medical first assistant to Robert Platt, later Lord Platt [*Munk's Roll*,Vol.VII,470] and Gurney Yates [*Munk's Roll*,Vol.VII,627]. The outbreak of war intervened and Edward Rhind served in the RAMC for a period of over four years, mostly in India; he was graded as a medical specialist and became officer in charge of the medical division in a military hospital with the rank of lieutenant colonel. Soon after his demobilization and return to Sheffield he was appointed physician to the Municipal Hospital, the city general hospital of Sheffield. So he became one of the consultants already in post when the National Health Service was founded, and was able to observe the impact of the new approach to medical care in a scene where the changes which resulted were undoubtedly at their greatest.

Only those who served through the early years of the NHS will find it credible that Edward Rhind was expected to look after both acute medical and many long stay geriatric patients in his own hospital, as well as performing outpatient clinics and responsible for other scattered groups of beds - some of them as far away as Bakewell, in Derbyshire. A staggering workload included responsibility for no fewer than 360 inpatients at that time. However, as the service evolved, his energies became concentrated in the City General Hospital, to which he was devoted, seeing it grow from its original role as a municipal hospital with very limited staff and resources into its present role as the second major teaching hospital within the Sheffield medical school, and now called the Northern General Hospital. In this transition Rhind was a key figure to whom the University could look for help at every turn.

Edward Rhind's manner was always quiet and unassuming but he invariably commanded immediate respect from patients, students and colleagues alike and his own high standards of behaviour in clinical practice brought out the best response from all associated with him. His character was outstandingly friendly and unselfish and these qualities, coupled with his professional skills and enthusiasm, made him a man in whom great trust was placed at all times. His personal qualities were perfect for the dual role of physician and teacher; at the same time, his warmth and friendliness allowed everyone to address him as 'Eddie' without any lessening of his authority.

His special interest in neurology led him to the chairmanship of the Sheffield branch of the British Epilepsy Association, and his concern for students to the responsibility for running the clinical committee of the faculty of medicine. He was also a loyal servant to the College, obtaining his membership in 1939 and his fellowship in 1964. He was the first College representative for the Sheffield region, a choice which recognized his ability to establish the important links between the College and the wider profession, particularly in relation to postgraduate training.

Edward Rhind had many interests outside medicine, most of which had become part of his life as he grew up in Aberdeenshire. It may be a surprise to learn that cricket is commonly played in the north east of Scotland and as a student he was both captain of the University XI as well as playing for his home team. He had a lifelong interest in shooting and salmon fishing, both of which he continued almost to the end of his life.

Rhind was twice married. His first wife, Mary Barker, a medical graduate of Sheffield, bore him three children, Anne, Margaret and John. She died in 1960. Five years later he married Kathleen McCombie, also a doctor and graduate of Aberdeen, who survived him.

After retirement, Edward Rhind found another special role in which he could help his hospital, by accepting the chairmanship of an outstandingly successful centenary appeal for the Northern General Hospital, which funded an important intensive care unit for neonates. In this post retirement work he found great satisfaction, as it allowed him to continue his association with the hospital in which he did so much of his life's work.

DSM

[*Brit.med.J.*,1986,**293**,1035; Photo]

RICHARDSON, George Oglethorpe *b.*31 Jan 1913 *d.*26 Mar 1984
MB BChir Cantab(1937) MA MD(1942) MRCP(1943)
FRCP(1959)

George Richardson was born in Newcastle-upon-Tyne. He was the fourth son of a prominent dental surgeon, who was chairman of the representative council of the British Dental Association. George died suddenly, though not altogether unexpectedly, for he had severe angina.

He was educated at Durham School, St John's College, Cambridge, and in clinical medicine at Newcastle. During this time he grew up to be a tall, fair, somewhat fastidious and well-dressed young man, a good tennis player and a keen cricketer.

After house jobs at the Royal Victoria Infirmary, Newcastle, and at Addenbrooke's Hospital, Cambridge, he trained in pathology in Dible's [*Munk's Roll*,Vol.VI,p.149] department at the Postgraduate Medical School, Hammersmith. On the outbreak of war in 1939 he was 'evacuated' to look after the pathology department at Hillingdon Hospital. Later he returned to Newcastle, where he married Jean Carrick. There he worked in the department of pathology under, and at times at odds with, Bernard Shaw. At this time he collected material for his MD thesis on the subject of renal artery occlusion in essential hypertension, but then felt he must join the armed forces. It was as well that he had written and submitted his thesis (with success) for when he left to join the RAFVR his renal artery specimens, along with others that he had carefully collected, were ordered to be incinerated. In the RAF he worked for and obtained in 1943 the MRCP London. He became a graded specialist in medicine and served in this capacity both at home and in India.

After he was demobilized he went as a medical registrar to the Newcastle General Hospital, and was soon appointed a consultant physician at that hospital. During his years there he ran a most efficient general medical unit, and also started a diabetic clinic which soon became the largest in the region. He made a number of notable contributions, mainly on the subject of diabetes mellitus. Essentially he was a first class general physician who was devoted to his patients, as they were to him.

Intensely loyal to his hospital, and determined to enhance its standing and reputation, he soon became involved in hospital and local medical politics, and his was an important influence in bringing the General Hospital fully into the University of Newcastle's group of teaching hospitals - *before* the disastrous reorganization of the National Health Service in 1974. For many years he was personally responsible for the teaching of internal medicine to the University's dental students.

Richardson was elected a Fellow of the College in 1959. After serving his apprenticeship as an examiner in internal medicine for the London Conjoint Board, he was appointed examiner by the College in 1968 and served as such until 1978. He was a councillor from 1969-72.

Many of those who had to negotiate with Richardson did not find him at all an easy man. He had decided views and not much time for those whose standards he considered lower than his own, or who were thought by him to be self-seekers. On the other hand he would support to the full those whom he admired. This search for the best, if not the perfect, was also exemplified in his domestic life by a continuous replacement of fireplaces and other heating appliances, and by a succession of new lawnmowers - he insisted on beautifully maintained and weed free turf. These foibles were endearing rather than the reverse, for he was apparently concerned for the general good, was not a zealot, and had a real sense of humour, evinced by his enjoyment when his leg was pulled about his various eccentricities.

At a comparatively early age he developed severe hypertension, in spite of which he led a rewarding and productive life. This was partly due to the efficacy of modern therapy when carefully and somewhat obsessively followed, but also to Richardson's ability to encapsulate this part of his life, for neither this nor any of his concomitant disabilities were allowed, to the limit of physical endurance, to interfere with his activities. Nobody who was not very close to him could have guessed how much he suffered from his very considerable handicaps.

George was a devoted family man, having two sons and a daughter - all themselves with children; very much the shepherd of his flock. He retired to a Yorkshire village and was active in local community affairs. He owned some two acres of land, which he and his wife tended with loving care. In return the land produced ample fuel for the sequence of wood burning appliances which he installed year by year in his sitting-room.

GS

[*Brit.med.J.*,1984,**288**,1315; Photo]

RIDDELL, Ronald William *b.*9 Nov 1913 *d.*21 Nov 1984
MRCS LRCP(1939) MB BS Lond(1947) MRCPE(1947) MD(1948) FRCPE(1951) FRCPath(1964) *FRCP(1972)

'Roland' Riddell was born in London, son of William Henry Riddell, a musician, and his wife Ada, daughter of John Littlemore, an engineer. Later in life he chose to modify his first name to Ronald. He was educated at William Ellis School, Highgate, and at St Mary's Hospital Medical School, qualifying with the Conjoint diploma in 1939.

The early part of his career, like that of most of his contemporaries, was largely determined by war service. In the RAMC from 1939-45, his first postings were in Scottish Command, where he started his training as a pathologist; later he served as pathologist to a military hospital stationed at Caserta, and finally as assistant director of pathology to the Central Mediterranean Force. On demobilization he was appointed senior lecturer in medical mycology at the London School of Hygiene and Tropical Medicine. In 1948-49 he was Commonwealth Fund travelling fellow in microbiology at Duke University, Carolina, and Columbia University, New York. During this time his interest in medical mycology was fostered, especially at Duke. On his return to London he was appointed senior lecturer in this subject at the Institute of Dermatology. His appointment as consultant bacteriologist to the Brompton Hospital in 1951 provided him with the main focus of his work for the rest of his career, but he maintained an interest in the mycology of the skin through part-time appointments at the Institute of Dermatology and St John's Hospital.

At Brompton the pathology services of the hospital were being expanded, and the newly formed Institute of Diseases of the Chest was developing academic departments in some laboratory disciplines. J R May was appointed to head the Institute's department of bacteriology. Relations between the clinical and academic sides were not always easy, but a *modus vivendi* was established by which research activities were divided; Riddell concentrating on the mycobacterial infections that were then prominent in the work of the hospital, and on his established interest in fungal infections, while May devoted himself to other bronchopulmonary bacterial infections, especially those complicating chronic bronchitis. For the hospital Riddell developed a comprehensive bacteriological service, and was active in the formulation and implementation of policies for the control of infection and the use of antibiotics; in research, he made important contributions to the early studies of the role of *aspergillus* in human lung disease. He became additionally responsible for bacteriological services at the London Chest Hospital in 1970, and at the National Heart Hospital in 1972, after the incorporation of the three hospitals into the National Heart and Chest Hospitals. His principal publications were contributions to textbooks, relating to fungus diseases of the lungs and to mycobacterial infections. In the College, he gave the Mitchell Lecture in 1974 on 'The modern chemotherapy of tuberculosis'. He served as a member of the microbiology committee on the International Union against Tuberculosis, and as chairman of the Acid-Fast Club in 1972.

Ron Riddell was of average height, always dressed in a conventionally professional style, even after less formality had become usual. His conversational manner was careful and considered. Although his working contacts with his colleagues were generally urbane, he seemed unduly sensitive to possible affront. In his younger days he played tennis, and was captain of the St Mary's team. He was fond of music, especially early Italian, and after he had come to live in Chelsea became interested in the history and conservation of that borough.

He was twice married. By his first wife, Sheila, daughter of John Douglas, a civil servant, he had two sons - both of which followed him into the medical profession through St Mary's Hospital Medical School and became general practitioners. The breakdown of this marriage unhappily led to his losing all contact with his sons. His second marriage was to Margaret Ann Lewis, a scientific officer in the haematology laboratory at Brompton.

JGS

[*Brit.med.J.*,1985,**290**,164]

RILEY, Michael John *b.*12 Oct 1935 *d.*1.9.87
MB ChB Cape Town(1959) MRCP(1968) FRCP(1980)

Michael John Riley was born in Coventry but in 1947, to avoid the expected evils of the National Health Service in England, his father, a general practitioner, brought his family by sea to Cape Town - en route for the opportunities of postwar Southern Rhodesia. Tragedy struck on the journey and, after a short illness which started at sea, Mike's mother died soon after disembarking. The father and two sons continued on their planned journey, and John Riley became a well respected practitioner in Salisbury for more than 30 years.

Mike completed his secondary schooling at St George's College, Salisbury, and his undergraduate medical training at Groote Schuur Hospital and the University of Cape Town, graduating in 1959. Returning to Salisbury, he became one of the first interns at Harare Central Hospital in 1960, working there for a further year before returning to England for training in internal medicine and rheumatology at the Hospital of St Cross, Rugby,1962-65, the MRC rheumatology research unit, Taplow,1965-67, and the Royal Berkshire Hospital, Reading, 1967-69. He took up the post of lecturer in the department of medicine at the then University of Rhodesia, becoming senior lecturer three years later. In 1977 he became dean of the faculty of medicine for three years, and in the same year was appointed inaugural professor and chairman of the department of clinical pharmacology, holding the latter post until leaving the country, now Zimbabwe, in 1983. He was appointed dean of medical studies at the Military Hospital, Riyadh, Saudi Arabia, in 1983. He first became ill while on holiday in Corsica with his family and returned immediately to England.

While working in Rugby, Mike met Kathleen, a little dark-haired Irish theatre nurse, and they married in 1963. They were an ideally contrasted couple, with Kay's vivacity and forthrightness affording an ideal foil for his quieter and more intense disposition. They had three children, two boys and a girl.

Although Mike's principal interest was rheumatology, a field in which many patients were helped by his particular expertise, kindly interest and sympathy, he was a true general physician - a necessity in the context of the developing country in which he spent the major part of his practising life. He undoubtedly had more than a passing interest in clinical pharmacology, but his moving into a new department specializing in this area probably had more to do with his ability to influence administrators to his own advantage and a desire to move out of a department in which he felt restricted, than by a prime concern for drugs and their clinical applications. Mike read widely, remembered well, thought logically, and could pass on his knowledge to others, to which many generations of students benefiting from his quiet and thorough persistence will bear witness for years to come. To his colleagues he was ever a haven of sound advice in troubled times, becoming one of the vital points of reference and stability to which younger, less experienced, or simply less confident, colleagues found themselves turning; a frequent happening in a young medical school. An able administrator and committee man, his agenda was often over-loaded. Nevertheless, he was always better prepared than most - if not all - and his knowledge and opinions demanded attention and found frequent agreement.

Mike's leisure interests were wide, and varied from his early passion for motor racing and an active involvement with the Territorial Army, as captain with the RAMC and 7th Royal Warwickshire Fusiliers, to his enjoyment of bridge, golf and trout fishing. As in his work, his devotion to his hobbies was deep and brooked no external interference. In times when personal safety could not be taken for granted, during the Independence War in Rhodesia, fishing isolated trout streams in the eastern districts was never given up in deference to guerilla activity in the area but simply required the addition of an Ouzi submachine gun to the more usual items of fishing equipment. The bag did not seem to suffer!

But, above all, Mike was a family man, who loved also the company of his friends. Relaxing weekends round the pool and barbecue would find him happily splashing with the youngsters, smoke-shrouded over the grilled food or languorously stretched, chatting away the afternoon sun in his soft drawl, his conversation spiked with wit and frequent reminiscences of his student days.

Following a year's illness and disabilty, remarkably borne and devotedly attended by his wife, he died in Leicester.

RID

[Photo]

ROBERTS, John Alexander Fraser *b.*8 Sept 1899 *d.*15 Jan 1987
CBE(1965) BSc Wales(1920) MA Cantab(1925) FRSEd(1927) DSc Edin(1933) MB ChB(1936) MD(1943) MRCP(1943) FRCP(1949) FRS(1963) FRCPsych(1971)

John Fraser Roberts, though he only qualified in medicine when he was 37, was a pioneer in medical genetics, one of the founding fathers of clinical genetics and, with a handful of others, one of the leaders in human genetics. He acquired an international reputation.

From early researches on animal genetics, where he concentrated on complex characters demonstrating a blend of inheritance and environmental effects , both of nature and nurture, he soon moved to work on similar characteristics in man. He became well known for his studies on measured intelligence, and his work on the genetic nature of hypertension ensured that his was a household name in medical circles. His studies on the relationship between blood groups and disease turned out to be equally important. His research spanned a period of almost 60 years which saw many major changes in genetics in general, and in human genetics in particular, and the practical application of genetics by the medical profession to the service of man. It found Fraser Roberts not only receptive to current advances but also eager to help in practical ways, and it was to the practical application of genetic knowledge that he dedicated much of his thinking and working time.

John Fraser Roberts was born at Fox Hall, Henllan, Trefnant, near Denbigh, in North Wales, the son of Robert Henry Roberts and his wife Elizabeth Mary Fraser, being the eldest of their three children. Henry Roberts was a dedicated farmer, and he and his wife had great hopes that their first-born would follow in his father's footsteps and choose farming as his future career. But this was not to be. As a young boy, living on a large farm, he displayed instead a keen interest in aspects of inheritance, especially in sheep - with which local farmers were preoccupied and which were largely defeating their attempts at selective breeding for treasured characteristics. It was this that first fired his lifelong interest in two parallel directions: the study of complex traits seemingly inherited, and the application of the findings and the principle of inheritance to practical purposes. Thus, he decided upon a career in genetics which, on his own admission, represented an 'honourable escape' from practical farming while still being relevant to animal husbandry - so fulfilling the desires of his parents in a roundabout way.

Fraser Roberts' brother, Robert Peter, practised at the Bar in London and, later, was dean of Corpus Christi, Cambridge. His sister, Marjorie, obtained a botany degree at Bedford College, London, and remained at the family farm to help out as her mother died of pulmonary tuberculosis at the age of 58, leaving their father to attempt to manage the farm single-handed.

John himself was educated at Denbigh County School, but for two years during his 'teens he was unable to attend school as he was 'delicate and ill'. However, he had excellent tuition at home at this time and so began his rather unconventional 'formal' education and studies, which often seemed to be dictated by opportunity or chance. In the autumn of 1916 he entered the University College of North Wales at Bangor, where he studied for his BSc degree - with a break of two years on war service, 1918-19, when he served as a second lieutenant in the Royal Welsh Fusiliers. In October 1920 he was admitted as an affiliated student from University College, Bangor, to Gonville and Caius College, Cambridge, obtaining his BA in the natural sciences tripos, followed by his MA in 1925, being first a Rhonodda student and then an exhibitioner. In 1923 the Ministry of Agriculture gave a grant to the agriculture department at University College, Bangor, - headed by R G White - for a study on the wool of Welsh mountain sheep, and John Fraser Roberts was appointed to do the work. But by the end of the year the scope of the research project had increased and the research was transferred to Edinburgh. So, while continuing to supervise experimental work at Bangor as an honorary lecturer in genetics, he was appointed research assistant in the Animal Breeding Research Department at Edinburgh (later the Institute of Animal Genetics) to work under the supervision of F A E Crew. He continued his studies there and was elected a fellow of the Royal Society of Edinburgh in 1927. His work on the wool of Welsh mountain sheep continued at Bangor and between 1925 and 1931 he was responsible for the joint abstracts on livestock for the *Welsh Journal of Agriculture.* In 1928 he was appointed head of the biology department of the Wool Industries Research Association at Leeds, a post which he held until late 1931 when he was granted a Macaulay research fellowship in human biology at the University of Edinburgh. During this period he obtained his DSc with his thesis 'Studies on the biology of sheep' in 1933, and immediately after this he enrolled in the faculty of medicine at Edinburgh, qualifying in 1936.

From 1942-46 he served in the RNVR as a surgeon lieutenant and later as a surgeon commander. He was a consultant in medical statistics to the Royal Navy.

In 1943 he obtained his MD for a thesis which was highly commended, 'Blood group frequencies in south-western England and North Wales: A study in racial variation, together with a search for evidence that the blood groups possess selective value; and other contributions to human genetics.' In the same year he obtained his membership of the College.

In 1933 Fraser Roberts had been appointed principal investigator at the Burden Mental Research Department, Stoke Park Colony, Bristol, and in 1940 he became its director, a post he held until 1957. During this time he had a chance of renewing his association with R A Fisher, whom he had met in 1924 while working with Crew, and with whom he had instantly been in tune. By then Fisher was Galton professor of eugenics at University College London and a member of the research advisory committee at the Burden. This relationship with Fisher was clearly material in refining still further John Fraser Roberts's statistical skill and in indicating the direction of some of his researches, particularly his work on the population variation of blood group frequencies and on measured intelligence; the latter being of special importance to him in view of his own deep interest in mental handicap. By this time his dedication to medical genetics was spurring him on to putting his genetic expertise to practical use.

He was appointed a consultant in medical genetics to the Royal Eastern Counties Hospital, Colchester, in 1946, where he started genetic counselling clinics which he continued until 1981, and he was director of research there until 1957. That same year, 1946, he had started a genetic counselling clinic at the Hospital for Sick Children, Great Ormond Street; the first ever recognized clinic in the British Isles, and the second or third in the world. Later, he started another at the Children's Hospital, Bristol. He was also appointed lecturer in medical genetics at the London School of Hygiene and Tropical Medicine. In 1957 the Medical Research Council established a clinical genetics research unit at the Institute of Child Health, Great Ormond Street, and Fraser Roberts was appointed director, a position he held until his retirement in 1964. The unit's major contribution was the execution of a whole series of important family studies carefully carried out and relevant to medical genetics both in theory and practice, especially directed at unravelling some of the genetic factors underlying abnormal development. After his retirement from the Institute he was appointed geneticist to the newly established paediatric research unit at Guy's Hospital medical school, and honorary consultant in medical genetics to the Hospital, where he continued with research and took an active part in the establishment of a model, comprehensive, population-based genetic service backed up by laboratory and genetic research facilities and general services; the first of its kind.

In 1949 he was elected a Fellow of the College. Through his work on blood groups he became president of the Royal Anthropological Institute of Great Britain and Ireland, 1957-1958, and his statistical skills led to his election as president of the Biometric Society 1960-1962. He was elected a fellow of the Royal Society in 1963, and received his CBE in 1965. In 1971, in recognition of his work on intelligence, mental abnormality and mental illness, he was elected a founder fellow of the Royal College of Psychiatrists. And for 37 years he was associated with the Eugenic Society, both as a member of council and of its various committees. From 1959-1970 he was adviser to Oxford University Press and series editor for the Oxford monographs on medical genetics, a function he shared with C O Carter (q.v.) from 1975.

His book *An Introduction to medical genetics*, London, Oxford University Press, 1940, deserves special mention because of its seminal influence derived from its timeliness, its approach, its style and content, all of which contributed to its popularity. It was more successful than any of its predecessors or successors which were written in a different vein and were not, as one physician said of them, 'bedside genetics', which Fraser Roberts' book clearly managed to be; therein lay the secret of its success. Originally conceived one summer evening in 1935 after his friend, Duncan Duthie, then a medical student in his final year, asked him to explain '...this genetics business...', it ran into eight editions; Marcus Pembrey being co-author of the last two books. Some 19 years later, with the confidence born out of experience, he was outlining clearly nine main pillars on which to rest professional genetic counselling, such as the importance of accurate diagnosis, knowledge of the literature, explanation in terms of odds, the use of empirical chances and the assessment of the size of risk against reasonable yardsticks.

During the second world war he served in the RNVR, first as a surgeon lieutenant and later as a surgeon commander. He was a consultant in medical statistics to the Royal Navy and concerned, among other matters, with the analysis of sickness statistics. Between the two wars he held a commission in the Territorial Army.

In 1941 John Fraser Roberts had married the actress Doris Hare, by whom he had two daughters. They were a remarkable couple, ostensibly very different; Doris Hare's ebullient personality, wit, gift for mimicry, quick turn of phrase and brilliant miming, contrasted vividly with Fraser Roberts' incisive but shy approach, spare gesturing and controlled neatness. He was always pleased with his wife's jests and perfectly at ease with her ability to entertain and amuse the most solemn crowd of scientists. It was sad when the marriage broke up, but he was fortunate in his second wife, Margaret Ralph, whom he married in 1975. She had been his assistant since 1948, from the days of the London School of Hygiene and Tropical Medicine. Having helped him in his work for so many years, she helped him in his final retirement with unfailing devotion.

By temperament John Fraser Roberts was mild and gentle, quiet and courteous, and - like the style of his book - somewhat Edwardian in his manners and beliefs. He was extremely precise both in the investigation of facts and the study of data, and also in the wording he used to describe his findings. He was also charming and helpful, sympathetic and attentive, and an excellent listener. He could not fail, given his wide knowledge and expertise, to be a model genetic counsellor, always ready to advise but never desirous of pushing his own views. But it must be said that for all his sympathetic stance and courtesy when it came to work he was very determined and held strong views, and he could be stubborn once he had made up his mind.

His principal recreation was walking, especially in the hills of his native land. Although always active physically and mentally, he found relaxation in his passion for computing on everything: the stock market, elections and by-elections, and anything else that came to hand. He was a religious man, brought up by his mother in a strict Scottish Presbyterian manner. As a scientist he had some doubts but, especially in later life and with failing health, he found comfort and relief in religion.

As a young man he had been treated for tuberculous lymphodentitis of the neck. In 1974 he suffered from an attack of ulcerative colitis which was successfully treated. By 1982 he had become practically blind from a progressive macular degeneration and a year later his heart had given occasion for anxiety, but with treatment he was able to continue with his favourite recreation, walking, right up to the end, which came suddenly and quietly while he was taking an afternoon nap.

PEP

[*Brit.med.J.*,1987,**294**,379; *Lancet*,1987,**1**,341-2; *The Times*,19 Jan 1987; *Guy's Hosp.Gaz.*,1976,**90**]

ROBERTS, John Clifford *b.*31 Oct 1905 *d.*28 Mar 1988
MRCS LRCP(1931) MB BS Lond(1936) MD(1939) MRCP(1941) FRCP(1969)

John Roberts was consultant physician at Harefield Hospital for many years until his retirement in 1970. He was brought up in Swansea, the only surviving boy in a family of girls,and was educated at Swansea Grammar School and the University of Wales. He was originally destined to follow his father into business as a builders' merchant, but it soon became clear after a trial period that he had no taste for business affairs. One of his sisters, a nurse, suggested that he might make a good doctor and therefore he went up to St Bartholomew's Hospital where he qualified in 1931. In later years it gave him particular pleasure that his daughter Marion followed in his footsteps and also entered Bart's to study medicine.

John returned to Wales after qualification and worked for a number of years in general practice not far from Maesteg. He had always been fond of children and decided to specialize in chest medicine with particular reference to its paediatric aspects. He took up a number of house posts in hospitals in East London, culminating in a post as resident medical officer at the Queen Elizabeth Hospital for Children in Hackney. From 1941-1943 he was medical officer for Middlesex County Council, then being appointed as consultant chest physician to Harefield Hospital where he soon distinguished himself as a shrewd diagnostician. At that time, tuberculosis and bronchiectasis were very common conditions both in children and adults. It is now commonplace for chest physicians to do their own bronchoscopies, but John Roberts was one of the early practitioners of the art among the physicians of his day.

John Roberts was devoted to his work and to his patients, in whom he took a fatherly interest. He would often go to the hospital on Sundays to see relatives, some having come from considerable distances to visit the sick patients. Many of the children suffered from extrapulmonary forms of tuberculosis, such as tuberculous spines, and required long periods of hospitalization. The links between John Roberts and his patients became very close and many of them kept in touch with him for years after their recovery, and even after he had retired.

He was dedicated to the principles of the National Health Service and steadfastly refused to undertake any private work. His opinion on difficult cases was eagerly sought by his colleagues, both senior and junior, and was always given freely and unsparingly. He moved some of his sessions away from the rarified atmosphere of the Chest Hospital and served as district health physician at Hemel Hempstead during the latter part of his consultant career.

In 1940 he married Mary Gertrude Tate, a medical practitioner, who died in 1963. They had one daughter. Some three years later he married again, and much enjoyed the music which his second wife brought into his life. John was always proud to be a Welshman and had a natural flair as a reconteur; he could illuminate any situation with a suitable story from his vast repertory.

After retirement, John Roberts began a new career as a ship's doctor, during the course of which he visited Brazil and also sailed up the Nile. He took up painting, attended classes in Spanish, and travelled extensively. He was greatly attached to his garden, where he particularly enjoyed tending roses. He was always willing to lend a hand to his fellow doctors, doing locums both in general practice and as a chest physician. He continued to read and discuss medical topics, and was a regular attender at meetings of the British Thoracic Society.

He was survived by his second wife and by his daughter.

JHA

[*Brit.med.J.*,1988,**297**,354; Photo]

ROBERTSON, Sandy Edwin John *b.*18 Oct 1915 *d.*14 July 1988
BM BS Sydney(1939) MRACP(1947) MRCP(1951) FRACP(1961) FRCP(1974)

Sandy Robertson was born at Collarenebri, New South Wales, Australia, where his father Frederick Gordon Robertson was a general practitioner. He was educated at Cranbrook College, Sydney, gaining his colours in rugby, rowing and athletics. He graduated in medicine from Sydney University, pursuing his clinical studies at the Royal Prince Alfred Hospital, and spending his first year after graduation as a resident medical officer at St Vincent's Hospital, Darlinghurst.

In 1940, following the outbreak of the second world war, Sandy enlisted in the Australian Army Medical Corps and was commissioned as a captain in the 2/10 field ambulance, 8th division. In the same year he was sent to Rabaul, where he was captured in December 1941 while endeavouring to escape by boat to Papua, New Guinea, with other officers and nurses of the field ambulance. All were taken by ship to Zentzuji Camp on Shikoku Island, Japan. During his three years as a prisoner of war, Sandy lost the sight of one eye due to trauma after capture, and developed diabetes mellitus. He suffered severe malnutrition and it was his unemotional view that the starvation policy in the prison camp helped to save his life.

On demobilization Sandy began his medical career in general practice, devoting considerable energies to the Karitane Mothercraft Movement. In 1947 he was appointed an honorary relieving assistant physician to the Royal Alexandra Hospital for Children, Camperdown. This was followed by the study of paediatrics both in Australia and the UK, where he obtained his membership of the College before returning to Australia. He rejoined the staff of the Children's Hospital on his return, being appointed honorary physician in 1963, elected to the board of management in 1972, and remaining active until his retirement in 1980, when he was appointed as emeritus consultant physician. He was also senior clinical lecturer in paediatric medicine at the University of Sydney.

Sandy was in the vanguard of the management of juvenile diabetes. He became very experienced in haemolytic disease of the newborn, and his knowledge and work in this field were highly acclaimed. He was honorary paediatrician to St Margaret's Hospital for Women and it was there he established the technique of exchange blood transfusion and performed more than 1000 such procedures. He became physician in charge of the haematology clinic at the Children's Hospital, in close liaison with V A Lovric, the hospital's haematologist.

Sandy Robertson was elected to the council of the Royal Blind Society of New South Wales, and served the Society for some 25 years as paediatrician and counsellor. In recognition of his important contribution, the Society named the Sandy Robertson Kindergarten, Burwood, in his honour.

The staff of the hospitals on which he served will remember his ready availabilty, his dedicated care of the sick, and his willingness to teach - both by his example as well as by his words.

In 1940 he married Gwenda Ashcroft, daughter of an electrical engineer, and they had two daughters, Edwina and Victoria, who gave them seven grandchildren. His interests outside medicine were golf, gardening and philately. His family all survived him. GW/VL

[*Aust.Paediats.J.*,1988,**24**,000-000; Photo]

RODGERS, Robert Ewing *b.*21 Mar 1911 *d.*2 Jan 1985
BA Cantab(1932) MRCS LRCP(1935) BChir(1935) MB(1938)
MD(1940) MRCP(1941) FRCP(1969)

Robert Ewing Rodgers was born in Sunderland, Co Durham, where his father, Albert Nathaniel Ewing Rodgers, was a doctor whose wife, Edith Isabel Jane Sim, was the daughter of a research chemist.

Robert was educated at Oundle School, Queen's College, Cambridge, and St Thomas's Hospital, London. After qualification he went into general practice. In 1939 he suffered an accident and was advised to give up medicine, which he ignored and proceeded to undertake various house appointments, including the Evelina Hospital for Children. He obtained his doctorate in 1940 and was then appointed registrar to the Sheffield Royal Infirmary under Robert Platt, later Lord Platt [*Munk's Roll*,Vol.VII,p.470]. After obtaining his membership of the College he entered the RAF Reserve as a medical specialist, with the rank of squadron leader, but was invalided out in 1943 with a prolapsed disc. He was chief assistant to James Livingstone (q.v.) at the Brompton

Hospital (evacuated to Horton) and then at the Royal Free Hospital under Theo Hoskin [*Munk's Roll*,Vol.V,p.201].

In 1946 Robert Rodgers was appointed consultant physician to Acton Hospital, and in the same year to the Bolingbroke Hospital,holding both posts until his retirement in 1976. He was elected a Fellow of the College in 1969.

The Bolingbroke Hospital was his centre of gravity and from the knowledge he had gained during a period in the USA in 1967, and his own enthusiasm, he set up one of the first coronary care units in the country. He was primarily a clinician and practical doctor, displaying considerable enthusiasm for his subject and defending his beloved hospital in committee.

Rodgers displayed a certain impatience with those who did not accept his point of view, and had little insight into the reason why others sometimes grew impatient with him - both in regard to his peers and occasionally with patients who refused to accept his advice. But he repaid loyalty and hard work among his juniors and secretarial staff by championing their cause if their position were threatened, and he took an interest in furthering their careers. He was a perfectionist for minor detail and time was of no importance if he wished to make a point.

He was an earnest and sincere physician, capable rather than brilliant, with a somewhat petulant manner if crossed but holding his patients' interests very much at heart.

He married Dorothy Joyce, née Riddle, in 1940 and they had three children, a son and two daughters. He was greatly devoted to his family and supported by a strong Christian faith.

MD

[*Brit.med.J.*,1985,**290**,476-7; Photo]

ROGERS, Howard John *b.*18 June 1943 *d.*30 Mar 1987
MRCS LRCP(1968) BChir Cantab(1968) MA MB(1969) PhD Lond(1972) MRCP(1973) FRCP(1985)

Howard Rogers, professor of clinical pharmacology at Guy's Hospital medical school, died at the age of 43. His early death was a tragic loss not only to his family but also to his subject and his medical school.

He was born at Crayford, the son of George Rogers and his wife Vivien Welch, and was educated at Sidcup and Chislehurst Grammar School. He won an open major scholarship in natural sciences to Downing College, Cambridge, and an open clinical scholarship to Guy's Hospital medical school. After qualification his first appointment was as house physician to Ralph Kauntze and Charles Joiner. He showed an early interest in the action and use of drugs and started laboratory research at Guy's while still a student. From 1969-72 he was Governor's research scholar in the department of pharmacology at

Guy's and was awarded a PhD for his work on the cellular action of psychotropic drugs. His colleagues in the laboratory rapidly became aware that Howard was a gifted research worker with a most valuable obsessive streak. He rapidly came to appreciate the finer points of such difficult techniques as amino acid analysis and the measurement of drug levels using high performance liquid chromatography.

Howard was firmly of the opinion that the practice of clinical pharmacology within a medical school required a firm grounding in general medicine, with practical experience in the use of drugs, and he was appointed junior lecturer in medicine and later medical registrar at Guy's.

In 1974 he returned to clinical pharmacology and became the Medical Research Council's Wellcome travelling fellow, in Paul Talalay's department at the Johns Hopkins Hospital in Baltimore. He fitted in well with the American scene and formed lasting friendships with several of his colleagues, in particular with Paul Leitman, professor and head of clinical pharmacology. On returning to Guy's he became, successively, lecturer, senior lecturer and reader, and - on the retirement of J R Trounce - was elected to the chair of clinical pharmacology, which he held for two and a half years until his death.

Howard Rogers combined to a remarkable extent a formidable intellect with an ability to deal with everyday affairs. After obtaining his doctorate his main scientific interests were in pharmacokinetics, where he achieved a considerable reputation, and at the time of his death his particular interest was in the kinetics of cytotoxic drugs and the relationship to cancer chemotherapy. In addition to his commitment to general medicine, he worked regularly in the oncology clinic and developed a considerable expertise in cancer chemotherapy. His interests, however, extended over a wide range of subjects. The institution of the Guy's drug research unit, where he was in medical charge, resulted in his becoming involved in a variety of Phase I studies, and his extensive knowledge and common sense proved invaluable in this difficult area.

He wrote fluently and easily and published, in his short career, well over a hundred papers as well as being the leading author of five books, including the definitive *Textbook of Clinical Pharmacology* 1981.

He was a good organizer and was very active in teaching at both undergraduate and postgraduate level. He instigated an intercalated BSc in clinical pharmacology at Guy's which proved both popular and successful, and in a relatively short time a remarkable number of postgraduate students passed through his hands.

His integrity enabled him to form a sensible and fruitful relationship with the pharmaceutical industry, at a time when some looked on such relationships with suspicion, and his ability to see reality in complicated situations led to his to being in increasing demand on committees and advisory bodies.

On first acquaintance he sometimes appeared a little brusque and retiring, but his humanity was never far beneath the surface. Although he disliked cant, hypocrisy and stupidity, and was very good at diagnosing all three, he respected human dignity, was exceptionally pleasant and gentle with his junior staff and never pulled the carpet from under their feet.

Howard's intellect illuminated everything to which he addressed himself, yet he was never cold and analytical. He was in fact a sensitive, kind and understanding person, who could take enormous trouble over those who came to him for advice or with problems, be they patients, colleagues or juniors. He had a considerable sense of humour, albeit on the dry side.

Like many other highly intelligent people, Howard Rogers had a sensitive appreciation of the arts. His knowledge of modern literature was extensive, his love and understanding of classical music, particularly Bach and Mozart, and - strangely enough - Prokofiev, was profound.

He married Moira O'Boyle in 1968, whose family came from County Mayo, and they had three daughters. Howard was very much a family man and outside his work most of his activities were centred on his home. He was not an enthusiastic traveller and, apart from the requirements of his work, limited himself to an annual holiday in Ireland, a country for which he had a great affection.

Howard was a religious man, as were his family. Although his faith was rarely apparent on casual acquaintance it was a major unifying factor in a complex personality.

JRT

[*Brit.med.J.*,1987,**294**,1425; *Lancet*,1987 **1**,932; *The Times* 9 Apr 1987; Photo]

RUGG-GUNN, ***Surg.Capt.*****Mark Andrew** *b.*2 Aug 1908 *d.*7 Mar 1988
BA Cantab(1930) MRCS LRCP(1933) MA MB BChir(1934)
MRCP(1940) MD(1947) FRCP(1958)

Mark Rugg-Gunn was born in Ullapool. His father, Andrew Rugg-Gunn, then in general practice was later to become an eminent ophthalmic surgeon. His mother, Gertrude Martha Smith, was the daughter of a business man.

Mark was educated at Sherborne, St John's College, Cambridge, and St George's Hospital - winning the Thompson medal for medicine in 1932. He qualified in 1933 and joined the Royal Navy the next year, serving in HM ships *Challenger, Repulse and Vasna.* He gained his membership of the College in 1940 and his doctorate in 1947, writing his thesis on nutritional neuropathy in prisoners of war returning from Japanese custody. He was elected a Fellow of the College in 1958.

As a physician his career centred on the naval hospitals and he served in Chatham, Haslar, Plymouth, Malta and Trinomalee. His skills as a physician were notable for an encyclopaedic knowledge of general medicine and a capacity for penetrating analysis of the difficult clinical problem. He contributed a number of articles on various subjects to scientific journals.

When he retired from the Royal Navy, as a surgeon-captain, he was appointed senior medical adviser at May & Baker Ltd. He retired from industry in 1975 and worked for a time with the National Blood Transfusion Service, and he was still working on disability pension assessments until shortly before his death.

Mark Rugg-Gunn had many interests outside his chosen profession. He was a good linguist, fluent in Gaelic and Urdu and with a sound knowledge of French, Latin and Greek. He was immensely proud of his Scottish ancestry and was the author of a history of the Clan Gunn, published in 1982, and an enthusiastic bagpiper. His conversation was always erudite and wide-ranging and his reminiscences, particularly of naval life, highly entertaining.

Mark married twice, first in 1934 Hilda (Jill) Rowell who died in 1951, and subsequently Patricia Cowan, a naval nursing sister, who survived him together with five children and 10 grandchildren. Two of his children entered the medical profession, and one grandchild. GJM-T

[*Brit.med.J.*,1988,**296**,1204; Photo]

RUSBY, Norman Lloyd *b*.26 Oct 1905 *d*.22 Nov 1988
MRCS LRCP(1931) MA BM BCh Oxon(1931) MRCP(1933)
DM(1941) FRCP(1948)

Lloyd was the son of Edward Lionel Macpherson Rusby, a general practitioner in Brixton, and Katherine Helen, née Wright. He was always known to his friends as Lloyd, and to others as Lloyd-Rusby. Lloyd was not a family name; his father, when registering his birth, forgot the agreed names and his mother strongly objected to the name Norman - so the baby was referred to as Lloyd, and Lloyd he remained.

Lloyd was educated at boarding schools; Winchester House preparatory school, Lancing College and St John's College, Oxford, where he took an honours degree in physiology in 1928. After graduation in medicine, he held junior appointments at St Thomas's Hospital in 1932, then started his training in thoracic medicine at the London Chest Hospital and, during a house appointment there, obtained his membership of the College.

In January 1934 he was appointed resident medical officer at the London Chest Hospital, and subsequently medical registrar. He was elected physician to the hospital in 1936, and combined this with the appointment of medical registrar and tutor at the Postgraduate Medical School, Hammersmith.

During the second world war he was directed to be resident medical superintendent at the London Chest Hospital, where he and others were subjected to many high explosive and fire bomb attacks. It was at this time that he met and married his house physician, Elizabeth Broadhead.

Lloyd was in charge of the hospital when, in 1941, it received a direct hit from enemy bombers, a whole wing was destroyed and the evacuation of patients and the treatment of casualties were his responsibility. At the same time, he acted as assistant physician at the West London Hospital in place of a colleague on active service. He was also editor of *Tubercle*, and a member of the standing advisory committee on tuberculosis.

On cessation of hostilities in Europe and the return of colleagues from active service, Lloyd was called up for service in the RAMC. This was unfortunate since many senior hospital appointments left vacant during the war were now being filled. He was posted first to Normandy, then Brussels, and subsequently as consulting physician Middle East Land Forces, with the rank of local brigadier.

On his return to civilian life in 1946 he was appointed assistant physician to the London Hospital; this was followed by appointments to Benenden, King Edward VII Hospital, Midhurst, and civilian consultant in diseases of the chest to the Royal Navy. He was Nuffield visitor to East Africa in 1950, and he lectured for the British Council in Europe and the Middle and Far East. He gave the Mitchell lecture at the College in 1967, and was a member of Council. He was also examiner for the College, and for the universities of London, Cambridge and Wales.

From 1957-77 he was vice-chairman of the Chest, Heart and Stroke Association, a member of the British Heart Foundation and of the Metropolitan Hospital Sunday Fund, a member of the board of governors of the London Hospital and of the Hospital for Diseases of the Chest and a member of the British Thoracic Society and of the Association of Physicians of Great Britain and Ireland. He was joint editor of *Recent Advances in respiratory tuberculosis*, 4th and 5th editions, 1948 and 1959, and contributed papers on pneumonia, pulmonary embolism and infarction to *Chest Diseases*, edited by K M A Perry and Sir Thomas Holmes Sellors, 1963. Lloyd's MD thesis on 'Dermoid cysts and teratoma of the mediastinum' was published in the *Journal of Thoracic Surgery*, 13,169-222, June 1944.

Lloyd Rusby was the old-style English gentleman. He was a family man whose home was well ordered; he liked company, and those who enjoyed his hospitality will always remember the small Georgian silver salver on which he passed drinks to his guests and the slim-stemmed pipe he smoked. He had no great athletic ability, though he played cricket for his school on occasions.

Lloyd's meticulous dress and scholarly appearance, with an unhurried calm and equable temperament, gave confidence to his patients. His sound judgement and firmness of decision were reflected in the number of bodies on whose councils he served. His opinion and advice were sought both by senior and junior colleagues, who found him easy to approach with their problems. He had a good sense of humour and appreciated a joke, even against himself. He used a monocle for close inspection and this, together with other little mannerisms, made him an ideal subject for mimicry in the students' Christmas shows.

In his younger days, Lloyd was a keen lepidopterist and he had a very respectable collection of British butterflies; he used to spend weekends tracking down uncommon species, with two of his colleagues. After the war, however, this interest declined. He had no other hobbies, but he developed an interest in cricket and became a member of the MCC. On retirement his visits to Lords became frequent. His main relaxation was English literature and history.

Lloyd Rusby was an excellent committee man and this was probably his main contribution to medicine.

He married Elizabeth Broadhead, a doctor and daughter of an architect, Frank Arthur Broadhead FRIBA, in 1941 and they had three sons. They all survived him.

EHH

[*Brit.med.J.*,1989,**298**,24; Photo]

SALMON, Harold William *b.*2 June 1909 *d.*28 Mar 1986
MRCS LRCP(1932) MB BS Lond(1932) MRCP(1944) MD(1948) FRCP(1969)

Harold William Salmon was born at Brighton, Sussex, the son of Albert Salmon, a medical practitioner, and his wife Mary Owen, daughter of a naval engineer. He was educated at Epsom College, Surrey, and Middlesex Hospital medical school where he won many prizes including the Broderip scholarship. After graduation he held house appointments at the Middlesex, the Brompton Hospital and the Samaritan Hospital, where he worked for such illustrious physicians as E A Cockayne [*Munk's Roll*,Vol.V,p.76], George Beaumont [*Munk's Roll*,Vol.VI p.33], Sir Robert Arthur Young [*Munk's Roll*, Vol.V,p.467] and Sir Harold Boldero [*Munk's Roll*,Vol.V,p.43]. In 1935 he began work with his father in general practice at Rickmansworth, Hertfordshire, where he remained until after the war. He gained his doctorate and membership of the College before commencing his compulsory military service, during which he served as a medical specialist, with the rank of major, at the Royal Herbert Hospital, Woolwich. In 1949 he was appointed consultant physician at the East Ham Memorial Hospital, St George-in-the-East Hospital and Queen Mary's Hospital, Stratford.

Salmon was particularly interested in cardiology and his *Handbook of clinical cardiology*, London, Harvey & Blythe, was published in

1962. He was an excellent diagnostician and his memory for rare medical facts was phenomenal. He delighted in trapping unwary candidates for the MRCP when on ward rounds. He was also a kindly and courteous man, and many patients had good reason to be grateful for his calm, unhurried manner and the sympathetic way in which he shared their problems.

In 1934 he married Blanche Percy Piper, daughter of a dental surgeon, and they had three children, two sons and a daughter. Both his sons are Fellows of the College; Paul is a consultant physician and senior lecturer in gastroenterology at the Middlesex Hospital and Michael is a consultant paediatrician at Stoke Mandeville Hospital.

Outside medicine Harold Salmon had many diverse interests: he had gained his 1st XV colours while at Epsom College and repeated this distinction at the Middlesex Hospital, taking part in three Cup Ties; he was passionately fond of his old school and followed its fortunes through the years; comparative religion, Egyptian archaeology and - latterly - painting were abiding hobbies.

He was survived by his wife and family.

MAS/GW/VL

[*Brit.med.J.*,1986,**292**,1087; *Lancet*,1986,**1**,925; Photo]

SANDERS, Eric *b*.25 Apr 1916 *d*.21 Mar 1987
MRCS LRCP(1939) MB BS Lond(1940) MRCP(1949) MD(1949) FRCP(1977)

Eric Sanders was born Eric Schlesinger, in Berlin, the son of a prominent Jewish banker. His family came to England as refugees from Nazi Germany, and he secured his British nationality in the name of 'Sanders' in 1940, the year he graduated in medicine from the Middlesex Hospital, University of London. After house appointments in the Middlesex sector of the EMS, he enlisted in the RAF and served as a medical officer in the UK and the Middle East. After demobilization in 1946 he obtained his doctorate in medicine and his membership of the College.

Following medical registrar posts at St Stephen's Hospital and Hillingdon Hospital he began specialization in chest diseases as senior registrar at Bevenden Hospital, Brighton, and in 1953 as chest physician to St Helier Hospital and Merton Chest Clinic. He continued to serve in this post, being upgraded to consultant in thoracic medicine in 1963, until he retired.

The writer worked alongside Eric Sanders for some 24 years and came to respect and appreciate his integrity as a compassionate physician, who was always concerned for the personal and social problems of his patients. In spite of his considerable ability and wisdom as a clinician, he was a man without ambition and the manoeuverings and pragmatic judgements of medical politics were anathema to him.

He never forgot a friendly act, and both in his professional and private life he never criticized anyone unkindly. He was a man truly without a trace of malice. Although an active participator in cooperative research projects, he was not a good initiator. He was, above all, a shining example of 'the good and caring doctor'.

Eric never married, but had a wide circle of friends, and in his private life was sustained by his deep love of classical music and, in particular, grand opera. He was a Friend of Covent Garden from its inception. Despite the trauma of his childhood in Germany, he remained a devout Wagnerian. For many years he was an active member of Kingston Cycling Club, and he missed the company and the camaraderie when the long club outings became too much for him, but he still enjoyed the Veterans' Run.

After his retirement, activity became more restricted by the onset of angina and he died from a coronary thrombosis on the day he was to enter hospital for surgical assessment. He was survived by his sister, Susie.

HFH

[Photo]

SANDERSON, Gerard *b*.14 Nov 1912 *d*.5 Feb 1987
MB ChB Liverp(1938) MD(1943) MRCP(1941) FRCP(1964)

Gerard Sanderson was born in Didsbury, Manchester. His father, Samuel Sanderson, a respected secretary of an approved Friendly Society and a justice of the peace for Southport, was awarded an MBE by King George VI and made a Knight of St Gregory by the Pope for his services to the Catholic community. He married Annie Howell and they had two sons one of whom was Gerard; the other died in infancy. Their eldest child, a daughter, became a nun.

At the age of 12 years Gerard went to the diocesan seminary at Upholland to study for the priesthood. But this was not to be his vocation. At the age of 16 he left Upholland, matriculated and took a course in optics, leading to a fellowship of the Spectacle Makers Company in 1931. He practised for a time in Manchester, mostly on insurance work, and had an appointment at one of the Liverpool hospitals dispensing spectacles to consultants' prescriptions. Following a period in hospital for an appendicectomy, he decided to study medicine and in 1932 he registered as a medical student at Liverpool University, where he qualified with a gold medal in obstetrics. He was appointed house physician to Wallace-Jones [*Munk's Roll*, Vol.V,p.433], and after a period as registrar was promoted to resident medical officer. In the latter post he came under the influence of Henry Cohen, later Lord Cohen of Birkenhead [*Munk's Roll*, Vol.VII,p.106], for whom he always maintained a warm affection and the greatest admiration; both were reciprocated.

Because of a delicate constitution he was not accepted for military service at the outbreak of war. But he made a valuable contribution through an attachment to the Liverpool school of tropical Medicine. ICI had given a number of antimalarial compounds to A R D Adams for clinical trial and Sanderson played a part in that work which led to the final product proguanil (Paludrine). He was co-author of publications on the trials but he declined to have his name associated with the definitive one on Paludrine; characteristically, he reckoned that his contribution did not merit it. In fact, his assistance to the work of the Tropical School was of great value at a time when medical manpower in Liverpool was stretched to the limit and hundreds of Service personnel were being sent home with tropical diseases.

In 1944 Gerard married Gwyneth Owen, the daughter of a Welsh farmer. She had been the nursing sister in charge of his ward when he was a houseman. There were no children of the marriage.

Sanderson was appointed consultant in 1946 to the Royal Southern Hospital, a branch of the United Liverpool Teaching Hospital, to which he gave devoted service until his retirement, and afterwards as chairman of the hospital authority archives committee. He was also consultant to the Waterloo Hospital and to the Ormskirk General and District Hospital. Often he would refer to the latter as 'The University of Ormskirk' and always took his 'firm' there for a taste of work in a country hospital. By his example, and the standard he demanded, he transformed the medical service in that town. In his retirement he founded a medico-literary club for medical and dental practitioners associated with Ormskirk. He named it after Joseph Brandreth, a physician of the town who had been appointed to the staff of the Liverpool Infirmary in 1780.

Sandy, as he was known, became one of the most popular and respected physicians on Merseyside. An exemplary doctor, never seeking the limelight, he was endowed with an innate compassion and kindness. He was a man of quiet strength and, in causes which interested him, of drive and initiative. He had neither the desire nor, in literal truth, the time to climb the ladders to professional acclaim, either in college, specialist association or research. Such paths, he judged, could be explored by those whom he considered more able - and certainly more interested. His professional life was dedicated primarily to his patients and, only just in second place, to his students and disciples of whom there were many. His strategy in diagnosis and therapy was holistic - the whole patient, the family and sometimes the friends, and always the environment, physical and psychological, ethical and religious, all were studied and encompassed in his treatment. He was conservative in that he did not embrace every medical advancement immediately it was introduced, especially if it seemed to involve risk or trauma to his patient, but once convinced of the value of a procedure he was enthusiastic for it.

His belief in apprenticeship was as firm as that of any 18th century physician. He, and only he, would choose the students for his 'firm' and his housemen; no dean or any other authority was allowed to stand in his path. Once selected, those boys and girls became his family, and equally that of Gwyneth. They were lucky, for he was an inspiring teacher and ever loyal and supportive to them.

Sandy gave outstanding service to the Liverpool Medical Institution; he was largely responsible for raising funds for an extension to the building, for achieving the presence of HRH Princess Margaret at its opening, and also instrumental in persuading the Privy Council to approve its Royal Charter. To be its president was his sole ambition for professional office. He held every other: librarian, secretary, treasurer and vice-president. But when invited to the chair, 13 years before his death, he had to decline on health grounds. He would never start anything which he felt he could not carry through meticulously and with dignity. For many years the presidents who were elected were second choice to Sandy, who continued to decline the honour.

Sandy was only of moderate height and of slim build. He had simple tastes and few personal wants - although he appreciated comfort, good wine and good food. He was without flamboyance and, although not easily provoked to laughter, had a dry sense of humour which could be very revealing about a situation or person. Such comments would be accompanied by a very characteristic, almost French shrug of the shoulders and raising of eyebrows. That he was a stickler for correct procedure showed in the perfection with which he arranged many professional functions; he was particularly in demand to plan memorial services for colleagues, often himself delivering the address: in character, he left detailed instructions for his own which was to be, preferably, on a Saturday so that his colleagues would not have to miss clinics or ward rounds.

All things of beauty mattered greatly to him, especially the countryside, and the sea and the shore by which he lived for the greater part of his active life. He loved books and music, and played the piano effectively and with delicacy, often accompanying his wife whose singing he enjoyed. Unfortunately he was an inveterate smoker. As a result, he was to endure for many years much suffering and immense frustration, but always with courage, fortitude, and a determination to fulfill his commitments - as he did almost to the end.

Gerard Sanderson was not a religious man in the conventional sense but he remained in the faith of his parents. His attitude in this respect is best summed up in the words of Pasteur, in his speech on his reception into the Académie Française, April 27,1882: 'Happy is he who bears a god within himself, an ideal of beauty, and obeys him: an ideal of art, an ideal of science, an ideal of the nation, an ideal of the virtues of the Gospel. These are the living springs of great thoughts and great actions. All are illuminated by reflections of the infinite.' *Familiar medical Quotations*, ed.Maurice B Strauss, London, J & A Churchill Ltd., 1968.

TCG

[*Brit.med.J.*,1987,**294**,713-4; *Lancet*,1987,**1**,639; Photo]

SARGANT, William Walters *b*.24 Apr 1907 *d*.27 Aug 1988
BA Cantab(1928) MRCS LRCP(1930) BChir(1931) MA MB(1933) MRCP(1933) DPM(1936) FRCP(1949) FRCPsych(1971)

Dr William Sargant was the most important figure in post-war psychiatry. He was a rebel *with* a cause. He had the build of a rugby forward, narrowly missed a blue at Cambridge, but went on to play for the Barbarians and represented Middlesex and St Mary's Hospital. He was over 6ft tall, huge by any standards in stature and charisma. He was born at home in Highgate. His mother was Alice, née Walters.His father was Norman Sargant, a Methodist whose hero was John Wesley, and Will became very interested in religious conversion early on in his psychiatric career. He had several uncles who entered the church, and one of his brothers became the Bishop of Mysore, Church of South India.

Will went to a boarding school from the age of 7 years, initially at St Wilfred's School, Seaford, where he was thrashed for his rebelliousness until he reformed and became head of house and captain of football. At Leys College, Cambridge, where he was also head of house and captain of most sports teams, he chose medicine as a career because he was so bad at Greek. At St John's College, Cambridge, he failed anatomy fairly regularly while captaining the rugby team and being president of the Cambridge University Medical Society.

Before deciding to take up psychiatry, Will was house surgeon to the surgical unit at St Mary's, where he worked for Dickson Wright and Lord Porritt. He later became assistant on the medical unit, and was St Mary's youngest medical superintendent in charge of all admissions and nursing and junior medical staff. He took charge of the biochemistry laboratory for the medical unit. Not surprisingly, after this whirlwind career, he became ill with tuberculosis, which was to affect him again in 1954.

After three months of inactivity, he did a locum job at Hanwell Mental Hospital, where he was horrified by the lack of treatment, the incarceration and suffering. He was later delighted when Edward Mapother [*Munk's Roll*,Vol.V,p.266], who was to become his mentor, invited him to join the staff of the Maudsley Hospital. His therapeutic enthusiasm soon became evident in psychiatric circles. He used amphetamines for depression in 1936; insulin treatment for schizophrenia in 1938; and then the convulsive therapy (ECT) for severe depression. He helped phase out bromides in favour of barbiturates for sedation and did research on anorexia nervosa and the physiological basis of depression.

In 1938 Will Sargant was awarded a Rockefeller fellowship to Harvard. He returned to England as soon as war was declared, to work at the Maudsley Hospital's neurosis unit, Sutton Emergency Hospital, later to become Belmont Hospital, with Eliot Slater [*Munk's Roll*, Vol.VII,p.541]. There he saw thousands of acutely disturbed shell-shocked soldiers, and developed acute sedation with pentothal and amytal and ether abreaction with Joe Shorvon. His methods of early treatment must have saved thousands of psychiatric casualties after the second world war. He published *An Introduction to physical methods of treatment in psychiatry*, with Eliot Slater, Edinburgh and London, E&S Livingstone, 1944. It was to have a profound influence on postwar psychiatry and ran into five editions, five translations and five reprints.

Will's hectic schedule was cut short by pneumonia complicated by infectious hepatitis. He spent his convalescence reading Wesley and Pavlov, and finally published *The Battle for the Mind*, London, Heinemann, 1957, which has had an extraordinary influence on psychiatric thinking to the present day. Eliot Slater, then editor of the *British Journal of Psychiatry*, said of Will: 'You have done so much to help patients, to teach, and to keep British psychiatry on a sane road.'

Will Sargant had a lovely sense of humour, which was infectious. He had a deep affection for dogs and understood them better then most, and he did more than anyone else to bring Pavlovian physiology into clinical psychiatry.

He was the outstanding clinician and teacher of his generation. Even Sir Aubrey Lewis [*Munk's Roll*, Vol.VI,p.284], the greatest intellect of postwar psychiatry, asked the question'How did Will Sargant do it?'. Will was an excellent communicator. Even if you could not speak a word of English and had to communicate with him through an interpreter, and your culture was totally alien to Anglo-Saxon ways, there was no bar to being healed.

His work at St Thomas's Hospital was legendary. When he was physician in charge of the department of psychological medicine, more medical students entered psychiatry than from any other medical school. He had massive self-confidence, sometimes amounting to recklessness, therapeutic fervour, stubborn persistence and clinical intuition. It was this last quality that was the most remarkable. He had am amazing sixth sense - a talent for taking short cuts to knowledge. So many of his former patients have said 'Dr Sargant saved my life.' A rare statement in the case of other psychiatrists, but a commonplace with Will. Most of all, he gave his patients hope.

Among Will's registrars were David Owen, who later became Foreign Secretary and Minister for Health; Jim Birley, president of the Royal College of Psychiatrists, and Peter Gautier-Smith, later to become dean of the Institute of Neurology, The National Hospitals, Queen Square.

In Will's early days, he met opposition and therapeutic nihilism.

He had to do what the *New Testament* calls 'Doing good by stealth'. He was driven by the fact that, as he put it, 'No suffering can compare with the suffering of the mind or what has been called the dark night of the soul. It is well, sometimes, to wake up haunted by what still has to be endured, by neurotically and mentally sick patients either in hospitals or in their own homes.' So he wrote in his autobiography *The Unquiet Mind,* London, Heinemann,1967.

As David Owen put it: 'He understood the awful pain of depression which drove him to take risks for those who were depressed - because he knew they would take risks with their lives. Yet for him the truest tribute was to be honoured by those who knew him best - the patients and physicians who worked for him.'

Hs wife Peggy sustained him through all his battles. In his autobiography he wrote: 'My thanks are due to my wife who didn't want to be talked about too much in the book but whose presence permeates every page since without her I could no longer be myself.' DK

[*The Times*,31 Aug 1988; *The Independent*,5 Sept 1988; *The Guardian*,2 Sept 1988; *Brit.med.J.*,1988,**297**,789-90; *Lancet*,1988,**2**,695-6,859; *St Thomas's Hospital Gazette*,Spring 1967,**65**,No.1; *World Medicine*,18 June 1975; Photo]

SATHE, Ramchandra Vishwanath *b*.28 Nov 1905 *d*.28 Aug 1983
MB Bombay(1927) MD(1931) MRCP(1933) FRCP(1963)

Ramchandra Sathe was born at Ahmedabad, India, the son of the late Vishwanath Sathe, who was in government service, and his wife Lakshmibai. He married Kashibai Govind Awate, a doctor, in 1935 and they had one son, Sadashiv, who became an orthopaedic surgeon, and a daughter, Prabha, an obstetrician and gynaecologist.

His early education was at the Municipal School and Elphinstone High School, Bombay, and he pursued his undergraduate medical training at Grant Medical College and Sir J J Hospital, Bombay, where he gained the gold medal for clinical medicine and the Hewlett prize for medicine. From 1936-38 he was honorary assistant physician at St George's Hospital, Bombay, and from 1938-41 he was honorary assistant physician and tutor at Grant Medical College and the Sir J J Hospital. From 1941-55 he was honorary physician at the Hospital and teacher in medicine at the College. In 1955 he was appointed associate professor of medicine at Grant Medical College and later became successively a full professor, honorary director of postgraduate studies in medicine, and finally emeritus professor of medicine. He was also visiting physician at Jaslok Hospital, Bhatia General Hospital, the National Hospital, Bombay, Karnataka Health Institute and Ghatprabha and Talegaon General Hospital.

For three years, 1963-66, he was vice-chancellor of the University of Bombay, being nominated by the State Government. He had been a member of the senate and the faculty of medicine since 1952. He served on the health survey and planning committee of the government; the *ad hoc* committee (import/export) on policies for drugs; the drugs equipment standards committee; the Indian pharmacopoeia committee and the national formulary committee. Since 1951 he had been a member of the medical council of India and for some years was a member of its postgraduate medical education committee. He also served on the Maharashtra medical council from 1961, and for six years was a member of its executive committee.

In 1962 he was elected president of the Indian Medical Association, having been vice-president from 1952-55 and president of its Bombay territorial branch from 1956-57. He was president of the World Medical Association 1962-63, and deputy president of the Third World conference on medical education in 1966. He was also president of the Diabetic Association of India, and president of the Association of Physicians of India. From 1932 he was a member of the British Medical Association and an examiner for the MD degree at many universities. Ramchandra Sathe was an admirable physician with a high standard of medical ethics, a keen medical educationist, and an able teacher and scientist. He had few interests outside his profession. SCS

[Photo]

SAYERS, *Sir* Edward George *b.*10 Sept 1902 *d.*12 May 1985
Kt(1965) CMG(1956) KStJ MB ChB NZ(1924) DTM&H(Eng)
MRCP(1935) FRACP(1938) FRCP(1949) MD(1959) FRS
NZ(1961) Hon FACP(1957) Hon FRCPE(1960)

Sir Edward Sayers was formerly dean of the medical faculty and professor of therapeutics at the University of Otago, New Zealand. As dean of the Otago Medical School he succeeded Sir Charles Hercus [*Munk's Roll*,Vol.VI.,p.236] when he retired in 1958.

Edward Sayers was educated at Christ's College, Christchurch, and Otago University and medical school, Dunedin, where he graduated in medicine in 1924. He then came to the UK to study for his DTM&H at the London School of Hygiene and Tropical Medicine before becoming a medical missionary. From 1927 to 1934 he served in this capacity in the British Solomon Islands. A new hospital was built, and time was found to send specimens of mosquitoes to the London School of Hygiene and Tropical Medicine and snakes to the British Museum. Unfortunately, the mission was closed in 1934 for economic reasons. In 1940 Edward Sayers was awarded the Australian Cilento medal for his outstanding work for native welfare in the Pacific.

He used up all his meagre savings to come to London in 1935, did a cram course, and passed his MRCP at his first attempt. Once his money ran out he returned to Auckland and set up in general practice, but his manifest clinical abilities were soon in great demand and after 1938 he became a consultant physician, holding a visiting appointment at Auckland Hospital.

On the eve of the second world war, Edward Sayers was called up into the New Zealand Medical Corps. He was drafted to the Middle East in 1940, as a specialist in tropical medicine at the 1st New Zealand General Hospital, which went to Greece. When the Pacific theatre of war opened he was transferred to the 3rd NZ Division and was in command of the 4th General Hospital at Noumea, New Caledonia. He did important work on malaria prevention, was promoted to the rank of colonel, and awarded the US Legion of Merit in 1944.

After the war he returned to consulting practice in Auckland and became a leading physician. His grand rounds at Auckland Hospital soon became known, and the hospital became a showplace for open discussion and informed criticism long before it was common medical practice. Indeed openness was one of Edward Sayers' endearing traits; he was always prepared to stand up and be counted.

Sayers was chairman of the Dominion Committee of the Royal Australasian College of Physicians and its nominee on the Medical Council of New Zealand. He was a founder fellow of the College and the first New Zealander to be elected president, 1956-58, which involved much travel to Australia.

In 1958 he became dean of Otago Medical School, but he retained his clinical interests and the weekly round at his medical unit was a star attraction. With the splitting of the University of New Zealand into independent universities, several special schools including medicine lost their semi-autonomous position, and the deanship entered a difficult phase. But the clinical basis of the medical school was strengthened by new chairs, and Edward Sayers never forgot that it was the education of the general practitioner which was paramount. He persuaded the Wellcome Trust in London to fund a new building for medical research and to endow a chair of experimental medicine for Sir Horace Smirk. Edward's own memorial is the Sayers Building in Dunedin, where he initiated the medical library and administration offices before his retirement.

He was president of the New Zealand branch of the BMA in 1963, elected a Fellow of the College in 1949 and a FRS(NZ) in 1961. He was also an honorary fellow of the American College of Physicians and of the Royal College of Physicians of Edinburgh.

He kept up with the latest developments in medicine and always knew where to get the best advice. He was much consulted by national bodies and was chairman of the scientific committee of the National Heart Foundation of New Zealand, 1968-79.

Edward Sayers was a gregarious man who made easy contacts with everyone he met. He encouraged promising juniors and most people were the better for his wise advice. GW/VL

[*Brit.med.J.*,1985,**290**,1819; *Lancet*,1985,**1**,1346; *Otago Daily Times*,NZ,13 Aug 1958; *Wellington Evening Post*,3 June 1960,*Auckland Star* 3 June 1960,12 Aug 1958]

SCARBOROUGH, Harold *b*.27 Mar 1909 *d*.22 Aug 1988
CBE(1976) MB ChB Edin(1932) PhD(1940) FRCPE(1947) MRCP(1955) FRCP(1960) FRS Edin

Harold Scarborough was born in Leeds, the son of a general practitioner, Oswald Lowndes Scarborough, and his wife Elizabeth Menzies. After education with a classical emphasis at Bridlington Grammar School he went on to the University of Edinburgh, where he graduated in medicine with honours in 1932. After junior posts in the Edinburgh Royal Infirmary, including the clinical tutorship in the department of therapeutics, and a demonstratorship in pharmacology, he took a PhD in biochemistry. It was evident in those early years that he was heading for a distinguished career - indeed, he and a number of his colleagues in Edinburgh, most notably John McMichael and Melville Arnott, were destined to have a major influence on British medicine in the post war years. Harold spent the war years as principal medical officer of South East Scotland's emergency blood transfusion service and after the war returned to academic medicine, obtaining a Beit Memorial research fellowship at St Mary's Hospital Medical School, London, 1945-47. A Rockefeller travelling fellowship then took him to Harvard Medical School, 1947-48, and in 1949 he rejoined Melville Arnott, now professor of medicine in Birmingham, as reader in medicine.

Scarborough went to Cardiff as professor of medicine in the Welsh National School of Medicine in October 1950, where he had the challenge of introducing scientific medicine to what was a clinically sound but academically undeveloped unit. In Cardiff, he soon established himself as a leader. The medical unit in particular, and the School in general, flourished under his influence - the effects of which can be clearly seen today. In medicine he formed a strong alliance and close friendship with the senior physician, William Phillips [*Munk's Roll*,Vol.VI.,p.377] and together they made a deep impact on medicine in South Wales. Harold Scarborough was a shrewd judge and the unit quickly attracted a number of first-class clinical academics. Particular mention should be made of Paul Fourman [*Munk's Roll*, Vol.VI.,p.181], who brought to Cardiff international distinction in metabolic medicine and endocrinology. Harold's own interests in purpura and capillary fragility, in haematology, and latterly in renal disease, were reflected in the growth of these subjects in Cardiff and in the careers of the young physicians he fostered. In later years he was quietly proud of the 14 professors who at one time or another had worked in the medical unit at the Cardiff Royal Infirmary. These included Peter Fentem who became professor of physiology in Nottingham, Peter Adams who became professor of medicine in Manchester, Keith Peters who became professor of medicine at the Royal Postgraduate Medical School and later regius professor of physic in Cambridge, and Eldryd Parry who was to make his career in Africa.

Harold Scarborough used his theatrical interests and aptitude (as an undergraduate he had contemplated a career in acting) to great advantage in his teaching, and his lectures were outstanding. His approach to medicine was to explain disease, wherever possible, in terms of disordered physiology. This was epitomized in the highly successful *Textbook of Physiology and Biochemistry*, Edinburgh, E & S Livingstone, 1950, which he wrote with George Bell and Norman Davidson, known to generations of medical students as 'BDS'. The book had seven editions written by the original authors. His students learned from his critical and balanced approach, and he showed them how to combine scientific and compassionate approaches to medicine.

In 1970, the next phase of his career began. He had developed a deep interest and affection for Nigeria as a visiting professor to Ibadan. He felt that after nearly 20 years in Cardiff it was time for change, both for him and for the Welsh National School of Medicine. He therefore eagerly accepted the post of dean in the newly established faculty of medicine at Ahmadu Bello University in Zaria, in Northern Nigeria. These were difficult days: the civil war was just ending; there was little money for higher education; recruitment was difficult; and few people from the south of Nigeria wanted to work in the north. Harold Scarborough set about his task with tact, hard work, and level-headed detachment. He quickly accumulated an excellent professoriate, including Eldryd Parry and James Lowrie, in medicine and surgery respectively, and in spite of the difficulties the new medical school flourished. In addition to his administrative load, he continued to teach, and his wise and balanced counsel won him the respect of his junior and senior colleagues. In its early years the medical school was predominantly expatriate. Harold Scarborough, with characteristic sensitivity, prepared the way for senior Nigerians to assume responsibility.

When the time came for him to retire, Harold returned to his home in Malta. However, he was quickly bored with the relative inactivity and he returned to Nigeria as a teacher of clinical physiology, of medicine in the field, and of bedside physical signs at the new faculty of health sciences of the University of Ilorin, where his former house physician, Eldryd Parry, was the foundation dean. At Ilorin he lived simply; he gladly went to live in villages when students were doing field work, and applied himself with diligence and care to problem solving teaching, and to the imaginative curriculum designed by John Hamilton. After three years Harold Scarborough decided to leave, yet his return again to Malta was not to be for long before the call came once more, this time for Maiduguri in the north of Nigeria where he was desperately needed to act as provost of the medical school, and medical director of the teaching hospital. The difficulties there were even greater than at Ilorin, or in Zaria, but he persisted and laid the foundation for a sound medical school.

Harold finally retired to Malta in 1984, retaining his interests in medical education and especially in medical education in under-privileged nations. He travelled widely, and every year returned to visit friends in South Wales.

Harold Scarborough was a man of great concern, whose clear and incisive mind got to the heart of a problem but without forcefulness or rancour. He was excellent company, with a deep interest in literature and a passionate love of the theatre, and he was an expert gardener. Although he was a shy and private man, he engendered great loyalty and affection from his students and junior doctors, many of whom became lifelong friends. He contributed greatly to medicine in the UK and in Nigeria, and helped many to achieve their professional goals. In short, he was a great physician.

EHOP/DKP

[*The Times*,1 Sept 1988; *Brit.med.J.*,1988,**297**,736,1124; *Lancet*,1988,**2**,635-6,805; *RCP Edin.Proceedings*, Jan 1989, No.1,108-9; Photo]

SCHLESINGER, Bernard Edward *b.*23 Nov 1896 *d.*25 Jan 1984
OBE(1945) MRCS LRCP(1923) MB BChir Cantab(1924)
MRCP(1925) MD(1930) FRCP(1934)

Bernard must have gone straight into the Army from his public school at Uppingham. He enlisted in the Queen's Westminster Rifles as a rifleman and served in France, Salonika and Egypt. He then went up to Emmanuel College, Cambridge, and got his BA in 1921. Following house appointments at University College Hospital and the Hospital for Sick Children at Great Ormond Street, he determined on a career in paediatrics. In 1927 he was appointed physician to outpatients at Great Ormond Street, and to the Royal Northern Hospital. He had just become a physician in charge of beds at Great Ormond Street in 1939 (which was when I first met him) when his career was again interrupted by war. He volunteered at once and had a distinguished career in the RAMC, ending as a brigadier. He was appointed OBE (military) in 1945. This did not end his RAMC activities for he became the first consultant paediatrician to the Corps and helped to found a paediatric service. He was an original member of the British Paediatric Association, of which he became president in 1953; of the British Cardiac Society, and of the Association of Physicians.

This curriculum says little of Bernard as a man or as a physician. As a man he was gifted with infectious enthusiasm for life in all its aspects, particularly for his profession but also for all the arts; and sport - particularly rugby football and squash racquets. He published many papers on rheumatic fever and juvenile rheumatoid arthritis. Later he published jointly the definitive paper on the severe form of juvenile idiopathic hypercalcaemia.

Bernard married in 1925. His wife, a notable musician, must have increased his love of music; any house they had had to have room for at least one grand piano. They had five children, but the family's life was not without tragedy for two daughters were lost at an early age. His own life was threatened by a carcinoma of the colon, which was successfully resected. Even this did not stop him playing squash, and he continued to play well over his sixtieth birthday.

Being a generous man by nature he made a splendid host. He was also compassionate. Few if any of his colleagues knew that in the 1930s he and his wife cared for and saw to the education of no less than 13 refugee children.

I first met Bernard in 1939 when I was a resident at Great Ormond Street but did not become his colleague at UCH, to which he was appointed in 1947, until 1948. He was the easiest of colleagues and I cannot remember a cross word between us during the 17 years we spent together. He started two memorable things at UCH at this time. The first was an intensive care unit for premature babies, which has now developed into an experimental unit with an international reputation. The other was to encourage the whole department to travel, ostensibly to see what other departments were doing but this also had a wider effect, knitting the UCH department more closely and consolidating Bernard's position at its head.

His love of life and his enthusiasm for everything he did infected everyone who had the pleasure of working with him. This made him an excellent teacher. There must be many paediatricians, including at least four professors, who were influenced by him in their early years.

REB-C

[*Brit.med.J.*,1984,**288**,494-5,649; *Lancet*,1984,**1**,352; *The Times*,10 Feb 1984; Photo]

SCHNEIDER, Jacob *b.*27 Feb 1923 *d.*23 Sept 1987
MB BCh Wits(1949) DTM&H Eng(1952) DMed Wits(1956) MD(1958) MRCPE(1958) MRCP(1958) FRCPE(1972) FRACP(1979) FRCP(1986)

Jacob Schneider (Jack) was born in Johannesburg and graduated in medicine and surgery from the University of Witwatersrand. This was the start of a lifelong commitment to general medicine, tropical medicine and infectious disease. He came to London to study at the London School of Hygiene and Tropical Medicine and obtained his DTM&H. He was later awarded a doctorate by the University of Witwatersrand for an original thesis on intestinal schistosomiasis. This was followed by membership of the College, and of the Edinburgh College, and he was subsequently elected a Fellow of both.

On his return to South Africa in 1961 he held appointments as consultant physician at the Natalspruit Hospital, as part-time gastroenterologist at the Johannesburg hospital, consultant venereologist to the Johannesburg City Council and lecturer in the subject at the University of Witwatersrand, as well as being lecturer in tropical medicine.

Jack Schneider's growing concern with social injustice and racial prejudice in South Africa led him and his family to emigrate to Australia in 1977. They successfully met the challenges of a new life in Perth. He was initially appointed as physician in extended care at the Sir Charles Gairdner Hospital, and shortly afterwards as a general physician. In 1983 he became director of postgraduate medical education at the hospital, while still retaining a special interest in infectious disease.

This brief history gives only the bare outline of a man who embodied all that is best in a physician and human being. He had knowledge, clinical skills and judgement in ample quantity, but his understanding of human nature, his empathy, warmth, humour and dedication to his patients were shining examples to his students and the medical and nursing staff. To medical students, in particular, he emphasized the importance of caring as a total commitment to the well-being of each patient, which included not only training in the essential scientific skills of medicine but also the application of those skills with a sensible concern for the physical, psychological and social circumstances of the individual.

Jack's substantial contribution to professional life was paralleled by his involvement in the wider community in which he lived. He had a great love of Jewish tradition and exemplified its values of justice and charity. Singing gave him great pleasure, and he was one of Perth's leading exponents of Jewish Liturgy and Yiddish folk songs. His deep baritone voice enriched many social gatherings, religious festivals and charitable functions.

Jack Schneider was a fine, undemanding and gentle man. His loving and supportive wife, Diana, and his two children, Jill and Mark, survived him together with his grandchildren Rachel and David. LJB

SCOTT, George Albert *b.*6 June 1911 *d.*21 Oct 1985
MB BCh BAO(1934) MD(1946) MRCP(1949) FRCP(1971)

George Scott was born at Clover Hill, County Cavan. His father was Richard William Scott, merchant and farmer. His mother, Elizabeth Matilda, was the daughter of William Crawford, also a farmer. When George was still a boy the family were driven from their home by the disturbance of the Irish rebellion.

George was educated at Coleraine Academical Institution and Queen's University, Belfast. Immediately after qualifying he went into general practice in Preston. In 1941 he gave up his practice and joined the Royal Naval Volunteer Reserve, reaching the rank of surgeon commander, and at the end of the war returned to Belfast for postgraduate study. He was appointed physician, the first such appointment, to the Moyle Hospital, Larne, in 1950; it was one of the early fruits of the National Health Service.

It is probable that the relatively small population of East Antrim did not give George all the opportunities of which his intellectual vigour could have made so much. Nevertheless, it was he who brought the specialty of internal medicine to East Antrim, and by his work and example transformed practice there to the benefit of its people. For many years he had no assistance other than a house physician, and he was on duty 24 hours a day except when he was on annual leave. He brought to light the silicosis of flint crushers in the area (which ran a rapid course to death), and the pneumoconiosis of bauxite crushers.

He was founder chairman of the Ulster Society of Internal Medicine and it was due to his friendly, sensible and commanding personality that the Society was a success from the beginning. He served in the RAMC Emergency Reserve, being lieutenant colonel in charge of the medical division of 4th General Hospital. He was president of the Larne branch of the British Legion, first chairman of the local banch of the British Empire Cancer Campaign, and an active worker for the Boys' Club.

The rocks of Antrim have long interested geologists, and George became a competent amateur geologist. In his youth he played rugby for his school, University and Ulster - and later played in Preston. He rowed on the River Bann at Coleraine and, later, at Queen's, where he also played the Scottish pipes in the OTC band. In his later years, fishing, golf, bridge and bowls gave him some recreation. After retirement, he cared for his garden and also revealed a remarkable gift for modelling in pottery.

George was a cheerful, kind friend and colleague, of a most firm and upright character, an enlightened physician, a good citizen and a true Ulsterman. In 1936 he married Zillah Elizabeth Todd and they had two sons. Zillah, who survived him, was the constant support of his life and work for nearly fifty happy years. One son, George, became an architect who designed surgery buildings for general practices; the other, Liam, entered general practice in Holywood, County Down. JSL

[*Brit.med.J.*,1986,**292**,1147; Photo]

SEATON, Douglas Ronald *b.*19 May 1912 *d.*30 Oct 1986
BChir Cantab(1936) MRCS LRCP(1936) MA MB(1937)
DTM(1939) MRCP(1939) DTH(1940) FRCP(1966)

Ronald Seaton was educated at Bradfield College, King's College Cambridge and Leeds University, graduating in 1936. His grandfather was president of Leeds Medical School and a surgeon at Leeds Infirmary, and his father Douglas Seaton was a general practitioner and anaesthetist in Leeds. His mother Constance, née Ingham, was the daughter of an engineer.

After qualifying, Ronald helped his father in his practice until 1939; as a child he had accompanied his father on foot on his medical rounds in Leeds. Having passed his MRCP and his DTM&H, he obtained an MRC junior fellowship and served his tropical research apprenticeship with Warrington Yorke. On the outbreak of war he joined the RAMC and was first posted to Freetown, Sierra Leone. In 1944 he was sent to India to take charge of the Army base typhus research unit in the Central Military Pathology Laboratory in Poona, with the rank of major and as specialist in pathology.

Under Ronald's leadership studies on the natural history of the various types of typhus, particularly in the Indian subcontinent, were undertaken. The team, which included Michael (later Sir Michael) Stoker, isolated strains of rickettsia, especially murine strains from rat fleas. During this time nearly all members of the research team, including Ronald, contracted murine typhus. The main contribution of the team was the identification of Indian tick typhus as one of this spotted fever group. At that time Ronald was one of a distinguished group at the Central Military Pathology Laboratory. It inclued Douglas Black, later Sir Douglas PRCP, Paul Fourman [*Munk's Roll*,Vol.VI, p.181], Michael Stoker and Bill Hughes - a founder of bacterial genetics. All of them were close friends of Ronald. In 1946 he was repatriated and joined the School of Tropical Medicine as an assistant physician and lecturer.

I first met him in 1949 and was privileged to be his senior house physician in the tropical ward of Sefton General Hospital, Liverpool. He was an astute, patient and humane person who showed me great tolerance in my early days. He published one of the earliest authoritative papers on the problem of resistance to antimalarial drugs, a subject much in the limelight these days. His lectures to generations of students were quite unique, mingling a dry sense of humour with erudition and a disarming smile. He was a compassionate, generous and completely self-effacing person, making it quite impossible to extract from him any information about his achievements no matter how well one knew him. Over my many years with him in Liverpool he treated me as a younger brother, frequently offering me the hospitality of his house in my early days and remaining a friend and confidant in my later ones.

Ronald was awarded his Fellowship in 1966, and characteristically said that he was elected because his son had got his MRCP by then.

Ronald Seaton was a man of great culture, superb intelligence, with an extraordinary memory. He could identify most of the visible stars and had a wide ranging knowledge of botany. He spoke French and German, the latter quite fluently. He regularly did *The Times* crossword puzzle in less than half an hour, retaining this faculty until the last week of his life. On one occasion he was a regional finalist in the crossword competition. He also played a good game of golf.

He married Julia, daughter of James Wilder Harrison, in 1937 and they had five children: four sons and a daughter. He was a devoted husband and father who put his family first, above all else. He was delighted that the medical dynasty continued, with two of his sons becoming consultant physicians, one of his 13 grandchildren a medical student, and his daughter Jane a nurse.

He was an undemonstrably religious man, and took his illness with a lot of quiet courage. He survived eight years after pneumonectomy for lung cancer, to die of left ventricular failure. His wife and daughter nursed him devotedly in his last illness. If he had been more ambitious Ronald Seaton would undoubtedly have been one of the great leaders in tropical medical science.

HMJG

[Photo]

SELLORS, *Sir* Thomas Holmes *b.*7 Apr 1902 *d.*13 Sept 1987
BA Oxon(1923) MRCS LRCP(1926) BM BCh(1926) MA(1927) FRCS(1930) MCh(1931) DM(1933) *FRCP(1963) Hon DSc Liverp(1970) Hon FACS(1971) Hon FRCSE(1972) Hon FRCSI(1975)

Sir Thomas Holmes Sellors was a past president of the Royal College of Surgeons of England, emeritus surgeon to the Middlesex Hospital and consultant surgeon to the National Heart and Harefield hospitals. He was the son of Thomas Blanchard Sellors, a family doctor, who first practised in Wandsworth and later, for health reasons, at Southend on Sea, and his wife Anne Oliver McSparron. Tom was born in Wandsworth; a vintage borough for RCS presidents since two recent presidents were born within a mile of Tom's birthplace, while a third attended a nearby primary school. Tom himself was educated first at Alleyn Court, Westcliff-on-sea and then at Loretto School and Oriel College, Oxford. At Oxford he was a contemporary of the late Derrick

Dunlop [*Munk's Roll*,Vol.VII,p.170] and Raymond Greene [*Munk's* Roll,Vol.VII,p.228]. He secured a university entrance scholarship to the Middlesex Hospital, where he qualified in 1926.

Following resident appointments at the Middlesex and Brompton hospitals, Tom Holmes Sellors became surgical registrar to Sir Gordon Gordon Taylor at the Middlesex. During this period, as first recipient of the G H Hunt award by Oxford University in 1928, he was able to spend a year at surgical centres in Scandinavia. After a thorough grounding in general surgery, he decided to specialize in chest work which, until then, had been a somewhat limited field and associated with considerable hazard. But he was an outstanding technician and the rapidly advancing anaesthetic techniques made him determined to pioneer this new specialty. At the tender age of 31 he had the temerity to publish a book on thoracic surgery, and it is reputed that this did not endear him to some senior members of the surgical establishment who had been more tardy in putting pen to paper. Tom's opportunity came when he was appointed to the staff of the London Chest Hospital in 1934, followed by appointments as thoracic surgeon to the Royal Waterloo Hospital and to Queen Mary's Hospital, Stratford. He also served various LCC hospitals and sanatoria, and he started chest units at the Radcliffe Infirmary, Oxford, and the Leicester Royal Infirmary, all of which entailed much travelling and an immense workload. Throughout his life, Tom was meticulous in his support and loyalty to those whom he had trained, and he retained their deepest respect and affection.

On the outbreak of the second world war he was appointed adviser in thoracic surgery to the north west metropolitan region, based on Harefield Hospital in Middlesex, where he worked most happily and productively until his retirement. By the end of the war he was engaged purely in cardiothoracic surgery. On appointment as thoracic surgeon to the Middlesex Hospital in 1947 he developed close and cordial relationships with the cardiologists Evan Bedford [*Munk's Roll*,Vol.VII,p.28] and Walter Somerville, to whose skilful assessment of patients needing heart surgery he always paid warm tribute. In 1957 he became consultant surgeon to the National Heart and Harefield hospitals. His anaesthetists, notably Parry Brown and the late Bobbie Beaver at the London Chest, and Brian Sellick at the Middlesex, worked most happily with him for many years. With Parry Brown he had some challenging and occasionally hair-raising experiences while operating in Madrid, Moscow and Guatamala. He had a devoted and enthusiastic band of loyal trainees, some of whom became internationally renowned and a number of whom predeceased him. To each of them, and to many others, he was fondly known as 'Uncle Tom'.

From the inception of the National Health Service, Tom was active in the medico-political field. He was chairman of the north west metropolitan regional consultants and specialists committee for some years, as well as a member of the central consultants committee and its chairman for five years. He was elected to the council of the Royal College of Surgeons of England in 1957, and in the following year he succeeded Lord Brain [*Munk's Roll*,Vol.VI,p.60] as chairman of the important joint consultants committee, linking the British Medical Association with representatives of the various royal colleges - an arduous task which he undertook for nine years. For this work and his services to surgery he received the well earned accolade of knight bachelor in 1963. He was elected vice-president of the RCS in 1968, and president from 1969-72. During this period he gave the Tudor Edwards lecture, the Gordon Taylor lecture, and was Hunterian orator in 1973. In 1972 he became president of the British Medical Association and was awarded its gold medal. He was a patron of the Royal College of Surgeons, an honorary fellow of its faculty of dental surgery, and a member of the board of trustees of the Hunterian Museum of which he was ultimately chairman.

Throughout a very busy surgical life he travelled widely abroad, lecturing and demonstrating in Europe, India and South America. He also visited the United States, Canada, Japan and South Africa. He was made an honorary fellow of the surgical colleges of South Africa and the USA, of the Royal College of Surgeons of Edinburgh, and of the Royal College of Surgeons in Ireland. He was elected a Fellow of the College in 1963. He was also an honorary member of the Academy of Medicine in Rome, the Royal Academy of Medicine in Belgium, and of the European Society of Cardiology. In the course of his travels Tom gave many eponymous lectures and was elected honorary MD Groningen, DSc Liverpool and MS Southampton. He received the Médaille de la Reconnaissance Française and was an officer of the Order of Carlos Finlay, Cuba. He had a keen sense of the unity of medicine, with a strongly international outlook and sympathies.

Throughout his working life Tom was up at daybreak, or earlier; often visiting a ward before the resident staff and day nurses were about, and sometimes guided to a patient by the ward cleaners. His gentlemanly style and good manners ensured excellent rapport with the nursing staff. His kindly, compassionate character, his good humour and recognition of the foibles of human nature, gave immense confidence to his patients. He worked with deceptive rapidity and economy of effort and seldom wasted time in idle chatter, so much so that an astute trainee - anxious to secure his shrewd counsel under rather pressing conditions - once hopped into Tom's car and took an unplanned trip from Harefield to London.

In the second half of his surgical life Tom set up no less than three open heart surgery units; at Middlesex, Harefield and the National Heart hospitals. Starting mainly with closed mitral valvotomies and Blalock operations in the early 1950s, he moved on to the closure of atrial septal defects under hypothermia. His results with this operation were unrivalled at the time. In those days he put the Middlesex and

Harefield firmly on the map, with surgeons from all over the country - and from abroad - coming to watch him operate. Some of his colleagues could hardly get into the surgeons' changing room because of the mass of visitors. Much as Tom came to enjoy explaining technical points to his visitors, he was essentially a modest and humble person; never one to take inordinate pride in his own achievements, nor to say disparaging and unkind words about his rivals.

As a technical surgeon he had few rivals, but his basic humanity and kindness made him reluctant to expose his patients to untried procedures and he was sometimes criticized for this. Whatever may have been his reluctance to subject patients to frankly experimental procedures, he did indubitably undertake the first direct operation for relief of pulmonary valve stenosis. Having set out to do a Blalock operation (which proved impracticable due to dense lung adhesions) he was impressed by the tightness of the valvular obstruction and the labouring right ventricle. Borrowing a tenotomy knife from a nearby orthopaedic theatre at Harefield, he calmly corrected the pulmonary valve narrowing and the patient was alive and well years later. News of this operation sped through the bush-telegraph and a rival surgeon quickly followed suit and was first to publish the procedure; not for the first time - Tom had also undoubtedly been the first to perform an 'unpublished' dissection lobectomy of the lung.

Although he never spared himself in the service of his patients and his profession, Tom had a pleasing capacity for gracious living. He lived in London for most of his life, but had been fortunate enough to acquire a country home, Spring Coppice Farm, at Speen in Buckinghamshire, before the war. This had been a strategic centre for his wartime work and was a marvellous peacetime retreat at the weekends when, except in utmost urgency, his senior registrars rarely disturbed him. He was a keen and knowledgeable gardener and a pleasing artist; rapidly painting delicate little pictures in watercolour and often leaving amusing sketches on programmes and menus.

Only those close to Tom were aware of the personal tragedies he suffered. In 1928, at the age of 26, he married Brenda Lyell but she died of acute appendicitis a few weeks later. In 1932 he had the good fortune to marry Elizabeth Cheshire and this was a happy and fruitful union. There were two children of the marriage, a son who is an ophthalmic surgeon and one time oculist to Her Majesty's household, and a daughter. She gave him all the love and support which built his confidence in the joy and security of their family but, sadly, while their children were still young, she suffered a stroke and hypertension. At that time, the early rising Tom would often tidy the house and lay the fire for his dear wife before going to work. Elizabeth died in 1953, when Tom was at the busiest and most demanding period of his life. In 1955 he married Marie Hobson, a union which was to last 30 years, and together they presided over the social and family affairs of the Royal College of Surgeons of England with grace and distinction during his outstanding presidency. Ironically, as the non-smoking spouse of a pioneer thoracic surgeon, Marie developed lung cancer and died nearly two years before him.

Tom was lonely during the last year or so, but he never lost his sense of humour or his exceptional courtesy. A year or so before his death he had a slight stroke. Visiting him a few days later, the writer found him chuckling with glee over a Saturday tabloid newspaper which was all he had been able to buy in hospital. Remarking on his unusual reading matter and enquiring the cause of his mirth, Tom replied: 'Just fancy. A house of ill-repute in Ambleside Avenue, Streatham. I had two respectable maiden aunts who lived there; Heaven knows what they would say if they could read this!'.

His gentle style and good manners characterised even his terminal illness. Only four days before his death, in a tired and weak voice, he did not omit to ask after those of my family that he knew. On the following day he took Communion before quietly lapsing into coma. He was survived by the children of his second marriage.

RM

[*Brit.med.J.*,1987,**295**,790; *Lancet*,1987,**2**,979; *The Times*,16 Sept 1987; *Address*,St Clement Danes,2 Dec 1987; *World Medicine*,1974,3 July 53-56; *Trans.Coll.Med.SAfr.*,Sept 1972, **16**,64; *Middx.Hosp.J.*,Dec 1970,**70**,(6); Photo]

SEMPLE, Robert *b.*21 June 1917 *d.*5 Mar 1988
MB ChB Aberd(1939) MRCP(1946) MD(1952) FRCP(1964) FRCPE(1979)

Robert Semple was born in Aberdeen, the son of Robert Semple, a general practitioner in the city and his wife Flora Mackenzie Ferguson, daughter of a civil servant. He was the eldest of six children. He was educated at Aberdeen Grammar School and Aberdeen University, where he graduated with first class honours. He won several gold medals and was first in his year. He was also an all round sportsman who gained a blue in golf, and was a member of the OTC. But the only distinction of which he spoke was that of passing his driving test in his father's Rolls Royce.

After a house physician post at the Aberdeen Royal Infirmary, he joined the RAMC and served from 1939-46. After a spell as RMO to the Argyll and Sutherland Highlanders at Stirling Castle, he was posted to the Gold Coast Regiment, serving with the 37th General Hospital at Accra, and later as medical officer to GHQ, West African Command. His next move was to Middle East Command where he became medical officer to a mountain artillery unit which, after training in the Lebanon, moved to Italy - complete with mule train - as part of the 8th Army. In 1944 he became a graded medical specialist in Italy and, later, in the United Kingdom.

After demobilization Robert was appointed registrar in medicine and pathology in Aberdeen, and obtained his membership of the College at his first attempt. In 1947 he was appointed second assistant, then first assistant in 1951, to the medical professorial unit at the Middlesex Hospital, London, and he was also a clinical assistant at the Brompton Hospital. He always maintained that his appointment to the Middlesex was due to his low golf handicap!

Robert could have continued in academic medicine but preferred, and was ideally suited to, the life of a general physician in the NHS. He also felt the call back to his native land and in 1952 he became consultant physician to Maryfield Hospital, Dundee, and to Stracathro Hospital, Angus. In 1974 he transferred to the new Ninewells Hospital in Dundee, and for a period served as chairman of its medical division.

His outstanding clinical skill and experience, and his acute mind, quickly established him as a highly esteemed consultant in eastern Scotland. He was tall and had a distinguished presence and natural authority which, together with his courteous manner, inspired trust and confidence in his patients. His dedication and compassion brought him warm affection. By his colleagues he was greatly respected as giving a first class opinion in difficult cases; to many of them he was a comforting support in personal illness.

Robert was involved in undergraduate teaching as honorary senior lecturer in medicine in the University of St Andrews (later of Dundee) and was very firm in ensuring a systematic groundwork in clinical examinations. He shared also in postgraduate training, and was noted at clinical meetings for his shrewd and perceptive comments. His years at Maryfield Hospital, in joint charge of a clinical teaching unit, were among his happiest, though when Ninewells Hospital opened he was delighted by his warm reception and respected status as senior physician in the medical professorial unit.

Robert lived a full and happy life. He was active in local professional organizations and was, at different times, president of Dundee Medical Club and of the Forfarshire Medical Association, and captain of Dundee Medical Golfing Society. Though essentially a private man he had a natural friendliness and ease of manner, and he had a wide circle of friends. He was a skilled fly-fisher, a good shot, and a knowledgeable gardener. But his chief recreation was golf, and he was certainly a fine player. He was a member of the Royal and Ancient Golf Club of St Andrews and a winner of perhaps its most difficult trophy, the Calcutta Cup. He continued to find pleasure and solace on his beloved Old Course up to the last few months of his life.

Robert showed remarkable courage and dignity in his last illness. He was determined to carry on his usual interests and activities for as long as possible and did so with grace, humour, and consideration for others. He was greatly blessed by the loving support of his wife, Alison, whom he had married in 1943, herself a doctor from a distinguished medical family her father was John Cruickshank, a professor of bacteriology; together with the support of their son Peter, a consultant physician in Glasgow, and their daughters Hilary and Rosemary.

JSF/RNJ/DE-S

[*Brit.med.J.*,1988,**296**,1203; Photo]

SEVER, Edmund Day *b.*1 Sept 1928 *d.*17 July 1986
MB ChB Edin(1951) MRCP(1957) FRCP(1974)

Edmund (Ted) Sever was proud of being a Yorkshireman, having been born in Scarborough where his father, Frank Sever, was a bank manager. His mother was Hilda Gertrude Day. He was educated at Oundle School and Edinburgh University, where he graduated in 1951. Following house officer appointments at Northallerton he entered the Army for his period of national service, serving in England and the Middle East. He received training in pathology while in the Army and, on demobilization , he took up a senior house officer post in pathology at Wigan. However, his heart was set on securing a consultancy in general medicine and he realized that he would have to work his way up from the bottom of the tree again, in prestigious appointments, if he were to have a real chance of fulfilling his ambition. Later in life he often gave similar advice to his junior staff, to correct their career plans towards a consultancy.

He returned to Edinburgh as house physician to Rae Gilchrist at the Royal Infirmary, and later worked with Ian Grant at the Northern General Hospital. In furtherance of his plan he moved south to London with house appointments at the National Hospitals Queen Square and Hammersmith Hospital. In 1959 he returned to Edinburgh Infirmary as registrar to J K Slater, an appointment which furthered his lifelong interest in clinical nuerology. It was at this time that he was observed helping one of the house physicians with her daily chores - and he married Jess Gordon in 1961.

Although he had set his eyes on a senior registrarship in neurology, fate took him to a position in rheumatology at the West London Hospital under William Copeman [*Munk's Roll*,Vol.VI,p.120] and Oswald Savage. With this comprehensive training he was appointed consultant physician, with an interest in rheumatology, at Northampton General Hospital in 1964.

Ted Sever quickly established his reputation as a very able general physician with sound clinical judgement. He retained his interest in neurology and delighted in making clinical neurological diagnoses. At Mansfield Hospital, Northampton, he worked in close collaboration with his orthopaedic surgical colleagues in the combined management of joint disease and, because of the availability of ancillary staff there, he was also able to establish a stroke unit.

Ted was a gifted and conscientious teacher who thoroughly enjoyed preparing candidates for the MRCP examinations. He devoted extra time to the preparation of tutorials, bedside classes, and journal clubs for the junior staff. Shortly after arriving in Northampton he was appointed the hospital clinical tutor, coinciding with the opening of the new Cripps Postgraduate Medical Centre. The immediate success of this centre was entirely due to his drive in presenting an intensive training programme for all varieties of medical staff. In particular, he participated in general practice activities and, later, in the vocational training scheme. He was a great stimulator of the junior medical staff in their preparation for examinations, and a ready source of advice for them.

He had a continuing interest in the College and served as Oxford regional adviser for a short period. He represented the College frequently on visiting committees for the assessment of the training of junior doctors and consultant appointments. He was a firm believer in election to the Fellowship depending entirely on merit and not on seniority.

Throughout his consultancy he remained a loyal supporter of the National Health Service. He continued as a full time physician and shunned private practice. He aimed to treat all patients individually, giving them his personal attention. He was as concerned with their social and emotional welfare as with their physical illness, and frequently held joint meetings with psychiatrists, psychologists, social workers, and all types of therapists. He was foremost, locally, in introducing problem orientated records, and in instructing his junior staff in this approach to patients.

Being a Yorkshireman, he naturally played cricket for his school, university and hospital, until ill health prevented him. He was also a golfer. He enjoyed reading, especially poetry and history, and was particularly interested in Greek culture. Through his regular commitment to his local Anglican church he became involved in the work of Amnesty International. He also took a great interest, initially, in the establishment of the Social Democratic Party and attended their mobile conferences. Becoming alarmed at the shortage of books in schools, he waged a largely single-handed protest against the local education authority.

Ted was happiest when working, and found it difficult to relax on holiday. Tragically he died alone, while walking the Zwölferhorn in Austria. He was survived by his widow, Jess, a consultant in child psychology at Northampton, and his three sons.

PCR²/GW

[*Brit.med.J.*,1986,**293**,1447; *Lancet*,1987,**1**,56; Photo]

SHANKS, Robert Andrew *b.*13 Sept 1917 *d.*25 Dec 1984
MB ChB Glasg(1940) FRFPS(1946) MRCP(1948) MD(1949)
FRCPG(1964) FRCP(1970)

Robert Shanks was born in Oslo, the son of a British father and Norwegian mother. He was educated at Dollar Academy, Glasgow Academy and the University of Glasgow, from where he graduated in 1940. His interest in paediatrics developed soon after qualifying when he became house physician to Stanley Graham. As with many other young men, his chosen career had to await the end of the second world war. He served with the RAMC from 1941-46, going to Madagascar and Burma. He played a prominent role in establishing a malaria research unit, achieving the substantive rank of lieutenant colonel.

On demobilization Shanks returned to his career in paediatrics in the Royal Hospital for Sick Children in Glasgow and, indeed, gave the rest of his professional life to this hospital, until early retirement due to ill health was forced upon him in 1981. A distinguished MD thesis earned him the Bellahouston gold medal of the University. His subject was 'The convulsions of infancy', and although Shanks was in all respects a general paediatrician he always had a special interest and expertise in paediatric neurology. His published works were not large in number but of high quality, and always written in beautiful English. While his principal commitments lay in the childrens' hospital, he also carried responsibility for the paediatric department of Glasgow Royal Maternity Hospital and he acted as consultant to the Eye Infirmary and the Ear, Nose and Throat Hospital. For many years he was physician superintendent of Eastpark Home for Infirm Children, where he amassed a wide experience of neurological and other chronic disorders of childhood. He was elected president of the Scottish Paediatric Society, and twice served on the council of the British Paediatric Association. Until the time of his death he served on the council of the Medical and Dental Defence Union of Scotland.

This factual account of Robert Shanks's career does not fully reveal the man, the doctor, or his interests outside medicine. He was a fine clinician and a stimulating teacher. Several of his juniors have achieved distinction in academic or NHS posts. He was possessed of an acute sense of humour, even of the ridiculous, and would often relieve a tense moment at a committee meeting with an extremely witty comment. He was a man of deep religious faith which seemed to enable him to bring particular comfort to his patients and their parents when they faced chronic or progressive disabilities. A senior lay member of St Mary's Episcopal Cathedral in Glasgow, Shanks once described himself as 'an enthusiastic Christian, but not much of a Calvinist'. As a member of the Episcopalian-Roman Catholic joint study group he was one of the few church leaders to be presented to the Pope during his visit to Scotland in 1982.

Shanks's other love was music; not surprising, as his mother had been a professional opera singer and his father was a talented amateur performer in opera. Shanks himself was a gifted pianist. For a long period he was a director of the Scottish National Orchestra Society and of Scottish Opera.

Severe emphysema forced early retirement and his last few years were burdened with increasing breathlessness, but it was a measure of the man that all his friends and colleagues who visited him came away uplifted by his cheerfulness and equanimity. He never married. JHH

[*Brit.med.J.*,1984,**290**,164; *Lancet*,1985,1,355; *The Times*,8 Jan 1985; Photo]

SHEEHAN, Harold Leeming *b.*4 Aug 1900 *d.*25 Oct 1988
TD MB ChB Manch(1921) MD(1931) MSc(1932) DSc(1940) MRCP(1941) FRCP(1947) FRCOG(1949) FRCPath(1964)

Harold Sheehan was born in Carlisle where his father, Patrick Sheehan, was a general practitioner who had built up a large working-class practice with over 4000 panel patients, the largest panel in the city. His mother had been Eliza Leeming before her marriage to Patrick Sheehan and Harold was one of their twelve children. Harold was educated at Carlisle Grammar School and the University of Manchester, graduating with honours in 1921. In the meantime his father had died, and Harold began his career as a doctor by joining his elder brother Gerald, who was now running the family practice in Carlisle. He worked as a general practitioner for the next six years.

The academic phase of his career began in 1927 when he became a demonstrator, and later a lecturer, in the department of pathology in the University of Manchester. At this time the professor of pathology in Manchester was John Shaw Dunn, whose chief interest was in renal disease, and he encouraged the young Harold Sheehan to undertake experimental work on renal function. The result was a thesis on the deposition of dyes in the mammalian kidney for which he was awarded a doctorate, with a gold medal, in 1931; the MSc degree followed a year later for a thesis on the renal elimination of injected urea and creatinine.

In 1934 he married Eve Suzanne Gertrude Potter, who was so dissatisfied with the names that her parents had given her that her real name became a sort of skeleton in the family cupboard. She would introduce herself as Susan Sheehan, while amongst intimates she was always known as 'Shooz'. Their married life began with a year in America, Harold having been awarded a Rockefeller medical fellowship to continue his studies on renal function in the department of pharmacology at the Johns Hopkins medical school in Baltimore.

The most important phase of Sheehan's professional career began in 1935 when he attained the post of director of research at the Glasgow

Royal Maternity Hospital, for it was during the next five or six years that he began to establish himself as an international expert on the pathology of pregnancy. He analysed the effects of obstetrical shock, he differentiated between the fatty liver of delayed chloroform poisoning and the condition of primary fatty liver of pregnancy, he demonstrated the reactivation of latent rheumatic heart disease that was induced by pregnancy, he clarified the effects of eclampsia upon the liver and kidneys, he identified the encephalopathy of hyperemesis gravidarum as Wernicke's disease, he recognized the association between concealed accidental haemorrhage and renal cortical necrosis and recognized that obstetrical shock and haemorrhage could induce necrosis of the anterior lobe of the pituitary gland.

The war interrupted his work on the pathology of pregnancy. He sometimes said himself that his most important work in the Army was the contribution that he made towards the solution of the problem of 'syringe jaundice'. His overseas service was in Italy, and by the end of the war he had been mentioned in despatches and was deputy director of pathology to the allied forces in Italy, with the rank of colonel. In 1946, after demobilization, he was appointed to the chair of pathology in the University of Liverpool.

In Liverpool he had to provide curricula for large classes of medical and dental students many of whom had served for several years in the forces. Such students were conscientious and were very discriminating about the instruction they received. They found Sheehan's teaching entertaining and memorable, for he had striking mannerisms - his eyes would bulge as he expressed horror at the enormity of some wrong answer. He was highly provocative and challenged much cherished dogma. He was at his best in the informal class on Saturday mornings, when specimens obtained during the previous week were demonstrated and discussed. The give and take of the 'meat class' reduced, for the regular attenders, any dread they might have felt for oral exams. Postgraduates also often attended these classes.

Although the students enjoyed his classes they knew that he was a demanding examiner and they suspected him of slyness: his nickname was 'Crafty Harry'. In his dealings with the juniors in his department, and with his colleagues, he liked to hold his cards close to his chest, and he kept his right hand in strict ignorance of what his left hand was up to. He obtained many advantages for his department by private deals rather than by formal negotiation at the faculty board.

In the post-war expansion of the laboratory services in Merseyside his influence was important but inconspicuous. His success, however, in establishing an endocrine unit within the region was of course something in which his vested interest was obvious.

He acted as a histopathological consultant for the region and his opinion on difficult sections was frequently sought by pathologists in Merseyside, North Wales and beyond. He arranged that he should be informed of any obstetrical death in the region and he would carry out the post mortem himself, even late at night, for he believed firmly that only fresh tissue was worth histological examination. His autopsy technique was chaotic but effective.

In Liverpool he consolidated and developed the work he had begun in Glasgow. In 1949, with V K Summers [*Munk's Roll*,Vol.VI,p.427] he published an important paper in the *Quarterly Journal of Medicine*, 1949,NS viii,277-309, on the syndrome of hypopituitarism. The paper showed convincingly that emaciation and premature senility, previously considered to be essential for the diagnosis of pituitary insufficiency, were not features of the syndrome. This paper, rather than Sheehan's earlier papers on pituitary necrosis, led to the syndrome of post-partum pituitary necrosis becoming known as Sheehan's syndrome. He continued to study renal cortical necrosis and his conclusions were published in book form, with H C Moore as co-author, *Renal cortical necrosis...*, Oxford, Blackwell Scientific, 1952. In addition he initiated much experimental work designed to elucidate the pathogenesis of both pituitary necrosis and renal cortical necrosis. He published, for example, a long series of papers with J C Davis on experimental renal ischaemia, and supervised various studies of the vasculature of the human pituitary gland by members of his staff. He never succeeded, however, in establishing that spasm of critically sized blood vessels, a mechanism that he had postulated to explain his findings in human material, had the importance he attributed to it.

The value of Sheehan's work appears to have been recognized abroad before it was recognized in Britain. He accepted many invitations to undertake lecture tours in both North and South America and in Europe, including both sides of the iron curtain. His wife always accompanied her husband on his foreign travels. Her linguistic skill was of value, especially in Spanish-speaking countries, but Harold himself could read a paper in French or German. A series of foreign visitors also came to Liverpool to undertake research in Sheehan's department.

In 1965 he was obliged by his age to retire from the chair of pathology but he continued his researches, first in the Liverpool School of Tropical Medicine and later in the University department of physiology. In 1973 he published a comprehensive book, with J B Lynch as co-author, *The Pathology of toxaemia of pregnancy*, Edinburgh, Churchill Livingstone. In 1982, with J C Davis, the whole of his work on post-partum hypopituitarism at last appeared as a book, *Post-partum hypopituitarism*, Springfield, Illinois, C C Thomas. That same year, with K Kovacs, he made an important contribution on the neuro-hypophysis and hypothalamus to the second edition of Bloodworth's book *Endocrine pathology*, 2nd ed. Baltimore, Williams and Wilkins, 1982,pp.45-100.

The Sheehans had no children. His wife died in 1986 and after her death Harold lived alone in their big, cold house in Liverpool until a fractured hip led, eventually, to his admission to a nursing home in

Kendal, where he died at the age of 88. He was a remarkable man and a memorable character, who well deserves the eponymous fame that has been bestowed upon him. AHC

[*Independent*,3 Nov 1988; Photo]

SHELDON, Joanna *b.*10 Mar 1931 *d.*31 Dec 1985
MRCS LRCP(1955) MB BS Lond(1955) MRCP(1962) MD(1966) FRCP(1975)

When she died, aged only 54, Joanna Sheldon was consultant physician to the Royal Sussex County Hospital, Brighton, and Hove General Hospital, and a distinguished figure in diabetes in Britain.

She was born in London, the daughter of Wilfrid Sheldon, later Sir Wilfrid [*Munk's Roll*,Vol.VII,p.531], and Mabel Winifred née Netherway, educated at St George's School, Harpenden, and graduated at King's College Hospital Medical School. After house appointments at King's and Brighton, she spent four years in the departments of medicine and endocrinology at Ann Arbor Michigan, where she was offered a staff appointment. However, she returned to the UK, where she became registrar to the diabetic department and then senior medical registrar at King's College Hospital. In 1968 she was appointed consultant physician at Brighton.

Joanna came from a remarkable medical family, Her father was the first Royal paediatrician, and her election to FRCP in 1975 marked the first time that a father and daughter had been Fellows of the College concurrently. An uncle, the late Joseph Harold Sheldon, had also been a Fellow [*Munk's Roll*,Vol.VI,p.402]. Another uncle, two cousins, a niece and nephew, were all doctors. Every one of them had been trained at King's!

In Brighton it did not take long for Joanna Sheldon to develop her exceptional talents as a clinician, teacher and researcher. But it was in diabetes that she achieved dominance both locally and nationally. She was one of the pioneers of evening clinics for working diabetics and, as long ago as 1968, started regular sessions, also in the evenings, with patients in order to educate them in their own disease. Her devotion and kindness to her patients were often disguised by a firmness and sometimes headmistress-like strictness, but always in their interests and never because of ill-temper or haste. Woe betide the obese diabetic who had failed to lose weight by his next visit. Her imposition of discipline for the sake of her patients was also illustrated by her insistence on quietness in the wards during her rounds, and that general practitioners must always attend her domicilary consultations.

During her early years in Brighton she continued laboratory research at the Univerity of Sussex, where she became a visiting research fellow, but eventually her clinical commitment as well as the load of her committee work and her involvement in local medical societies, including the presidency of the Brighton and Sussex Medico-Chirurgical Society from 1984-85, forced her to abandon this in favour of clinical research. Her publications totalled no fewer than 42 articles and chapters in books. She did a formidable amount for the British Diabetic Association; was secretary of its medical advisory committee from 1971-77 and chairman of the U.100 syringe subcommittee from 1980-85, during which time she played a major part in the smooth transition to the U.100 system in this country.

Joanna Sheldon never married and what little time she allowed herself away from medicine was filled by her other loves - her family, classical music, gardening, birdwatching and mountain walking, epecially in the Dolomites.

In 1980 she developed a malignant tumour which was successfully treated by surgery. However, four years later she learned that she had a second primary which this time was widespread. She endured operation and very unpleasant chemotherapy with great fortitude, and later, when recurrence became evident, she chose to have yet another and even more extensive operation followed by radiotherapy. Between each of her numerous hospital admissions she returned to work and saw her last patient only three weeks before her death. In the fifteen months during which she knew her time was limited her primary concern was to minimise distress to her family and friends and, with her courage and optimism, based on her firm Christian belief, she succeeded. PS[1]/GW

[*Brit.med.J.*,1986,**292**,209; *Lancet* 1986,**1**,116; Photo]

SHONE, Samuel *b*.22 Feb 1906 *d*.1 Jan 1986
OBE(1948) MRCS LRCP(1928) MB BS Lond(1929) MRCP(1931) MD(1932) FRCP(1968)

Samuel Shone was born in Warsaw, Poland, where his father, Isaac, was a tailor. His mother was Stephanie Leshgold. In 1912 the family moved to London where his father re-established himself as a merchant tailor. While Sam was undoubtedly a diligent student who displayed great ability, his achievements owed a great deal to the sacrifices made by his indulgent parents.

Sam Shone was educated at Sir George Monoux' Grammar School, Walthamstow, and at Guy's Hospital medical school. After qualification, his clinical appointments between 1929 and 1934 at Guy's Hospital - in neurology, dermatology and venereology - were combined with general practice. He obtained his doctorate in 1932.

In 1935 he entered the Indian Medical Service and was additional professor of medicine at the Madras Medical College from 1939-40. However, in August 1939 he had volunteered for military duties and he was posted overseas in 1940. He served successively in the Sudan,

Egypt, and in Italy as the officer commanding the medical division, eventually attaining the rank of lieutenant colonel. He returned to India in 1944 as officer commanding a medical division at Lucknow.

In June 1945 Sam Shone returned to civilian duties as professor of medicine at Andhra College and physician in charge of the King George V Hospital at Vizagapatam. For this work he was awarded the OBE. After Indian independence, he worked for a time as chief medical officer of the Kailing Mining Administration in China but was obliged to return to this country when the communists took control.

In 1949 he was appointed deputy senior administrative medical officer to the newly created Sheffield Regional Hospital Board. In 1962 he became senior administrative medical officer, a post which he held until his retirement in 1971. Initially, his assistant medical officers were all either ex-IMS or British ex-servicemen and, because they knew a thing or two about work avoidance, the department was for many years operated on quasi-military lines; heated exchanges and barked orders were commonplace. But the launch of Enoch Powell's hospital plan for England and Wales provided Sam with the opportunity to appoint additional staff of his own choosing. They were younger men, not set in their ways, who actually wanted to help in refashioning the hospital services. Sam brought an analytical approach to problem solving although some of his critics, who had a tendency to rely on intuition rather than reason, used to say he could always find six answers to every question.

Sam Shone exhibited a contradictory character. Sometimes he was the archetypal small, corpulent, choleric bully - and many lived in fear of arousing his wrath. At other times he could be utterly charming and unbelievably patient, especially so with pretty young women and small children. Indeed, he was patient with anyone who showed a willingness to learn, for he was an able teacher. He had a masterly command of written and spoken English, and was well versed in Latin prose from which he frequently quoted. Memoranda and letters drafted for him were often returned with endless corrections, and only the absence of a mark out of 10 distinguished them from school work.

He was a highly articulate man who loved debate, especially with his intellectual equals. On one celebrated occasion he and the late Richard Crossman, Secretary of State for Social Services, had an embarrassing slanging match at a meeting held to discuss services for the elderly. Later that day Crossman summoned him to his room, where it was agreed that they were both strong-willed and the argument was resolved over a drink.

Within the Zuckerman Committee on the future of the scientific services in the NHS, of which he was a member, he was known behind his back as the 'Dalek', partly on account of the similar physical attributes but mainly because of a shared propensity to emit expressions like 'exterminate, exterminate'; he might have been flattered.

After retirement, he moved to Nottingham to take up the chair created for him in the University medical school as special professor of administration. Here he played an important part in the life of the new school, and also pursued his interest in teaching consultants about the management of the NHS. A severe stroke in 1974 compelled him to abandon further work; mercifully he was not deprived of either his speech or his reason and was able to occupy himself in reading and conversation.

In 1943 he married Mary Isabel Forbes, the daughter of a Scottish farmer. He had one son, who has lived in South Africa for many years. He had few interests outside medicine except photography at which he excelled, and occasionally he was persuaded to exhibit his works but, like many artists, he was very self-critical.

Of his many achievements during his time at Sheffield, three are particularly worthy of mention. He was the main architect of the plans for the refurbishment of the hospital and specialist services throughout the region. Much of the new building was not finished before he retired, but he had brought about very substantial improvements in medical staffing levels. He was highly influential in persuading the Department of Health that Nottingham and Leicester could provide ample clinical experience to support the proposed new medical schools. Finally, he was one of the pioneers in introducing short courses on management for consultants, which he did with the help of a grant from the Nuffield Provincial Hospitals Trust. These courses always proved popular and continue to this day; in many ways he viewed them as a form of briefing to his troops before they went into battle.

Sam survived for eleven years after his stroke, thanks to the selfless care and devotion of his wife.

JAS

[Photo]

SIMONS, Peter Josef Gabriel *b.*10 Aug 1929 *d.*25 Apr 1988
MB BS Lond(1956) MRCP(1964) FRCP(1986)

Peter Simons was born in Hamburg, Germany. His parents, Walter Simons and his wife Edith Alice, then came to live in England, and Peter was educated at Stanmore Park School, Hertford, and Bedford School. After national service in the Army he entered St Thomas's Hospital medical school, where he won several prizes. After graduation he held several junior appointments including that of house surgeon at Plymouth Hospital, and then house physician at St Thomas's. He was medical registrar at St Luke's Hospital, Guildford, the Gordon Hospital and the Brompton Hospital, London, and subsequently senior registrar at Crumpsall Hospital and Manchester Royal Infirmary. In 1971, following some locum appointments he was appointed consultant physician, with a special interest in diabetes and metabolic medicine, to Medway Health Authority.

Peter was a skilful and caring physician, with a warm manner which inspired trust and confidence in his patients. He expected high standards from his junior staff, but he showed them great loyalty and did all in his power to further their careers. He was chairman of several district committees and had a gift for reconciling differences of opinion by his persuasive arguments, but on occasion he could show great tenacity when supporting causes in which he believed deeply. He played a major part in the development of undergraduate teaching at a local level.

Peter Simon's most outstanding qualities were his honesty and integrity. He was a sensitive and cultured man, with a passion for music, art and literature. He was not married, but enjoyed the company of a wide circle of friends. He was survived by his mother and brother, Martin.

GW/VL

[*Brit.med.J.*,1988,**297**,1604; Photo]

SIMPSON, Cedric Keith *b.*20 July 1907 *d.*21 July 1985
CBE(1975) MRCS LRCP(1930) MB BS Lond(1930) MD(1932) MRCP(1958) FRCP(1964) FRCPath(1964) Hon DMJ(1962) Hon MA Oxon(1964) Hon MD Ghent(1973) Hon LLD Edin(1976)

Keith Simpson was the son of a doctor. He was educated at Brighton and Hove Grammar School and entered Guy's Hospital as a student in 1924. As a student he was outstanding. He won the prizes for his year in anatomy, physiology and pathology, and gold medals in clinical surgery and bacteriology. He also won three research scholarships. After qualification in 1930 he proceeded to MD in pathology in 1932. He had already decided to become a forensic pathologist but his entry into the field was delayed by the early and sudden death of G W de P Nicholson, professor of pathology at Guy's Hospital. Keith Simpson was then the lecturer in the department and took over its running, although still under 30 years of age, until Payling Wright [*Munk's Roll*,Vol.V,p.461] was appointed to succeed Nicholson. Keith was always proud of the fact that his efforts had been recognized by the School of Governors, who presented him with an honorarium. He left the department of pathology in 1937 when he became lecturer in forensic medicine and a full time forensic pathologist.

In 1937 forensic pathology was in the hands of individual police surgeons and general practitioners with an interest in pathology. It was only in special cases that the services of a recognized consultant pathologist were called upon. Keith Simpson rapidly demonstrated to the coroners and the police the value of the trained pathologist when dealing not only with criminal cases but also with routine coroners' autopsies upon sudden or unnatural deaths. He wrote a series of articles in the *Police Journal* called 'Studies in Reconstruction' Based upon his own cases, showing the importance of forensic pathology in the solution of certain crimes. From the start of his career he was a prolific and lucid writer. In 1947 his undergraduate textbook *Forensic Medicine*, London, Edward Arnold & Co., became the best seller in the field, and was not only used by undergraduates but was widely read by general practitioners and police surgeons. Although forensic medicine was dropped soon after the war as a separate paper in the final examination, the popularity of the book never waned and during his lifetime it achieved eight editions and six reprints. The ninth edition appeared shortly after his death. The third edition had gained him the Swiney Prize of the Royal Society of Arts. In 1956, with Sir Sydney Smith, he edited the eleventh edition of *Taylor's principles and practice of medical jurisprudence*, and in 1965 he edited the centenary edition. He was always delighted that the editorship had returned from Scotland to Guy's Hospital. He also edited two editions of *Modern Trends in forensic medicine*, London, Butterworth, 1953 [1952]. Besides these well known publications he also wrote two editions of *A Doctor's guide to Court*, London, Butterworth, 1962, and a series of cases under the pseudonym of Guy Bailey.

Keith was in great demand as a lecturer, not only in the British Isles but also worldwide. He visited many countries and established contacts between his department at Guy's Hospital and forensic and police departments throughout the world. Every year he welcomed numerous overseas visitors in the department. Many were social calls, but not a few were to seek advice which he was always willing to give, however busy. He was a well liked and fair, but thorough, examiner at both undergraduate and postgraduate levels. If postgraduates fared badly through lack of experience he would often invite them to spend a fortnight at Guy's Hospital.

In 1946 he was appointed the first reader in forensic medicine in London University, at Guy's, where he began to build the first department of forensic medicine in the University. He was given a personal chair in 1962 and in the same year was appointed lecturer at Oxford University. He was a founder member and later president of the British Association of Forensic Medicine; a founder, fellow and council member of the Royal College of Pathologists, and a president of The Medico-Legal Society. He received honorary degrees from Oxford, Ghent and Edinburgh, and was appointed CBE in 1975. He played a major role in the rapid advancement of forensic pathology and science after the war. He was a member for many years of the Home Office Scientific Advisory Committee and he served on the Committee of the Medical Protection Society. He retired from Guy's in 1972, but continued both to work and write until his death. His autobiography *Forty years of murder...*, London, Harrap, was published in 1978, and was among the top ten best sellers for many months.

In spite of his great commitment to forensic medicine, Keith liked to relax. At his country cottage in Tring he frequently entertained guests over the weekend. At one time he carried out much work on the cottage. He was very musical, a keen gardener, and a one time president of the local horticultural society. He always took an interest in student activities, having been chairman of the Clubs' Union Council and president of Guy's Hospital Boxing Club. He also took an active role in many hospital committees.

He married Mary Buchanan in 1932 and they had three children, including a son now practising medicine. After his wife died he married his secretary, Jean Scott Dunn, in 1956. Left a widower a second time, he married Dr Janet Thurston, widow of Gavin Thurston the former coroner for Westminster, who had been his fellow student at Guy's. AKM

[*Brit.med.J.*,1985,**291**,416; *Lancet*,1985,**2**,399; *The Times* 23 July 1985,28 May 1988; *Guy's Hosp.Gaz.*,Aug.1985,**99**, No.2355,282-284; *J.roy.Soc.Arts*,Sept 1985,**5350**,733-34; *Doctors*,14 July 1988,54; *Trans.Mem.address*, 23 Dec 1985; Photo]

SIMPSON, John *b.*23 May 1917 *d.*25 Apr 1987
MB BS Durh(1939) MD(1946) MRCP(1947) FRCP(1965)

John Simpson was born at Tynemouth, Northumberland, into a medical family. His father, Robert Wilfrid Simpson, was in general practice at Heaton. His mother, Bessie Ainsley, was the daughter of a business man.

John was educated at Newcastle Royal Grammar School and the University of Durham, undertaking his clinical studies at the Royal Victoria Infirmary, Newcastle. He graduated with second class honours, obtaining house appointments at the Royal Victoria, where he recalled with special gratitude the influence of J C Spence, later Sir James Spence [*Munk's Roll*,Vol.V,p.386], and other teachers there. He then joined the RAF medical branch and was posted to Singapore just before the invasion. Over the next four years, as a prisoner of war in Java, Malaya, Burma and Siam, he managed to keep records of the nutritional deficiencies as they developed in the various camps. These formed the basis for his MD thesis soon after liberation, which was followed by his membership of the College.

In 1949 he became the first consultant physician to be appointed to the remote West Cumbria area, where he found much ill health from high unemployment and a decaying mining industry. Based on two small hospitals in the main towns, and with the nearest university centre 100 miles away, he found full scope for his clinical expertise, his capacity for sustained hard work, and his teaching ability.

Over the next 25 years John Simpson had the satisfaction of building up a medical unit based at the new West Cumberland Hospital, as well as maintaining close contact with all the general practitioners in the area and teaching housemen, who came from all parts of the world, as well as students from Newcastle. He was elected a Fellow of the College in 1965 and served on the committee which considered the functions of a district general hospital.

When he was 55 years old he went to Uganda on a two-year secondment, where he found his interest in tropical disease and nutrition of so much use that, realizing how much he could contribute to developing countries, he resigned from the NHS and thereafter worked in Zambia and the Persian Gulf until far beyond normal retiring age. Apart from his personal contribution, he was also able to arrange for promising medical students to come to the United Kingdom for further education.

In his free time, John was an active fell walker and interested in birdwatching. He also enjoyed gardening, music and the arts.

In 1945 he had married Monica Mary, daughter of Patrick O'Grady, and they had three daughters. His family survived him. CBW

[*Brit.med.J.*,1987,**294**,1557; Photo]

SKIPPER, Eric Walter *b.*11 Sept 1904 *d.*1 Oct 1987
MRCS LRCP(1926) MB BS Lond(1927) MD(1929) MRCP(1929) FRCP(1950)

Eric Walter Skipper was born in London where his father, Frank John Skipper, was a business director. His mother, Minnie Alice Beale, was the daughter of a furniture manufacturer. He was educated at Parminter's School and the University of London, pursuing his clinical studies at the London Hospital. In a succession of junior resident posts he came under the stimulating influence of such giants as Russell Brain, later Lord Brain [*Munk's Roll*,Vol.VI,p.60], Donald Hunter [*Munk's Roll*,Vol.VII,p.288] and William Evans. After the glamour of the London Hospital, succeeding appointments in Cambridge and south-east London proved dull and disenchanting. Indeed he was not sorry to join the RAMC as a medical specialist at the outbreak of war in 1939. He served as a lieutenant colonel in Northern Ireland and West Africa, and then in France during the Normandy landings, where he was engaged in blood transfusion work. He said afterwards that he would not have missed it for anything but would not like to endure it again. Demobilized in 1946, he was appointed consultant physician at the Royal Infirmary, Sheffield, and Barnsley Beckett Hospital. He became president of the Sheffield Medico-Chirurgical Society - founded more than 100 years ago.

I first met him in Sierra Leone. We were both posted to similar jobs at adjacent West African military hospitals, and struck up a friendship at once. Of serious mein, he was well able to laugh at himself, and other people too if the occasion warranted it. He was, though, a dedicated physician. His special skill and interest lay in the field of diabetes mellitus - while I was particularly interested in the new sounds I heard with my stethescope. But our most entertaining pursuit had nothing to do with medicine or the Army. We examined the new environment, for the ecology of Sierra Leone was different from anything we had known. The mile upon mile of evergreen, leafy forests with exotic blooms aloft, and the wildlife they supported, never failed to interest us. Chattering monkeys, acrobatic to a degree; fast-darting gaily coloured lizards; vibrant and tuneful mosquitoes, and sinuous and often sinister snakes - they were all our constant neighbours. We climbed the hills together, but were disappointed not to have achieved the summit of the Sugar Loaf (2494ft); the undergrowth defeated us. On one of these expeditions, with sweat exuding from every pore, we paused at an unexpected pool which lay at our feet. We would have loved to enter the water but knew of the danger of contracting schistosomiasis from such freshwater snails it might contain. Instead, I introduced the topic of his joining me in Sheffield when the war was over. Sheffield had its points. The neighbouring terrain, on our very doorstep, was none other than the picturesque Peak District. More to the point, perhaps, we had four longstanding teaching hospitals under the benign governance of the University. Little more was said, and in any case the vacancy had yet to arise. Furthermore, I felt also that the prospect of life in Yorkshire did not appeal to him.

Our ways parted in Sierra Leone as a result of the RAMC system of postings, whereby we were scattered to the various theatres of active service and never with the same personnel. It was, therefore, a pleasant surprise, four years after our last meeting, to hear his voice over the telephone. He had been demobilized, was looking for a suitable job and wanted to know all about the Sheffield appointment he had seen advertised in a recent *BMJ*. Within three or four short weeks, after a brilliant interview, he had the job...

At the Royal Infirmary, some distance away from my hospital, we saw less of each other, but met at clinical meetings and at the Philharmonic Concerts at the City Hall. Socially we made the acquaintance of each other's wives and families. We each had three children who soon became boisterous friends. After dinner Eric would play the violin, to my accompaniment on my lovely Bluthner. He played with a skill and interpretation that I had never heard or seen in any other amateur. I liked to see him, instrument on high and bow asweep, at the conclusion of a sonata by J S Bach. And then he gave it all up for the pleasure of painting Derbyshire landscapes - an indulgence for which I never forgave him! He was a lover of nature, though he would not have approved of that designation. It gave him pleasure to feed a squirrel out of his hand; it gave him pleasure to converse with the resident owl at the bottom of his garden. He was, indeed, a person.

Many of his former patients in Sheffield, Derbyshire and Barnsley remember the help he gave unstintingly in times of trouble and illness. They speak of him still.

TEG

[*Brit.med.J.*,1987,**295**,1492; *Lancet*,1987,**2**,1348; Photo]

SKOTTOWE, James Stuart Ian *b*.2 Mar 1902 *d*.2 Mar 1984
MB ChB Glasg(1923) DPM Lond(1926) MD Glasg(1930)
MRCP(1951) FRCP(1958) Hon FRCPsych(1971)

Ian Skottowe was in that classical mode of Scottish psychiatry which, stemming from Sir David Henderson [*Munk's Roll*, Vol.V.,p.188], inevitably embraced precise diagnostic formulation on the basis of detailed history and careful mental examination; yet this school has allowed for such diverse idiosyncracy of psychiatric personality as those of T A Munro, Arthur Spencer Paterson, Angus McNiven and Robin McInnes, as well as his own notable individuality; with Robin, who was by contrast far more of a dynamic psychotherapist and less of an exact nomothetist than Ian, long standing friendship and neighbourly collaboration during Ian's Buckinghamshire days led to a remarkable balance of approach being offered to all who worked under them both, from the time Ian joined Robin as second consultant at Warneford Hospital, Oxford, in 1951, until his retirement to Winchester in 1964.

Born in Dumbartonshire, and educated at Sevenoaks School, Kent, and Morrison's Academy, Crieff, he received his psychiatric education as an undergraduate at the Western Infirmary, Glasgow, becoming MD with commendation, and as a postgraduate at Boston, USA, which afforded him fascinating insights into American psychiatry.

His natural ability to put a point of view over clearly and firmly ensured that he would combine the roles of clinician and administrator: it was in 1927 that he crossed the border to become medical superintendent of Cef'n Coed Hospital where he gained a knowledge of Welsh that, I suspect, about equalled that of his native Gaelic, but which he used for rapport with his Welsh patients - producing occasionally the first smile in depressive misery.

It is intriguing that his death was within a few weeks of that of his wife Phyllis, whom he married in 1927, and within a month of that of S R Tattersall, his deputy and most long standing colleague during the period when he was medical superintendent at Royal Bucks Mental Hospital, Stone, Aylesbury; to be renamed St John's during his tenure. There he set about an extensive programme of modernization, less of its structure than of the clinical approach: his success led to, and was itself further fostered by the association of St John's with the teaching of psychiatry at the Middlesex Hospital during the war time evacuation. This collaboration in teaching continues at St John's till today, now mainly with the Oxford department of psychiatry. Michael Shepherd can be instanced among the top psychiatrists in whose formation Ian played a part. His postgraduate students, particularly his own junior staff on the road to consultantship, kept in loyal communication with him from all over the country: he once remarked that en route for a Scottish holiday he and Phyl could stay with former students at almost any point along either the eastern or western route.

During Warneford's private hospital days he seemed to vie with Tattersall - referred by either, the patient was always interesting and could be helped - in the length of letter of referral, the detail of background and work-up and the exactitude of the problem which the Warneford needed to resolve. Since they both felt that Warneford was the most suitable for certain patients, even after it entered the NHS in 1948, referrals continued on the same meticulous basis until Ian himself moved over.

By this time, he was not only psychiatric member of the Oxford Regional Hospital Board, serving from 1948-55, but was also engaged in writing his short textbook of psychiatry. He would appear at clinical meetings apparently absorbed in his proof reading, only to produce a comment on the presentation so devastingly trenchant as to make it plain that he had not missed a relevant word. The book, which ran into two editions, can be considered the most felicitously succinct modern textbook of psychiatry in the English language - ultimately a collector's piece. Take his initial description of the essential schizophrenic process, of the clinical elucidation of which he was a master:

'Schizophrenic disorders, as the name implies, are characterized by a splitting of mental functions from one another, so that the mental activity of the individual no longer appears as an integrated whole function, but is fragmentary, inconsequential, freakishly aberrant and out of touch wih the immediate material environment, while the component sub-functions of the mind, for example, emotional functions and thinking functions, are liable to be grotesquely incongruous with one another. This produces a kind of bizarre caricature of the patient's personality as it was before he became ill and, in many cases, in the end leaves nothing but a dilapidated fragmentary relic of the personality as it once was.' *Clinical Psychiatry for practitioners and students*, London, Eyre & Spottiswoode, 1953; 2nd edition, London, Churchill, 1964.

It is tempting to suggest that the sophisticated diagnostician follows far more closely Ian's diagnostic methods than relying on the presence or absence of 'front-rank' symptoms. Ian made his juniors connoisseurs of the earliest manifestation of thought disorder, to which he preferred the term 'dyssymbole', less clumsy and more suggestive of possible psycho-physiological aetiology.

In formulating delusional activity, Ian taught 'Avoid being over categorical. "He believes that his sex is being changed by emanations from freemasons in Gravesend" conveys far more than "He harbours paranoid delusions."'.

His second clinical connoisseurship lay in the extraction of the essential depressive state from reactive and neurotic overlays. In the field of minor psychiatry: 'Common sense and wise good will rather than depth analysis were his psychotherapeutic weapons. Sceptical of psychiatric over encroachment into the field of behaviour and personality disorders, he often talked of the need to 'extend the boundaries of normality.' [SS.*Brit.med.J.*,1984,**288**,866]. Tolerant of his dynamically-minded juniors, he would allow time for psychotherapy, recognizing the value of rapport while sceptical of the specific, as opposed to the general effects of the schools of dynamic psychotherapy: at the moment when therapeutic enthusiasm dwindles - 'Let us now see what we can do with some ECT.', Ian would intersperse; and within a fortnight the patient was better.

Not that he was over enthusiastic for mechanical methods. If only means could be found for comparing the effect of his combination of an amphetamine and a carefully monitored mixture of tinct opii with that of today's antidepressants and tranquillizers! Certainly many a depressant was maintained in the community, in the pre-antidepressant days, on this combination and without electroplexy. Resultant opiate addiction? Never.

He had no time for electronarcosis and was extremely reserved over cerebral surgery, though he rejoiced in the meticulous approach of Sir Hugh Cairns. For Ian, the really prolonged narcosis, using barbitone, provided curative respite for the agitated depressive and a few restless manics. To show how readily such narcosis once successfully constituted could be maintained, he would wheel narcotized patients from their wards into clinical demonstrations without disturbing their sleep which he might continue for up to a month.

Ian was much less of a researcher than clinician; but he collaborated enthusiastically with Richard Parnell (q.v.) in his studies correlating somatometry with psychiatry, joined with him in a number of articles and wrote, in collaboration with him, a mental health handbook.

When his move to Oxford, and his subsequent retirement from the Oxford Regional Hospital Board, allowed him respite from local administration he gave his wisdom and foresight increasingly to the Royal Medico-Psychological Association, to its parliamentary committee - of which he was chairman for years - and to its reformation into the Royal College of Psychiatrists. He was himself president of the RMPA, a well justified elevation which he encompassed with marked distinction from 1964-65.

Retirement to Winchester meant no cessation of work until he was about 80, as he became psychiatric adviser to both Wessex Regional Hospital Board (to become Wessex RHA) and Winchester College, as well as working in private consultancy.

Happily married to Phyl (their devoted daughter Gillian lives in Cornwall) he spent himself caring for her in her last illness and probably did not wish to survive her; his survival was in fact only five weeks.

'She first deceased
He for a little tried
To live without her
Liked it not, and died.'
[Adapted from Sir Henry Wotton,
'Upon the death of Sir Albert Morton's wife.']

Socially, he will be especially remembered at the piano, playing riotous improvised solos, or more measured duets with Robin McInnes. A natural extrovert and party man, he was delightful to have as a guest or to visit as a host. With his particular mixture of Scottish courtesy and impish humour - so that nothing, certainly not himself, was taken too seriously or too pompously - and with his wisdom as administrator, and his clinical expertise, Ian had a place among the illustrious psychiatrists of his day.
SJGS

[*Brit.med.J.*,1984,**288**,866; Photo]

SMART, ***Maj.General*** **Robert Arthur** *b.*29 Apr 1914 *d.*6 Nov 1986 CBE(1958) MB ChB Aberd(1936) DPH(1948) MRCP(1965) FRCP(1977)

Robert Arthur Smart (Sam or Robin to his friends) served in the Royal Army Medical Corps until 1972 and, on retirement from the Army, became medical officer to the Tower of London and chief medical officer of Esso Petroleum.

Robin was born in Aberdeen, where his father Arthur Francis Smart was in business as an importer. He was educated at Aberdeen Grammar School and Aberdeen University, where he graduated in medicine, being commissioned in the RAMC that same year.

During the 1939-45 war he served in the Middle East and in Germany, first as a regimental medical officer and later with medical units, and in medical staff appointments. He was mentioned in despatches on three occasions.

At the end of the war, he decided to specialize in preventive medicine. In 1948 he elected to make a study of some medical problems encountered when living and working in extremes of cold, for the dissertation then required by the DPH examination. This choice of subject was to have considerable influence on his subsequent career, as the post war situation dictated a new requirement for the Army to train in cold climates, in addition to the temperate and tropical zones which had dominated prewar service. In consequence, during 1948-51, while on the staff of the Army School of Health, he was appointed to a winter warfare study team and spent several months at Fort Churchill in

Canada. On return he made good use of this experience in his teaching, and many officers and soldiers were introduced to and encouraged to consider hazards to health in the Arctic and the necessary preventive measures. Many medical officers, both young and old, at the School learned for the first time in any detail about hypothermia, frost bite and snow blindness; subjects which heretofore had not received serious attention in the medical curriculum of the undergraduate.

In the latter part of 1951 the exigencies of the service dictated that he should discard cold weather problems for a time and return to sub-tropical and tropical climes in Tripoli and East Africa. In both these locations he was a much respected counsellor in the maintenance of health and prevention of disease.

In 1956 he again became involved in the subject of cold weather conditions, when he was seconded to the Royal Society to be leader of their International Geophysical Expedition to Antarctica, and spent a year at Halley Bay. While there he fell on the ice, with a camera strapped to him, and sustained internal abdominal injuries which, luckily, in a situation where there could be no surgical intervention, responded to rest in Fowler's position. On his return he was awarded the Polar Medal, and appointed CBE in 1958.

He was posted to Supreme HQ Allied Powers Europe in 1960, then in Paris, and while there he was awarded the Silver Medal of Honour of the French Military Medical Services. Following that appointment, and until his retirement from the Army in 1972, he was sucessively: deputy director Army Health Rhine Army, director of Army health at the Ministry of Defence, director of medical services Rhine Army, director of medical services Far East Land Forces and director of medical services HQ Army Strategic Command. He was honorary surgeon to HM The Queen from 1968 to 1972.

On leaving the Army he took up the post of medical officer at the Tower of London, and during the same period became senior and, later, chief medical officer of Esso Petroleum.

Robin Smart was a man of wide interests and an avid reader who could pick up a book or paper and give it his full attention whatever his surroundings. It is no apocryphal story that he would read when walking the dog, and on one occasion (on his own admission) continued to read and walk when the dog, still on the lead, had sat down.

His ability to achieve complete concentration was equally apparent in situations calling for speed of thought and action. When he was pulling away from London Docks for the Antarctica, with a very distinguished farewell party on the quay and the media well represented, his wife suddenly called 'Robin - the car key?'. Without a qualm he paused, took careful aim - and the key landed at his wife's feet.

Of quiet voice and reserved manner, he was happier in a 'round the table' discussion than when speaking from a lectern. He was always good company, with a sharp sense of humour, yet was not an easy man to get to know; those who did remained lifelong friends, and throughout his life he remained in touch with friends of his student and early service days. He could - but not often - be awkward to the point of cussedness. During the last years of his life he insisted, despite medical and family protests, on going alone from Aberdeen to London to attend a meeting at the Polar Club. It was later learned that it was his turn to take the chair. He returned without apparent harm. This incident highlights another facet of his career in that he always seemed to have great good luck and emerged from difficult situations relatively unscathed.

He much enjoyed hill walking and skiing; when in Aberdeen he was a frequent visitor to the Cairngorms, and when in Germany to the Harz mountains. He was also very fond of music and was an able pianist whenever he could be persuaded to play in company.

He married Josephine von Oepen, daughter of Bernard-Heinz von Oepen, in 1947, and they had one child, a daughter. On his final retirement they settled in Aberdeen, but his social life and outdoor pursuits were soon to be limited by progressive ill health which denied him the full enjoyment of the Deeside country which he loved. He was survived by his wife, his daughter and three grandchildren.

TWC

[*Brit.med.J.*,1987,**294**,188; *The Times* 10 Nov 1986; Photo]

SMITH, Bernard Hooper *b.*22 Nov 1917 *d.*24 Jan 1985
MB ChB Aberd(1940) MRCP(1947) DPM(1950) MD(1956) FRCP(1965)

Bernard Smith was professor of neurology at the University Medical School, Buffalo NY, USA, and head of the neurological department at the Edward J Meyer Memorial Hospital, now the Eric County Medical Center.

Bernard was born in Scotland at Peterculter, Aberdeen, where his father was a carting contractor. He attended the local school, going on to Aberdeen Grammar School and subsequently to Aberdeen University, where he graduated in medicine with first class honours. After qualification he held house posts at Aberdeen Royal Infirmary until 1941 when he joined the RAMC, serving in India and Burma until 1946. He was promoted to the rank of lieutenant colonel and became assistant director of nutrition at GHQ India Command.

On demobilization he was affiliated to the National Hospitals, Queen Square, and the Maudsley Hospital, for three years. He became interested in psychiatry and was appointed registrar to Eliot Slater [*Munk's Roll*,Vol.VII,p.541] at the National. But Bernard's primary interest was neurology, and in the years immediately after the war there were very few posts in this country. He decided to go to the United States, and spent a year in Cincinnati. From there he moved on to the Montreal Neurological Institute, Canada, under the direction of Wilder

Penfield, and was later appointed lecturer in neurology at McGill Hospital, Montreal. In 1953 he returned to the USA, becoming the first consultant neurologist to be appointed in Buffalo. In January 1955 he was appointed professor of neurology at the University of Buffalo Medical School. During this time he built up a fine department at the Meyer Memorial Hospital and instituted a trainng programme for residents. He was an excellent teacher, a first class organizer, and an outstanding diagnostician.

He wrote three excellent books on neurology: a textbook *Principles of clinical neurology*, Chicago, Year Book Medical Publishers, 1965; a monograph *Cervical spondylosis and its neurological complications*, Springfield,Ill.,Thomas, c.1968, and *Differential diagnosis : neurology*, New York, Arco Pub.,c.1979.

Bernard was a very likeable man with a good sense of humour; he loved people, enjoyed life fully, and travelled extensively after his retirement. He had married Ruth Hanna, daughter of James Sherman Hanna, in 1959. There were no children of the marriage. He and his wife were very hospitable and enjoyed entertaining many English neurologists at their home near Buffalo. His wife survived him. GW/VL

[*Lancet*,1985,1,709; Photo]

SMITH, Kenneth Shirley *b.*23 Jan 1900 *d.*29 Jan 1987
BSc Lond(1923) MB BS(1927) MRCP(1927) MD(1929)
FRCP(1933)

Kenneth Shirley Smith was born in Hendon. His father, Edward Shirley Smith, was a hotelier. His mother, Gertrude Water, came from a medical family; her father and two brothers were doctors. Kenneth was educated at Haberdashers Aske's. He obtained a senior scholarship to the Middlesex Hospital medical school and graduated BSc with first class honours. In 1925 he married Alice Hoogewerf, the daughter of a Dutch lawyer. He qualified in medicine in 1927, proceeding MRCP in the same year, and was awarded a doctorate in 1929. He became medical registrar at the Middlesex after his house appointment there, and then resident medical officer at the National Heart Hospital. He was appointed assistant physician to Charing Cross Hospital in April 1930, becoming full physician in 1935. At that time, consultant physicians combined general medicine with their specialty, as had done the great physicians of the past such as Sir Gordon Holmes [*Munk's Roll*,Vol.V, p.195] and W J Adie [*Munk's Roll*,Vol.IV,p.596] at Charing Cross Hospital. His contemporaries were also superb general physicians, and yet made lasting contributions to their subjects of special interest, such as Anderson Hickling's work on myelosclerosis which is still quoted today.

He served with the First Army in North Africa during the war, and was also with the Central Mediterranean Force in Italy, Greece and Austria, rising to the rank of lieutenant colonel RAMC.

Kenneth Shirley Smith made many advances in cardiology, but his work was mainly done before the explosion of knowledge produced by cardiac catheterization. Much of his earlier published work was concerned with the interface of cardiology and other specialties such as diabetes. He coined the term 'linked angina' to describe angina precipitated by other disease. He was particularly concerned with the prevention of impending myocardial infarction and the treatment of cardiogenic shock. His recogniton of minor episodes of myocardial infarction with their good prognosis enabled many patients to return early to a normal life at a time when prolonged bed-rest, and permanent entailment of subsequent activity, was the advice so often given to a patient. His work led to the widespread recognition of the relatively benign nature of right bundle branch block. He published many of the early papers on the treatment of hypertension with ganglion blocking drugs and their more specific successors. He was a pioneer in the study and radiological appearances of pulmonary infarction. He studied the application of kymography in the radiological diagnosis of patent ductus arteriosus at the London Chest Hospital. He provided a very active service in cardiology, both at Charing Cross Hospital and the London Chest Hospital. He was also consultant physician to the Samaritan Hospital for Women. He was such a good teacher, his outpatient clinic so active and his inpatient rounds so interesting, that each year he tended to attract the best house physicians.

At a time when academic departments in medicine were being set up in London there was some conflict between the academic units, rapidly increasing in size, and the part-time London teaching hospital consultants. It was claimed that teaching, NHS patient care, and research must suffer if time was spent on private practice. Kenneth Shirley Smith fulfilled all his commitments as generations of students, research associates, and all his patients can testify. There was never any conflict between his private practice and his NHS commitments. Reports on all his patients, to their general practitioners, were full and detailed. If a patient with an obscure heart disease had a small, reducible, indirect inguinal hernia, this would be included in the report. He hated slovenly speech and writing, believing that they denoted slovenly thinking. No student could say 'fast atrial fibrillation' when he meant 'atrial fibrillation with a fast ventricular response'. This abbreviation, repeated continuously at the bicentenary celebration of William Withering in Birmingham recently by members of his beloved British Cardiac Society, would have saddened him greatly.

He was secretary of the first European Congress of Cardiology held in London in 1952. He had already become associate editor of the *British Heart Journal,* to Maurice Campbell [*Munk's Roll*,Vol.VI,p.86], and editor in 1959. The cardiologists had made a wise choice; papers were rejected for ambiguity, but not for lack of literary style. He was very critical of clichés and pointless padding, and swept out jargon, giving the *Journal* a high international reputation. He was elected president of the British Cardiac Society in 1964. It was the same year that his brother, Sir Hubert Shirley Smith, became president of the Institute of Civil Engineers.

Kenneth Shirley Smith was so modest that many of his colleagues at Charing Cross Hospital were unaware of his standing in the world of cardiology. At the same time, some in the hospital considered him autocratic; he used to state that democracy did not work in hospitals. The consultant was in total charge and responsible for the patients under his care. He uncannily anticipated many of our present day NHS problems. The multidisciplinary teams and the prolonged case conferences held on individual patients, so common now, would have been anathema to him. Yet there was no person too unimportant for their opinion to be disregarded, and he treated all people with great courtesy. His kindness and reticence prevented many from realizing that he had an underlying will of steel. A man of strong convictions, he overcame his shyness to take a very active part in the medical world. He was prominent in the Assurance Medical Society of Great Britain, and became president 1955-56. His diffidence allowed him to be overshadowed to some extent by those on the other side of the watershed produced by the introduction of cardiac catheterization. He gave tremendous support to his successor, who had been trained in the modern techniques. His own work was not always appreciated by those who came after him. He was a very different type of cardiologist from Paul Wood [*Munk's Roll*,Vol.V,p.456], whose daughter married Shirley Smith's son.

Kenneth Shirley Smith was a gifted, sensitive, artistic man and it is no surprise that his son became a famous artist. His daughter obtained a degree in physiology, and his grand-daughter is reading physiology at Cambridge.

Kenneth Shirley Smith was stricken low by the death of his beautiful, warm-hearted wife, but his self-discipline and courage did not allow his grief to affect his work. He had a subtle sense of humour, complete integrity, a sensitivity and loyalty that made those who worked with him love him and feel privileged to have known him.

PBSF

[*Brit.med.J.*,1987,**295**,338-9; *Charing X Hosp.Gaz.*Summer 1966,**64**(2)86-87; Photo]

SMITH, Louis William *b.*12 July 1916 *d.*27 Dec 1984
BA Cantab(1937) MB BChir(1940) MRCP(1942) FRCP(1970)

Louis Smith was born at Barnard Castle, North Yorkshire. His father, also Louis, was an engineer and his mother, Mary Dent, the daughter of a local farmer. He was educated at Barnard Castle School, going on to Cambridge where he obtained first class honours in Part I of the natural sciences tripos. He proceeded to clinical studies at St Bartholomew's Hospital where he became house physician after qualification. Later, he worked as a house physician at the London Chest Hospital before going to the Bradford Royal Infirmary. On the inception of the NHS he became consultant physician at the Infirmary, and at St Luke's Hospital, Bradford.

As a physician he was conscientious to a fault. He took endless trouble, but would lie awake at night worrying about his patients and was sometimes reluctant to discharge them if any uncertainty remained. This prompted him to abandon private practice and take a full time contract; it was the same character trait - in addition to later spells of depression - which precipitated his early retirement at the age of 60. He was happiest in his large and busy diabetic clinic where he was rightly confident of his ability to deal with anything that presented itself. Had the system permitted, he would have liked to devote all his time to his chosen specialty of diabetes.

He had a small circle of friends, but few could claim to know him well. He was a retiring man, who disliked public occasions, and his colleagues were beginning to regard him as a confirmed bachelor when, at the age of 42, he abandoned his single life in a small residential hotel and married Eileen O'Mahoney who was at that time working as a senior house officer at St Luke's Hospital. Both were devout Roman Catholics, and they moved into an attractive house with a large garden which became his major leisure interest. He loved the Yorkshire Dales, and was a keen fell walker and a knowledgeable bird watcher. He also liked to visit country churches, on which he was something of an expert. He enjoyed his retirement and remained active up to the time of his sudden and unexpected death. His wife survived him, but they had no children.

PCR[1]/GW

[Photo]

SMITH, Marion Cecilia *b.*15 Jan 1915 *d.*19 Oct 1988
BSc Zool Glasg(1937) MB ChB(1944) MD(1955) MRCP(1965) FRCPath(1965) FRCP(1971) DSc(1985)

Marion Smith (Mrs Stephen Sherwood), known to a large number of people as Mai, made important contributions to the anatomy and pathology of the human nervous system, and to the histological methods of studying that system. She spent most of her life working for the Medical Research Council at the National Hospitals for Nervous Diseases at Queen Square. When she retired she was given facilities to continue her unfinished work at the Radcliffe Infirmary in Oxford.

Marion was educated at the Notre Dame Montessori School and Notre Dame High School, Glasgow, and went on to Glasgow

University, where she took an honours degree in zoology. On graduation she became a demonstrator in zoology, and then lectured in biology at Mount Pleasant Training College, Liverpool. In 1939 she returned to Glasgow to study medicine, supporting herself by demonstrating at the university and at the Anderson College of Medicine, and running coaching classes in zoology for her fellow medical students. She qualified in 1944, and after junior hospital appointments she went on to the National Hospitals for Nervous Diseases, Queen Square, London, to be trained in neuropathology by J G Greenfield [*Munk's Roll*,Vol.5, p.164].

Her research work has contributed to our basic knowledge of the human brain, cerebellum and spinal cord. As her work was on man, it was immediately useful to neurosurgeons carrying out pain-relieving operations and stereotactic operations for Parkinsonism and other movement disorders. She was consulted by veterinary, and human, neuropathologists for her opinion of anatomy, and was often asked to try and find degeneration of nerve fibres within the central nervous system, that must have occurred but could not be demonstrated. She developed and made great use of the Marchi method of staining degenerating nerve fibres, so that it became an excellent tool for investigating the human nervous system with both surgical and naturally occurring lesions.

Marion Smith was married to Stephen Sherwood, a neurosurgeon who was mainly interested in research; he predeceased her. They had two daughters, one of whom is a dermatologist and the other an architect.

One of her main interests was the British Neuropathological Society, of which she was a founder member, then secretary from 1960-68, and later president from 1977-79. Her wise counsel during a period of considerable changes was invaluable to the society. She was also secretary-general of the International Society of Neuropathology, 1970-78, and an enthusiastic member of the Harveian Society, of which she eventually became secretary and president.

She was known in private life as 'Dr Sherwood' and was an amusing and warm-hearted woman. She had a great number of friends to whom she meant a great deal. She kept up a large correspondence with those who lived abroad and many of them, of different generations, would often stay with her in her London home and at the small house she had in Wiltshire. At the latter, she worked strenuously in the garden up to and including the last few months of her life. Her affection extended to her dogs and cats, of which there were usually at least four. In a very busy - perhaps one should say 'hectic' - life, she also found time to read and listen to music.

Marion died from pulmonary embolism following a shunt operation for low pressure hydrocephalus at the National Hospital, where she had worked for most of her life.

PWN

[*The Independent*,8 Nov 1988;*The Times*,2 Nov 1988; *Lancet*,1988,2,1376]

SNEDDON, Ian Bruce *b.*6 Mar 1915 *d.*10 Oct 1987
CBE(1974) MB ChB Sheff(1937) MRCP(1941) FRCP(1956) Hon MD Sheff(1980)

Ian Bruce Sneddon was born in Sheffield and lived there for most of his life. His father William was a Scottish general practitioner who had set up in practice in the city, and his mother was Kathleen Hield. His father died when Ian was still a schoolboy; he always remained proud of his Scottish ancestry.

Ian was educated at Uppingham School and graduated in medicine from the University of Sheffield. His first medical appointment was as house physician to Edward Wayne, later Sir Edward, and Rupert Hallam [*Munk's Roll*,Vol.V,p.169] at Sheffield Royal Infirmary. The dermatological side of this post was to influence the rest of his life, and after other house jobs he returned to the skin department in 1938 as clinical assistant. This was followed by appointment as registrar to Robert Platt, later Lord Platt [*Munk's Roll*,Vol.VII,p.470], a post which was interrupted by the outbreak of the second world war.

As Sneddon was in the RNVR he was called up immediately, and after serving two years at sea he was made a skin and medical specialist. Part of the rest of his naval service was spent in the Pacific, some of the time in Australia, and his love of sailing started in dinghies on the waters of Sydney Harbour. A fellow sailor was C A Clarke, later Sir Cyril and a former president of the College, and at that time they produced a paper and film on nutritional neuropathy in prisoners of war.

Sneddon returned to Sheffield when he was demobilized in 1946 and intended to specialize in general medicine, but he obtained the post of supernumerary registrar in the skin department at the Royal Infirmary and later that year accepted a consultant post in the department. At the same time, he was appointed honorary clinical lecturer in the medical school. He was a popular and enthusiastic teacher, and his interest in and concern for young people led to his becoming clinical dean in 1950. While holding this appointment he initiated a tutorial system for the medical students which provided them with a member of the consultant staff to whom they could turn for advice.

Sneddon's description, with Wilkinson, of subcorneal pustular dermatosis (Sneddon-Wilkinson Disease) brought him eponymous fame and gave him a special interest in blistering eruptions, about which he wrote numerous papers and chapters in books. But his interests in dermatology were wide, jointly with R E Church he wrote a small textbook, *Practical Dermatology*, London, Arnold, 1964, which went into four editions, and about 100 articles on various aspects of skin disease. A fluent and witty speaker, he was in constant demand as a lecturer and gave the Dowling Club oration in 1957, the Watson Smith lecture in 1963, and the Thom Bequest lecture in Edinburgh in 1968. He also gave numerous invited lectures all over the world.

Ian Sneddon was president of the British Association of Dermatologists at the time of its 50th anniversary in 1970, and of the section of dermatology of the Royal Society of Medicine 1980-81. He was subsequently made an honorary member of both the Association and the Section. At various times he was president of the North of England Dermatological Society and of the Sheffield Medico-Chirurgical Society. He was awarded the CBE in 1974, and when he retired in 1980 Sheffield University gave him an honorary MD.

In 1946 he married Joan, daughter of George Simon, an engineer, and they had three daughters and two sons. Joan, also a Sheffield graduate in medicine, was co-author of some of his papers on the psychiatric aspects of skin disease, and one of his sons qualified in medicine.

Small in stature, and bald from an early age, his appearance seemed almost unaltered by the passing years. Like many of his build, what he lacked in height he made up for in energy and determination. Brought up in the northern work ethic, he continued in private practice and in the writing of articles until the end. A coronary thrombosis in 1968 caused him to relinquish the office of clinical dean and impeded his interest in gardening, but made no difference to his enthusiasm for dermatology or for participation in its affairs. Sailing remained his main interest outside medicine; he graduated from dinghies to a 10-tonner which he kept at Cowes, and he sailed to France and back for the last time shortly before he died. In his last year he passed the examination for Yachtmaster's Certificate, an achievement of which he was very proud. Fly-fishing and painting were among his other interests.

In every branch of medicine each generation produces one or two figures of outstanding ability, to train and advise the young and set the standards of the specialty. In his day Ian Sneddon was such a figure in dermatology.

REC

[*Lancet*,1987,**2**,1039; *Brit.med.J.*1987,**295**,1356; Photo]

SPILLANE, John David *b*.7 May 1909 *d*.23 July 1985
BSc MB BCh Wales(1933) MRCP(1936) MD(1939) FRCP(1950)

John David (Jerry) Spillane died at his home in Dyfed, West Wales. Born the son of a schoolmaster, Thomas Spillane, and his wife Margaret, daughter of John David Condon, a farmer, he was educated at High School in Cardiff, University College Cardiff, and the Welsh National School of Medicine, graduating with a gold medal and distinction in medicine. He was a University of Wales champion boxer.

After obtaining his membership of the College he travelled in the USA as a Commonwealth Fund fellow, studying under Paul White at the Massachusetts General Hospital and at the New York Institute of Neurology, before returning to proceed MD; his thesis being in cardiology. In 1939 he married Joan Rees, daughter of John Calwalader Rees, a distinguished Cardiganshire doctor. Shortly afterwards he enlisted in the RAMC and was posted first to Aldershot, then to the Middle East, and ended his war service as adviser in neurology, Middle East Forces, with the rank of lieutenant colonel.

On demobilization he spent a short period in Queen Square before being appointed to the Cardiff Royal Infirmary, first as assistant physician and then as consultant neurologist and honorary lecturer in neurology in the Welsh National School of Medicine, a post which he filled with the utmost distinction until his retirement in 1974. During this period he earned the Cardiff department a national and international reputation by dint of his energy and enthusiasm for clinical research, writing and teaching. He travelled widely on a variety of fellowships, keenly observing and recording all that he saw. His contributions to his subject were legion; outstanding among them cervical spondylosis, craniovertebral anomalies, the encephalopathy of boxers, iatrogenic neurology, 'Tegretol' in the treatment of trigeminal neuralgia, and tropical neurology.

In later years his interest turned to books. *An Atlas of clinical neurology,* London, Oxford University Press, 1968, ran to three editions in five languages. It was a clinical commentary on his already vast selection of photographs embracing the whole range of clinical neurology, being hailed as 'the very essence of one of the greatest teachers of our day' and based on his weekly clinical demonstrations, known affectionately as 'The J D Show' to generations of undergraduates.

Jerry's travels inspired him to edit *Tropical Neurology,* London: New York, Oxford University Press, 1973, and also provided the title 'The Geography of Neurology' for the subject of his Bradshaw Lecture and his presidential address to the section of neurology at the Royal Society of Medicine in 1972. His most scholarly work *The Doctrine of the nerves,* London:New York, Oxford University Press, 1980, appeared to critical acclaim and was followed by *Medical Travellers,* London, Oxford University Press, 1984. A further work, on his travels in America before the second world war, was in the publisher's hands at the time of his death.

Jerry Spillane was elected president of the Association of British Neurologists in 1974. Speaking after dinner he recounted how, as an aspiring young neurologist who had studied in detail the nutritional disorders of prisoners of war, he had presented his findings to the

Association. Sir Francis Walshe [*Munk's Roll*,Vol.VI,p.448] commented, in discussion: '... clearly someone will have to look into all this.' Understandably somewhat disgruntled he was subsequently consoled by Lord Cohen [*Munk's Roll*,Vol.VI,p.106]: 'Don't worry, Spillane, he was smiling as he said it.' Spillane gloomily rejoined: 'But his smile was like a glint of wintry sunshine on the brass handle of a coffin.'

Retirement did not still Jerry's energies, and in 1976 he greatly enjoyed a year as Sandoz visiting professor in Dartmouth Medical School, New Hampshire.

Throughout his life Jerry Spillane's burning ambition was to contribute significantly to his specialty, and this he pursued with literacy and zeal. Apart from his work, his interests were in his family, reading, and an informed love of music. In his retirement he enjoyed walking in the countryside and conversation with his friends. He had an impish sense of humour, little patience with fools, and respected those who held their ground when under critical attack.

Jerry was a superb clinician, especially astute in spotting the non-organic, and a magnificent teacher. His written works are his lasting memorial.

He is survived by his wife Joan, daughters Gillie and Susie, and his son John, also a consultant neurologist and responsible for the revision of the third edition of *An Atlas of clinical neurology*. JGG/GW

[*Brit.med.J.*,1985,**291**,549; *Lancet*,1985,**2**,677; *The Times* 13 Aug 1985; Photo]

SPRIGGS, Edmund Anthony *b.*30 Mar 1918 *d.*28 Dec 1988
BA Oxon(1939) BM BCh(1942) MRCP(1943) DM(1952)
FRCP(1966)

Edmund 'Tony' Spriggs was born in Banff, the son of a distinguished physician, Sir Edmund Ivens Spriggs KCVO [*Munk's Roll*,Vol.IV,p.468]. His mother, Alice Mary, née Watson, was the daughter of a farmer. 'Tony' was the first of two sons, who effectively provided a second family for parents tragically bereaved by the loss of two older daughters in a drowning accident. His father, Sir Edmund, had been a physician at Guy's prior to becoming dean at St George's Hospital medical school, but his professional life in London was cut short by a pleural effusion, presumably tuberculous, and the family moved to Banff, Scotland, in the hope of aiding his recovery. This did indeed follow and Edmund Spriggs Snr returned to medicine with considerable vigour to found, or help to found, two new diagnostic clinics - firstly in Banff (Duff House) and secondly in Denbigh (Ruthin Castle). The latter was particularly successful, received royal patronage and led to Sir Edmund's knighthood. The early years of Tony Spriggs and his brother Arthur were consequently privileged, and greatly influenced by the beauties and pastimes of rural Britain - an influence which was reinforced by their mother's farming background.

Both sons followed their father's prescribed course of education - Oxford preparatory school, Winchester, and Oxford University - and both duly entered the medical profession. For Tony, this meant an initial house physician post in 1942 at the Postgraduate Medical School, Hammersmith, and war service in the Army soon afterwards. He served as a regimental medical officer in India and France, as well as in Britain, and eventually returned to civilian life and the Postgraduate Medical School in 1947. He developed an interest in gastroenterology and, after further house and registrar appointments, undertook a fellowship at the Mayo Clinic in 1950-51. His plans to visit the United States coincided with the start of his long and happy marriage to Sheila More, although the plains of Minnesota proved to be rather less attractive to the bride than the groom.

They returned to Britain in 1951 when Tony became a physician to the Ruthin Castle clinic, his father having died in 1949. By that time the NHS had taken over most of the country's hospitals and clinics, and the privately endowed Ruthin Castle entered an inevitable decline prior to closure and sale. Tony completed his DM thesis concerning mobility of the colon and rectum, became an enthusiast of the new order of medicine, changed his allegiance to chest diseases, and in 1953 moved to the north east of England. In 1955 he was appointed physician to the Barrasford Tuberculosis Sanatorium, the Northern Counties Chest Hospital and Newcastle General Hospital, which was to become one of three teaching hospitals of the medical school of the University of Newcastle upon Tyne. With the sequential closures of the sanatorium and chest hospital he came to work wholly at Newcastle General Hospital by the time of his retirement in 1983.

His 28 years on Tyneside were busy and eventful. He made his greatest impact as a deeply caring physician, though he contributed his share to administrative duties and wrote modestly, if not prolifically, on a wide variety of subjects. He was instrumental in founding Northern ASH (Action on Smoking and Health), chaired the consultant staff committee, became postgraduate tutor, and served as treasurer and president of the North of England Thoracic Society.

Tony Spriggs was a keen member of the British Society for the History of Medicine, and much of his writing centred on this interest: '...the best reason for reading the history of medicine is the fun of it. But it also puts into perspective the fashion of the day.' His collaborative work with Bryan Gandevia, a fellow chest physician and historian from Sydney, on John Hutchinson, a Tynesider by birth, E A Spriggs, *Med.Hist.*,**21**,Oct 1977,357-64 and B Gandevia, *Med.Hist.*, **21**,Oct 1977, 365-83; on spirometry, E A Spriggs, *Brit.J.of Diseases of the Chest*, 1978, 165-80, and Hutchinson's colourful experiences in the Antipodes, were particularly notable. Spriggs was also a man of some wit, and as recently as 1979 managed to publish within the columns of *The Lancet* an account of his imprisonment within a faultily locked toilet at Newcastle's newly built Freeman Hospital.

Away from his work he was a man of many parts. His marriage and children were sources of great contentment. The family medical traits were preserved, and his two daughters became nurses, while one of his sons became a physician. Shooting, fishing and natural history were continuing passions from childhood, while in later life he took up gardening, Persian rugs and book binding. Retirement was celebrated by a university refresher course in Latin, and increased service to his village church. It lasted a little over five years and must have been as fulfilling as ever.

His former secretaries and nurses must have the last word: 'Dr Spriggs was quite the finest gentleman we ever knew. He was kind, compassionate, warm, human and above all humourous. He made everybody - staff, patients, colleagues, friends and neighbours - all feel as if they were important to him. He always remembered to ask after one's family, always by name, and was interested to know how they were getting on. His patience and kindess were an inspiration to all who knew him - he was truly a Christian man.' Indeed; in the traditional sense, he was a scholar and a gentleman.

DJH

[Photo]

STAHL, Jules *b*.6 Oct 1902 *d*.14 Mar 1984
MB Strasbourg(1931) Hon MAAP(1957) Hon FACP(1962)
*FRCP(1970)

Jules Stahl, one of the founders and first president of the European Association of Internal Medicine (AEMIE), was born at Barr in Alsace where his father, also Jules Stahl, was a chemist. He studied medicine at the University of Strasbourg, entering the Clinique Médicale B in 1926 where he undertook his clinical work.

From the beginning Jules Stahl was greatly attracted to research and involved in the early work on insulin therapy. In 1930 he began experimental studies on a hyperglycaemic factor which was ultimately shown to correspond to glucagon. Subsequently, he collaborated in various basic studies relative to carbon monoxide intoxication, thyroid stimulating hormone determinsation in biological fluids, pituitary-thyroid feed-back, and pathophysiological problems of hypochloraemia.

In 1934 he spent a year as a Rockefeller Foundation fellow at the Columbia Presbyterian Medical Center in New York, working with R F Loeb and D W Atchley, who became his lifelong friends. This period in the United States strengthened his own views on the essential role played by pathophysiology and clinical investigation in the training of internists. They remained his guidelines when organizing clinical research at a later stage of his career: the pathogenesis of oedema, hepatic effects of glucagon, and blood ammonia and liver diseases were the three main areas of his research. He wrote numerous papers, and gave many lectures, on the results of this research. He became professor of internal medicine at Strasbourg in 1946, and in 1948 was appointed head of the Clinique Médicale B where he remained until his retirement in 1974. Even after his retirement he continued to participate in the life of the department by clinical presentations of patients to students, by revising papers for publication, enabling his colleagues to profit from his wide knowledge of medical bibliography, and by undertaking all the translation work required by the secretariat of the AEMIE.

An accomplished scientist, but also a humanist in the true sense of the word, Jules Stahl was above all a highly talented and renowned physician, an impartial and selfless professional who gained the admiration of all his patients. His attitude towards his patients reflected his personal view of internal medicine: he never behaved as an omniscient super-specialist, but endeavoured to create a doctor/patient synthesis which would permit the harmonious integration of clinical data for the ultimate benefit of the patient.

Jules Stahl was a brilliant teacher able to impart his knowledge to students in a lively and stimulating manner. During ward rounds his students greatly appreciated the way in which he would instruct them in the art of taking case histories and making physical examinations. He was also an excellent linguist and received invitations to lecture in many countries, and to act as chairman at various congresses. He was a convinced European, believing that his own concept of internal medicine would be strengthened and more widely applied if it could be developed within all the countries of the European Community.

He married Annette Simon, daughter of Maurice Simon, a leather manufacturer, in 1936. They had one son and two daughters. Outside his work and his family, Jules's main interest was listening to chamber music.

JMB

[Photo]

STAMM, *Air Vice-Marshall* William Percivale

*b.*27 Aug 1909 *d.*10 Mar 1986

CBE(1960) MRCS LRCP(1932) MB BS Lond(1933) DCP(1946) DTM&H(1947) MRCP(1951) FRCP(1956) FRCPath(1963)

William Percivale Stamm, know as Bill, qualified with the Conjoint in 1932 at Guy's Hospital, and after house appointments there he obtained his MB BS in 1933. In January 1934 he started his 33-year career in the Royal Air Force Medical Branch. Before the second world war he served in RAF hospitals in the UK and the Middle East. During the war he commanded the RAF hospital at Takoradi from 1942-43, and then served at RAF hospital Cosford until early 1947, during which time he obtained his DCP. He was then posted to the RAF Institute of Pathology and Tropical Medicine at RAF Halton, where he was to serve for the next 20 years. He was elected a Fellow of the College in 1956 and made a founder fellow of the Royal College of Pathologists in 1963. He was an honorary surgeon to HM The Queen from 1959-69, was made a CBE in 1960, and promoted to air vice-marshal in 1966. He was consultant adviser in pathology and tropical medicine to the RAF from 1952-69. From 1966-67 he was president of the Association of Clinical Pathologists, vice-president of the Royal Society of Tropical Medicine and Hygiene from 1969-71, and an external examiner for the DTM&H at both Liverpool and London Universities. After his retirement in 1969 he became the director of the Amoebiasis Research Unit.

It was due to his efforts that the pathology and tropical medicine services of the RAF were put on to such a sure footing after the war, when new pathologists had to be trained. He was a very good manager and worked out methods of assessing workloads of laboratory technicians for establishment reviews. He introduced standardization of equipment and methods in both overseas and UK RAF hospitals, and devised systems for quality control. A true clinical pathologist, he was an acknowledged world expert on amoebiasis and a very good teacher. A man of forthright personality, he developed the pathology services for the RAF at a time of crisis and served it with flair, vigour, enterprise and loyalty throughout his long and distinguished career.

He was survived by his children, Pippy, John and Peter.

JGD

[*Brit.med.J.*,1986,**292**,1283,1467; *Lancet*,1986,**1**,1047; Photo]

STEWART, Ian McDougall Guthrie *b.*23 Dec 1914 *d.*9 June 1988
MRCS LRCP(1939) MB BS Lond(1940) MRCP(1947) MD(1949) FRCP(1966)

Ian Stewart was born at Hutton in Essex. His father, James Guthrie Stewart, was a civil servant and his mother, Katherine Mary, née Mundle, was the daughter of a Congregational minister.

Ian was educated at Berkhamsted School, Hertfordshire. He trained for medicine at Guy's Hospital, London, qualified in 1939, and almost immediately joined the RAMC. After the war he continued his education, holding house posts at Guy's, the Brompton Hopital, and Bristol Royal Infirmary, before being appointed consultant physician to the Victoria Hospital, Blackpool, in 1953, a post he held until his retirement from the NHS in 1980.

Ian Stewart had remarkable physical and intellectual energy, addressing many clinical and research problems with insight and careful analysis. His main clinical interests were in diabetes mellitus and hypertension. He was instrumental in the development of a major diabetic service for the district, and made many important observations on the management of hypertension, in particular the association between hypertension, its treatment, and ischaemic heart disease. He was in much demand as a lecturer, both nationally and internationally, and was always welcomed with respect and affection.

He was a member of the Council of the British Heart Foundation, and a member of the cardiology committee of the College. He also served on the advisory committee of the *Prescribers' Journal* for many years. Ian was very interested in medical education and he planned the development of the excellent postgraduate medical centre in Blackpool, where he was tutor for many years. After retirement he remained active in private practice and, in addition, was invited to take up a year's appointment as professor of medicine in the University of Malta.

Almost as if his professional life were not enough to fulfill him, Ian had a wide range of interests and commitments outside medicine. At medical school he played rugby for Guy's, where his career was described as 'brilliant and outstanding' - interrupted only by a series of fractures. His reluctance to shirk a challenge, be it intellectual or physical, was evident even at that time.

At the beginning of the 1939-45 war he was deeply affected by the loss of a brother. He served in the RAMC, being posted to the 1st Welsh Regiment. He was twice mentioned in despatches and was taken prisoner in Crete. During this period his normally restless nature was channelled into supporting his fellow prisoners, who came near to worshipping him. For this he was punished by his captors who segregated him from his fellows, confining him to the company of prisoners who spoke no English. He responded, characteristically, by learning not only survival techniques but also French, in which he became fluent, and German. He wrote a definitive history of *The Struggle for Crete*, London, OUP,1966, which was widely praised as one of the best campaign descriptions ever published. His short contribution to *Beyond Survival*, a symposium edited by Guthrie Moore, provided those who knew him with an unexpected insight into his philosophy of life.

After the war, he maintained and developed his interest in history, particularly in military history and he was very widely read. His interest in sport was also maintained; at Bristol he began a second career - parallel to medicine - as a sports commentator. Over the next 20 years he gave over 500 radio broadcasts, many on rugby union football.

Throughout his professional career Ian also showed an interest in medico-legal matters. He thrived on the cut-and-thrust of courtroom debate and contention and at one stage, when somewhat disenchanted with the National Health Service, he toyed with the idea of taking up law as a first profession. However, he conceded that although he would have enjoyed it thoroughly, he might have lost as many cases as he won.

Ian Stewart will always be remembered, by those who knew him, for his wit, his kindness and lack of malice which, with his extreme determination, provided an endearing combination. His first marriage, to Betty Boff in 1947, ended in divorce in 1972. They had a son, Simon, who is a consultant gynaecologist. He was survived by him, and by his second wife, Marilyn, and their children, Sarah and Adrian. FIL

[*Brit.med.J.*,1988,**297**,613; *Guy's Hosp.Gaz.*,Oct 1988,167-8; Photo]

SUMMERLY, Roy *b.*17 Apr 1932 *d.*2 Apr 1984
MB CHB Birm(1955) MRCP(1960) MRCPE(1961) FRCPE(1972) FRCP(1974)

Roy Summerly was born at Rothwell, Northamptonshire, and educated at Kettering Grammar School. He read medicine at the University of Birmingham, where he qualified in 1955. After house appointments at the Queen Elizabeth Hospital, Birmingham, and the North Staffordshire Royal Infirmary, he served in the RAMC for two years and became clinical officer for dermatology. After leaving the Army in 1958 he completed his general medical training at the North Staffordshire Royal Infirmary.

Roy decided to devote himself to dermatology and, after two years as a registrar in Stoke-on-Trent, was appointed first as registrar and later as senior registrar in dermatology at St Thomas's Hospital. During this time he was also honorary clinical assistant at the Institute of Dermatology, St John's Hospital for Diseases of the Skin, London. In 1964 he became a clinical research fellow at the Medical Research Council's unit of experimental pathology of the skin in the medical school, Birmingham, where he developed an apitude and interest in research which he maintained throughout his career.

He was appointed consultant dermatologist to the North and Mid-Staffordshire Health Districts in 1966, being based at the North Staffordshire Hospital centre, Stoke-on-Trent. He rapidly established a reputation not only as a superb dermatologist but also as an excellent physician, and he played a major role in building up a first class department of dermatology. He was a gifted teacher, and the number of junior staff rotating through his department who subsequently took up a career in dermatology is a testament to his enthusiasm and inspiration.

Roy continued to devote himself to research at the MRC skin unit in Birmingham which subsequently moved to the University of Keele when its department of postgraduate medicine was established. He was appointed senior research fellow in 1979, and senior lecturer in dermatology in 1980. His more recent research was concerned with studies on epidermal phospholipase A^2 activity in psoriasis and other epidermal hyperproliferations, and the monitoring of tropical steroid action by the inhibition of phospholipase A^2. His scientific contributions to dermatology were extensive and he published over 15 papers in the last five years of his life.

Roy Summerly was much in demand on local and regional committees, where his wise counsel was highly valued. He was a member of the Dowling Club, and was asked to give the Dowling Oration in 1973, the mark of a distinguished young dermatologist. He was a very active member of the British Society for Investigative Dermatology, and had been on the executive committee of the British Association of Dermatologists, a vice-president of the section of dermatology of the Royal Society of Medicine, and president elect of the North of England Dermatology Society in 1983. He was due to be elected a member of the Association of Physicians of Great Britain and Ireland in the month in which he died; a rare honour for a dermatologist and a testimony to his standing among physicians. He was also president of the North and Mid-Staffordshire division of the British Medical Association.

Roy was a quiet, extremely hardworking person, highly respected in his specialty, perhaps particularly because of two unusual attributes. Firstly, he continued active and productive basic research while carrying a busy clinical load, and secondly because his extensive knowledge of general medicine - quite apart from its relevance to dermatology - made him a dermatologist of the highest calibre.

He was a delightful person, always courteous, with many friends in all walks of life. He was renowned for his professional skill, integrity and intellectual honesty, his charm and his sense of humour. He enjoyed fell walking, cricket, and classical music - especially church music. He was also a campanologist. More recently he had developed an interest in historic battlefields. But most of the little spare time he had was devoted to his family; his wife Myrtle, a physician in community health, and his daughter and son, who all survived him. WvH

[*Brit.med.J.*,1984,**288**,1388; *Lancet*,1984,**1**,919; Photo]

SVARTZ, Nanna Charlotta *b.*25 July 1890 *d.*3 Apr 1986
Med.Lic.Stockholm(1918) MD(1927) *FRCP(1970)

Nanna Svartz was born in Vasteras, a provincial city west of Stockholm. At that time this city made no provision for educating girls to university entrance standard. However, young Nanna had made up her mind to study medicine and she went to Stockholm (later her family followed in order to support her), where she passed her university entrance examination in 1910 and immediately began her medical studies at the Karolinska Institute. She qualified in 1918 and commenced her hospital career in internal medicine at the Seraphimer Hospital, the old University Hospital in downtown Stockholm. After gaining her MD in 1927 she was appointed assistant professor of medicine. Ten years later she became a full professor of internal medicine at the Karolinska Institute, and chairman of the department of medicine at the Seraphimer Hospital. In 1940 she was transfered to the new Karolinska Hospital, where she was chairman of medicine until her retirement in 1957.

During the latter period she created a new department of rheumatology at the hospital and contributed in a major way to the establishment of the King Gustav Vth Research Institute, built within the grounds of the Karolinska Hospital with money collected in Sweden to celebrate the King's 80th birthday. Nanna was appointed the first head of the Institute - a position she held for some time after her official retirement from the academic chair. She continued to work in her laboratory until a few years before her death; up to 25 years after her retirement.

At an early stage in her career, two apparently unrelated areas of medicine caught her interest and remained major fields of activity throughout her long life: ulcerative colitis and rheumatoid arthritis. Her MD thesis had dealt with intestinal bacteria and had stimulated her interest in ulcerative colitis. She early understood the importance of good nutrition for patients suffering from this condition; on arrival in hospital they were often found to be suffering from near starvation. And in rheumatology she began to define and purify the rheumatoid factor; by surprisingly simple methods, such as euglobulin precipitation in distilled water, she achieved a major step in the process of purification. She also developed a sheep cell agglutination test for the determination of the factor in patient sera.

Perhaps her most important contribution was the creation of a link between her two fields of research: the sulphonamides had been introduced towards the end of the 1930s and Nanna Svartz decided to try and combine a sulphonamide with acetylsalicylic acid which, until that time, had been almost the only effective drug against rheumatic disease. The result was azulfidine ('Salazopyrin'): sulphapyridine linked to acetylsalicylic acid by an azo bond. Nanna had planned this new drug specifically for the treatment of rheumatoid arthritis (at that time called primary chronic polyarthritis), but she found that the granulated tissue in ulcerative colitis had definite similarities to that in rheumatoid arthritis and decided to try the drug on patients with ulcerative colitis - with immediate effect. This was before the steroid era and Nanna Svartz used 'Salazopyrin' for treatment of the active stage of the disease. Almost 50 years later it is being used all over the world, mainly to prevent recurrence of the colitis, with excellent results. But the link has not been broken, there is renewed interest in 'Salazopyrin' as an immunosupressive drug also in rheumatoid arthritis - especially in the UK.

Nanna combined her hospital work with private practice; she took very good care of all her patients and they all loved her. She also received very substantial financial support for her continuing research.

Nanna was an excellent clinician. I trained with her during my first years in hospital medicine and learned much; a knowledge that many years later I tried to pass on to my students when I myself became chairman of medicine at the Karolinska. She loved to travel and even in old age she liked to take part in international meetings as an active participant, often submitting papers. She was also active in the creation of the International Society for Internal Medicine.

Nanna was not the first woman doctor in Sweden; when she began her career there were a score already in practice. But she was the first woman within any university faculty to be appointed a full professor - after evaluation by an official Faculty Committee, according to the Swedish system, followed by appointment by the King. She was very proud of this fact, and rightly so. Nowadays there are many women professors within medicine but even today there are very few women who head large departments of internal medicine.

Nanna Svartz loved her work - and she loved working. She published a very large number of scientific papers. I well remember how she used to arrive at the hospital in the morning with a bunch of official documents under her arm, stating that she had read them all in bed during the night. She repeatedly said that she knew nothing better than work; that science, which reveals the truth step by step, was more exciting than the theatre, and that clinical work was always an adventure.

Nanna's long and very active life is now over; her contribution to medicine was made so long ago that there are many today who do not know who she was. Those who had the good fortune to work with her will always remember her with gratitude and admiration. LEB/GW

[Photo]

SWANN, Peter Geoffrey *b.*18 Feb 1921 *d.*15 Oct 1988
CBE(1981) OBE(1973) MRCS LRCP(1942) MB BS Lond(1943) MD(1949) MRCP(1949) FRCP(1969) FFOM(1978)

Peter Swann was born at Ilford, Essex, the son of Arthur Swann, an industrial chemist, and his wife Annette Watkins, whose father had an engineering background. He was educated at the Park School, Ilford, and Chigwell School, from where he obtained admission to the London Hospital medical college. He qualified in 1942 and, after house appointments, joined the RAFVR medical branch, where he served until 1946. After the war, following a period in which he was registrar and senior registrar in general medicine at the London and Oldchurch hospitals, he decided to enter the newly emerging specialty of occupational medicine. He was appointed senior medical officer to Esso Petroleum in 1954 and was promoted to chief medical officer in 1961. In 1975 he was appointed director of medical services, Esso Europe Inc., a post he held until retirement in 1984.

Swann served the oil industry well. He was a fellow of the Institute of Petroleum and served on its medical advisory committee. He was a member of the health advisory committee of the oil companies International Study Group for Conservation of Clean Air and Water, Europe, the Working Party of Medical Officers in the Chemical Industry in the EEC, and the Board of Offshore Medical Support Ltd.

Peter's outstanding quality was an ability to see and seize opportunities. Combined with this was an exceptional flair for working with others, and for being hospitable. It made him an exciting and delightful colleague with whom to work; with Peter something new was always about to happen. In contrast to most people, it generally *did* happen.

His achievements outside the company were many and distinguished. He was a member of the standing committee on occupational medicine at the College, and attended College meetings and functions regularly. Through his influence, Esso was very generous to the College. His periods as honorary secretary of the Society of Occupational Medicine, member of the occupational health committee of the BMA, examiner in the conjoint diploma of industrial medicine, president of the occupational medicine section of the Royal Society of Medicine, member and latterly chairman of the specialist advisory committee on occupational medicine, and a member of the standing medical advisory committee of the DHSS, convinced him that occupational medicine would not achieve adequate recognition by the profession, or by industry, unless its academic standing was on a par with other branches of medicine. He recognized that the support of the College was crucial in this. He worked assiduously for this support and, with the help of like minded Fellows, this led to the formation of the College working party, of which he was appointed chairman, in 1976. In turn, this led to the formation of the Faculty of Occupational Medicine within the College in 1978. It was fitting that Peter Swann was appointed the first dean of the Faculty, and served for three years in that capacity. He was, therefore, also a member of Council and of the finance committee of the College, and of the Conference of Medical Royal Colleges and their faculties. It would seem likely that without his persistence and his ability to take people with him, the formation of the Faculty would have been long delayed. He helped to set the high academic and ethical standards which the Faculty proposed and the College endorsed. In 1981 he was created CBE, having already received the OBE in 1973.

Peter found time to be honorary medical adviser, and later deputy president, of RoSPA, and also to sail and play golf. He was a charming and convivial host; in appearance short, rotund, balding, with a left facial tic but always affable and good humoured. No one minded being manipulated by him for a good cause.

In 1954 he married Ruth Audrey, daughter of Reginald John Stapleton, a chartered accountant, and they had two children, a son and a daughter.

Peter's early retirement from Esso in 1984, after 30 years' of distinguished service, was much regretted. It was sad that he cut himself off from all his old friends, and was not seen at College functions again. In December 1987 he was diagnosed as having a serious illness and died, as he wished, at his home in Frinton. JB[1]/PABR

SYMONS, Cecil *b.*16 Dec 1921 *d.*9 Oct 1987
Chevalier de l'Ordre National du Mérite,France MB BS Lond(1945) MRCP(1947) MD(1948) FRCP(1969)

Cecil Symons was consultant physician and cardiologist to the Royal Free Hospital, New End Hospital, St Dunstan's and the Dispensaire Français.

He graduated from the Middlesex Hospital medical school and, following house officer posts at Barnet General and the Royal Northern Hospitals in 1946, was appointed casualty medical officer to Hampstead General Hospital, part of the Royal Free group. His association with the Royal Free was to last, barely broken except for short periods in his training, until his death in 1987 nine days after he had been appointed an honorary consulting physician and cardiologist on his retirement.

He was appointed senior registrar to Hampstead General Hospital in 1951, he was an American Heart Association fellow and Fulbright grantee, at Cornell University medical college from 1953-54, and in 1956 became first assistant to the cardiac department at the Royal Free. In 1961 he was appointed consultant physician at New End Hospital, and consultant physician and cardiologist to the Royal Free Hospital in 1968.

Cecil was dedicated to all who were patients or staff at the Royal Free. He was especially concerned that when the Royal Free moved from its old site in Gray's Inn Road to Pond Street it should lose none of its identity or historical sense of purpose, and he commissioned a series of drawings depicting all the hospitals in the Royal Free group. To ensure that the new Royal Free Hospital would present a relaxing atmosphere, he set up a works of art committee in 1974, which he chaired until his retirement. He conceived the idea of an art gallery in the main concourse of the hospital - which is now named after him - where local artists could exhibit their work. The Works of Art committee also commissioned the magnificant mural in the Rowland

Hill Street entrance of the medical school, depicting the history of the school and the hospital. Together with the board of governors, he instituted the annual Marsden lecture commemorating the founder of the hospital.

In 1987, as a singular honour, he was the first member of the hospital staff ever to be invited to give the lecture. Sadly, he died before he could deliver his lecture 'Self care and health care in William Marsden's district', but the lecture was read posthumously by a BBC newsreader, Robert Dougall, assisted by Cecil's wife Jean.

Over many years he collected medical instruments, particularly from the Georgian period, which form the basis of the Symons Collection. In 1981 his Samuel Gee lecture at the College was entitled 'Invalids in the Georgian era'.

Symons published many scientific papers on thyroid heart disease, and as a scientific fellow of the Zoological Society he was interested in comparative medicine.

He was a governor of the Royal Free, chairman of the medical executive committee, and a special trustee. In Hampstead, he was a trustee of the Campden Wells and Hampstead Trust and the Camden Arts Trust.

Cecil Symons had a particular interest in France, the French people and their language, and enjoyed promoting collaboration between British and French colleagues. He was the founder and president of the Société Clinique Française which held regular meetings with the proceedings entirely in French. He was also a trustee of the Alliance Française de Londres. For his services to France he was honoured by his appointment as Chevalier de l'Ordre National du Mérite.

Cecil was the perfect colleague, a man of absolute integrity, great charm and wit. He is survived by his wife Jean; there were no children of the marriage.

TRE

[*Brit.med.J.*,1987,**295**,1288; Photo]

SZEKELY, Paul *b.*16 Aug 1910 *d.*15 Nov 1985
MD Prague(1934) MB BS Durh(1946) MD(1947) MRCP(1970) FRCP(1971)

Paul Szekely was the son of a Czech physician, Vojtech Szekely. He studied medicine in Prague, where he graduated MD in 1934. After graduation he worked in the medical clinic at the University of Prague until 1938, when he moved to Paris. There he spent two years at the cardiac clinic of the Hôpital Broussais, and when France was invaded he escaped through Bordeaux to England.

In 1941 Paul Szekely became house physician to the medical and cardiovascular department of Newcastle General Hospital where he came under the influence of William Hume, later Sir William [*Munk's*

Roll, Vol.V,p.206], who encouraged his career and became a close family friend. Paul was medical registrar at the Newcastle General from 1942-48. In 1948 he was appointed consultant cardiologist to the regional cardiovascular centre of Newcastle General Hospital and he remained in this post until his retirement in 1975, when he was appointed honorary consultant cardiologist.

Szekely was a highly respected and widely consulted clinical cardiologist with a particular interest in arrhythmias, rheumatic heart disease, and heart disease in pregnancy. He first developed his interest in research in Prague and later in Paris, where he worked with Laubry, who became a lifelong friend. It is for his contribution to the care and understanding of cardiac disease in pregnant women that he will be long remembered. His interest dated from 1946 when he developed a close working relationship with the obstetrician Linton Snaith. Over the succeeding 30 years they shared the management of a large number of affected women, and their experience formed the basis for a joint work: *Heart disease and pregnancy*, Edinburgh, Churchill Livingstone, 1974, now a standard textbook. Paul followed up the patients he had seen during pregnancy over the rest of their lives, and not surprisingly they were devoted to him and were greatly distressed when he retired in 1975. Although at that time he virtually gave up clinical work, he continued to be consulted by his colleagues and his last papers on his favourite topic were published in 1985. He participated in research work, and in investigations for the Committee on the Safety of Medicines, until the end.

Paul Szekely was liked by all those who came into contact with him. He seemed to be personally interested in everyone and his gentle charm and unfailing courtesy were always evident. He was a fluent linguist in French, German and English, and thoroughly enjoyed travel. He got particular pleasure from his several visits to India whose culture, history and food he relished. In his earlier days he was an accomplished skier and tennis player, and in later years he turned rather reluctantly to golf but preferred to avoid hitting the ball when accompanying his wife round the course. His other interests included archaeology, opera and art. He was a member of the Society of Antiquaries and a friend of the Laing Art Gallery.

In 1941 he met his future wife Peggy Lynch, daughter of a journalist, when she was also working at Newcastle General Hospital. Their subsequent marriage undoubtedly formed the cornerstone on which he built the rest of his full and rewarding life.

DGJ

[*Brit.med.J.*,1986,**292**,67; Photo]

TANJEA, *Surg. Rear Admiral* **Bashi Lal** *b.*18 Feb 1903 *d.*25 Jan 1976
MB BS Punj(1926) DTM&H(1928) MRCP(1930) MRCPE(1930)
FRCP(1968)

A kindly man of enduring charm and absolute integrity, people had great faith and trust in Bashi Lal Taneja and his advice was willingly accepted. 'Bashi', as he was popularly called, became a household word in most families uprooted from Pakistan and he will long be remembered with gratitude, respect and affection by a large circle of friends, students, colleagues and patients.

Bashi Taneja was born in Dera Ismail Khan on the North West Frontier of India, now Pakistan, where his father Thakar Dass Taneja was a business man dealing in sugar and general merchandise. He was educated at the V B High School and Punjab University, Lahore, undertaking his clinical studies at the Mayo Hospital, Lahore.

After completing his house appointments, Bashi came to London and obtained his membership of the College, subsequently being elected a Fellow. On his return to India in 1930 he joined the Indian Medical Service and became a recognized specialist in medicine and pathology. During the second world war he served in the RAMC, being assistant director (pathology) to the 11th Army Group Allied Land Forces, South East Asia Command from 1944-45 and a member of the British Scrub Typhus Commission. He was commended by Sir Oliver Leese, commander in chief Allied Land Forces, for his investigations on scrub typhus. He won an international reputation as a research worker in tropical diseases, and held the rank of colonel on demobilization.

In 1945, Bashi Taneja was appointed professor of pathology and principal, Glancy Medical College, Amritsar. Two years later he became director of pathology to the Irwin and Safdarjang Hospitals. In 1950 he was appointed professor of pathology at the Armed Forces Medical College, Poona. In 1956, with the country's independence, he was appointed director of Medical Services (Navy), 1956-58. In 1958 he was instrumental in the founding of Maulana Azad Medical College, New Delhi, paying personal attention to the construction of the College building. His efforts ensured that the professorial posts were filled in time and the first session became operational without delay. The medical college is now of international repute.

Six years later Taneja was appointed director-general of the Indian Council of Medical Research, an assignment he held with distinction for five years. He then became executive director of the Indian Academy of Medical Sciences, of which he was a founder member.

Bashi Taneja was the author of a number of scientific papers on medical practice, hospital administration and medical education in India. His name will always be associated with field trials of cholera vaccine, carried out under his guidance at the Cholera Research Centre, Calcutta.

He was a member of the scientific advisory committee of the

Cabinet; a member of the planning commission panel on scientific research and national science policy, Research Advisory Committee, Armed Forces Medical Service; a member of the governing body of the National Institute of Health Administration and Education, and of the Central Family Planning Institute. For many years he was a member of the WHO Advisory Panel on the organization of medical care. He was also a member of the Indian delegation to the UNESCO conference on the application of Science and Technology, South East Asia, and a member of the national committee in the International Council of Scientific Unions.

In 1959, Bashi Taneja had been selected for promotion to the rank of major general, but he continued to serve in the civil sector in order to complete the task he then had in hand - that of building up the Maulana Azad Medical College and associated hospitals. By refusing promotion, and the consequent rise in status, he was true to himself and his dedication to duty. A man of sterling qualities and outstanding personality, and of few words, he was endowed with a gift for grasping essentials. He believed in and encouraged applied research in preference to fundamental studies.

In 1927 Bashi had married Karna Devi, daughter of Seth Chuhar Lal, a banker. She was a woman of rare courage, calm and patience, and devoted to her husband. They had four sons.

YPM/GW

[Photo]

TANNER, Selwyn Edward *b.*3 Oct 1898 *d.*22 Nov 1988
MRCS LRCP(1923) MB BS Lond(1923) MD(1925) MRCP(1926) FRCP(1944)

Selwyn Tanner was consultant physician to Leicester Royal Infirmary, and later also to the Leicester General Infirmary and small hospitals at Hinckley and Market Harborough. He was the son of a mine manager and was born at Crofty in South Glamorgan.

Selwyn was sent to Gowerton Intermediate School from where he won a county scholarship to Aberystwyth University to read physics. After a year he changed to medicine, taking his second MB at Cardiff. His clinical training was at University College Hospital, London, where he qualified. After resident posts, including one with Wilfred Trotter at University College Hospital, he went to Leicester in 1926 and entered a partnership with N I Spriggs and T W Allen. In 1930 he became honorary physician at Leicester Royal Infirmary; he abandoned general practice in 1936 and devoted himself entirely to consultancy. With the advent of the National Health Service in 1948 he became absorbed into the new system until he retired in 1963.

During house posts at University College he became interested in endocrinology, an interest he kept up throughout his career. He had many original ideas, but at Leicester his laboratory work was torpedoed by a hostile laboratory service. Many of his theories, particularly on Addison's disease, were ahead of his time. His main work, however, was as a general physician. He was the leading physician in Leicester of his day. From 1954-55 he was president of the Leicester Medical Society.

Selwyn Tanner was straight-laced and honourable in all his dealings, and a rather private man. He was greatly respected both as a person and for his academic knowledge. He had some interest in Christian societies, probably at the prompting of his wife, because he seemed to be something of a doubter. He believed firmly in ethical medical behaviour and was critical of those whose ethics he thought to be less strict. On a point of principle, he would not mix with osteopaths or other fringe practitioners. At times he could be dogmatic, but he was very supportive of all those who worked for him.

In retirement he abandoned medicine and devoted himself to his garden, and to reading. He had married Gwen Burgess and they had two daughters; after he retired he and his wife moved house to be nearer their children.

FDR

[*Brit.med.J.*,1989,**298**,379-80; Photo]

TAYLOR, Peter John *b.*24 Sept 1929 *d.*6 Jan 1987
BSc(1951) MB BS Lond(1954) MRCP(1956) DIH(1962) MD(1966) FRCP(1971) FFCM(19760) FFOM RCP(1978) Hon FFOM RCP(1982) O St J(1982)

The name of Peter Taylor is bound up with the origins and development of the Faculty of Occupational Medicine within the Royal College of Physicians of London, and with the emergence of occupational medicine as a distinct discipline within the main body of British medicine. The recognition of occupational medicine has been growing slowly in Britain since the work of Charles Turner Thackrah and it is interesting to see in the previous volumes of *Munk's Roll* how many distinguished physicians in this field are represented. Peter Taylor takes his place among them as an influential figure who commanded widespread respect, not only by virtue of his authoritative contributions to the literature, especially to that on sickness absence which was a subject that he made his own, but also because of his standing as a leader and someone to whom one might turn for good advice and commonsense.

His father, Commander Sam Taylor, was a submariner and his mother, Alice Storm, was partly of Danish origin. Peter, an only child, was educated at Clifton College and was originally destined for a career in the Navy but, having failed the eyesight test, decided on medicine instead and in 1947 entered St Thomas's Hospital which was close to his home. He took a BSc in physiology and graduated MB BS with honours in pathology. This was followed by appointment as house physician at St Thomas's. Between 1956-58 he was a medical specialist and acting squadron leader in the Royal Air Force, serving in Cyprus. There he met his wife, Josephine Marian Hetherington (Jo), a sister in the Princess Mary's RAF Nursing Service who had also trained at St Thomas's. They were married in 1959 and had two children, a son and a daughter. Later in his Service career, Peter was medical specialist at the British Military Hospital at Tripoli; in both these appointments he impressed others by his clinical skills and meticulous and untiring efforts on behalf of his patients.

Peter Taylor returned to civilian life as a medical registrar at St Thomas's, in 1958, but two years later he became a medical officer to Shell International at Balikpapan in Indonesia where he spent three years during which he learnt to speak fluent Pasar Malay. He returned to the UK in 1963 to be a senior medical officer at Shell Haven in Essex where he began his classical studies of sickness absence, which he made the subject of an MD thesis, for which he was given the BMA occupational health prize in 1967. This work formed the basis of an important paper on 'Personal factors associated with sickness absence' published in the *British Journal of Industrial Medicine* in 1968 (25., pp.106-118). His investigations emphasized the importance of a man's attitude to his work and to his own health in relation to sickness absence.

In 1968 he was appointed deputy director of the Institute of Occupational Health at the London School of Hygiene and Tropical Medicine where he set up an information and advisory service while continuing his work on sickness absence, shift work and chronic disability, contributing to the teaching programme and writing 23 papers.

In 1971 he was invited to become the chief medical officer to the Post Office where he was responsible for an occupational health service for 422,000 people, which employed 26 doctors, 75 nurses and an occupational hygiene group, on a budget of 2 million pounds. He filled this appointment with distinction, using his expertise in sickness absence studies to the full on the accumulated health records which became available to him. He moved in 1981 to another major post in the occupational health field, international medical adviser to Unilever, which inevitably involved a heavy programme of foreign travel in Europe, North and South America, Africa, Australia and the Far East, which further enhanced his international reputation and brought him into contact with occupational medicine as it is developing worldwide.

In addition to his exacting job, Peter Taylor gave much time to professional organizations concerned with occupational medicine. He was an active member of the Society of Occupational Medicine and served on its research and education panels, and the executive committee, before becoming president of the Society in 1974. In the same year he was made secretary of a RCP working party on industrial medicine which produced a discussion paper that October. This group was replaced by a standing committee of the College whose terms of reference were to explore how the College could provide a professional focus for occupational medicine and, among other tasks, formulate training requirements. Peter Taylor put forward arguments for a faculty within the College, and with the support of the standing committee he drafted standing orders and raised the possibility of recognition of a membership qualification by the General Medical Council. The latter was achieved in 1985. The first Faculty board meeting took place on 18 May 1981 with Peter Taylor as vice-dean, and he became dean of the Faculty a year later. He was the first chairman of the Faculty's ethics committee and played a major part in the production of *Guidance on ethics for occupational physicians* published by the Faculty in 1980; two further editions appeared respectively in 1982 and 1986 under his supervision. This booklet has had a wide circulation in the United Kingdom and has attracted interest abroad. His published work, beginning when he was in the RAF, amounted to some 60 papers on a variety of subjects, particularly on absence attributed to sickness, shift work, the disabled and stress. He was joint author, with A Ward Gardner, of *Health at work*, London, Associated Business Programmes, 1975, and contributed chapters to a number of books on occupational medicine.

Peter Taylor became a member of the International Congress on Occupational Health in 1969, and of its absenteeism and shiftwork subcommittees. He was visiting professor at McMaster University, Canada, in 1979-80. Other activities included part-time honorary appointments as consultant/senior lecturer to Guy's and King's College Hospitals in London; membership of the medical advisory committees of the Health and Safety Commission, the Industrial Society and the Migraine Trust, and appointment as civilian consultant in occupational medicine to the RAF in 1984.

He acted as examiner to London University for the MSc, PhD and Diploma in Occupational Medicine; the Diploma in Industrial Health of the Conjoint Board 1973-77; as an external examiner to Dundee University from 1977, and for the MSc to Singapore University in 1979. He was chief examiner for the Associateship of FOM from 1984 until his death.

Fully committed in everything he did, Peter Taylor was a tireless worker, an accomplished lecturer, and had the quiet authority that enabled him to represent occupational medicine effectively at home and abroad. He was easily approachable and ready to give help, especially to those in training or involved in research. He encouraged women to take up occupational medicine, feeling that they were under-represented, and became involved in a project on 'Women in Occupational Medicine' based in Lucy Cavendish College, Cambridge, and supported by the Leverhulme Trust.

Peter Taylor's recreations were music - he was especially fond of Haydn and Mozart operas - gardening and donkeys; he was vice-president of the East Anglia Donkey Show in 1984. His son is now a schoolmaster and his daughter is married.

In 1985 he was treated surgically for achalasia of the cardia, an ordeal about which he made no fuss and which he appeared to take in his stride. His death from a dissecting aneurysm of the aorta was totally unexpected and cut short a life which had still a great deal to contribute to occupational medicine and many other activities.

A thanksgiving service was held for Peter Taylor in All Souls Church, Langham Place, London on 31 March 1987, at which the dean of the Faculty of Occupational Medicine, John Aldridge, gave an address; the service included a reading of *The Donkey* by G K Chesterton, reflecting Peter's interest in the breed.

The Faculty, together with the Society of Occupational Medicine and the College, combined to establish a Peter Taylor Medal to commemorate his contributions to occupational health and to all three organizations. It marks outstanding academic achievement by trainee occupational physicians and will be awarded to the candidate submitting the best dissertation in the Faculty's membership examination. The medal is in silver, with the College arms and a figure of Charles Turner Thackrah. In recognition of Peter Taylor's work as a civilian consultant the Royal Air Force has established the Peter Taylor Chair of Occupational Medicine, to be held by a serving medical officer nominated by the Director General and approved by the Faculty. RIMcC

[*Brit.med.J.*,1987,**294**,255; *Lancet*,1987,**1**,342; *The Times*,9 Jan 1987; *FOM Newsletter*,Feb 1987,No **23**; *J.Soc.occ.med.*, 1987,**37**,65; *Address*,All Souls Church, Langham Place, Mar 31,1987; Photo]

TAYLOR, *Lord* **Stephen James Lake** *b.*30 Dec 1910 *d.*1 Feb 1988
(1st Baron Taylor of Harlow)
BSc MRCS LRCP(1934) MB BS Lond(1934) MRCP(1937) MD(1938) FRCP(1960) FRCGP(1960) FFCM(1979)

Stephen Taylor had a curiously split career in medicine and politics. His father was a civil engineer but his uncle, James George Taylor was a Fellow of the College [*Munk's Roll*,Vol.V,p.406] and his paternal grandfather was a surgeon. He married Charity Clifford, herself a successful doctor who, from the Prison Medical Service, became Governor of Holloway Prison.

Stephen James Lake Taylor had a very full and varied life and contributed significantly to general practice, industrial medicine, psychiatry, medical research, the formation of a medical school, the design of health services in two countries, party politics and government. He was educated at Stowe School and St Thomas's Hospital medical school, where he held junior posts after graduation. He first specialized in psychiatry, going on to the Bethlem Royal Hospital, the Royal Free, and the Maudsley. Briefly, before the second world war, he was an assistant editor of the *Lancet*. He had obtained his doctorate and his membership of the College before seeing service in the RNVR as a specialist in neuropsychiatry. He had also written on 'Suburban Neurosis'(*Lancet* 1938;1,759). In 1941 he was invited to serve in the Ministry of Information as director of home intelligence. In this capacity he set up the wartime social survey, forerunner of other national surveys which continue today. After joining the Labour Party, he contributed a number of articles to *The Spectator* on health service matters which influenced the shape of the National Health Service, and in 1945 he resigned from the Ministry to stand as Labour candidate for Barnet and was elected. He was not directly involved in the preparation of the NHS, though he had some indirect influence, and he served as Parliamentary Private Secretary to Herbert Morrison from 1948-50. He lost his seat to Reginald Maudling in the 1950 election and returned to industrial consultancy, and to the Harlow New Town Corporation of which he was vice-chairman. It was at Harlow that he founded the Industrial Medical Service and he acted as its first director from 1955-67. He was also responsible for the important pioneer development of health centres, one for each neighbourhood, at a time when the NHS was inactive in this important area. No other 'New Town' even planned for such an important initiative and, although Stephen was later to recant some of his original belief in health centre development by public authorities, he deserves much credit for this insight into the needs of new housing developments.

In 1951 he was invited by the Nuffield Provincial Hospitals trust to carry out a survey of general practice, provoked by the very critical report of J S Collings (*Lancet* 1950,1,555-890). Stephen Taylor's book *Good general practice,* London, Oxford University Press, 1954, is perhaps his most important contribution to the future of British medicine coming, as it did, at a time when general practice was depressed and under attack, and much needing signposts for the future. If the evolution of better primary health care is, at least, one of the two most important features of the first 40 years of the NHS, Stephen Taylor deserves some of the credit.

Allied with the health centre development at Harlow, Stephen had initiated an industrial health service for smaller undertakings in combination, and he was much concerned in this field in succeeding years.

Stephen Taylor was created Lord Taylor of Harlow in 1958, as one of the first group of life peers, taking his title from Harlow to which he had contributed so much, and thereafter he often led for the Labour Party in the House of Lords in debates on health, housing, science and higher education. In 1962 he chaired the Labour Party's study group on higher education, which produced *The years of crisis*, known as the Taylor report, and in the same year he was invited by the provincial government of Saskatchewan to mediate in its dispute with the doctors of the province over the terms of its revised health service. He succeeded in gaining the trust of the doctors and produced the nucleus of a solution within eight days, The resulting system of Medicare formed a pattern which became general in Canada and had some influence on aspects of medical care in the United States.

Stephen's political activity diminished but in 1964, with the new Labour administration he became Under-Secretary of State for the Colonies in the Commonwealth Office. He was a member of the British delegation to the first Commonwealth Medical Conference in Edinburgh, but did not long continue after that. He resigned his post in 1967 when he received another call from Canada - to be president and vice-chancellor of the Memorial University of Newfoundland, at a time when the medical school there was about to be formed. He remained there until 1973 when, sadly, he was struck with ill health and was obliged to retire. But he became a visiting professor and maintained his links with the university. Stephen was highly successful during his time in Canada and it represents the highlight of his career. He continued to take the Labour Whip in the Lords until 1981, when he resigned from the Labour Party to become an independent and sit on the crossbench. He did not consider himself a 'good party man' and found it '...impossible to adhere mentally to the decisions of democracy when they are manifestly wrong'.

Stephen Taylor was not notably active in College affairs and his political career was disappointing in spite of his considerable influence behind the scenes. The Canadian contribution was substantial but his outstanding achievement was in the early revival of general practice - and it came at a crucial time. Industrial medicine was less important, but nevertheless an original innovation. If he does not emerge as a major figure, either in politics or in medicine, he was nonetheless someone who played a distinct personal part in medical progress. He was very much his own man, and a leader rather than a follower. His wife, two sons and a daughter survived him.

GEG

[*Brit.med.J.*,1988,**296**,578,1010; *Lancet*,1988,1,425-26; *The Times*,3 Feb 1988; *The Independent*,4 Feb 1988; *Bull.roy.Coll.Psych.*,Sept 1988,396; *World Medicine*,19 July 1966; *Western Morning News*,9 July 1962; Photo]

THOMAS, Gerald Trevor Griffith *b.*26 Apr 1916 *d.*27 May 1987
MRCS LRCP(1940) MB BChir Cantab(1940) MRCP(1944) FRCP(1966)

Gerald Thomas was born in Richmond. His father, Frederick

Ernest Thomas, was a company director and his mother, Ida Marsh, was the daughter of a postmaster. He was educated at Colet Court Preparatory School, Berkhamsted School and Caius College, Cambridge. His clinical studies were undertaken at Guy's Hospital, where he graduated in 1940, winning the Treasurer's prize and the gold medal in medicine. During the latter part of the second world war he was medical tutor and registrar at Guy's, then joined the RAF medical division from 1945-47. He had early shown an interest in cardiology and on leaving the RAF he entered the Institute of Cardiology, working with Paul Wood [*Munk's Roll*,Vol.V,p.456] as senior medical registrar until, 1949 when he became chief clinical assistant to the National Heart Hospital and physician to the special unit for rheumatism research at Taplow, based at the Canadian Red Cross Memorial Hospital which in 1957 became the MRC's rheumatism unit. He gained his membership of the College in 1944, followed by his election to the Fellowship in 1966. He was also a member of the British Cardiological Society and of the Association of Physicians.

Gerald Thomas was essentially a clinical cardiologist, serving the Windsor area in that capacity from 1949 until his retirement in 1981, well trusted by general practitioners and consultant colleagues. In addition, as cardiological consultant to the rheumatism unit working in collaboration with the local staff and, through our weekly National Heart Hospital visits, with Paul Wood, Wallace Brigden and Aubrey Leatham, he greatly contributed to a long-term research study of the natural course of rheumatic fever and its several aspects. This included the international multicentre controlled trial of aspirin, cortisone and ACTH, equally shared between 11 centres in Great Britain and 10 centres in the USA. From 1945-65 he was author or part-author of some 18 publications in this field, including four classics, but never stepping beyond his essential role as a wise and responsible cardiologist and clinician. Before rheumatic fever disappeared, about 1965, and during the later follow-up period, patients at Taplow, as well as those in the Windsor area, owed him a great deal. On retirement he continued to work as a locum consultant, examining medical officer, and occasionally in private practice, to the great benefit of the NHS and his patients, and his own immense satisfaction. He was a retiring and most modest person who accomplished a great deal in the areas he worked in; and as a recording angel always accurate and caring during the death throes of rheumatic fever in the western world. He acquired a love for crème brulée during his years at Caius College and never lost it; it was, he said, his College's 'favourite duff'.

In later life Gerald moved to Padstow in Cornwall. He married Annabella Doogan, daughter of a farmer, in 1945 and they had three children, a son and two daughters: Richard, Geraldine and Caroline. None of his children went into medicine: Geraldine became an actress, married with one daughter; Caroline, also married with one daughter, runs her own public relations company; and Richard is managing director of a music company. They all survived him, together with his wife Patsy.

EGLB

[*Brit.med.J.*,1987,**295**,278]

THOMAS, Harold Ernest *b.*24 Oct 1919 *d.*24 Mar 1986
MRCS LRCP(1942) MB BS Lond(1943) MRCP(1947) MD(1949) FRCP(1972)

Harold Thomas was born in Middlesborough, the son of Ernest Thomas, a medical representative for a pharmaceutical company, and his wife Holly Rosaline Amesbury whose father was in the Merchant Navy. He was educated at Dulwich College and won an open scholarship to the University of London. His clinical undergraduate training was at University College Hospital, where he won the Fellowes gold medal in clinical medicine and graduated in 1943. He subsequently gained his membership of the College and was elected a Fellow in 1972.

After house appointments at UCH and the Brompton Hospital, he served in England and Germany as a captain in the Royal Army Medical Corps from 1944-47. After his Army service, he became registrar at UCH, and in 1951 was appointed consultant chest physician to the Birmingham regional hospital board, in which post he remained until his retirement in 1984. Initially he provided a service at the Yardley Green Sanatorium (later incorporated into the East Birmingham Hospital), Selly Oak, Solihull and West Heath Hospitals, but he later gave up his commitment at Selly Oak and West Heath as the need for beds for tuberculosis patients fell and the East Birmingham Hospital, together with Birmingham Chest Clinic, became a major centre for tuberculosis, chest surgery and chest medicine.

Harold's professional lifetime saw the introduction and widespread use of anti-tuberculous drugs. When he began, tuberculosis was his main concern; by the time of his retirement it occupied only a small part of the chest physicians' day. He developed the Birmingham tuberculosis drug resistance register, and won the research prize of the British Tuberculosis Association in 1958 for a paper entitled 'Retrospective survey of tuberculosis patients developing drug-resistant bacilli.' Further papers on this topic followed,the last published posthumously.

Away from the wards, he was active in local medical politics and was president of the Solihull division of the British Medical Association. He also served on the East Birmingham district health authority. He was the British member of the sub-committee on statistics and epidemiology of the International Union Against Tuberculosis from 1964-67, and was president of the Midland Thoracic Society from 1982-85.

Harold was almost invariably good humoured and his presence in a building was often quickly announced by his loud and distinctive laugh. He tended to come to work early and to make a point of bidding a hearty good morning to everybody he met. These traits were occasionally wearing to those less lively early in the day. He was a keen student of the group psychology of committee meetings, believing that rather than telling a difficult colleague the right thing to do it was better to approach matters obliquely and allow the colleague to discover and suggest the right course of action for himself. Harold's contributions in committee where, therefore, often enigmatic and occasionally incomprehensible but his listeners, on later reflection, sometimes saw the light as he saw it.

He married Elizabeth Jean Anderson in 1947 and they had two children, Anne Elizabeth and Peter Angus. He was a devoted family man. His wife, children and two grandchildren survived him.

Harold enjoyed robust health until he suffered a stroke, from which he never fully recovered, at the summer meeting of the British Thoracic Society in York in 1985. Ironically this presented as a dysphasia while he was asking a question in the discussion period following a paper on drug-resistance in tuberculosis.

CS

[*Brit.med.J.*,1986,**293**,274; Photo]

THORBURN, Ian Oriel *b.*21 Aug 1903 *d.*6 May 1985
Kt St John(1969) MB BS Melb(1928) MD(1931) MRCP(1933) FRCP(1959)

Ian Thorburn was born and died in Perth, Western Australia. His father, Ernest Smet Thorburn, died in early life and he was raised by his mother, Laura Louisa Growse.

Ian was educated at Guildford Grammar School and proceeded to Melbourne University, where he graduated in 1928 and took his MD in 1931. After resident appointments at Melbourne (afterwards Royal Melbourne), the Children's (afterwards Royal Children's) and Women's Hospital in Melbourne, he returned to Western Australia as medical registrar at the Perth (afterwards Royal Perth) Hospital in 1932. Between 1933-34 he was a postgraduate in London, taking courses, and later he worked at the Royal Chest Hospital. He returned to Western Australia as medical superintendent of the Royal Perth Hospital for a two year period, ending 1936; shortly afterwards being appointed honorary physician in general medicine, and honorary physician in infectious diseases, at the same hospital. He held these appointments until his retirement in 1963.

It was his role in infectious diseases that made him the beloved physician, in every sense, to hundreds of poliomyelitis and diphtheria patients. He was physician to the infectious diseases branch of the

Royal Perth Hospital for at least three decades; his tenure including three epidemics of poliomyelitis between 1948 and 1956. His dedication to acutely ill patients was without peer. Those were the days of the iron lung, tracheostomy and terrible tragedy amongst young people and children with diphtheria. He would spend countless hours at the bedside of acutely ill and dying patients. A succession of his former registrars, now senior physicians on the staff of the Royal Perth Hospital, remember him as a conservative physician completely absorbed by his work. He was often observed to be ill at ease on social occasions and avoided them whenever possible, sometimes even prolonging clinical rounds to achieve this purpose. His outpatient clinics and ward rounds were long standing and painstaking, and often did not finish until 10 p.m. The profound clinician, imbued with the value of repeated clinical observation, he was distrustful of laboratory fads but welcomed tested developments. A ditty composed about him in a residents' sing-song in 1952 reflected not only his conservative practice of medicine but also the affection and trust in which he was held by patients and colleagues alike:

In the ward of 42 there's a happy little throng,
If you want to hear about it just listen to our song,
You've got to watch your ulcer if you're put in under Thor,
You'll still be holding up a bed in 1984.

He believed strongly in honorary service, and saw the introduction of sessional payment to specialists working in public hospitals as a disaster for the independence of the medical profession. He adhered strongly to his belief in principles. He could not stomach bureaucracy. He was a man of great faith and a convinced Anglican. He was always hospitable in his own home, where he used to have monthly meetings and discussed physiology with his junior medical staff. He was popular as a clinical teacher and for many years participated in the teaching of the introductory course in clinical methods to medical students in the University of Western Australia, even after his retirement from hospital practice. He held the first and principal appointment in teaching medicine in the University Dental School.

In 1936 he married Patricia Margaret Lilian May, a nursing sister he met in Melbourne, and this was an extremely close and lifelong relationship.

Thor's only recreation was golf. His main extra-curricular interest was in the St John Ambulance Association, to which he gave nearly 50 years of service. He was made a knight of St John of Jerusalem in 1969. He gave generously to educational charities, notably to his old school, Guildford Grammar, and he was president of Bush Church Aid which provided support for Anglican communities in isolated areas.

In 1959 the conversion of the infectious diseases branch of the Royal Perth Hospital into the Royal Perth Rehabilitation Hospital was established, with specialist services in other hands, but Thorburn never lost his dedication to the institution and its affairs, contributing wise counsel and support to both patients and colleagues. In 1977, amongst the last of the former honorary consultants at Royal Perth Hospital, he was appointed emeritus consultant physician. Even though he discontinued attending the medical outpatient clinic at the Hospital, he was still to be seen at postgraduate meetings - recognizable by his lean build and the whiskers sticking out of his ears. He reluctantly closed his private consulting rooms in 1983.

At the time of Ian Thorburn's death in his 82nd year he had weakened in his mind and was fortunate to have been continuously cared for by a physician who had one time been his registrar. He gave long and dedicated service to the Royal Perth Hospital and his name is proudly displayed on the Memorial Boards.

KS

[*Uninews*,Univ.Western Aust.,**4**,7,May 1985; *Servio News*, Roy.Perth Hosp.,**11**,6,May 1985; *The Brigade*,St John Ambulance Assoc.,West.Aust.,Aug 1985; *Annual Report* 1984-5,Roy.Perth Hosp.; Photo]

THORNE-THORNE, Bezly *b.*20 July 1910 *d.*13 Mar 1986
MRCS LRCP(1934) MB BChir Cantab(1935) MRCP(1937) FRCP(1968) DO Oxon(1943)

Bezly Thorne-Thorne was born in Woking, Surrey, the son of Richard Thorne-Thorne, a medical practitioner, and his wife Constance Houghton, daughter of a ship owner. His grandfather was Sir Richard Thorne KCB, a Fellow of the College [*Munk's Roll*,Vol.IV,p.236]. Bezly was educated at Marlborough College and Caius College, Cambridge, and qualified in medicine at St Bartholomew's Hospital. At Marlborough, Bezley Thorne-Thorne was lucky to be a pupil of A G Lowndes, one of the great biology masters of that time, who trained both scholars and fellows of the Royal Society. In the Easter holidays Bezly accompanied the biology class to the Marine Biological Stations at Plymouth, or Cumbrae, where they trawled for plankton and other specimens. Once, when he felt sick, Lowndes suggested he ate a raw clam as a 'kill or cure' remedy - with disastrous results.

At Cambridge, Bezly was a conventional, average undergraduate of his time - playing games, doing the required amount of work and enjoying May Week balls. He had a lot of good friends among medical students, was excellent company, and enjoyed comfortable living, but he did not belong to the Union, political clubs, or indulge in field sports. Yet underneath his easy going nature he had a strong will, which was useful in later life when he had to contend with awkward colleagues.

From 1940-45 Bezly served in the RAFVR as a medical specialist, with the rank of squadron leader, and spent two years in India. On return to civilian life he had to make a choice between London practice or a rather more secure life in the provinces. A consultant appointment at Brighton Eye Hospital (now the Sussex Eye Hospital) came vacant at the right time and he opted for the provinces - a decision he did not regret. As an ophthalmologist he had a strong medical bent and obtained his membership of the College. He examined for the College and for the Society of Apothecaries of London, and he brought his wide knowledge of medicine into the realm of ophthalmology in Brighton.

Clinically Bezly was very sound, but he had his own ideas about things and he stuck to them. He did not suffer fools and was sometimes irascible in the clinics, and in committee. He was a member of the Faculty of Ophthalmologists and of the Oxford Ophthalmic Congress, a specialist in ophthalmology to the DHSS, and ophthalmic surgeon to the East Sussex Association for the Blind. Particularly close to his heart was his work in connexion with St Dunstan's at Brighton. A most sociable person, he regularly attended meetings of the Society of Apothecaries and of Bart's Fountain Club, being master in 1976.

Although small in stature, he was never small in ambition or confidence; he knew precisely where he was going and never appeared to have trouble with his work or innumerable exams. He was an entirely unpretentious person, a caring man, generous and courteous, who epitomised all that is meant by that strangely misunderstood phrase - an English gentleman. He was a man of standards; of behaviour, judgement, and integrity. He saw things clearly, and was without prejudice. He was unfailingly kind, gentle and a marvellous listener. He was indeed a very special person and much loved. The 150 letters received by his widow after his death testify to the high regard in which he was held.

After retirement, he was at his happiest pottering about his garden and cultivating seeds in his greenhouse. He loved a good party and had a tremendous sense of humour; perpetrator of several practical jokes on his friends in his youth. He enjoyed sport; playing squash for the Jesters Club, and was good at both tennis and golf. Music also played a prominent part in his life; he was particularly fond of opera, and enjoyed both concerts and the theatre. His married life was singularly happy. His wife, Ierne, shared many of his activities and theirs was a most cultured and hospitable home. His wife and two daughters survived him.

AST/LTT/AHO

[*Brit.med.J.*,1986,**292**,1284,1679; Photo]

TODD, *Sir* Geoffrey Sydney *b.*2 Nov 1900 *d.*24 Dec 1987
KCVP(1951) CVO(1947) OBE(1946) MB ChB Sydney(1925)
MRCP(1929) FRCP(1949) FRACP(1960)

Geoffrey Sydney Todd was born in Melbourne, Australia, the son of George William Utley Todd, a newspaper editor and his wife Amy Louisa, née Webb. He was educated at King's School, Parramatta, and at Sydney University where he graduated in medicine.

After an appointment as resident medical officer at the Wagga District Hospital he came to London for postgraduate study. His intention was to take up surgery, but in 1929 the course of his career was determined when he became house physician at the Brompton Hospital and obtained his membership of the College. He was then appointed assistant resident medical officer. The title resident surgical officer would have been more appropriate for the post, and was later adopted, but at that time thoracic surgery was such a recent development that the house officer who worked for the thoracic surgeons at the Brompton was a trainee physician. It was a demanding post, carrying sole responsibility for post-operative care as there was no surgical registrar. Todd worked for a year for the two leading thoracic surgeons in London, J E H Roberts and Tudor Edwards. Most of the surgery was for pulmonary tuberculosis and for broncho-pulmonary and pleural suppurative diseases, with some operations for intra-thoracic tumours. He remained at the Brompton for another three years, as resident medical officer, a post with important administrative responsibilities as well as clinical duties comparable to those of a senior medical registrar today. He became involved in all aspects of the activities of the hospital and staff, and was as capable of giving clinical advice to consultants as of contributing a conjuring act to the residents' Christmas show.

Towards the end of 1933 the post of medical superintendent of the King Edward VII Sanatorium, Midhurst, became vacant on the resignation of Richard Trail [*Munk's Roll*,Vol.VI,p.437]. At this time surgical procedures, mainly thoracoplasty, were widely used in the treatment of selected cases of pulmonary tuberculosis and plans for facilities for these to be carried out at Midhurst were well advanced. With his wide experience of current medical and surgical practice in respiratory diseases and tuberculosis, and his proven administrative ability, Todd was the obvious candidate for this post and was appointed to it at the early age of 33.

Todd started at Midhurst in January 1934. Just as at Brompton, but with more autonomy, he quickly assumed control of and became omniscient about all activities at the sanatorium. Rest followed by controlled return to normal activity was the standard treatment for pulmonary tuberculosis and success in its implementation required not only good organization but sympathetic, firm and open dealings with patients; many old patients have attested to Todd's qualities in this respect. In the introduction of surgery at Midhurst his surgical experience was invaluable and the first major operations were performed within a few months of his arrival. During the second world war half the beds at Midhurst became a surgical chest centre, a development which similarly owed much to his special abilities. After the war he steered Midhurst through the changes imposed by the revolution in treatment of tuberculosis brought about by antimycobacterial agents, leading to its evolution into a chest hospital as the need for tuberculosis beds diminished. The links with the medical services of the Army and Royal Air Force, forged during the war, continued and Todd became civilian consultant to both.

Todd retained close contacts with Australia and many young Australian graduates seeking specialist study in chest diseases were welcome and made a distinctive contribution at Midhurst under his tutelage.

In spite of its origin as a royal foundation the King Edward VII Sanatorium faced difficulty in defining its role with the coming of the National Health Service in 1948. At first arrangements with regional hospital boards made it possible for a substantial number of NHS patients to be admitted; in this context Todd became a member, and later vice-chairman, of the South West Metropolitan regional hospital board. His links with the Brompton Hospital were renewed when this board nominated him to the board of governors of the hospitals for diseases of the chest.

As the number of patients with respiratory diseases referred to Midhurst diminished, clinical activities expanded in other fields. This was recognized in 1964 when the name was changed from 'Sanatorium' to 'Hospital' by supplemental charter. After this a wide range of medical and surgical problems were dealt with. Some required specialized facilities; for instance, John Charnley started hip replacements at Midhurst in 1969. Todd supervised the administrative and individual problems of this transition with skill and success, and in 1970 retired after 36 years as medical superintendent. At the time a large benefaction had made possible the development of the Midhurst Medical Research Institute in the hospital grounds and Todd became administrator of the Institute until 1981. He thus remained a key figure at Midhurst for another 11 years.

His outstanding services to the royal foundation at Midhurst were recognized when he was created CVO in 1947 and KCVO in 1951. He served as president of the British Tuberculosis Association from 1951-53.

He gave the first Marc Daniels memorial lecture in the College in 1956, on 'Chemotherapeutic control in the treatment of fibro-nodular pulmonary tuberculosis', and the Tudor Edwards lecture in 1971 on 'Tudor Edwards and his time'.

Geoff Todd remained purposefully Australian in style throughout his long career in Britain. His customary directness of speech hurt nobody's feelings, since it could not conceal obvious concern for the welfare of other people. He was an able and informed clinician in his chosen field; he was not drawn to laboratory research, his contributions to journals being those of an acute clinical observer. His concern for the success of the institution to which he devoted his professional life was an overriding consideration with him. He guided it successfully through the transition from a tuberculosis sanatorium to a hospital with wide medical and surgical activities, and in this he was probably unique. At the end of his life he must have been disappointed when the difficulties of integrating a research institute with a small isolated hospital proved to be intractable.

In his youth he played lawn tennis to a high standard, being singles champion of Sydney University in 1921. He was also a good performer at golf.

In 1955, when he was aged 55, he married Margaret Alan, daughter of Frank Alan Sheen, and engineer. There were no children of the marriage.

JGS

[*Brit.med.J.*,1987,**294**,517; *Lancet*,1987,1,459,639; *The Times*,9 Jan 1987 Photo]

TRENCHARD, Hugh John *b*.8 Dec 1913 *d*.3 Feb 1988
MB ChB Birm(1936) MRCP(1940) DMR(1946) FRCP(1966)

Hugh Trenchard was born in Hackney, London, the son of an itinerant schoolmaster who settled shortly afterwards in Somerset, where Hugh was brought up. His education commenced at Christ's Hospital, Horsham, but his father - by this time headmaster at Dr Morgan's School, Bridgewater, - thought that Hugh was not working hard enough, an accusation which could never again be levelled against him, and removed him to attend his own school so that he could keep a closer eye on him. Hard work was clearly the rule, and Hugh matriculated at the age of 14. He went on to study medicine at Birmingham University and graduated MB ChB in 1936. After house posts he spent some time in general practice but because he was convinced that war was inevitable he elected to join the RAF medical service and was commissioned in 1939.

He married Elizabeth Mallinson, who came of a farming family, in 1940, and there were six children of the marriage although, sadly, one suffered from Down's syndrome and died at an early age. The eldest is a general practitioner.

He gained his membership of the College while serving in the RAF, and it was during this period that his interest in chest diseases was stimulated when he joined the late Richard Trail [*Munk's Roll*,Vol.VI, p.437] to set up the first Mass Miniature Radiography unit (MMR). Latterly, he served in India and reached the rank of wing commander before being demobilized in December 1945.

On demobilization he was appointed registrar at the London Chest Hospital and during his tenure there he found time to study for and obtain the diploma of medical radiology (DMR). In 1947 he was appointed physician to the Harrow Chest Clinic and, on the introduction of the NHS, he felt that he could combine this post with that of assistant senior medical officer at the North West Thames Regional Health Authority - in charge of tuberculosis services. This combination of posts enabled him to see the then tremendous problem of providing beds for the care of the tuberculous. In 1949 he resigned his post with the RHA and moved to the Edgware Chest Clinic where his drive and enthusiasm soon made it one of this country's renowned centres, attracting a constant stream of visitors and scholars from all over the world. He was in the forefront of the move to treat tuberculosis on a domiciliary basis and this, together with his constant alertness to the principles of prevention and control of infection, soon resulted in his making deep inroads into waiting lists, well before the advent of chemotherapy.

His training in radiology proved to be a great advantage in that it enabled him to have his clinic fully equipped with its own X-ray department, and it was on his advice that the use of the Odelca camera using 70mm films was introduced at a time when the provision of full size X-ray plates was becoming increasingly difficult and costly. This meant that he could continue with a programme of selective radiography, concentrating on groups at risk, a technique he showed to be cost effective when compared with MMR.

As tuberculosis waned, Hugh's energy and enthusiasm were directed towards the problems of chronic bronchitis and bronchial carcinoma. Again he realized the importance of care within the community, with the provision of effective support from the clinic and the hospital. These things were only possible because he was such an excellent physician and, above all, because he was such a remarkable man. He would never spare himself in his efforts for others, be they patients or colleagues, and he was not afraid to fight for the things he believed in - and at times he had to fight very hard indeed, as those who served on committees with him knew only too well.

Although his main concern was always patient care, his numerous papers over a wide range of subjects, during most of his working life, are indicative of his constant search for excellence.

Medicine was Hugh's life and after retirement he continued with locum work, and in addition he found great satisfaction in helping in his own general practitioner's surgery - he was in fact seeing patients the week before he died. He was also heavily involved in activities outside medicine: he had inherited an interest in geneology from his father and, with more time on his hands, he was able to trace and develop correspondence with kinsmen far and wide, and he reaped infinite pleasure from each addition to the family tree. This interest extended beyond the immediate family, and he was a member of the Somerset and Dorset Family History Society and a member of the Cromwell Society.

These are the bald facts of his life - none tell of his nature. It requires few words to sum up Hugh as a man; he was kind, gentle, caring and generous. Self always came last. His abiding interest was in people and how he could help those who needed it. He was especially caring for the elderly, but the young were not to be neglected and he devoted his Sundays, for many years, to teaching at his local Methodist Sunday School. He gave much and asked for little, and as a result was adored by his patients and revered by his colleagues. To have known him and worked with him was a privilege. WCH

[*Brit.med.J.*,1988,**296**,869; Photo]

TUNBRIDGE, ***Sir*** **Ronald Ernest** *b.*2 June 1906 *d.*12 Jan 1984
Kt(1967) OBE(1944) BSc Leeds(1927) MSc(1929) MB ChB(1931) MD(1933) MRCP(1933) FRCP(1944) FFOM(1983) FCP Ceylon(1966) Hon DSc Hull(1974) Leeds(1975) Warwick(1979) Hon F Leeds Polytechnic(1983) JP(1958)

The son of the Rev W J Tunbridge, a Methodist minister, and his wife Norah Young, Ronald Tunbridge became a consultant physician and professor of medicine whose life was characterized by a conscientious devotion to the care of the elderly and the disabled, combined with a dedication to the training of doctors to acquire greater technical skills without loss of compassion.

Ronald went to school at Scarborough College, Ashby-de-la-Zouch Grammar School, and Kingswood School, Bath. He developed into a tall, strongly built man, who played cricket and hockey and captained the medicals' rugby team at Leeds University. The author's first memory of him is as a vigorous player of male and mixed hockey in days of early military service. Ronald obtained first class honours, BSc and MSc at Leeds, in physiology; winning a university research scholarship, British Association and BMA studentships. Prophetically, in addition to sport and study, he also made time to be secretary and president of the student medical society, chairman of the Rag committee, joint foundation editor of the medical school magazine, founder and leader of a local boys' club and an active member of the Yorkshire Association of Boys' Clubs.

After graduating MB ChB with honours, and following house appointments at the General Infirmary at Leeds, Ronald held an important formative post as clinical assistant to Francis Fraser at St Bart's [*Munk's Roll*,Vol.V,p.141]; during this time he acquired his MD and MRCP before returning to Leeds as RMO (1933-35) and then becoming, successively, medical tutor (1935-37), reader in medicine

(1937-39) and consultant physician to the Public Dispensary and Hospital, St James's Hospital, Leeds, and Staincliffe and Dewsbury Hospitals in the West Riding.

The RAMC claimed him as a medical specialist early in 1941 and by the end of that year he was officer in charge of the medical division, Malta, command adviser in medicine and psychiatry, and physician to the Governor. He was appointed OBE in January 1944.

In the invasion of Europe he was for a time in charge of the infectious diseases unit in Normandy, then served in Belgium and Germany, being twice mentioned in despatches, and finally became consultant physician to the British Army of the Rhine with the rank of brigadier.

Ronald returned to Leeds in 1946 and was appointed to the newly established full time chair of medicine at the University of Leeds and consultant physician to the United Leeds Hospitals; posts he filled with distinction until his retirement in 1971. At the same time, he willingly accepted an astonishing range of responsibilities, locally, nationally and internationally. Most doctors dislike, or affect to dislike, all committee work; Ronald never lost faith in the combined wisdom, good sense and practicality of most of his colleagues, and as chairman of many committees often achieved more than might have been expected from the sum total of the individual talents of the members.

He was a member of the Regional Hospital Board 1948-52, and of the board of governors of the United Leeds Hospitals, 1952-71; chairman of the medical advisory committee to the Ministry of Health, vice-chairman of the Central Health Services Council, adviser in medicine to the Platt survey in the Newcastle and Bristol Regions, chairman of the Ministry's standing committees on cancer and immunization, and chairman or deputy chairman of departmental committees on accident and emergency services, general practice, tissue transplantation, medical records, medical care of hospital staff, rehabilitation, remedial services and psychogeriatric provision of care. He was also joint chairman, with Lord Knutsford, of a national working party on liaison between health, social, and employment services for disabled people.

For the College, Ronald was a member of Council, an examiner, and chairman of a committee on social and preventive medicine. He was president of the Association of Physicians, and he became a Fellow of the British Medical Association after long and valued service, including the chairmanship of committees on rehabilitation and on the future policy of the General Medical Council, the first chairmanship of the board of science, membership of Council, and from 1974-75 the presidency.

His almost inexhaustible list of other responsibilities can only be indicated here by some principal examples. In diabetes, he became vice-president of the British Diabetic Association and later honorary president, chairman of its research committee, Banting Memorial lecturer, and he was joint founder of holiday camps for diabetic children. In geriatrics, he became president of the British Gerontological Society, chairman and organizer of the second international congress of gerontology, vice-president of Age Concern, helped the writer to organize three international symposia on biological aspects of ageing, and won the Bobst award for international services to gerontology. In rheumatology and rehabilitation, he was a strongly supportive member of the British Arthritis and Rheumatism Council and chairman of its medical, education, and research committees, orator and president of the Heberden Society, and president of the British Spas Federation. He received a gold medal from the Montreal Institute for Rehabilitation and the Osler award of the Canadian Medical Association.

The Tunbridge Building at Cookridge Hospital, Leeds, houses the new cancer research laboratories and was named after Ronald Tunbridge as a tribute to his work for many years for the Yorkshire Cancer Research Campaign.

Ronald's passion for improvement in medical education and his concern for young people in training led to his being in great demand as examiner and educational adviser in many places in the UK, and in Ceylon and Malta. He went on special missions to Iraq, Iran, Tunisia, Ethiopia, Kenya and the Cameroons, as consultant in medical education to WHO. He was president of the Council for International Organizations of Medical Sciences (CIOMS), chairman of the training committee of King Edward VII Hospital Fund, and for the Chaplaincy Board he was responsible for report on the work of hospital chaplains. He was a governor of several schools, president of the Yorkshire Science Masters' Association, served as a magistrate for the city of Leeds, and also even as first chairman of the BBC Leeds radio station.

The mind boggles at his tireless, dutiful, manifold and effective participation and leadership in these and many other activities. It might well be supposed that his patients, his students, and his family could hope to see him only rarely and briefly, but Ronald's patients demonstrated on his death that they had never been in doubt of his continuing concern for each and every one of them, and his students and housemen knew that he was as anxious for them to become competent, caring doctors as they were themselves, and that they could approach him freely in his office or in his home. In his home life and his friendships he was singularly happy. He married Dorothy Gregg in 1935, and for their many friends in England and abroad it was impossible to think of the one without the other; their love and companionship gave joy to all who knew them. To their delight, both their sons, Michael and David, achieved distinction in medicine and were elected Fellows of the College.

In his generation, Ronald was a tower of strength to British medicine.

GW

[*Brit.med.J.*,1984,**288**,333; *Lancet*,1984,**1**,294; *The Times* 19 Jan 1984; *Daily Telegraph* 17 Jan 1984; Photo]

VASEY, Ronald Houston *b*.11 June 1913 *d*.3 Oct 1986
MB BS Durh(1936) DRCOG(1938) MD(1939) MRCP(1947) FRCP(1969)

A native of the north-east of England, Ronald Vasey spent the greater part of his life in that area. Born at South Shields, he was educated initially in Edinburgh but his later studies were at the Royal Grammar School, Newcastle. His medical studies were also at Newcastle, which at that time was part of Durham University, where - having gained an entrance scholarship - he distinguished himself by being awarded the Gibson prize. After qualifying, his junior appointments were at the Royal Infirmary, Newcastle. Then came World War II and Ronald served with the Indian Medical Service. He spent the next six years in India and Burma as a medical specialist and became acting lieutenant colonel.

On demobilization Ronald returned to England and the north-east, initially back to the Royal Victoria Infirmary but subsequently to Sunderland where he spent the rest of his life. His first appointment was to the General Hospital, transferring as senior consultant physician to the Royal Infirmary. As such he became a very well known and highly regarded physician in addition to being a popular personality. He did not confine his interests solely to clinical medicine but also served the medical community as a member of the Hospital Staffs Association, the Medical Advisory Council and the British Medical Association. In each of these associations his abilities were recognized when he was made chairman.

Outside the field of medicine his interests were wide. A family man, being the proud father of a son and a daughter, he nevertheless found time to enjoy golf and bridge, at both of which he was adept. One of his great loves was his garden, which gave him enormous pleasure until the end of his life. He gave meticulous dedication to all he undertook and his death left a gap in the community.

VJM

[Photo]

VINES, Robert Harold *b.*10 Jan 1921 *d.*25 Dec 1986
MB BS Sydney(1943) MRCP(1950) DCH(1951) FRACP(1963) FRCP(1974)

Robert Vines was born in Armidale, New South Wales, the son of a Presbyterian clergyman. His childhood schooling took place in various country centres in outback New South Wales, graduating from Dubbo High School where he was school captain and Dux of the year. He obtained an exhibition to the University of Sydney, and a scholarship to St Andrew's College, graduating with honours in 1943. After a twelve months' residency at the Royal Prince Alfred Hospital, he enlisted in the Royal Australian Army Medical Corps,1944-1947, and served in Bougainville. Following discharge from the Army he obtained further training, gradually moving into paediatric medicine.

In 1950 he came to England and was house officer to Reginald Lightwood (q.v.) at the Great Ormond Street Hospital for Sick Children and no doubt this association kindled his interest in growth and hormone disorders. On return to Australia in 1952 he was appointed physician to the Royal Alexandra Hospital for Children, and practised consultant medicine there until illness forced his resignation in January 1986. In 1965 Robert Vines journeyed to Baltimore, USA, and furthered his endocrine knowledge under the great Lawson Wilkins, undoubtedly the doyen of modern paediatric medicine. Vines, in his turn, was the father of this discipline in Australia and was instrumental in the development of the Australian human pituitary programme, with the first child being treated with extracted growth hormone in 1963. His contributions to the development of growth hormone and other hormone therapy in Australia have been enormous, and done on a honorary basis. He was chairman of the Human Pituitary Advisory Committee for seven years, and served for many more on the growth hormone subcommittee. He was the founding president of the Australasian Paediatric Endocrine Group, and a director of the Diabetes Youth Foundation.

Robert's family life was a happy one. He married Lorraine Halse Rogers in 1948, the daughter of Percival Halse Rogers, Justice of Supreme Court of New South Wales and one time chancellor of the University of Sydney, and they had four children with whom he had an excellent rapport.

Robert Vines was a man whose energies led him into many other pursuits. He enjoyed the challenge of learning French and Italian, both of which he spoke with enthusiasm to the consternation of native French and Italian speakers. His love of exercise and the outdoors was seen in his sailing, sailboarding, skiing and bushwalking, all of which he did until forced to stop by illness. A man of ready wit and a love for the English language, he proved to be a wise counsellor in many disputes.

Robert faced the multiple challenges of his terminal illness with great courage, arriving back at work after performing impossible feats, such as climbing Ayers Rock, within weeks of major surgery. When diplopia developed as a consequence of his illness, he wore an eyepatch and even with his piratical appearance his patients continued to come in increasing numbers. A literary man of impeccable integrity, having an encyclopaedic knowledge of medicine, he is best remembered as a quiet achiever who never sought the limelight, but despite that has been honoured by having a research fellowship and an annual oration named after him.

MS

[Photo]

WALKER, William *b.*1 Jan 1920 *d.*18 Dec 1984
MA MB ChB St And(1946) MRCP(1950) FRCPE(1964)
FRCP(1971)

Bill Walker was born in Dundee, the son of William Sharp Walker, who was engaged in the jute industry, and Joan Strachan. His intellectual precocity was evident at the Harris Academy in Dundee, and in 1936 he was awarded a Harkness residential scholarship in the faculty of arts in St Andrews where he read English literature, history and philosophy. His undergraduate career was distinguished and he was elected president of the Literary and Philosophical Societies. But he was also a robust all-rounder, enjoying the cut and thrust of interuniversity debate, membership of the Students' Union Committee and captaincy of the boxing team (he was only knocked out once in his boxing career - in a Brigade championship in the British Expeditionary Force in 1940). In the summer vacations he travelled extensively in Europe and he could sing romantic songs of several nations. Foreseeing the coming war he worked out an elaborate code with a friend in Czechoslovakia and throughout the war they kept up a correspondence in French.

At the outbreak of war he was commissioned in the Royal Scots and in 1940 he was badly wounded in Belgium and evacuated in coma via Dunkirk. The bullet that lodged in a temporal bone left him with permanent unilateral deafness, and he was ataxic for quite some time. Invalided from the Army in 1941, he returned to St Andrews to graduate MA, and then entered the faculty of medicine, graduating with commendation. He was house physician to Adam Patrick [*Munk's Roll*,Vol.VI, p.368] whom he resembled in his broadly based education and erudition. He then held junior appointments in the department of pathology in Dundee, and with C C Ungley [*Munk's Roll*,Vol.V,p.427] in Newcastle. There he worked on the treatment of pernicious anaemia with concentrated liver extract and developed what was to be a life-long interest in haematology. He returned to Dundee as a lecturer in the department of therapeutics under R B Hunter, where he established an efficient haematological service and carried out research on disorders of blood coagulation and anticoagulant drugs. Walker then spent a year at Boston University and the Massachusetts Memorial Hospital, and in 1955 was appointed senior lecturer in therapeutics and consultant physician at Maryfield Hospital, Dundee.

Bill Walker had a wide range of extra-mural interests. He was a co-opted member of Dundee Education Committee, 1957-62. He championed the cause of secular, scientific humanism in a notable Sunday morning television series on religion and philosophy, 'Seek the Truth' 1962-63, and later he became president of the Edinburgh Humanist Group. An internationalist in outlook, he was co-founder of the Anglo-German Medical Society and for six years held its presidency. In 1963, as Colombo Plan adviser to the Burmese government, he helped set up a school of paramedical services.

In 1964 he was appointed consultant physician to the City Hospital, Aberdeen, where, as a general physician, he had charge of the wards dealing largely with infectious diseases. His first challenge was the Aberdeen typhoid epidemic, in which over 500 cases were diagnosed, and his hard work and enthusiasm over these months were in no small measure responsible for the low morbidity as well as the eventual containment of that outbreak. His profound understanding of drug therapy was particularly useful to him in developing a first-rate medical unit, and he was also able to continue his interest in haematology. His excellence as a teacher, both at the bedside and in the formal lecture, was recognized in his appointment as reader in clinical medicine. Indeed, such was his academic and professional recognition in Aberdeen that he was a natural choice to fill the Regius Chair of Materia Medica in 1973, following the untimely death of Alastair Macgregor [*Munk's Roll*,Vol.VI,p.311].

Despite all his academic and extra-mural interests Bill Walker maintained that his first duty was to his patients, and he was a dedicated and compassionate doctor much loved by his patients and highly regarded by his clinical colleagues.

He had a genuine interest in the problems of medical students and young doctors, and it was very appropriate for him to be elected president of the Aberdeen University Medical Society. He readily accepted that committee work was an increasing part of a clinician's activity, and his erudition and diplomacy were utilized to the full as chairman or member of many health service, university and national bodies. He was a member of the old board of management in Aberdeen and subsequently of the Grampian Health Board. As chairman of the Grampian Medicines Committee he fought for the introduction of a hospital formulary as a step towards rational prescribing. He was a member and subsequently vice-chairman of the committee on Review of Medicines.

Bill Walker's eloquence, combined with his knowledge and love of history and the arts, attracted regular requests for impromptu or after-dinner speeches, many of which were spiced by references to his deeply felt political ideals, his commitment to the National Health Service, and his abhorrence of injustice. He often delighted his audience by ending a speech with a sonnet written for the occasion or with an appropriate operatic excerpt. He had been an accomplished sonneteer since his schooldays and it is hoped his poetry may be published posthumously. In 1931 he was the first Aberdeen based clinician to be elected president of the Harveian Society of Edinburgh. His oration, published in the *Scottish Medical Journal*, can be read and re-read with pleasure and, of course, it ends with a sonnet!

Ever since the earliest realization of the devastating capabilities of nuclear weapons he campaigned passionately against their existence and development, and believed that the medical profession, more than many others, could help prevent nuclear biocide. He was a founder and recently chairman of the Aberdeen branch of the Medical Campaign Against Nuclear Weapons.

In 1982, because of the necessity of cuts in university expenditure and in a characteristically unselfish manner, he took early retirement from his Regius chair and accepted a part-time teaching commitment as professor of clinical medicine. Again characteristically, he continued to teach as much as before but he began to envisage more opportunity to pursue his extra-mural interests - gardening, poetry, literature, music, philosophy and political controversy. Alas, it was not to be. He bore his final illness with courage and dignity, not deviating in any way from his lifelong philosophy. It is sad that his untimely death prevented the late flowering of his literary talent that his friends and admirers had expected.

He married in 1948 Mary Cathleen Kenny, a nurse trained in Dundee Royal Infirmary. She supported him loyally throughout his career and especially, along with their adopted children, Kenneth and Katrina, in his terminal illness.

PDB/KGL

[*Brit.med.J.*,1985,**290**,641; Photo]

WALLACE, Hugh John *b*.21 Feb 1909 *d*.11 Sept 1984
MRCS LRCP(1935) MB BChir Cantab(1936) MA(1938)
MRCP(1938) MD(1943) FRCP(1951)

Hugh Wallace was born of farming stock in Bedfordshire, where he completed his schooling. In childhood he was afflicted with a severe kyphoscoliosis, which left him a dwarf with a deep, rumbling voice. He started medicine at King's College, Cambridge, where he distinguished himself academically with ease and coxed the College boat. He continued at St Thomas's Hospital, where he was again recognized as a remarkably competent and enterprising student. After qualifying, he took up dermatology in close association with G B Dowling [*Munk's Roll*,Vol.VII,p.163] with whom his professional and social life became intimately connected. During the war his physical disability prevented active service, so he was stationed at Woking in the Emergency Bed Service, which he turned into a centre of happy organization and professional enthusiasm. After the war he was appointed to the staff of St Thomas's Hospital, London, and St John's Hospital for Diseases of the Skin, in each of which he became a governor. In the College he gave the Watson Smith lecture and was a councillor,1967-70, and an examiner. His private practice was outstandingly successful.

The key to Wallace's professional life lay in his character and his association with G B Dowling, to whom he was a complement; being first a disciple, then colleague and finally successor. His friends tended to allow him the licence so often given to a cripple, and he would say and do things which needed to be said and done without offence. In every group his enthusiasm, friendliness and wise tolerance, made him the acknowledged centre. This was shown particularly in the Dowling Club, which he founded to bring together colleagues at home and abroad and to encourage the young, and which ended by making British dermatology an energetic and powerful force which attracted those with ambition and enterprise.

He was survived by his wife, also a doctor, and by a doctor son and nursing daughter.

JBH

[*Brit.med.J.*,1984,**289**,1011,1150; *The Times*,1 Oct 1984; *St Thomas's Hosp.Gaz.*,Winter 1984,95-96; Photo]

WALLACE, William Davidson *b.*3 July 1909 *d.*16 Jan 1984
MB ChB Sheff(1934) MSc(1937) MRCP(1944) FRCP(1967)

William Wallace was born at Horncliffe, Northumberland, the son of James Wallace a Scottish minister of the Presbyterian Church of England. He was educated at Firth Park Grammar School, Sheffield, and Sheffield University; and the Royal Hospital and Royal Infirmary. Sheffield. After house appointments at Sheffield and Leicester, he joined the RAMC in 1940 and was posted first to India and then to Burma. He served until 1945, but in spite of the interruption to his career, and his wartime commitments, he undertook a correspondence course for the membership examination of the College and obtained it in 1944. On demobilization he was appointed an honorary physician to the Royal Salop Infirmary in 1947, and later became consultant physician to the Shrewsbury Group of Hospitals. He moved to Copthorne Hospital in 1953 and worked there until his retirement. His brother-in-law T E Gumpert, and his nephew, E J W Gumpert, were also Fellows of the College.

To Bill, medicine was a way of life. His workload was enormous and his free time sparse. At Shrewsbury he rose early, often between 6am and 7am and worked until 6pm or 7pm, came home for dinner, and then went out and worked until 10pm or later. An 'evening' of relaxation would then follow, and when the rest of the household had retired to bed he would happily sit down to 2-3 hours' reading, often about medical matters. His capacity for sustained work was enormous, but in spite of his workload he never appeared rushed or harassed, and never failed to give every patient his undivided attention, kindness and patience.

He married Jean Margaret Coates in 1940, and they had four children. He was essentially a private man, with an able and enquiring mind and a delightfully dry sense of humour. The time that was not devoted to medicine was reserved for his wife and family, and their close friends. He enjoyed lawn tennis and competed regularly with friends. His other interests were snooker and hill walking, and in later years his holidays were based on travel and caravanning. PLB/GW/VL

WARD, Donald Jason *b.*21 June 1931 *d.*6 June 1988
MB ChB St And(1955) MRCP(1960) FRCP(1979)

Donald Ward was born in Brighouse, Yorkshire, the son of a trade inspector. There were no family links with the medical profession.

He was educated at the Hipperholme Grammar School and St Andrew's University medical school, qualifying in 1955. After house appointments to Professor (later Lord) Hunter at Maryfield Hospital Dundee, and at the Bradford Royal Infirmary, he spent 1957-61 at the Royal Hospital Sheffield in junior medical posts. In 1961 he moved to the Canadian Red Cross Memorial Hospital at Taplow as senior registrar to Eric Bywaters. He remained there for four years, at a time when this internationally renowned unit was in its heyday. He then spent a year as a lecturer in rheumatology in the department of medicine at Bristol University before moving to the Robert Jones and Agnes Hunt Orthopaedic Hospital Oswestry in 1966.

The hospital at Oswestry had been founded by Dame Agnes Hunt, with the help of Sir Robert Jones, and developed through the efforts of a series of eminent orthopaedic surgeons, including Sir Reginald Watson-Jones and Sir Henry Osmond-Clark. The problems posed by rheumatoid arthritis and related disorders formed an increasing portion of the clinical workload at the hospital following the conquest of tuberculosis and poliomyelitis, and the need for a rheumatologist was recognized. As a physician heavily outnumbered by orthopaedic surgeons the early years at Oswestry were not always easy for Donald Ward as he struggled to set up a rheumatology unit. Nevertheless, he succeeded in obtaining his own ward and undertook lengthy outpatient clinics, with patients travelling long distances from north and mid-Wales, Cheshire and Herefordshire, as well as from Shropshire itself. With the help of a biochemist, D J Lea, he set up a rheumatology laboratory which not only carried out routine laboratory tests but also produced a series of papers on immunological topics related to rheumatology. Facilities in the rheumatology unit and in the laboratory were developed with the help of funds provided by the Oswestry Rheumatology Association, which Donald had founded soon after arriving at Oswestry to provide a self-help group for patients with rheumatic disorders.

Donald Ward was a rather shy man, who disliked large meetings and rarely travelled far from his home or hospital; consequently his achievements were not widely recognized. In small groups, however, he was witty and talkative, and he was widely revered by his patients.

Donald met his wife Mildred Rose Turner (Penny) when she was nursing at the royal hospital Sheffield, and they and their two sons formed a close family unit. having played the piano in his youth, Donald retained an interest in classical music. but perhaps his favourite form of relaxation was working in his delightful garden overlooking the Shropshire plain and the Welsh hills.

RCB

[*Lancet*,1988,2,805]

WATERFIELD, Reginald Lawson *b. d.*10 JUNE 1986
FRAS(1916) MRCS LRCP(1925) MB BS Lond(1932)
MRCP(1933) FRCP(1955)

Reggie Waterfield, as he was known to all, was the son of a remarkable father, the Very Reverend Reginald Waterfield, Dean of Hereford Cathedral, who had been Chaplain to Queen Victoria's Household, a master at Rugby School, and Headmaster of Cheltenham College. His charming and caring mother, Mary Lawson, was the daughter of a farmer. The Dean died in his hundredth year, a short while after his wife. Though different in many ways, Reggie and his father respected one another, and his mother was very supportive.

Reggie was educated at Temple Grove Preparatory School and Winchester College, and he graduated in medicine from Guy's Hospital Medical School, University Of London. His interest in astronomy had started at the age of 10, when he was still at preparatory school and saw the daylight comet of 1910. In 1914, while at Winchester, he joined the British Astronomical Association. He had a special interest in comets, was a skilled photographer and took pictures of the principal eclipses in various parts of the world, as well as carrying out innovative astrometric work. His planetary observing books were a model and full of information, particularly on Mars, Jupiter and Saturn. He soon had an international reputation as an astronomer and the Royal Astronomical Society elected him a Fellow at the age of 16, while still at school. In 1942 they awarded him the

Jackson-Gwilt Medal, and in 1966 the British Astronomical Association awarded him the Walter Goodacre Medal - their highest award. He was also the first recipient of a medal for work on comets awarded by the Astronomical Society of the Pacific. He was later to become vice president of the Royal Astronomical Society, and president of the British Astronomical Association, having been director of its Mars section from 1931-42. He usually observed from Cheltenham, and would join the late Rev T E R Phillips at his observatory in Headley, Surrey, where eventually Reggie himself went to live, building his own observatory.

On leaving school, during the first world war, Reggie joined the Army and saw service at the front before the end of hostilities. On demobilization he decided to study medicine and entered Guy's Hospital medical school. After qualification he spent the next two years, 1927-29, as lecturer and assistant physician at the Johns Hopkins hospital, Baltimore, USA, returning to Guy's as a medical registrar. Unfortunately he became ill and had to spend some time in Switzerland undergoing treatment for severe chronic thyroiditis. On recovery, he returned to Guy's to work in general medicine, with an increasing interest in haematology. The advent of the second world war found him at first in the Emergency Medical Service carrying out valuable work with sector haematology, and in air-raid resuscitation at Guy's. In 1941 he joined the RAMC, in the Army Blood Transfusion Service, and formed a blood supply base unit which operated in North Africa, advising and caring for the needs of the mobile field transfusion units under his command.

After the war Reggie returned once more to Guy's and began to build up the department of haematology. His enthusiasm was infectious and the department soon produced outstanding detailed laboratory work. He was meticulous, precise and thorough, insisting on very high standards both for himself and for those who worked with him. Reggie was also extremely artistic and he produced beautiful illustrations and slides in order to demonstrate detailed morphology of the disorders he was investigating or demonstrating while teaching. He carried out much original work, and he owed a great deal to the excellent assistance given him by T H Newman, his chief technician, a fine laboratory organizer and trainer of technicians, and always a sound and much valued opinion in haematology, who is now retired with his wife and family near Vancouver, Canada.

In 1949 Reggie developed an unusual and severe form of poliomyelitis which left him a paraplegic. His determination to maintain as much independence as possible, aided by a fine brain and quick wit, enabled him to continue to pursue his career as a haematologist and as an astronomer. Despite his handicap, he continued his day to day diagnostic work and the teaching of undergraduates and technical and medical staff. He had already written papers on abnormal blood pigments, aspects of blood sedimentation in relation to anaemia, and changes in plasma proteins. He worked extensively on coagulation both in haemorrhagic disorders and thrombosis, and was able to organize a first class anticoagulant unit for the treatment of venous thrombosis, and acute and after care for coronary thrombosis.

An example of Reggie's attention to detail in his search for perfection was his concern when he found that it was apparently impossible to draw out thin, smooth films for staining and microscopic examination. They were uneven, streaky, and on examination showed excessive red cell rouleaux formation. Further investigation showed raised fibrinogen content in the plasma proteins. He then noticed that increased blood sedimentation rate was also present and varied according to the degree of film streakiness. After this correlation of streaky films to blood sedimentation rate a Waterfield report on a blood picture often included 'streaky films indicating a high ESR'. A useful pointer to the presence of infections, or perhaps malignancy, and many a keen student who learned to make good slides through Reggie's tuition scored marks by following up 'streaky film' appearance with an ESR estimation.

Reggie was greatly interested in red cell measurements, where he found many similarities to photography in astronomy. He built the Waterfield Halometer, an accurate machine for determining red cell diameter. The Halometer was of great value in differentiating macrocytic, microcytic or normocytic anaemia, and was particularly useful in early diagnosis of pernicious anaemia. A more sophisticated Waterfield Spherocytometer was developed which added cell thickness and diameter thickness ratios, greatly facilitating investigation into haemolytic anaemia. He followed this branch of research, and for two years after his retirement Guy's granted him laboratory space to complete his work. Automated cell counting machines may have superseded his lovely instruments with the advent of large scale haematology, but in many ways they are based on Reggie's early work.

During his distinguished career in haematology, Reggie continued his work in astronomy and more detailed information about his achievements in this field can be obtained from the Royal Astronomical Society.

Always elegant in dress and manner, when away from work Reggie was excellent company. He enjoyed playing darts, over a beer, with students and colleagues, and he excelled in off-board play; very amusing for all, except the unfortunate marker at risk from the straying dart. He was a splendid raconteur with delightful stories; perhaps of his yearly visit to the Hobby Horse Fair at Padstow in Cornwall, where his many friends, from dukes to dustmen, would wheel him precariously from one place of entertainment to another; or of the time when he entertained a famous visiting professor of medicine and his wife to dinner at the Athenaeum Ladies Section. The dinner was splendid in all respects, but when leaning forward in the drawing-room afterwards Reggie became unbalanced in his chair, sending the coffee and liqueur trolley hurtling to the other side of the room, passing bishops and their ladies, and other guests. After apologising profusely came the cry 'Oh, my God, I hope they won't tell the Lord or my father.'

Reggie was a very fine pianist, with a love of Bach and Beethoven, and enjoyed playing to friends. He often said, with pride, that after suitable priming he was persuaded to accompany Einstein, a very good violinist, for a short recital. His modest comment was 'Surprisingly, it seemed to go very well.'

Reggie Waterfield never married, but after his parents' death he enjoyed the help and companionship of his widowed sister, Patsy, who survived him. On retirement, Reggie returned to Somerset, where he set up his observatory with an electrically operated roof. He remained there for the rest of his life. Many friends came to see him; some were visiting astronomers who observed with him and during the winter of 1985 he viewed with delight Halley's comet, which he had first seen in 1910. He continued to work in his observatory until shortly before his death.

Reggie Waterfield was a unique and talented man who, in spite of severe physical handicaps, ran an excellent department of haematology while producing a wealth of new and important knowledge. He brought inspiration to many, and will be remembered and missed by all who knew him. He was indeed a remarkable genius.

FCS

[*The Times*,27 June,2 July 1986; *Brit.med.J.*,1986,**293**,214,397; *Lancet*,1986,**2**,114; *Guy's Hosp.Gaz.*,January 1987; Photo]

WETHERLEY-MEIN, Gordon *b.*30 June 1918 *d.*24 Mar 1987
MA Cantab(1939) MB BChir(1942) MD(1953) MRCP(1965) FRCP(1971)

Gordon Wetherley-Mein, professor of haematology at St Thomas's Hospital, was one of the farsighted few who founded the discipline of haematology in this country.

He was born at Cosham, Hampshire, where his father William Archibald Wetherley-Mein was a surgeon. His mother Elizabeth, née Taylor, was the daughter of an Edinburgh businessman. Gordon was educated at Wychwood Preparatory School, Loretto and St John's College, Cambridge, and graduated in medicine at St Thomas's Hospital medical school.

After graduation he joined the RAMC as a regimental medical officer with the 50th Infantry Division in North Africa. He served in Sicily and in the Normandy landings, and during his service in the field handled blood transfusions and organized a laboratory to deal with malaria.

On demobilization he joined the staff of the clinical pathology laboratory at St Thomas's, and fell under the spell of Joe Bamforth [*Munk's Roll*,Vol.VI,p.25], a unique teacher of pathology. Gordon

Wetherley-Mein found the atmosphere of the laboratory stimulated his own interests and he increasingly specialized in haematology. But he was not really content to spend his life behind a laboratory bench; he was a man who preferred to be among his patients. His major interests were in proliferative diseases of lymphoid tissue and bone marrow, especially the polycythaemias and leukaemias.

In 1954 he became physician to the laboratory, and ten years later was the first occupant of the chair of haematology. He recruited staff with wide ranging interests so that the entire haematological spectrum was represented. He expected a total commitment from his team, equal to his own, and worked hard to support and encourage them.

He cut an unusual figure in the wards, with his Pickwickian appearance and his gales of laughter, but he was man of profound common sense and had little time for either waffle or red tape. He was an extremely shrewd doctor and his own straightforward teaching methods were popular with students.

Wetherley-Mein was elected a Fellow of the College in 1971, and after his retirement in 1983 he became Wolfson Research Fellow and set about organizing a countrywide survey of polycythaemia and its complications. It produced fascinating results which are expected to change the treatment of this condition.

He married Elizabeth McBride, daughter of an Australian sheep farmer,in 1945, and they had two children, Colin and Karen. In retirement he was able to spend more time sailing, which was a lifetime passion, and also to indulge in his other hobbies of reading and painting. His family survived him.

GW/VL

[*The Times*,27 Mar 1987]

WHELAN, Robert Ford *b.*22 Dec 1922 *d.*21 Nov 1984
MB BCh BAO(1946) MD(1951) PhD(1955) MD(Ad Eundem) Adelaide(1958) DSc Belf(1960) FRACP(1963) FAA(1966) FACE(1974) *FRCP(1980)

Bob Whelan inherited from his father, a civil servant in Northern Ireland, a rigorous self-discipline and an administrative ability of a high order. In his youth it was the open-air that called him. He became a scoutmaster and organized camping and canoe trips almost as if they were antarctic expeditions. His troop had to remain on Lough Neagh in all weathers because that was what the timetable indicated, and when ashore modern means of lighting fires were scorned. Whelan had intended to enter the Civil Service, but in 1941 he became a medical student at Queen's. He graduated in 1946, and after being resident medical officer at Belfast City Hospital spent a year as a ship's surgeon with the Holt Line.

Fired with enthusiasm for physiology by Henry Barcroft, he joined the department of physiology at Queen's. There he was one of a group of talented young men and he, Shepherd, Roddie, Shanks and Love, under the leadership of David Greenfield, investigated the control of the peripheral circulation in man. Their experiments, often on themselves, demanded steady nerves and a high level of mutual trust. How else could men breathe 30% CO_2, or investigate the effect of strong emotion on skin blood flow by telling the head of department - while he was on a couch with a plethysmograph on his arm in a waterbath - Don't worry. Everything is all right. There's been a fire in your house, but it's out now. Everything is all right, just lie still'.

Whelan's work earned him an MD and a research fellowship with Henry Barcroft at St Thomas's Hospital Medical School in 1951. In that year he married Helen Elizabeth MacDonald Hepburn and they returned to Belfast where he was appointed lecturer in physiology in 1952, and presented his thesis for the PhD in 1955. His work on neurotransmitters, and on chemical agents that modify blood flow to the skin and muscles, was widely recognized as outstanding, and in 1958 he was appointed to the chair of physiology and pharmacology at the University of Adelaide. When he arrived the department was languishing, but despite the poor facilities, he organized on-going experimental work within a month of his arrival and his infectious enthusiasm attracted many able graduates to his discipline. The department was soon identified as a major centre of cardiovascular research. Whelan was awarded a Carnegie travelling fellowship in the USA in 1962, and in 1967 he published his authoritative monograph on *The control of the peripheral blood flow in man,* Springfield, Ill.,Thomas.

Besides leading his research team, he was a fine teacher. He firmly believed that undergraduates were being taught so much that they did not have time to learn. He therefore concentrated his teaching in the laboratories, for he was convinced that the main contribution physiology should make to medical education was to teach young men to observe and record accurately. They needed to learn that it was what they found that mattered, not what they believed they ought to find.

He was elected dean of the medical school in 1966, was on the board of directors of the Eliza and Walter Hall Institute for Medical Research, and was an inaugural member of the drug evaluation committee of the Commonwealth Department of Health. He played a major part in the establishment of the Australian Physiological and Pharmacological Society, and of the Australasian Society of Clinical and Experimental Pharmacologists.

His administrative talent was widely recognized and in 1971 he was appointed vice-chancellor of the University of Western Australia. For this office he had just the right temperament, and he and Betty made a great impression on the staff and students. Bob gave the university firm and wise leadership at a time when it was expanding and developing rapidly. He was elected a member and was later on the council of the

Australian Academy, and he played an important advisory role in the development of the Flinders University of South Australia.

These were happy years for Bob, Betty and the children - two boys and a girl - and he may have needed some persuasion to return to Britain in 1977 to become vice-chancellor of the University of Liverpool. The move was a great success both for the Whelans and for the University. On his return he had to pass the driving test but this was achieved under the tuition of Mr Apter, the university chauffeur, - 'You are meant to hold the steering wheel with both hands, Vice-Chancellor'.

It was not an easy time to be a vice-chancellor, for there was retrenchment of the government's funding of the UGC. But Whelan thrived on it. He was not a man to be confined in a vice-chancellor's office. He visited departments. Staff were invited to his home. He was always accessible to students. Secretaries, technicians, and the uniformed staff of the university respected him. He was always impeccably dressed, unruffled and courteous. He was firm and indeed tough, but he would listen and was always ready to be persuaded by sound argument. During his term as vice-chancellor the university celebrated its Centenary, and the new University Hospital was commissioned. He and Betty worked hard for the university and for the city. When someone asked him what he did when he was not working his reply was one word: 'Sleep'.

His services were in demand nationally. He was a member, and later vice-chairman, of the Committee of Vice-Chancellors and Principals, and chairman of its medical advisory committee. He became an important link man in many delicate negotiations between the UGC, the DHSS and the BMA about medical manpower and clinical academic staff salaries, and he represented the UK on medical matters at the EEC.

In the last week of his life, Bob Whelan was in great form. He had had a letter from the Prime Minister informing him that she was minded to recommend that he be knighted in the new year. Naturally, there followed good natured badinage in the privacy of the Vice-Chancellor's Lodge.

On the last day of his life, the students had a 'day of action' - a day of inaction might be more exact - to protest against the financial cuts imposed on the university by the government. The students' president had wanted the vice-chancellor to speak on TV, because he knew Whelan was sympathetic to the student concern, but Bob declined, saying: 'I am not good-looking enough.' It was while he was addressing the students that he collapsed and died.

His premature death was a sad blow for his family, and his many friends and colleagues in Belfast, Adelaide, Perth and Liverpool; although there was perhaps some envy that he had successfully avoided the anxieties, fears, illnesses and indignities that so often trouble most of us as we age.

A week later, a letter to *The Times* predicted that his death heralded a big mortality among vice-chancellors due to the strain they were experiencing in handling students and staff at a time of severe financial retrenchment. If Bob had been around still, his comment on that letter would have been caustic.

OLW

[*Brit.med.J.*,1984,**289**,84,1701-2; *Lancet*,1984,**2**,1407-8; *The Times* 24 Nov 1984; Photo]

WHITFIELD, Allan George Williams *b.*30 Jan 1909 *d.*18 Feb 1987 CBE(1974) MB ChB Birm(1931) MRCP(1946) MD(1950) FRCP(1953) PhD(1955) FFCM(1978)

Most people, when they first met George Whitfield, concluded that he could not be as good as he appeared. As time passed, they realized that he could. His warmth, his courtesy, his charm, his friendliness, his willingness to help others, no matter how much it demanded from him, became legendary.

George Whitfield was born at Ironbridge, Shropshire, where his father was a general practitioner, and for five generations many members of his mother's family had practised medicine in that part of England. He was educated at Wellingborough School and the University of Birmingham, where he graduated in medicine with distinction in pharmacology and therapeutics at the early age of 22. After house appointments he went into general practice. He had joined the RAMC Territorial Army in 1933 and during the war he served as second in command of a field ambulance in France, being mentioned in despatches. In 1942 he was promoted lieutenant colonel and became assistant director of medical services to the First Army. He was invalided out with a severe chest infection in 1943, and returned to general practice. Four years later he relinquished general practice and returned to academic medicine, becoming medical registrar to the new professorial department of medicine at Queen Elizabeth Hospital, Birmingham. In 1948 he was elected to the staff as consultant physician and lecturer in medicine to the University, subsequently becoming director of the board of graduate medical studies at the university 1955-74, and professor of medicine 1966-74.

George Whitfield gained his membership of the College in 1946 and was elected a Fellow seven years later. He gave unstinting service to the College: he was a councillor, examiner and second vice-president. He gave the Lumleian lecture in 1975, the Croonian sermon in 1983 and the Harveian oration in 1986. An accomplished orator, he was able - without notes - to hold an audience spellbound. He was also assistant director of the College's research unit from 1977 until his death.

He was editor of the *Quarterly Journal of Medicine* for 13 years; a member of the Association of Physicians of Great Britain and Ireland, the British Cardiac Society and the Thoracic Society. He was secretary and treasurer,1950-58, of the West Midlands Physicians Association and was largely responsible for it becoming such an active body. He was its president in 1970.

Postgraduate education was one of his interests and he examined for many universities. During his 19 years as postgraduate dean in Birmingham he supervised the development of postgraduate medical education and the building of postgraduate centres in the West Midlands region. He took a prominent part on a national level as chairman of the advisory committee of postgraduate deans of the Council for Postgraduate Medical Education. He was elected a fellow of the Faculty of Community Medicine in 1978, and had been convenor of the Midlands and South Western inter-regional training scheme in community medicine. He became deputy lieutenant for the county of the West Midlands in 1975.

George Whitfield's chief research interests were in lung volume, radiation damage to thoracic tissues and cardiomyopathy, but he contributed over 100 scientific papers to the literature on a wide variety of subjects. He was also interested in Victorian history and published a number of historical papers and two books *Beloved Sir James*, Sutton Coldfield 1982, and *The First thirty-seven registrars* of *the College*, Sutton Coldfield 1981, both published privately. A brief period in his youth as a part-time member of Sir Barry Jackson's Company at the Repertory Theatre in Birmingham led to a lifelong interest in the theatre, and among his other leisure activities were cricket, racing and golf.

It was George's training in general practice which helped him to be such an outstanding consultant. He understood minor problems as well as major ones, and was an inexhaustible source of sympathy and understanding. His politeness and consideration for others were a striking part of his character, but behind his polished manners was tremendous drive and an application which only those who knew him well could appreciate. Having decided on a paper, he would retire to the library and the first draft would appear within days. This, coupled with the fact that he knew everyone worth knowing in medicine in this country was of inestimable value to the RCP research unit - he held the unit together as no one else could have done.

The fact that he knew what it was like to have frequent ill health contributed to his excellence as a doctor, but made not the slightest difference to his application to work. His terminal illness was already causing him distress when he gave his Harveian oration, but he gave the lecture without a note and the applause emphasized the quality of his performance. Soon afterwards he was obliged to retire to his bed, but this did not stop him working; he edited and produced manuscripts almost to the end. George hated being a nuisance to anyone, but never thought of anyone who needed his help as being a nuisance to him. As a young colleague said: 'He was a perfectionist in all things, but most of all in the spirit of being a gentleman.'

In 1937 he married Barbara Franks and they had one daughter. Barbara looked after him devotedly through his many health setbacks, with courage and exemplary cheerfulness, particularly so during his last few weeks.

George Whitfield was typical of all that is best in general physicians, and he could sum up people admirably, as evidenced by the frequency with which he was asked to write obituaries for colleagues. Just before he died he had submitted a chapter to a book on medical writing, still to be published. His contribution dealt with writing an obituary for a colleague: 'That,' he said to a friend, with a smile, 'will give you something to chuckle about after I've gone.'

CAC/RH

[*Brit.med.J.*,1987,**294**,584; *Lancet*,1987,**1**,577-8; *The Times*,2 Mar 1987; *Univ.Birm.Fac.of Med.Bull.*Autumn 1974,**40**,37; 41-3; *Times of Malta*,23 May 1964; Birmingham Post,28 May 1966; *Address*,Service of Thanksgiving,Edgbaston Old Church,1 May 1987; Tape recording; Photo]

WHITTLE, Claude Howard *b.*2 May 1896 *d.*1 Mar 1986
MRCS LRCP(1920) MB BChir Cantab(1921) MA(1923)
MRCP(1924) MD(1926) FRCP(1945)

Claude Howard Whittle, a pioneer in dermatology, died in Cambridge just two months before his 90th birthday. His very long career had been spent almost entirely in Cambridge. He was born in Felixstowe, Suffolk, where his father Tom Whittle was a schoolmaster. His mother, Edith Annie, was the daughter of Samuel Thompson, also a schoolmaster.

After his early education at Eastward Ho, Felixstowe and the Masonic School at Bushey, in Hertfordshire, he went up to Cambridge as a scholar at Queen's College in 1915. After medical training at King's College Hospital, London, he qualified in 1920 and spent the next two years in his teaching hospital successively as house physician in general medicine, in clinical pathology, as casualty officer and house surgeon. The physician in charge of the dermatology department at that time was Arthur Whitfield (q.v.). Howard Whittle was particularly attracted to clinical pathology and in 1923 he returned to Cambridge as assistant clinical pathologist at Addenbrooke's Hospital. Although he retained his involvement in various aspects of pathology for many years, and was responsible for blood transfusion in Cambridge through-out the 1939-45 war, his skills in the emerging laboratory services led to his becoming a consultant physician with special interests.

In 1929 he was appointed honorary assistant physician to Addenbrooke's and physician in charge of the skin department. In this latter appointment he succeeded John Aldren Wright, who had taken a special interest in diseases of the skin so that they were gradually concentrated in his outpatient clinic. For many years Howard Whittle was the only dermatologist to Cambridge, Bedford, Peterborough, Wisbech, Bury St Edmunds and Newmarket Hospitals, serving an area which now keeps six dermatologists fully occupied.

Howard Whittle really established dermatology as a specialty in East Anglia. In the years before the NHS it was only in large centres of population that it was possible for a man to earn a reasonable living in a practice restricted to dermatology, and it was not until 1948 that Whittle began to confine himself to dermatological cases. He owed much to the support and influence of the dermatology section of the Royal Society of Medicine, of which he was secretary for many years and president 1961-62, and the British Association of Dermatologists of which he was president in 1954.

Whittle's early interest in bacteriology was followed by close attention to fungal diseases and, after retirement, he helped to found the British Society for Mycopathology, became its president 1973-76 and for years after was its most durable supporter. He published a considerable number of papers, mainly on bacteriological and fungal infections.

Howard married Phyllis Lena Fricker, daughter of a manufacturing chemist, in 1923, and they had three sons. His wife died in 1979. All three sons survived them.

His life was a full one and his interests outside medicine were legion. He was commodore of Cambridge Sailing Club, vice-president of the Cambridge Drawing Society, a gifted painter in watercolours and a regular exhibitor at the Medical Art Society, he sang for years with the Cambridge Philharmonic and never missed a church choir practice after his confirmation at the age of 83. He loved Cambridge and Addenbrooke's. He was a doggedly determined campaigner in the Cambridge Preservation Society. In his 90th year he continued to attend the consultant staff council meetings at Addenbrooke's and readily acknowledged that the hospital of the '80s, not least the dermatology department, was a very different place from the hospital of his day.

AJR/RHC/GW

[*Brit,med,J.*,1986,**292**,1086; *Lancet*,1986,**1**,1107; Photo]

WIG, Khushwant Lal *b.*30 Sept 1904 *d.*8 June 1986
MB BS Punj(1929) MRCS Eng(1930) DTM&H(1930)
MRCP(1931) FCCP(1950) FRCP(1961)

Khushwant Lal Wig was born at Gujranwala, Punjab, India, the son of Mohan Lal Wig, a landlord, and his wife Dhan Devi Chib, daughter of Kanahya Lal Chib, also a landlord. He was educated at the U P

Mission High School, Gujranwala, and the Government College and D A V College, Lahore. He studied medicine at the K E Medical College, Punjab University, Lahore, pursuing his clinical studies at the Mayo Hospital, Lahore. He did postgraduate work at the Charing Cross Hospital medical school, London, obtaining his membership of the College and subsequently being elected a Fellow.

From 1941-46 he was assistant professor in the department of medicine at the King Edward Medical College and the Mayo Hospital, Lahore. In 1947 he was appointed professor of medicine at the Medical College and senior visiting physician to the V J Hospital, Amritsar, a post he held until 1958; he was also dean of the medical faculty at Punjab University. He was invited to become the first professor of medicine at the All India Institute of Medical Science, New Delhi, a post which he took up in 1958 and later became director of the Institute until 1969.

In 1959 Khushwant Wig was awarded a travelling grant by the Rockefeller Foundation, to study modern methods of medical education in the United States and United Kingdom. On his return to India he made a significant contribution to the teaching methods being adopted at the All India Institute of Medical Sciences, both at undergraduate and postgraduate levels. He also introduced many improvements in training methods at the medical colleges attached to Punjab University, of which he was a fellow and member of the senate.

He contributed some 35 articles to scientific journals and was president of the board of editors of the *Punjab Medical Journal* for some years.

Wig was a member of the board of the medical faculty of Lucknow University, and of Delhi University, and a member of the clinical trials sub-committee of the Indian Council of Medical Research. In 1961 he was president of the annual conference of the Indian Association of Chest Dieases. He was elected a fellow of the American College of Chest Physicians in 1950 and was the governor of the Northern India Chapter. He was an authority on tuberculosis and lung diseases.

Before the partition of India, Khushwant Wig was an examiner for the MB BS at Punjab University, Lahore, and after partition to Punjab University, Chandigarh. He also examined for the universities of Lucknow, Patna, Delhi, and Madras.

In his earlier days Khushwant Wig had an extensive private practice but he gave this up to concentrate on the advancement of medical education in India. He received the prestigious Dr B C Roy national award for eminent medical men, and as a recipient of the Padma Bhushan in the early '60s he attended the 6th Congress of International Physicians for Prevention of Nuclear War at Cologne, West Germany, as head of the Indian delegation.

He married Shanta Puri, daughter of a lawyer, in 1934 and they had a son and two daughters.

He died at Bern, Switzerland. His wife and children survived him.

GW/VL

[Photo]

WILKINS, Eric Gordon *b.*30 Oct 1907 *d.*31 Mar 1988
MB BS Lond(1930) MD(1935) DTM&H(1935) MRCP(1948)
FRCP(1971)

Eric Gordon Williams was born at Cuttack, Orissa, India, the son of Rev Gordon Smedley Wilkins and his wife Ellen Lucy, daughter of Rev John Gregory Pike. His father, his maternal grandfather and a maternal uncle were Baptist missionaries in Orissa, India.

He was educated at Eltham College and the Middlesex Hospital medical school. His intention, based on the strong religious convictions derived from his family background, was to return to Orissa as a medical missionary. He graduated MB BS in 1930, having been awarded the Leopold Hudson prize in the final year. Towards the end of his time as a student he met Honor Elizabeth Cooper Harvey, the daughter of an architect; she was studying medicine at the Royal Free Hospital, shared his purpose in life, and graduated at the same time. They became engaged, and determined to work together in the medical missionary field. Having in mind the probable needs of his intended work, he applied for surgical rather than medical house appointments, and was house surgeon and casualty officer at Willesden General, and house surgeon at Nottingham General Hospital.

Seeking an appointment in an existing mission hospital in India, they found that the Baptist Society had no vacancies, and were accepted for a 5-year posting to a Methodist mission hospital at Sarenga in West Bengal. They married, and went to Sarenga in 1932. After two years Gordon was invalided home to London with a diagnosis of tropical sprue, for which he was treated at the Hospital for Tropical Diseases. In spite of the prevailing advice of the time, that sufferers from tropical sprue should not return to the tropics, he was determined to work in India and while convalescing studied for and obtained the Diploma in Tropical Medicine and Hygiene, and the London University MD in tropical medicine.

An opportunity arose when the Baptist Missionary Society proposed a new hospital in the Kond Hills, an area in Orissa which was without medical facilities, as a memorial to a former medical secretary of the Society, Fletcher Moorshead. Gordon and Honor Wilkins undertook this task, and returned to India early in 1936. In retirement 50 years later he wrote a book, which was published privately, describing the 15 years they devoted to the establishment of this hospital in the Kond Hills. The title '*By Hands, bullocks and prayers*' reflects the pioneering nature of their work. They started from scratch, planning and supervising building, improvising when necessary everything from drainage and water supply to medical equipment; he taking responsibility for administration and for medicine and surgery, including eye surgery - especially for cataract, and she for obstetrics and gynaecology. This work continued throughout the second world war, and was recognized to be of such importance that he was exempted from service with the armed forces.

They returned to England in 1946, with the three sons who had been born during their stay in India. Gordon was successful in the MRCP examination in 1948; a remarkable achievement for one who had never had a house physician appointment and whose clinical experience had been as much surgical as medical, and in medicine principally tropical. At the end of 1948 they returned to the hospital in the Kond Hills for a final three years, having decided that for the sake of their three sons this should be the limit of their service abroad. When they finally returned from India in 1951, they left an established hospital of 140 beds, serving a population of mainly aboriginal hill tribes, numbering about half a million, and attracting patients from long distances. At the silver jubilee celebrations of this hospital in 1964 they were honoured guests.

Gordon's first post after return to England was at Dorking General Hospital, in the senior hospital officer medical grade as deputy medical superintendent, with clinical duties in general medicine, chest medicine and the care of the chronic sick. In 1956 he was appointed to the new post of consultant physician in geriatrics for the York area, and the second part of his professional life began. He once again did a pioneering job. Starting with little more than an old poor-law infirmary which needed adaptation to modern requirements, he built up medical services for the elderly so that when he retired, in 1972, he left an integrated geriatric service. The high regard in which his clinical colleagues held him was shown when they elected him chairman of the medical division in York.

Gordon Wilkins (he never used his first name) remained throughout his life quietly true to the ideals of his youth. Rather above average height, slim and upright, he gave those meeting him for the first time an impression of being somewhat precise and formal, but his quiet sense of humour soon became evident in conversation. He was a Rotarian and served as president of the York Club in 1969. He was an accomplished flautist, playing in the York Symphony Orchestra.

After his retirement, he and his wife moved to Colwall, near Malvern. He continued to be active in fund raising for medical aid to the third world; among these was a typically ingenious scheme for collecting unused medical samples, sorting them and sending them to places where they might be of use.

His wife died in 1986. Her final illness began when he was writing the book in which he described the work they had done together in establishing the hospital in the Kond Hills; he dedicated it to her memory, and was pleased that she had been able to discuss each chapter as it was written and had seen the final one. Of their three sons, one followed his parents into medicine and became a consultant anaesthetist.

JGS

[*Brit.med.J.*,1988,**297**,418; Photo]

WILLCOX, Richard Robert *b.*10 May 1912 *d.*2 May 1985
MRCS LRCP(1936) MB BS Lond(1937) MD(1951) MRCP(1970) FRCP(1976)

Richard Willcox had a conventional upbringing but a most unusual medical career. He was educated at Cranleigh School and at St Mary's Hospital Medical School, London, where he qualified in 1936. He won the Cheadle gold medal in 1937. After a number of junior posts at St Mary's, and other hospitals in the home counties, his interest in venereology began in the Army, where he eventually became command specialist in Eastern Command, command venereologist in West Africa 1943-45, and adviser in venereology to the War Office with the rank of lieutenant colonel. On demobilization in 1946 he was immediately appointed consultant venereologist to St Mary's Hospital, and in 1947 he became consultant venereologist to the King Edward VII Hospital, Windsor. He held both these posts, and also managed a busy private practice, until his retirement in 1976.

For many years after retirement he continued to see patients, to conduct clinical trials, and to travel widely. He was a very heavy smoker for much of his life and was well known to his colleagues for inhaling strong pipe tobacco. In addition, he was of portly stature and took no exercise whatsoever. Not surprisingly, he developed severe obstructive airways disease and although he gave up smoking he had a number of episodes of respiratory failure, from which he was rescued more than once by the intensive care unit at St Mary's. Despite this severe handicap he continued to travel all over the world in the interests of his specialty, and his courage and tenacity were greatly admired by all his colleagues and friends. Eventually he died in St Mary's Hospital, having requested that modern techniques of resuscitation should not be used again in his case.

Dick was an indefatigable traveller, writer and rapporteur: he used to say that his hobbies were writing, travelling and sun-bathing. He wrote a large number of papers on a wide variety of subjects, but mostly reporting clinical trials of new antibiotics. So great was his output that his friends called him 'the Edgar Wallace of venereology'. Moreover, his papers were published in journals all over the world, and during the 1950s and '60s it was rare for a new antibiotic to be put on the market without, among others, a paper from Richard Willcox describing his experience with it. Some of his colleagues regarded him as a compulsive writer.

His textbook *A textbook of venereal diseases*, New York, Grune and Stratton, 1950, was widely read throughout the world; a second edition of this book entitled *A textbook of venereal diseases and treponematoses*, Springfield, Ill., Thomas, was published c.1964. He also wrote *Progress in venereology*, London, W Heinemann Medical Books, 1953, and *Treponema pallidum*, published by the World Health Organization in 1966. A series of reviews of the world literature in venereology from 1948-76 appeared in the *British Journal of Venereal Diseases*.

In the medical world Dick Willcox not only had an international reputation as a great writer but also as a great traveller, who was always willing to undertake consultantships in any area of the world. As a result, he might be found in places as remote as Papua, New Guinea, or as sophisticated as Washington DC. No journey was too tiring for him and he contributed greatly to the reputation already established by a small number of physicians; that the British were the leaders and trendsetters in this expanding field of medicine.

He will be remembered by his colleagues principally for his skill, tact and prodigious hard work as a rapporteur at all sorts of international meetings throughout the world. Since the second world war the post of rapporteur has assumed increasing importance in international medical meetings, since the rapporteur guides the direction of the discussion, decides what should be included in the report and, in most instances, writes the drafts of the final report. Dick had a special ability to put together documents which produced rapid agreement among the delegates, and many established international reports owe their very existence to his diligence, diplomacy and hard work. Some of the highlights of his work as a rapporteur included the well known travelling seminars in the USSR and USA, and many surveys in Africa, South East Asia, India and the USA. Special praise was given for his two surveys in New Guinea, and for his collaboration with Thorstein Güthe of WHO in Geneva.

In recognition of his contributions to medicine he was elected president of the Medical Society for the Study of Venereal Diseases for the years 1966 and 1967, and was made an honorary member of many overseas medical societies.

At his riverside home in Barnes, he and his wife 'Sadie', née Alma Caffery, were always most hospitable to people from all over the world. He derived great satisfaction from the fact that his son Jeremy took up the same medical specialty and was appointed consultant physician in genito-urinary medicine at Plymouth. In what must be an unusual father/son collaboration, he and Jeremy jointly wrote a book entitled *Venereological medicine*, London, Grant McIntyre, 1982. The team of Willcox pére et fils produced here an excellent volume which has been widely read and greatly appreciated in their specialty.

Dick bore the burden and responsibility of his worldwide reputation with modesty and charm. Everyone who has attended international meetings overseas will be familiar with the anxious demands of foreign colleagues: 'Is Willcox coming?'; 'Have you seen Willcox yet?'; 'Is Willcox giving a paper?' and 'Where is Willcox?'. He was a great ambassador for British medicine and a good companion. RDC

[*Brit.med.J.*,1985,**291**,1131; *Lancet*,1985,**2**,902; *St Mary's Hosp.Gaz.*,1966,**72**,78-79; Photo]

WILLIAMS, Cyril James *b.*22 Apr 1915 *d.*26 April 1985
MB ChB Liverp(1938) MRCP(1946) MD(1949) FRCP(1969)

Cyril James Williams was born in Dinorwic on the Menai Strait, where his father was headmaster of the local school. His mother, Charlotte Anne Clee, was the daughter of a slate mine manager in Ffestiniog, where Cyril was later to receive his secondary school education. Young men from North Wales usually chose London or Liverpool for their medical education, and Cyril Williams journeyed north to enter Liverpool Medical School. He qualified in 1938 after a distinguished undergraduate career which included the gold medal in medicine. Because of the sudden death of his father he was forced into general practice in Standish, south-west Lancashire, in order to provide a family budget. Within twelve months his two senior partners, who were both in the Territorial Army, were called up and he was left running a large practice single-handed throughout the war years. In spite of his workload he found time to hold a commission in the Home Guard and, in the later war years, assisted the senior visiting physician in out-patients at Wigan Infirmary. In 1946 he passed the MRCP at his first attempt and, to his great surprise, was called up into the RAMC in the next year.

Williams served with distinction in Malaya as a medical specialist, and during this time he collaborated with an American research team introducing chloramphenicol in the treatment of typhoid and typhus. In 1949 he returned to Liverpool and joined the staff of Walton Hospital, where he proceeded from registrar to senior registrar to consultant within an 18 month period. He was known affectionately by patients and all members of the staff as 'CJ' and he lived in a hospital flat, even after he married his ward sister, Phyllis Baker. Throughout these years he devoted his whole time to clinical medicine, and his nightly visits to the wards were a reflection of his dedication to patient care. More than once the writer was embarrassed to be second to reach the bedside when on call. Williams' reputation as a diagnostician and teacher was established at this time, and doctors from Merseyside came to enjoy his ward rounds. Although never a gossip, he had an innate sense for the social life within the hospital; stories being retold with a little Welsh lilt. He remained fluent in his native tongue in spite of being away from the Principality for so long.

As the years went by 'CJ' found great difficulty in coming to terms with the changes in hospital medical practice, the concept of a unit of medical time being quite abhorrent to him. Following major surgery, he retired at the age of 60 on health grounds. He had moved to surburban Liverpool when his children had reached school age and the last ten years of his life were spent in quiet retirement, devoted to his house and garden. He was survived by his wife and daughter, and a doctor son of whom he was very proud.

IKB

[*Brit.med.J.*,1985,**291**,148; Photo]

WILLIAMS, David Aelwyn *b.*2 Feb 1907 *d.*14 Mar 1986
MRCS LRCP(1930) MB BCh Wales(1935) MD(1937)
MRCP(1941) MSc Wales(1944) FRCP(1955)

David Williams was born in Cardiff and spent most of his life there. In retirement he lived in Barry, near Cardiff. His father was a mining engineer in South Wales and his mother was the daughter of a tin plate manufacturer.

Williams was educated at Cardiff High School and then went to the University of South Wales and Monmouthshire, at Cardiff, for the preclinical years of medical study. His clinical years should have been spent at Cardiff, but in 1927 the historic dispute closed the clinical school there and most of the students went to University College Hospital, London, where they were made welcome. It was there that Williams took the MRCS LRCP in 1930. He returned to Cardiff Royal Infirmary for posts as house physician and house surgeon, going on to City Lodge Hospital (now St David's Hospital), Cardiff, as resident medical officer. He was to retain an association with his hospital for many years.

The Welsh National School of Medicine came into being by Royal Charter in 1931, with Alex Mills Kennedy [*Munk's Roll*,Vol.V,p.228] already there as full time professor of medicine; a man of wide culture and devoted to his students. He made David Williams an assistant in the department of medicine in 1932, which began a life-time friendship.

Williams suffered from asthma and Kennedy encouraged him to study allergy and asthma. In 1933 the medical scene in Cardiff was changing with the opening of the Municipal Hospital at Llandough, near Cardiff. In order to expand his teaching facilities, Kennedy took over some of the medical beds with Williams as a full time physician in charge of the patients. In 1934, D Greenwood Wilson opened an asthma clinic at St David's Hospital with David Williams in charge. Thus encouraged, he set up an asthma research programme - later to become the Asthma Research Unit, St David's Hospital, and now at Sully Hospital. This led to a lifetime of research, to the setting up of pollen counting in Cardiff in the 1940s in cooperation with Mr Hyde, keeper of the department of botany in the National Museum, and to the study of mould spores as a cause of allergy and asthma. Research at the asthma clinic resulted in many publications from 1950-76 on the natural history of asthma and its treatment.

During the second world war, part of University College Hospital came to Cardiff; the department of medicine with Max Rosenheim, later Lord Rosenheim of Camden [*Munk's Roll*,Vol.VI,p.394], and Sir Thomas Lewis [*Munk's Roll*,Vol.V,p.531] arrived at Llandough Hospital. Lord Rosenheim later remarked that on his visit to arrange the move he met a new kind of physician - one who willingly shared his beds with another. He and 'DA' became firm friends. Sir Thomas Lewis and David Williams developed a mutual respect and had a common interest in research topics.

Williams became consultant physician to the Cardiff teaching hospitals, clinical teacher in medicine, and examiner in the MB Wales. Teaching was important to him and his ability was recognized at all levels. He instituted postgraduate rounds which were popular with aspirants for higher qualifications. To the select few, his guidance and advice on medical research were invaluable. He was a founder member of the British Society of Clinical Allergy, the European Society of Allergy and of the Asthma Research Council, and was proud of his fellowship of the American Academy of Allergy.

'DA' was a family man, delighting in his four children and proud of their ability at work, in sport, and in field events. Two are doctors: a daughter in practice and a son who is a consultant in rheumatology. 'DA' was a keen fisherman and loved his country cottage, with its fishing, near Brechfa in Carmarthenshire. He was also a music lover and as his physical handicap increased music played an important part, especially in his retirement years.

David Williams was soft spoken, quiet and modest, and deeply interested in people and their welfare. His old students and colleagues held him in highest regard, and he was always pleased to have good news of them. He kept in touch with his specialty and his department until the end. He died at the home of his daughter in Barry.

AJT

[*Brit.med.J.*,1986,**293**,60-61; Photo]

WILLIAMSON, Norman Bruce *b.*[?] 1893 *d.*3 Jan 1984
MB ChB Edin(1923) MD(1925) MRCP(1926) FRCP(1939)

Bruce Williamson was born in South Shields. He was one of eight children, five of whom chose to enter the medical profession. Bruce was strongly influenced by his elder brother, Scott Williamson. It was Scott who founded the Pioneer Health Centre at Peckham, which was an outstanding success in social medicine in the 1930s and '40s.

Bruce served throughout the first world war and after demobilization,at the age of 25, he commenced his medical studies at Edinburgh University, financed by damages he received for war injuries. He became an active member of the Edinburgh University Club, a founder member of the Scottish Medical Golfing Society, and from 1921-22 was senior president of the Royal Medical Society in its 185th session, when still a student. Shortly after qualifying he moved to the south of England, eventually becoming senior consultant at the Royal Northern, Prince of Wales, and Barnet General hospitals.

Williamson was an accomplished writer and his handbook *Diseases of children,* Edinburgh, E&S Livingston, 1931, ran to nine editions, being published in Britain and the USA. *Vital cardiology: a new outlook on the prevention of heart disease,* Edinburgh, Livingston, 1934, followed and in his later years he wrote *The executive and the seventies: men only,* London, Lloyd-Luke, 1973. His style had a clarity which appealed to generations of medical students and to many laymen interested in matters of health.

On 2 August 1958 Bruce Williamson wrote to the president of the College, Robert Platt, later Lord Platt of Grindleford (*Munk's Roll,*Vol.VII,p.470], expressing his pleasure that no back street site near Harley Street had been chosen for the new home of the College, and suggesting that the authorities might be persuaded, with Royal support, to surrender the site of Someries House 'provided the design carried on the Nash facade to some extent'.

He was survived by his only son Lorne, and his second wife Yvonne.

GW/VL

[*The Times,*16 Jan 1984; Photo]

WILSON, Brian Douglas Raymond *b.*13 Nov 1915 *d.*30 Apr 1988
MRCS LRCP(1939) MB BS Lond(1939) MRCP(1946) DCH(1948) FRCP(1960)

Brian Wilson died without warning in his beloved Woking garden among the shrubs and roses he had tended for thirty years or more. From 1958 until he retired in 1979 he was physician in charge of the childrens' department at St Thomas's Hospital, London - an institution with which he was associated as man and boy for half a century. The department then included wards at the old Royal Waterloo and Lambeth Hospitals, in addition to which he had consultant sessions at St Andrew's, Bow, and also the Mile End Hospital. With furled umbrella, his measured strides in one or other direction were a familiar sight, pausing only to say a few kind words to any passing 'salt of the earth' locals whose children he had looked after, or to peer through the view-holes of hoardings around building sites - a temptation he could never resist. He also visited the Cheyne Spastics Centre, a charity he supported staunchly and of which he became chairman.

His association with Woking was almost as long as that with St Thomas's. It was perhaps more than a coincidence that having gone to a preparatory school in Woking he was to make it his home town, sending all three sons to his old school and then on to Sherborne, where he himself was educated. He believed in traditional values and preferred to build on solid ground and proven soil. He had the patience and perseverance to wait for results and was never one to change course lightly.

Brian entered St Thomas's in 1933 as a student, to return after the war as paediatric registrar, then becoming chief assistant and finally, in 1950, physician. Eight years later he took charge of the department which, like the garden he cultivated with green fingers, took seed and bore fruit. His first task was to appoint Dennis Cottom [*Munk's Roll*,Vol.VI,p.122] as his junior. Between them the department was to enjoy its halcyon years, setting some 25-30 consultant paediatricians (and Fellows of the College) on their future careers. The wasteful early death of Dennis Cottom in a road accident in France was a bitter blow, coming as it did at a time when he himself needed arterial surgery for his legs, and followed not long after by the tragic death of his middle son at the early age of 19. Brian was not one to give up easily. With characteristic patience and fortitude, he rebuilt what he could and few outsiders ever guessed the truth.

He was an unerring judge of ability, with a rare instinct for picking winners, but never sought the limelight for himself. He had an eye for good prose and shrewd observation, yet only wrote three or four papers himself. Instead, he gained great satisfaction from the ever-growing number of important contributions by past and present members of his department. He had a talent for bringing out the best in others without the slightest pressure, appeal or reproach. If there was a secret in this it was the ability to inspire a natural filial devotion. With his neat white, clipped moustache and fruity voice, he looked the epitome of paternal benevolence, and his appearance never changed in thirty years. He evidently had the same effect on mothers and children, with whom he was an instant success.

As a paediatrician he was an excellent no-nonsense clinician who kept himself well-informed and achieved speedy results with the fewest investigations. He had the wisdom and foresight to be among the first to foster the development of neonatal intensive care, eclectic child psychiatry, refresher courses for general practitioners, and many other points which were to relieve the burdens of sick children. He instantly knew a good thing when he saw it, but eschewed passing 'bandwagons' and had scathing things to say about some paediatric fashions of the moment.

Brian married Evelyn Marsden, the daughter of a parson and a fellow medical student who willingly relinquished her studies to make a supremely happy home for Brian, their three boys, and some 10-20 animals of various shapes and sizes - from stick insects to a python. Almost as soon as he qualified, Brian was recruited into the RAMC and sent to India and Burma. It was there that he presumably acquired his endearing capacity for light-hearted derogatory remarks against the Far East which, after the death of Dennis Cottom in France, grew to encompass all nationals east of Calais, except for Gurkhas, punka wallahs, and anybody who could make a good curry.

His favourite hobbies were cricket, trout fishing and gardening. He staunchly supported the hospital cricket team and, at home, revelled in the sporting exploits of his boys. For his fishing, he flogged the Surrey waters, often for no better reason than to bring back a briefcase full of trout for everyone in the department. He enjoyed nothing better than to holiday in Scotland with his family, and his fishing rods and flies.

When Brian cut the Christmas pudding on the children's ward at St Thomas's for the last time he had, as always, surreptitiously contrived to put a sixpenny bit in every slice, making sure his successor knew how it was done. A traditionalist to the last, his legacy to paediatrics can be found in the many consultants and professors who grew out of his department.

JWS/HB[1]/GW

[*Lancet*,1988,**1**,1237; *The Times*,20 May 1988; *Guardian*,13 May 1988; Photo]

WILSON, ***Sir*** **Graham Selby** *b.*10 Sept 1895 *d.*5 Apr 1987
MRCS LRCP(1916) MB BS Lond(1916) MD(1919) DPH(1920) MRCP(1920) FRCP(1930) Hon LLD Glasg(1962) Hon FRCPath(1972) FRS(1978)

Graham Wilson, known all over the world as 'GS', was born in Newcastle but, his family having moved south, was educated at Epsom College and Charing Cross Hospital medical school, where he came under the influence of W W C Topley [*Munk's Roll*,Vol.IV,p.561] who was to be his inspiring colleague from 1919 to 1941. Soon after qualification in 1916, Wilson joined the Army and worked in the laboratories at Kasauli, India. He returned to England to work first as a demonstrator in bacteriology at Charing Cross Hospital medical school, from 1919-22, and then moved, with Topley, to Manchester as lecturer in bacteriology; together they moved to the London School of Hygiene and Tropical Medicine in 1927, where Wilson later became professor of bacteriology applied to hygiene. Having been intimately concerned with the planning and establishment of the Emergency Public Health Laboratory Service, he succeeded as director when W M Scott was killed in an air-raid in 1941. He directed the emergency service and its peace time successor until 1963.

Wilson's contributions fell into three parts: his prewar work on the gradings and pasteurization of milk, his authorship with Topley of what is arguably the best textbook of microbiology in the world, and the effective creation of the Public Health Laboratory Service.

The studies on the hygiene of milk were of crucial importance in justifying the general introduction of pasteurization, and were thus instrumental in the near total eradication of childhood tuberculosis.

The first edition of Topley and Wilson's *Principles of bacteriology and immunity*, London, E Arnold & Co., was published in 1929 and set standards in clarity and critical argument that can rarely have been equalled. The book soon established itself as the major bacteriological textbook in English; even after six subsequent editions it has few real rivals. Wilson continued to edit successive editions, first with A A Miles (q.v.) and later with M T Parker. Until the last few months of his life he was sending comments and material for the next edition, often - as M T Parker writes - '...on the backs of old letters or circulars. Many were of daunting erudition and all - right up to the most recent - beautifully composed.'

With Topley, Wilson was involved in the establishment of the postgraduate diploma in bacteriology courses, both in Manchester and London, which played a very large part in establishing high standards for the practice of medical microbiology in Britain, and indeed in many other countries.

The idea of the emergency public health laboratory service, to be established in the event of war, was almost certainly Topley's - who saw the need for such a service irrespective of war and who had, with Wilson, been drawing attention to that need in his teaching at the London School of Hygiene and Tropical Medicine. Wilson inherited and developed Topley's plans and, with the administrative support of A Landsborough Thomson in the Medical Research Council, he created a network of laboratories designed to help in the control of wartime epidemics and, when they did not occur, in control of the infectious diseases normally present in the community. Wilson's belief that public health bacteriology was different from hospital bacteriology, with which he was never really familiar, meant that the Public Health Laboratory Service developed somewhat separately from the rest of the NHS laboratory services, with both advantages and drawbacks. But Wilson's insistence on the highest technical standards and his abilty to establish a coherent team out of a collection of individualists, most of whom he had selected himself without reference to committees, but with the use of a personal interview of awe-inspiring duration and austerity, did create a remarkably successful organization, not only in providing the routine services but also in conducting large scale collaborative research. His whole administration was essentially personal, and included annual visits to almost all of the 50-60 laboratories, often reached by bicycle.

Wilson's achievements were recognized by many prizes and honorary degrees from institutions worldwide. He was knighted in 1962, and elected FRS in 1978.

Wilson did not reveal much of his private life to his professional friends, and indeed he cannot have had much time for non-professional activities. He married Mary Joyce Ayrton in 1924 and they had one son and adopted another. His wife died in 1976.

He was a successful gardener, a knowledgeable connoisseur of claret and of church architecture, and a determined cyclist. Once, when a colleague offered to take the Underground in order to see him, he replied: 'I'll take my bicycle to you; it will be much quicker.'

Wilson was also sustained by a strong religious faith and was held in great respect and affection by many with whom he came into contact, and those whom he had helped in various ways.

Like his father, Ralph Graham Ward Wilson, he achieved the age of 91 and for the last 13 years of his life he was senior Fellow of the College and had the privilege of thanking the president on the occasion of his annual address, a task he accomplished with enviable eloquence.

RW

[*Brit.med.J.*,1987,**294**,1039,1238; 1988,**296**,279-80; *Lancet*,1987,**1**,992; *The Times*,6 & 18 Apr 1987; *Biog.Mems.of Fellows of the Roy.Society*,1988,**34**,887-919; *J.roy.Coll.Path.*,Sept 1985,5-6; MOH *Public Health Lab.Service Year Book 1964*,15; *The Evening News*,1 Nov 1962; Photo]

WINNER, *Dame* Albertine Louise *b.*4 Mar 1907 *d.*13 May 1988
DBE(1967) OBE(1945) BSc Lond(1929) MB BS(1933) MD(1934) MRCP(1935) FRCP(1959) FFCM(1974)

Albertine Winner was born in London. Her father Isidore Winner (formerly Weiner) came originally from the Netherlands; her mother Annie Stonex from this country. She was educated at the Francis Holland School in Clarence Gate, where she was head girl, and at University College London, where she took her BSc with honours in physiology, and then at University College Hospital, qualifying with the University gold medal.

In her early medical career she worked with and was much influenced by Sir Thomas Lewis [*Munk's Roll*,Vol.IV,p.531] and Sir Francis Walshe [*Munk's Roll*,Vol.VI,p.448]. Walshe gave her an abiding interest in neurology, which was finally of great help in developing palliative treatment of motor neurone disease at St Christopher's Hospice when she returned to clinical medicine as a final end to her varied and most distinguished career.

After a time as physician to the Elizabeth Garrett Anderson Hospital and the Mothers' Hospital, Woolwich, she joined the RAMC in 1940, becoming lieutenant colonel and later consultant to the Women's Services. This experience revealed her liking and aptitude for administration, and after a brief return to the Elizabeth Garrett Anderson Hospital she joined the Ministry of Health. Here she kept her intense interest in patients' welfare, particularly in the chronic sick. She became deputy chief medical officer, the first woman to hold that office.

At this time, Sir Max (later Lord) Rosenheim [*Munk's Roll*,Vol.VI, p.394] persuaded her to become the College's Linacre Fellow. She was not quite the first person to occupy this post, since Charles Newman briefly doubled up as Harveian Librarian and Linacre Fellow, but Albertine was the first to serve for any length of time - eleven years - and she certainly set the standard for her successors. As, again, she was the first woman to serve as a College officer, she might be regarded as a feminist ice-breaker; she did break ice, not as a feminist but simply as an extremely competent and reliable administrator who succeeded in getting to the top in what was at the time very much a man's world.

As Linacre fellow, Albertine was responsible for developing the training posts that were to be recognized by the newly-formed JCHMT. She visited most of these posts herself until their number grew too great. She then appointed visiting teams to carry out the task, holding herself in reserve to lean on recalcitrant colleagues who were reluctant to accept the College's recommendations.

She became DBE in 1967 as she retired, but this was but the beginning of yet another career as deputy medical director of St Christopher's Hospice. This interest began in 1961 when she heard of the plans for St Christopher's, the first research and teaching hospice, and joined in a ward round at St Joseph's Hospice. Her clinical acumen and personal interest in people in need was at once aroused and she carried on a quiet campaign of advice and support for the Hospice from then on. It was typical of her that she should go on a refresher course in medicine before returning to clinical duties. As she said: 'The moment I began to examine a patient my hands took over and I was back.'

Albertine worked at St Christopher's part-time from 1967-73, helping to establish both its clinical and administrative standards and fill the rest of her time with many other interests. She was on the council of the Disabled Living Foundation; was one of a group of physicians who travelled to assess Issel's controversial cancer clinic in Bavaria, and was an active member of the Medical Women's Federation and its president in 1971. She retained her interest in the expansion of hospice work as part of a generally accepted clinical challenge. She was fond of describing it as '...just good medicine in a special setting.'

She became chairman of St Christopher's in 1973 and was elected president in 1985, and was also much involved in founding the North London Hospice, with its special Jewish interest and inter-faith connexions.

Brought up in a Jewish home, she joined the Christian foundation of St Christopher's as a 'sympathetic agnostic'. In later years she returned to the faith of her fathers, fitting in as well as ever in the Hospice 'community of the unlike'.

Dame Albertine's interests were as wide as her medical career. A perceptive observer of art, she built up a valuable collection of Japanese prints, and she was an enthusiastic opera lover. She played tennis for the university, and retained her interest in sport and travel.

To some she appeared formidable, but to all she was kind, giving the best in sensitive and utterly reliable moral support. A superlatively good clinician, she finally became a cooperative and unfailingly courteous patient, much loved by all who had the care of her. Of her nursing staff she said: 'They will always come'. Her patients and her many friends would say the same of her.

CMS/RH

[*Brit.med.J.*,1988,**296**,1545; *Lancet*,1988,**1**,1292,1349; *The Times*,17 May 1988; *The Independent*,24 May 1988; Memorial Service *Address*,President RCP; Dame Cicely Saunders; St Christopher's Hospice, *Annual Report & Year Book* 1987-88; Photo]

WOOD, *Sir* Ian Jeffreys *b.*5 Feb 1903 *d.*1 Sept 1986
Kt(1976) MBE(1942) MB BS Melb(1927) MD(1929) MRCP(1932) FRACP(1938) FRCP(1944)

Ian Jeffreys Wood was born in Melbourne and educated at Melbourne Church of England Grammar School and the University of Melbourne. After the usual house appointments, he became medical superintendent of the Royal Children's Hospital, Melbourne, in 1930. Subsequently he spent a year as house physician to the Great Ormond Street Hospital for Sick Children, London, in 1931; he was Marion Carty research fellow to the Walter and Eliza Hall Institute of medical research in 1935, physician to outpatients at the Royal Children's Hospital in 1936 and to the Royal Melbourne Hospital in 1939. He served in the Australian Imperial Forces during the second world war, as physician specialist with the rank of major, being promoted to colonel, OC 2/7 Australian General Hospital. He was awarded the MBE.

In 1945 Ian Wood was appointed head of the newly created clinical research unit at the Walter and Eliza Hall Institute of Medical Research and the Royal Melbourne Hospital, and became assistant director to Sir Macfarlane Burnet (q.v.) of the Hall Institute until his retirement in 1963. During this time he achieved outstanding success in medical research. After 1963, Ian became consultant physician to the Royal Melbourne Hospital and to the Repatriation General Hospital, Heidelberg, Victoria. In 1974 he received the Neil Hamilton Fairley medal awarded jointly by the College and the Australasian College, and in 1976 was honoured with a knighthood for his services to medicine.

Ian Wood, in his professional life, was continually motivated by his quest for discovery and healing: his autobiography, fittingly entitled *Discovery and healing in peace and war*, Toorak, Australia, was published in 1984.

Prior to the second world war, Ian Wood carried out much of the essential research on which the establishment of blood banking in Australia was based. He joined a Red Cross committee dealing with the extension of the donor panel in 1938, and on the afternoon of the day when war broke out he accompanied Lucy Bryce on a visit to the chairman of the Red Cross, John Newman Morris, to obtain his approval for the implementation of the Red Cross Emergency Blood Transfusion Service. Ian Wood was one of the leaders in the organization of blood for the Australian Armed Forces, and organized the introduction of blood transfusion in the Middle East. On leaving the Army in January 1946, he rejoined the Red Cross Blood Transfusion Service committee and played an important role in the postwar development of the service.

Ian Wood is acknowledged as the founding father of clinical research in Australia. The major effort of his clinical research unit was in gastroenterology. Wood's approach was observational, intuitive and descriptive rather than reductionist, and he relied strongly on histopathological and clinical correlation. He was also a great believer in technical innovation. These principles were applied to two major diseases which received particular study in his unit and brought it a substantial reputation: hepatitis and gastritis. He used liver biopsy to study chronic active hepatitis which, in 1948, was a virtually unknown disease, and his foresight led to the recognition that autoimmunity was an important determinant of this disease.

In the course of aspirating gastric juice, Ian Wood fortuitously observed that gastric mucosa as well as gastric secretions could be sampled by per oral suction. This led to the design, with Arthur Hughes, an engineering colleague, of a thin biopsy tube with a sampling head into which a small fragment of gastric mucosa was drawn by a suction pump. Using the 'Wood tube' his unit examined close on 2000 gastric mucosal specimens and this experience enabled him and his colleagues to delineate the natural history of gastritis, to differentiate the 'simple' from the atrophic or 'pernicious-anaemia' type of gastritis, and correlate biopsy appearances with gastric secretory function. These studies culminated in his monograph, with Leon Taft, on *Diffuse Lesions of the Stomach...*, London, Arnold, 1958.

Later Ian Wood became a strong promotor of Burnet's concepts on autoimmunity and enthusiastically urged his unit to examine various chronic diseases, in addition to hepatitis and gastritis, for autoimmune associations. He gave unfailing encouragement to Mackay and Burnet during preparation of their monograph on *Autoimmune diseases...* Springfield, Ill., Thomas. 1963.

Ian Wood served medicine in many ways other than research. He was a most successful administrator and, apart from the organization of his own unit, he served Burnet as deputy director of the Walter and Eliza Hall Institute from 1948-63; Burnet greatly relied on his wise counsel and shrewd advice. Ian was a kind and patient as well as a skilful clinician, and this was reflected by the large and busy service in his ward and clinics at the Royal Melbourne Hospital. He was also a dedicated teacher of medical students and postgraduates, and went to an extreme degree of preparedness for lectures and tutorials whatever the size and nature of the audience.

He enjoyed writing and taught innumerable acolytes the principles of preparing a scientific paper. Whatever he wrote, there was never a lack of emphasis, so that the point to be made could not be missed. His writing reflected his punctilious devotion to a style based essentially on *Fowler's Modern English Usage*, with more often than not a touch of grandeur creeping in. His autobiography, written late in life, expressed this to a degree which evoked great nostalgia among his many protégés.

Ian Wood was a gracious and sociable man who enjoyed his many friendships. Many of them went back to his younger days when he was a talented sportsman; he recalls in his autobiography how he matched skills on the cricket field with players of international standard. He had a great 'will to win' at work and play, but this was allied with a generous humanism which deflected his competitive edge. He will be especially remembered for his rare sense of social devoirs. If he accepted an invitation, or received a reprint or report, or listened to a good lecture, there invariably followed a punctilious handwritten letter of appreciation and congratulations.

It was my privilege to read the funeral oration for Ian Wood which closed with some stanzas from Tennyson's *In Memoriam*, including:

I climb the hill: from end to end
of all the landscape underneath,
I find no place that does not breathe
Some gracious memory of my friend.

IRM

[*Brit.med.J.*,1986,**293**,1035,1248; *Lancet*,1986,**2**,760; *The Times*,11 Sept 1986; RACP *Coll.Newsletter*,Feb 1976; Photo]

WOODFORD-WILLIAMS, Eluned *b*.12 Sept 1913 *d*.25 Nov 1984
CBE(1978) BSc Wales(1933) MB BS Lond(1936) DCH(1938)
DRCOG(1940) MD(1940) MRCP(1940) FRCP(1964)
MRCPsych(1978) FRCPsych(1983)

Eluned (Lyn) Woodford-Williams was born in Liverpool, the eldest of four daughters. Her father, John Woodford-Williams, was a dental surgeon and her mother, Edith Mary Stevens, was the daughter of a horticulturist. From her Lyn inherited her interest in gardening, especially fruit and rose growing. Her father had always wanted his first born to be a son and it is possible that Lyn's great drive to succeed in medicine sprang from this fact.

Lyn began her education at Liverpool College, but the family moved to Wales and she attended the Cardiff High School for Girls. Always keen to become a doctor, she entered the Welsh National School of Medicine, where she obtained a BSc in 1933. She moved to London to University College Hospital for her clinical years, qualifying in 1936 and becoming house physician to Sir William Pearson. Her first interest was in children and she went on to become RMO at the Alder Hey Children's Hospital in Liverpool, obtaining the DCH in 1938. She then returned to University College Hospital as house physician to Sir Thomas Lewis [*Munk's Roll*,Vol.V,531] who, she was proud of telling her friends, diagnosed the mitral valve disease, from which she was to die almost half a century later.

In 1939 she became RMO to Redhill Hospital, Surrey. While holding this post she obtained the DRCOG, the MRCP and the London MD, with gold medal in medicine, all in the same year, 1940; the year of the Battle of Britain and the air raids on London.

In 1942 she moved to Manchester Royal Infirmary as chief assistant to T H Oliver [*Munk's Roll*,Vol.V,p.312], and became for a time temporary assistant physician to the Manchester Northern Hospital.

She married Dennis Astley Sanford in 1939, who also became MRCP in 1940. He subsequently became a surgeon and, after the war, was appointed to a consultant post in Sunderland. Lyn moved there with him. She had two young children and for a time had no regular appointment. But in 1950 Oscar Olbrich arrived in the town with revolutionary ideas about the way to run a geriatric service, and she became his first senior registrar and then his first assistant physician. She assisted him not only to develop the service but also to pursue his researches.

Olbrich, a refugee from Austria, spent the war in Edinburgh where he developed the idea that old people required the same high standards of medicine as were offered to younger people, and that if this were done many might escape joining the ranks of the chronic sick. He also believed in research into the problems of old age. One of his first acts when he inherited 600 chronic sick beds in Sunderland was to establish a research laboratory to study renal function in the elderly. Olbrich was a creative genius but he was also a tyrant in the mould of the continental professor. He was devoted to Lyn, but he abused her unmercifully, and in public. He always called her 'Villiams', never by her Christian name. 'Villiams, you fool', spoken in a teutonic accent, was something heard more than once in the wards. But when he had a heart attack he would have no one to attend him but Lyn.

When Olbrich died, Lyn was the obvious person to succeed him, and the Sunderland unit went from strength to strength under her direction. Lyn worked all hours, but in addition found time to serve as a JP. She established the first day hospital in the north of England, and published papers about it which are still quoted. Sunderland rapidly gained an international reputation for geriatric medicine. In the early years of the NHS it was one of only two or three units in Britain which could present geriatric medicine as an exciting form of medical practice with a great future. The unit attracted many trainees and she took pains to place them in good jobs. The people that she trained have developed geriatric services around the country and proved the validity of the Sunderland philosophy. She was proud of them.

One of her achievements was to establish geriatric medicine as an age related discipline. In Sunderland all medical emergencies in patients over 65 went to her department. She had more beds than all the general physicians put together, and insisted on the highest standards of medicine. She even developed her own coronary care unit. Bernard Isaacs recalls seeing Lyn sitting besides a hugh pile of case notes representing the week's discharges, reviewing them in the presence of her staff and pouncing on any errors or omissions. He added '...she required in her staff the same pace of work she demanded in herself. Her then senior registrar told me that if he wanted to get his hair cut, he had to take a week's holiday.' Her intensity was somewhat daunting, and when she peered at you through her glasses it could be a frightening experience.

As well as drive, Lyn had courage. In 1958 the new European Clinical Section of the International Association of Gerontology, having elected Olbrich as secretary, planned to hold a meeting in Sunderland. Soon after his election Olbrich died suddenly, and the entire responsibility for organizing an international meeting in what had lately been a workhouse, in a town which had never before held more than a local medical meeting, fell on Lyn's shoulders. The light burned in her office far into the night. Needless to say, the conference was a great success and launched Lyn on an international career. She knew everyone in Europe, on both sides of the iron curtain, who was interested in gerontology, and became, with Norman Exton-Smith, co-editor of the new international journal *Gerontologia Clinica*

In 1973 she was appointed to the directorship of the Health Advisory Service, a post which she held with distinction for five years and which she regarded as the pinnacle of her career. She was one of the few doctors to get on well with Barbara Castle, at that time Minister for Social Services. This helped her to affect government policies in ways which assisted the development of geriatric medicine. Through her influence many local geriatric services were improved, and her contributions as a speaker at King Edward's Fund courses helped to spread knowledge of good management in geriatric medicine and psychiatry.

Her concern for the development of the psychiatry of old age was recognized by the award of the MRCPsych in 1978 and the FRCPsych in 1983. During this time she was a regular attender at meetings of the College's section for the psychiatry of old age. There are plenty of psychiatrists among the Fellows of the Royal College of Physicians, but there are very few physicians among the Fellows of the Royal College of Psychiatrists, a fact which gave her great pleasure.

Within our own College Lyn also made her mark. With her UCH connexions, she was friendly with Max Rosenheim [*Munk's Roll*,Vol.VI, p.394] as well as with Norman Exton-Smith, and with their help and encouragement persuaded Comitia to establish a geriatrics committee within the College. She would have been delighted that within a year of her death the College had established a Diploma in Geriatric Medicine.

After her time at the Health Advisory Service, which was rewarded by a CBE, she went for a year back to Manchester to stand in as professor of geriatric medicine while Brocklehurst was in Canada. Her power to transmit enthusiasm for geriatric medicine once again made its mark.

Lyn then retired to her cottage at Abersoch in Wales. She loved the Welsh mountains and Welsh culture. She was a regular supporter of the National Eisteddfod. She acquired many paintings by modern Welsh artists, which she bequeathed to the National Museum of Wales. Her retirement was marred by ill health and by loneliness, her marriage having broken up at about the time she went to the Health Advisory Service. She survived an attack of bacterial endocarditis, but eventually her heart failed. At the very end Dennis reappeared in her life and took her to Sunderland, where she died peacefully. She was survived by a son and a daughter, both medically qualified. Her son John became a member of the College and followed his mother as a consultant in geriatric medicine, a 'first' of which she was naturally very proud.

Norman Exton-Smith summed her up well in his *BMJ* obituary: 'Lyn directed indefatigable energy towards all her pursuits - clinical work, research, teaching, editorial work, and as a committee member of many international bodies concerned with the health and welfare of the aged. All who met her were impressed with her tremendous enthusiasm.'

REI

[*Brit.med.J.*,1985,**290**,163,405; *Lancet*,1985,**1**,119,233; *The Times* 10 Dec 1984; Photo]

WOODS, Arnold Winkelried William Belmont

*b.*4 May 1915 *d.*13 Nov 1986
BA Oxon(1937) BM BCh(1940) MA(1940) MRCP(1943) FRCP(1965)

Arnold Woods was the son of W W Woods, pathologist at the London Hospital, and a nephew of Sir Robert Stanton Woods [*Munk's Roll*, Vol.V,p.458]. He was educated at Kings College School, Wimbledon (where he was later a governor until his death), Exeter College, Oxford, and the London Hospital. He graduated in 1940 and obtained his membership of the College in 1943. From 1943-45 he was EMS physician at the Connaught Hospital, Walthamstow, and later served in the RAMC in India and East Africa, reaching the rank of lieutenant colonel and becoming adviser in medicine to East Africa Command from 1946-47. On demobilization and return to England he was appointed senior registrar at the National Heart Hospital and St George's Hospital, prior to his appointment as consultant physician to Epsom District Hospital in 1955. He held this post until 1980, and was also honorary consultant physician to Epsom College.

Arnold Woods was among the most kind and generous of men. He was a man of absolute integrity. His opinion, whether as a physician or a member of committee was widely sought and universally respected. During his 25 years as a consultant he guided the development of his hospital from little more than a poor-law infirmary to a modern district general hospital. This was undertaken with total altruism.

His junior collleagues, whether houseman, registrar or consultant, came to rely very much on the gentle yet unerringly sound advice of Arnold Woods. He was greatly in demand for committee service, be it at local or regional level, and after his retirement he continued to serve his district well as a member of the local health authority. From 1971-75 he was an examiner for the College.

Arnold's interests outside medicine included languages, travel and cricket. His Norwegian ancestry gave him the ability to master foreign tongues and he enjoyed travelling which gave him the opportunity to put his mastery in languages to good use. He had a lifelong love of cricket and was no mean player himself. 'Many who shared his companionship on the field with the "Jesters", or in the Members Enclosure at Lords and the Oval, will miss his bright-eyed enthusiasm, his sportsmanship and sense of fun...' [extract from a tribute offered at his funeral.]

In 1948 Arnold married Marthe Emilie Isely, 'Ninette', daughter of a Salvation Army officer, and they had one son. The marriage was an exceptionally happy one and in his home life Arnold was complete. Their son is now a surgeon.

GHR

[*Brit.med.J.*,1987,**294**,188-9]

WOOLLAM, David Henry Morgan *b.*12 Aug 1921 *d.*19 Nov 1986
BA Cantab(1941) MA(1945) MRCS LRCP(1948) MB BChir(1949) MRCP(1949) ScD(1964) FRCP(1972)

David Woollam was born at Caldecote, Northamptonshire. His mother was Mollie Duncan Smith, daughter of an Australian landowner, and his father, Harry Morgan Woollam, was a company director. After early education in Australia, he went to the John Lyon School at Harrow and later, with a classical exhibition scholarship, to Peterhouse, Cambridge, where he got first class honours in the natural sciences tripos. His clinical work was undertaken at St George's

Hospital, where he was a senior scholar. After some house appointments he devoted his life to anatomy, at Cambridge University; first as university demonstrator 1950-53 and then as lecturer 1953-82. He became a fellow of Emmanuel College in 1959, and was subsequently domestic bursar, tutor, and director of studies in medicine. He also held the latter post at his old college, Peterhouse.

During his time at Cambridge he was the author of numerous papers on neuroanatomy, teratology, embryology, cytology and medical history. He edited *Advances in teratology* from 1966-72 and *Experimental embryology and teratology* in 1964. He was co-author with J W Millen of *The Anatomy of the cerebrospinal fluid*, London, OUP, 1962 and wrote a number of annotations and leading articles for the *British Medical Journal* and *Lancet*. He was Arris and Gale lecturer at the Royal College of Surgeons, 1958; Alex Bourne lecturer at St Mary's Hospital, 1979; Milroy lecturer at the RCP in 1979, and Lichfield lecturer at Oxford in 1984.

His most notable research was probably his work, initially with J W Millen, on teratology - when he demonstrated the effects (including hydrocephalus) of vitamin A deficiency in rabbits, and later studied the action of thalidomide on experimental animals, and the foetal alcohol syndrome. After the thalidomide disaster he was consultant to the WHO and leading drug companies.

David was a man of acute intelligence who seemed to have an encyclopaedic knowledge of anything which interested him. He was very widely read and could produce apt quotations to fit almost any situation. He also had a penetrating and sometimes irreverant sense of humour, and a remarkable ability to sense what was going on around him almost before it happened. He was essentially an 'ideas man' and tended to rely on others for the more manipulative aspects of research. In latter years he became somewhat frustrated by not getting the funding required to put his ideas into practice, which may also explain why he never got the readership which he undoubtedly deserved.

He had friends among London consultants, and deans of medical schools, particularly through his membership and later vice-presidency of the Cambridge Graduates Medical Club. He annually hosted lunches at Emmanuel College on the day of the Club dinner, at which valuable discussions took place between his London and Cambridge friends concerning the problems of admission of Cambridge graduates to the London clinical schools.

In 1943, while still a student, he married Margaret Leonora Harries and they had five children: two sons (one of whom died in infancy) and three daughters. He was a devoted family man and derived much strength and happiness from his close, and increasing, family circle, and the successful careers of his children. Perhaps the most fitting memorial to him is the Elizabethan Herb Garden in Emmanuel College, the creation of which was entirely due to his inspiration. EHRF

[*Brit.med.J.*,1987,**294**,1237; Photo]

WYNDHAM, Cyril Henry *b.*7 May 1916 *d.*3 Feb 1987
MB ChB Wits(1940) FRS(SA 1966) DSc(1968) MRCP(1946) FRCP(1973)

Cyril Wyndham was born in Maraisburg, South Africa, of a mining family. His father, Herbert Henry Wyndham, was a metallurgist and came from Bristol, England. Cyril received his secondary education at Maritzburg College, where one of his form masters was Alan Paton. He gained a reputation as an athlete, and then went on to study medicine at the medical college of Witwatersrand University, where he became the top student of his class, graduating with several prizes and awards.

He intended embarking on a career in cardiology but the second world war intervened and in 1942 he joined the South African Medical Corps, being seconded to the RAMC, 1943-46, serving in Italy at the 64th British General Hospital.

After the war, with the encouragement of Sir Basil Schonland, he carried out postgraduate work in London and Oxford, where the legendary Sir Wilfrid le Gros Clark set him in the direction of human applied physiology. On his return to South Africa he was appointed director of the Human Sciences Laboratory of the Chamber of Mines Research Organization. From 1962-63 he served on secondment as professor pf physiology at the University of Witwatersrand.

Through his work on solving the physiological problems faced by miners, Wyndham became the leading international expert on human thermal physiology. The fact that nowadays one-third of a million mineworkers can work in safety under stress in deep-level and often hot environments underground is largely due to the work of Cyril Wyndham's team. By 1975 he had published over 250 papers on applied physiology, participated in numerous international meetings, lectured widely by invitation from the USA to India and from Sweden to Malawi, and launched the careers of numerous young scientists, many of whom were later to hold chairs at universities in South Africa and overseas.

Cyril retired from the Chamber of Mines at the age of 60 and took up a second career, which had previously been a hobby; he joined the Medical Research Council as an epidemiologist at the Institute for Biostatistics. In the second phase of his research career his major contributions were to cardiovascular disease and its epidemiology in South African populations. His investigations confirmed the serious extent of coronary heart disease among the economically active white and Indian communities, and these observations provided strong support for the launching of a national heart effort by the MRC which led to the establishment of the Heart Foundation of Southern Africa.

Subsequently he evaluated leading causes of death among children under five years of age in the different ethnic groups. His last publications contained comparisons of mortality rates for cancer in various populations. He particularly regretted the inadequacy of death certification for blacks, a misgiving now felt universally with regard to death certification data in general. His observations raised important challenges regarding the means, commensurate with resources, that could be employed to combat diseases of underprivilege, especially among the young and, conversely, diseases of destructive lifestyles as they prevail among the prosperous.

Cyril Wyndham received many honours and accolades, including the MRC's gold medal, 1985, and an honorary DSc from his university.

Cyril's total absorption and personal participation in experimental studies were impressive, as was his familiarity with and acceptance by black and white miners alike. Another of his early and sustained interests concerned the nutrition of blacks. He would make house to house visits in Soweto enquiring, and seeing, just what foodstuffs there were in the family pot. No one had done this previously - and who has done it since? Cyril had deep feelings about the splendid devotion of physicians and helpers in black rural hospitals; on occasion he would stand in to allow a tired and harassed medical officer a weekend off. Indeed, man's humanity to man is a matter for angel's rejoicing.

Cyril Wyndham will be remembered not only for his exceptional contribution to medical research but also for his kind, fatherly nature which endeared him to everyone with whom he worked; his feeling for and understanding of the problems of others; his total commitment and unswerving devotion to his work - but never at an unfair cost to others. In later years he faced up to the ravages of a terminal disease, accepting each new onslaught with fortitude, a cheerful smile and no word of complaint.

In 1942 he had married Winifred Francis and they had four children, three daughters and a son. They all survived him. ARPW/GW

[*SA Medical Journal*,21 Mar 1987,409; *MRC News* 21 Mar 1987,2]

YOUNG, Carmichael Aretas *b.*19 Aug 1913 *d.*11 Aug 1986
MRCS LRCP(1936) MB BS Lond(1936) MRCP(1939) MD(1940) FRCP(1950)

Carmichael Young, always 'Car' to his friends, was born in Adelaide, South Australia. He was the eldest of three sons, in a family of five children, of Aretas Henry Young, a supervisor with Cable and Wireless Limited, and Isabella Young, née Wilson, of Parattah,

Tasmania. The family roots were in England and many of his forbears had been distinguished soldiers in the British Army,but in 1848 his great-grandfather, Sir Henry Edward Fox Young, was appointed governor of South Australia and the family moved to Adelaide. Subsequently Port Augusta was named after Sir Henry's wife, Lady Augusta Sophia Marryat Young. Car was related by marriage, through his sister, to Sir John Frew (q.v.), who was president of the Royal Australasian College of Physicians from 1972-74.

Car was educated at Carey Grammar School, Kew, Victoria, leaving at the age of 15 to work as a clerk in an office for a year, before accepting a generous offer from his aunt, Helen Young, who had qualified in medicine at Edinburgh in 1924, to come to Britain to study medicine. Helen Young lived in Harley Street, next door to Sir William Willcox [*Munk's Roll*,Vol.IV,514], and took students in as lodgers. She used to attend Sir William's ward rounds and consult Sir Alexander Fleming at St Mary's Hospital. It was thus that in 1931, after passing his first MB externally, with these introductions, Car began his medical studies at St Mary's. He qualified in 1936 and, after house appointments at St Mary's, became resident medical officer at the Brompton Hospital. After obtaining his MRCP in 1939 he was appointed assistant to the professorial medical unit at St Mary's, and gained his London MD the following year. He joined the RAMC in 1941 and was mentioned in despatches while serving as a medical specialist to the 10th (British) casualty clearing station in Tobruk. Later, he was appointed officer in charge of the medical division at the 43rd General Hospital in Beirut with the rank of lieutenant colonel, where for a short period he commanded the hospital.

On demobilization and return to England, he spent a brief period as medical registrar to the Prince of Wales' Hospital before being appointed, in 1948, as general physician to St Mary's where he also took charge of the diabetic clinic. In 1950, when the dean of the medical school, D H Brinton (q.v.), made arrangements for St Mary's students to do some of their clinical clerkships at the nearby Paddington General Hospital, he was also appointed general physician there. Car's firms, both at St Mary's and at Paddington, were very popular with the students. He was sub-dean of the medical school from 1952-53. At the College he served as Pro-censor 1973-74 and as Censor 1974-75. He was also an examiner for the Conjoint Board and London University.

Car was an outstanding character; he had a quick mind and was a shrewd physician with diagnostic flair. His views were clear, succinctly expressed and, not infrequently, provocative. He was a forthright teacher of essentials, much appreciated by his students and medical assistants. He successfully combined an insight into people with an understanding of their pathology and a warm approach to their problems. As a consultant he was much in demand and worked extremely hard. Unfortunately, opportunities for research did not come his way; the years when he might have been involved were spent in the

RAMC. But someone must apply the fruits of research at the bedside, and this he did with a skill and kindness which made him a much respected and widely trusted physician.

Car loved all sports, but especially swimming, rugby football and golf, all of which he supported with the same energy and zest that he applied to his profession. He worked hard and he played hard. He was captain of the Otter swimming team, swam for English universities and played water polo for Hertfordshire. But the apogee of his swimming prowess was his presidency of the Otter Club 1964-67. He wrote a history of the Club to commemorate its centenary: *The Otter Club centenary 1869-1969*, which was published in 1969. He played in the St Mary's team which won the seven-a-side rugby competition in 1940, and was president of the rugby club 1970-75. There were times when he spent more energy supporting on the touchline than did some of the players on the field. As a golfer he was indefatigable. Some would misread his bubbling vitality as brusqueness; in fact, it concealed a sensitive, kind and sympathetic nature.

His family life was extremely happy. In 1939 he married Marie, a physiotherapist at St Mary's with a keen interest in all sports. She was the daughter of W H Lewry of Botley, Hampshire, a butcher. They had three sons, who all qualified at St Mary's, and a daughter who was a nurse there. His brother, A W O Young, became Minister of Health in Tasmania and represented Tasmania in the Australian Parliament. In summary, Car's life was his family and St Mary's.

TAK

[Photo]

YOUNG, ***Sir*** **Frank George** *b.*25 Mar 1908 *d.*20 Sept 1988
Kt(1973) BSc Lond(1929) PhD(1933) FRSC(1933) DSc(1939) FRIC(1946) MA Cantab(1949) FRS(1949) Hon FRCP(1974) Hon LLD Aberd(1965) Hon DSc Rhodesia, later Zimbabwe(1975)

Frank Young was born in London, the son of Frank Edgar Young and his wife Jessie Eleanor, née Pinkney. He was educated at Alleyn's School, Dulwich and University College London, where he graduated with first class honours in chemistry in 1929, obtaining a PhD in physiology in 1933. His postgraduate education and research for his PhD thesis were carried out under J C Drummond, later Sir Jack; C Lovatt Evans, later Sir Charles [*Munk's Roll*,Vol.VI,p.166], and A V Hill. In 1932 he was awarded a Beit Memorial fellowship for medical research, which he held at University College and also, at separate times, in the laboratories of the two discoverers of insulin, J J R Macleod [*Munk's Roll*,Vol.V., p.259] - in Aberdeen,1933 - and C H Best [*Munk's Roll*,Vol.VII,p.35], in Toronto, 1934-35.

Frank Young became a member of the scientific staff of the Medical Research Council in 1936 at the National Institute for Medical Research under its director, Sir Henry Dale [*Munk's Roll*,Vol.VI, p.130]. During the six years that he spent there Frank made his most important discovery - that permanent diabetes could be produced in experimental animals by the injection of an anterior pituitary gland extract. He later demonstrated that the active material in these extracts was growth hormone. In this work he was able to apply his basic knowledge and skills as a chemist to the complex problems of endocrine physiology and medicine. It was therefore entirely in keeping with these interests when he was appointed professor of biochemistry at St Thomas's Hospital medical school in 1942 at the early age of 34. In 1945 he was appointed professor of biochemistry at University College London, and in 1949 became Sir William Dunn professor of biochemistry in the University of Cambridge. He held the latter position until his retirement in 1975.

Frank Young's contributions to endocrine and diabetes research remain to this day as foundation stones of modern knowledge in these fields. In addition to being elected a fellow of the Royal Society, the importance of his scientific achievements was recognized in a variety of ways too numerous to list. Suffice it to mention the awards of the Banting Memorial medals of both the British Diabetic and American Diabetes Associations and the Upjohn award of the Endocrine Society of the USA. Amongst many named lectures given were: the Stirling lecture, Yale University, USA; the Jacobeus lecture, University of Oslo, Norway; the Richardson lecture, Harvard University medical school, USA; the Litchfield lecture, University of Oxford; the Croonian lecture, the Royal Society, and the Linacre lecture, St John's College, University of Cambridge.

While continuing with some research at Cambridge, Frank's major interests and achievements during this period lay in fostering the role of biochemistry in medical education and research. He was a person who both demanded and imposed order in his own activities and those of others. Consequently he was an outstandingly effective administrator and also an excellent editor. Indeed, few pieces of paper that passed before him escaped the almost reflexively operated editorial pencil. The list of Frank's administrative responsibilities both nationally and internationally would lead someone who did not know him to conclude that few of these could have been satisfactorily executed. On the contrary, Frank was - by virtue of long hours of work - always well briefed. As head of the Cambridge department of biochemistry he was a patient and fair-minded chairman.

The department of biochemistry flourished under his leadership, but this in turn brought its own problems. Large numbers of students applied in their final year to read Part II Biochemistry. The intake grew steadily to 40, but still could not accommodate the applicants who frequently exceeded 100. The prospect of a new building to accommodate a larger class was a great attraction to him. Probably the only major setback which he suffered in the university, through no fault of his own, was the failure to gain the new building which would have crowned his development of biochemistry into one of the strongest biological sciences in the university.

Happily there are two other major monuments to Frank's achievements in the university: Darwin College and the Clinical School. The way in which Frank worked, steadily and unobtrusively to bring these two institutions into existence illustrates both the nature of his personality and his extreme effectiveness. He clearly perceived the need for a graduate college. In an appendix to the report of the syndicate on the interrelationships between the university and the colleges, under the chairmanship of Lord Bridges, he argued the case. When Darwin College was established in 1963 it was entirely appropriate that he should be the first Master. The critical role which he played in establishing the procedures and ambience of Darwin are well known. From these beginnings the intake of students grew steadily and with a minimum of fuss it became the first mixed college in the university.

From his early days in Cambridge, Frank strongly supported the establishment of a clinical school. Equally, he recognized the powerful internal opposition to this proposal. True to character, his reaction to this conflict was to build a basis of trust and friendship between the key people involved upon which a clinical school could be securely founded. In 1955 a dining club was established, with Frank as a founder member, which rejoiced under the name of The Carphologists. Frank was greatly fascinated by words, and this is probably an example. 'Carphology' is a terminal sign in typhoid fever, being a plucking at the bedclothes by the unfortunate patient. One can imagine that Frank saw the dining club as a terminal sign in the opposition to the Clinical School. This club, comprised of a membership drawn equally from the preclinical and clinical subjects, flourishes to this day and continues to foster friendship and academic exchange.

In 1969 Frank became chairman of the Cambridge Clinical School planning committee and the founding of the Clinical School began. Without Frank's wise leadership the fruition of these plans would have been long delayed, if indeed the Clinical School would have been established at all. Thus, the beginning of Frank's retirement coincided with the culmination of one of his most consistently pursued plans. In retirement he continued with a number of national and international commitments. Sadly, decreasing mobility eventually forced him to give up one of his greatest enjoyments - international travel.

He married Ruth Turner, who practised as a doctor, in 1933 and they had four children; three sons and a daughter. The strengh and love of Ruth and his family underpinned his professional life, and made his personal life equally fruitful. He was survived by his wife and three sons; his daughter predeceased him.

CNH

[*The Times*,21 Sept 1988; *The Independent* 26 Sept 1988; *Brit.med.J.*,1988,297,913,1040; *Address* by Prof.C.N.Hales,Great St Mary's Church,Cambridge,27 Nov 1988]

COLLEGE OFFICERS

The following lists are continued from Vol.VII.p.632.
* indicates that a life of the officer is included in this volume.
Civil honours, where indicated, are those to which the College officer was entitled at the time of holding the appointment. By tradition, only university degrees in medicine are included.

PRESIDENTS

93	1983	Sir Raymond Hoffenberg KBE MD Cape Town

VICE PRESIDENTS

51	1984	Thomas Ernest Oppé MB Lond
52		Sir Richard Ian Samuel Bayliss KCVO MD Cantab
53	1985	John Richmond MD Edin
54		Ian Ramsay Gray MD Lond
55	1986	Peter Albert Emerson MD Cantab
56		John Allen Davis MB Lond
57	1987	Maurice Hart Lessof MD Cantab
58		Barbara Mary Ansell CBE MD Birm
59	1988	Harold Philip Lambert MD Cantab
60		Margaret Elizabeth Harvey Turner Warwick DM Oxon

TREASURERS

57	1970 *	Nigel Dean Compston CBE MD Cantab
58	1985	Anthony Michael Dawson MD Lond

REGISTRARS

40	1975	David Alan Pyke CBE MD Cantab

ASSISTANT REGISTRARS

16	1981	David Gwyn Williams MD Wales
17	1987	Carol Anne Seymour DM Oxon

HARVEIAN LIBRARIANS

11	1979	Sir Gordon Ethelbert Ward Wolstenholme OBE MB Cantab

LINACRE FELLOW

3	1978	John Melville Malins MD Birm
4	1987	John Lister MD Cantab

HANS SLOANE FELLOW

1	1987	Sir John Badenoch DM Oxon

EDITOR OF THE JOURNAL

1	1988	Robert Frederick Mahler MB

LECTURES AND LECTURERS

The following lists are continued from Vol.VII p.633ff.
* indicates that a life of the lecturer is included in this volume.
** indicates that the lecture has not been published.
♦ indicates that the manuscript is in the College library.

Complete information about the subjects of lectures and their publication is in the College library. Subjects are also given in the College List.

Only university degrees in medicine and science, and fellowship of the Royal Society are indicated.

LUMLEIAN LECTURES

1984	Colin Terence Dollery MB
1985 ♦	Maurice Hart Lessof MD
1986 ♦	Desmond Roger Laurence MD
1987 **	John Stewart Cameron MD
1988	Reginald Hall MD

GOULSTONIAN LECTURES

1984 **	Andrew Jackson Rees MB
1985	Christopher Bunch MB
1986	Adrian Llewellyn MB
1987 **	Alan Michael McGregor MD
1988	David Christopher Linch MB

HARVEIAN ORATIONS

1984 ♦	Antony Clifford Dornhorst MD
1985 ♦	Dame Sheila Sherlock MD
1986 *♦	Alan George Williams Whitfield MD
1987 ♦	Sir James Learmonth Gowans MB FRS
1988 ♦	Paul Emanuel Polani MD FRS

CROONIAN LECTURES

1984 **	David John Weatherall MD FRS
1985	David Arthur John Tyrrell MD FRS
1986 ♦	Christopher John Dickinson DM
1987	Michael Anthony Epstein MD FRS
1988	David Hull MB

BRADSHAW LECTURES

1984	David Keith Peters MB
1985	Michael David Rawlins MD
1986 ♦	William Ian McDonald MB
1987	John Douglas Swales MD
1988	Magdy Habib Yacoub FRCS

MILROY LECTURES

1984	Adetokunbo Olumide Oluwole Lucas MD
1985 ♦	Peter Stanley Harper DM
1986 ♦	Archie Young MD
1987 ♦	Roy Goulding MD
1988	John Everard Cotes DM

FITZPATRICK LECTURES

1984	Sir Gordon Wolstenholme MB
1986	John Melville Malins MD
1987 ♦	Alex Sakula MD
1988	Sir Abraham Goldberg MD

OLIVER-SHARPEY LECTURES

1984 **	Cesar Milstein PhD FRS
1985 ♦	Charles Michel BM
1986 ♦	William Richard Keatinge MB
1987	Geoffrey Burnstock PhD FRS
1988	Lucio Luzzatto MD

LLOYD-ROBERTS LECTURE

§ Indicates that the lecture was given at the College

1984♦ §	HM King Hussein of Jordan
1985	Dame Helen Mary Warnock
1986♦ §	Lewis Wolpert ScD FRS
1987	Denis Noble PhD FRS
1988	Rt Rev David Jenkins, Bishop of Durham

BERTRAM LOUIS ABRAHAMS LECTURE IN PHYSIOLOGY

1984 **	Leslie Lars Iversen MD
1986	Colin Brian Blakemore PhD
1988	Ian Michael Glynn MD FRS

CHARLES WEST LECTURES

1986	Alexander Stewart McNeish MB

HUMPHRY DAVY ROLLESTON LECTURES

1986	Howard Christopher Thomas MB

FE WILLIAMS LECTURES

1985 ♦	Francis Irvine Caird DM

ERNESTINE HENRY LECTURES

1987	Robert Ian McCallum MD

LANGDON-BROWN LECTURES

1984	Paul Emanuel Polani MD FRS
1986 ♦	Peter Julius Lachmann MB FRS

TUDOR EDWARDS LECTURES

Lectures given at the College

1985	Margaret Elizabeth Harvey Turner-Warwick DM
1987	Kenneth Barrett Saunders MD

WATSON SMITH LECTURE

1984 ♦	Neville Robinson Rowell MD
1985	Brian Ernest Kendall MB
1986	Howard Saul Jacobs MD
1987	John Gerald Garvin Ledingham DM
1988	Robert Williamson PhD

MARC DANIELS LECTURES

1984	Harold Philip Lambert MD
1987	George Stewart Kilpatrick MD

PARKES WEBER LECTURE

1985	Ronald Marks MB
1988 ♦	John Lloyd Burton MD

PHILIP ELLMAN LECTURES

1984 ♦	Timothy John Hayes Clark MD
1987	Robert Alastair Livingstone Brewis MD

TEALE LECTURES

1985 **	Robert David Hugh Boyd MB
1987	Cyril Chantler MD

SAMUEL GEE LECTURES

1985 **	Rt Hon Michael Foot MP
1986 ♦	John Michael Newsom-Davis MD

LILLY LECTURES

1984	Lewis Thomas MD
1985 ♦	Jesse Roth MD
1986 ♦	John Robert Evans MD
1987	Robert Charles Gallo MD
1988	Arnold Seymour Relman MD

OSLER ORATIONS

1984 ♦	Henry Harris MD
1987	Charles Eric Stroud MB

JOHN THORNTON INGRAM LECTURES

1987 ♦	Ronald Edward Church MD

LOCKYER SCHOLARSHIP AND LECTURE

1984	Neville Roy Butler MD
1988	Albert Aynsley-Green MB

WILLIAM WITHERING LECTURE

1985	Colin Terence Dollery MB
1988	Michael John Stratton Langman MD

Key to Contributors' Initals

ACK	AC Kennedy
ADP	AD Perrett
AFC	AF Conchie
AG	A Ghareeb
AGB	AG Bearn
AGS	AG Stewart
AGW	AG Watkins
AGWW	AGW Whitfield
AH[1]	A Hollman
AH[2]	A Horwich
AHC	AH Cruickshank
AHO	AH Osmond
AHTR-S	AHT Robb-Smith
AJR	AJ Rook
AJT	AJ Thomas
AKG	A Kerr Grant
AKM	A Keith Mant
ALG	AL Graber
ALW	Dame Albertine Winner
AMH	AM Halim
APG	AP Goldman
APN	AP Norman
ARM	AR Morrison
ARPW	ARP Walker
ASM	AS Mitha
AST	AS Till
ATR	AT Richardson
BC	B Clarke
BDB	BD Bower
BG	B Gandevia
BGP-S	BG Parsons-Smith
BH	B Hudson
BIH	BI Hoffbrand
BJF	BJ Fairhurst
BLH	B Hazleman
BM	B Mullally
BML	BM Laurance
BP	B Perry
BW	Sir Brian Windeyer
CAC	Sir Cyril Clarke
CACP	CAC Pickering

CAK	CA Keele
CBS	CB Schoemperlen
CBW	CB Willey
CDA	CD Anderson
CDC	CD Calnan
CGU	CG Uragoda
CHE	CH Edwards
CHG	CH Gray
CJE	CJ Earl
CMS	Dame Cicely Saunders
CN	C Nourse
CNH	CB Hales
CS	C Skinner
DAKB	Sir Douglas Black
DAJT	DAJ Tyrell
DAP	DA Pyke
DC	D Coakley
DD	D Daley
DDG	DD Gibbs
DDM	Dowling D Munro
DEM	D Elsdon Myers
DEMS	DEM Speed
DE-S	D Emslie-Smith
DF	D Francis
DGG	DG Gibson
DGJ	DG Julian
DJG	DJ Gee
DJH	DJ Hendrick
DK	Desmond Kelly
DKO'D	DK O'Donovan
DKP	DK Peters
DMTG	DMT Gairdner
DNP	DN Phear
DSB	DS Browne
DSM	DS Munro
DTDH	DTD Hughes
DVB	DV Bates
DVH	DV Hamilton
ECBK	ECB Keat
EDC	Sir Dick Caldwell
EGK	EG Knox
EGLB	EGL Bywaters
EHH	EH Hudson
EHOP	EHO Parry
EHRF	EHR Ford

EN	E Neil
EW	E Williams
FDH	F Dudley Hart
FDR	FD Rosenthal
FFF	FF Fenech
FGS	FG Stoddart
FH	F Hobbiger
FIL	FI Lee
FJA	FJ Aumonier
FJM	FJ Milward
FL	F Lunn
FPAC	FPA Capps
FPM	FP Marsh
GDF	GD Forwell
GEG	Sir George Godber
GHduB	GH du Boulay
GHR	GH Robb
GLA	GL Asherson
GML	GM Little
GJM-T	Surg. Vice Admiral GJ Milton-Thompson
GPB	GP Baker
GRS	GR Steed
GS	Sir George Smart
GSC	GS Crockett
GW	Sir Gordon Wolstenholme
HAD	HA Dewar
HB[1]	H Barcroft
HB[2]	H Barrie
HEH	HE Holling
HFH	HF Harwood
HKL	HK Lucas
HMJG	HMJ Gilles
HRR	HR Rollin
HRV	HR Vickers
HS	H Saxton
ICSN	ICS Normand
IDC	ID Chisholm
IHM	IH Mills
IKB	IK Brown
IM	IP Munro
IRM	IR Mackay
IW	Sir Ian Wood

JAL	JA Lowden
JAP	JA Parrish
JAS	JA Scott
JB[1]	Dame Josephine Barnes
JB[2]	J Brothwood
JBB	JB Barlow
JBH	J Bishop Harman
JBP	JB Penfold
JC	J Colover
JCH	JC Haworth
JDC	J Donaldson Craig
JDD	JJ Daly
JDH	JD Hunter
JFPS	JFP Skrimshire
JGD	Air Marshal Sir John Donald
JGG	JG Graham
JGS	JG Scadding
JH	J Hannah
JHA	JH Angel
JHH	JH Hutchison
JHP	Sir John Peel
JKW[1]	JK Wagstaff
JKW[2]	JK Wales
JL	J Ledingham
JMB	JM Brogard
JML	JM Ledingham
JMP	JM Parkin
JMS	JM Smellie
JNF	JN Fell
JOF	JO Forfar
JPHD	JPH Davies
JR	Surg.Vice Admiral Sir John Rawlins
JRB	JR Billinghurst
JRF	JR Forbes
JRT	JR Trounce
JS	J Strong
JSF	JS Frew
JSL	JS Logan
JSP	JS Pippard
JWS	JW Scopes
KBT	KB Taylor
KFR	KF Russell
KGL	KG Lowe
KMcC	K McCarthy
KR	K Russell
KS	K Somers

KWGH	KWG Heathfield
LB	L Brent
LBS	LB Strang
LEB	LE Böttiger
LHC	LH Collier
LJB	LJ Beilin
LJR	LJ Rowley
LMB	LM Blendis
LS	L Sinclair
LTT	L Thorne Thorne
LW	L Wollner
MA	Sir Melville Arnott
MAES	MAE Symonds
MAS	MA Salmon
MB	M Baraitser
MC	M Caplin
MD	M Dulake
MFO	MF Oliver
MGW	MG Whiteside
ML	M Linnett
MLS	ML Slevin
MR	Sir Martin Roth
MRC	MR Clements
MRP	MR Parsons
MS	M Silink
NA	N Amarasekera
NAB	NA Black
NRR	NR Rowell
ODF	OD Fisher
OLW	OL Wade
OPG	OP Gray
PAE	PA Emerson
PABR	PAB Raffle
PAJB	PAJ Ball
PBC	PB Carey
PBS	PB Storey
PBSF	PBS Fowler
PCR[1]	PC Reynell
PCR[2]	PC Robertson
PDB	PD Bewsher
PDBD	PDB Davies
PEP	PE Polani

PG	P Ghalioungui
PHA	PH Adams
PJL	PJ Lachmann
PLB	PL Boardman
PLM	PL Mollison
PM	P Marsden
PRE	PR Evans
PS[1]	P Sharpstone
PS[2]	P Sleight
PT	Sir Peter Tizard
PWN	PW Nathan
RBS	RB Salter
RCB	RC Butler
RD	Sir Richard Doll
RDC	RD Catterall
RDM	RD Montgomery
REB	RE Beamish
REB-C	RE Bonham-Carter
REC	RE Church
REI	RE Irvine
REK	RE Kelly
RGG	RG Grainger
RGW	RG Welch
RH	Sir Raymond Hoffenberg
RHB	RH Balme
RHC	RH Champion
RHSM	RHS Mindham
RID	RI Dent
RIMcC	RI McCallum
RJT	RJ Terry
RJW	RJ Walker
RL	R Lawrie
RLH	RL Himsworth
RM	Sir Reginald Murley
RNJ	RN Johnston
RRH	RR Henderson
RW	Sir Robert Williams
SA	S Alexander
SC	S Cohen
SCS	SC Sheth
SG[1]	S Gold
SG[2]	S Goulston
SGB	SG Browne
SH	S Howarth
SJGS	SJG Spencer

SMW	SM Watkins
SO	S Oleesky
SP	S Padmavati
SRS	SR Smithers
SSS	SS Singhal
TAK	TA Kemp
TBSD	TBS Dick
TCG	TC Gray
TEG	TE Gumpert
TEO	TE Oppé
TF	Col.TM Fowler
TL	Sir Thomas Lodge
TRE	TR Evans
TS	T Schneider
TWC	Maj.General TW Carrick
VJM	VJ Metcalf
VL	V Luniewska
VR	V Ramalingaswami
WAG	WA Gillespie
WB	W Brigden
WBA	WB Armstrong
WCH	WC Harris
WEG	WE Gibb
WGH	WG Harding
WHJ	WH Jopling
WJEMcK	WJE McKee
WRG	WR Gauld
WSM	Brig.Gen.WS Millar
WvH	W van't Hoff
WW	W Walker
YA	Y Armstrong
YPM	YP Munjal

CUMULATIVE INDEX

CUMULATIVE INDEX

CUMULATIVE INDEX

CUMULATIVE INDEX

CUMULATIVE INDEX

CUMULATIVE INDEX

CUMULATIVE INDEX

CUMULATIVE INDEX

CUMULATIVE INDEX

CUMULATIVE INDEX

CUMULATIVE INDEX

CUMULATIVE INDEX

CUMULATIVE INDEX

CUMULATIVE INDEX

CUMULATIVE INDEX

CUMULATIVE INDEX

CUMULATIVE INDEX

CUMULATIVE INDEX

CUMULATIVE INDEX

CUMULATIVE INDEX

CUMULATIVE INDEX

CUMULATIVE INDEX

CUMULATIVE INDEX

CUMULATIVE INDEX

CUMULATIVE INDEX

CUMULATIVE INDEX

CUMULATIVE INDEX

CUMULATIVE INDEX

CUMULATIVE INDEX

CUMULATIVE INDEX

CUMULATIVE INDEX

CUMULATIVE INDEX

CUMULATIVE INDEX

CUMULATIVE INDEX

CUMULATIVE INDEX

CUMULATIVE INDEX

CUMULATIVE INDEX

CUMULATIVE INDEX

CUMULATIVE INDEX

CUMULATIVE INDEX

CUMULATIVE INDEX

CUMULATIVE INDEX

CUMULATIVE INDEX

CUMULATIVE INDEX

CUMULATIVE INDEX

CUMULATIVE INDEX

CUMULATIVE INDEX

CUMULATIVE INDEX

CUMULATIVE INDEX

CUMULATIVE INDEX

CUMULATIVE INDEX

CUMULATIVE INDEX

CUMULATIVE INDEX

CUMULATIVE INDEX

CUMULATIVE INDEX

CUMULATIVE INDEX

CUMULATIVE INDEX

CUMULATIVE INDEX